T0201475

THE PHILOSOPHY OF COSMOLOGY

Following a long-term international collaboration between leaders in cosmology and the philosophy of science, this volume addresses foundational questions at the limits of science across these disciplines, questions raised by observational and theoretical progress in modern cosmology. Space missions have mapped the Universe up to its early instants, opening up questions on what came before the Big Bang, the nature of space and time, and the quantum origin of the Universe.

As the foundational volume of an emerging academic discipline, experts from relevant fields lay out the fundamental problems of contemporary cosmology and explore the routes toward possible solutions. Written for physicists and philosophers, the emphasis is on conceptual questions, and on ways of framing them that are accessible to each community and to a still wider readership: those who wish to understand our modern vision of the Universe, its unavoidable philosophical questions, and its ramifications for scientific methodology.

KHALIL CHAMCHAM is a researcher at the University of Oxford. He acted as the executive director of the UK collaboration on the 'Philosophy of Cosmology' programme. His main research interests are in the chemical evolution of galaxies, nucleosynthesis, dark matter, and the concept of time. He has co-authored four books and co-edited ten, including *From Quantum Fluctuations to Cosmological Structures* and *Science and Search for Meaning*.

JOSEPH SILK FRS is Homewood Professor at the Johns Hopkins University, Research Scientist at the Institut d'Astrophysique de Paris, CNRS and Sorbonne Universities, and Senior Fellow at the Beecroft Institute for Particle Astrophysics at the University of Oxford. He has written seven popular books on cosmology, including *The Big Bang, A Short History of the Universe*, and *The Infinite Cosmos*. His research areas include dark matter, the formation of the galaxies, and the big bang theory. He has received numerous awards and prestigious international fellowships.

JOHN D. BARROW FRS is Professor of Mathematical Sciences at the University of Cambridge and Director of the Millennium Mathematics Project to improve the appreciation of mathematics amongst the general public, teachers, and school students. The recipient of many distinguished prizes, his research interests are in cosmology, gravitation, and the interface between particle physics and astronomy. He is also a prolific author; the most recent of his 22 books are *100 Essential Things You Didn't Know You Didn't Know about Maths and the Arts*, and *The Book of Universes.*

SIMON SAUNDERS is Professor of Philosophy of Physics at the University of Oxford, and a Tutorial Fellow of Merton College, Oxford. He is the lead editor of *Many Worlds? Everett, Quantum Theory, and Reality* and the author of more than 60 articles in philosophy of physics, with special emphasis on the foundations of quantum mechanics, quantum field theory, and symmetries. He also works on structuralism in philosophy of science and metaphysics, focusing on the logic of identity. He is president-elect of the British Society for the Philosophy of Science.

THE PHILOSOPHY OF COSMOLOGY

Edited By

KHALIL CHAMCHAM
University of Oxford

JOSEPH SILK
University of Oxford

JOHN D. BARROW
University of Cambridge

SIMON SAUNDERS
University of Oxford

CAMBRIDGE
UNIVERSITY PRESS

CAMBRIDGE
UNIVERSITY PRESS

University Printing House, Cambridge CB2 8BS, United Kingdom
One Liberty Plaza, 20th Floor, New York, NY10006, USA
477 Williamstown Road, Port Melbourne, VIC3207, Australia
4843/24, 2nd Floor, Ansari Road, Daryaganj, Delhi – 110002, India
79 Anson Road, #06-04/06, Singapore 079906

Cambridge University Press is part of the University of Cambridge.

It furthers the University's mission by disseminating knowledge in the pursuit of
education, learning, and research at the highest international levels of excellence.

www.cambridge.org
Information on this title: www.cambridge.org/9781107145399

First published 2017

Printed in the United Kingdom by TJ International Ltd. Padstow Cornwall

A catalogue record for this publication is available from the British Library.

Library of Congress Cataloging-in-Publication Data
Names: Chamcham, Khalil, editor.
Title: The philosophy of cosmology / [edited by] Khalil Chamcham, University
of Oxford, Joseph Silk, University of Oxford, John D. Barrow, University
of Cambridge, Simon Saunders, University of Oxford.
Description: New York : Cambridge University Press, 2017. | Includes index.
Identifiers: LCCN 2016045219 | ISBN 9781107145399 (hardback)
Subjects: LCSH: Cosmology.
Classification: LCC BD493.P49 2017 | DDC 113–dc23
LC record available at https://lccn.loc.gov/2016045219

ISBN 978-1-107-14539-9 Hardback

Contents

Contributors

David Z. Albert is Professor at the Department of Philosophy, Columbia University, USA.

Frank Arntzenius is Professor of Philosophy at the University of Oxford, University College, Oxford, UK.

Tom Banks is Professor at the New High Energy Theory Center and the Department of Physics, Rutgers University, USA.

Luke A. Barnes is a postdoctoral researcher at the Sydney Institute for Astronomy, University of Sydney, Australia.

John D. Barrow is Professor of Mathematical Sciences and Director of the Millennium Mathematics Project at Cambridge University. He is also a Professorial Fellow of Clare Hall, UK.

Claus Beisbart is Professor of Philosophy at the Institut für Philosophie, Universität Bern, Switzerland.

Kimberly K. Boddy is a Postdoctoral Research Fellow at the Physics Faculty, University of Hawaii, USA.

Bernard Carr is Professor at the School of Physics and Astronomy, Queen Mary University of London, UK.

Sean M. Carroll is Professor at the Department of Physics, Caltech, USA.

Cian Dorr is Professor at the Department of Philosophy, New York University, USA.

George F. R. Ellis is Professor at the Mathematics and Applied Mathematics Department, University of Cape Town, South Africa.

James Hartle is Professor at the Department of Physics, University of California – Santa Barbara, USA.

Thomas Hertog is Professor at the Institute for Theoretical Physics, Leuven University, Belgium.

Cormac O'Raifeartaigh is a researcher at the School of Science, Waterford Institute of Technology, Waterford, Ireland.

Don N. Page is Professor at the Department of Physics, University of Alberta, Edmonton, Alberta, Canada.

J. Brian Pitts is a Senior Research Associate of Philosophy at the University of Cambridge, UK.

Jason Pollack is a student at the Department of Physics, Caltech, USA.

Joel R. Primack is Professor at the University of California – Santa Cruz, USA.

Carlo Rovelli is Professor at the Centre de Physique Théorique or the Aix-Marseille Université, Marseille, France.

Svend E. Rugh is a researcher at the Symposion, "The Socrates Spirit", Section for Philosophy and the Foundations of Physics, Copenhagen, Denmark.

Martin Sahlén is a researcher at the Department of Physics and Astronomy, Upsala University, Sweden.

Joseph Silk is Homewood Professor at the Johns Hopkins University, USA, Research Scientist at the Institut d'Astrophysique de Paris and Sorbonne Universities, France, and Senior Fellow at the Beecroft Institute for Particle Astrophysics at the University of Oxford, UK.

Chris Smeenk is Professor at the Rotman Institute of Philosophy, Department of Philosophy, Western University, Ontario, Canada.

Ward Struyve is a researcher at the Department of Physics, University of Liège, Belgium.

Daniel Sudarsky is Professor at the Instituto de Ciencias Nucleares, Universidad Nacional Autónoma de México, México.

Roderich Tumulka is Professor at the Department of Mathematics, Rutgers University, USA.

Jean-Philippe Uzan is Professor at the Institut d'Astrophysique de Paris, CNRS, Université Pierre et Marie Curie, Paris, France.

Francesca Vidotto is a researcher at the Institute for Mathematics, Astrophysics and Particle Physics, Radboud University, Nijmegen, The Netherlands.

David Wallace is Professor of Philosophy at the University of Southern California, USA.

Henrik Zinkernagel is Professor at the Department of Philosophy, University of Granada, Spain.

Preface

Cosmology is the unifying discipline par excellence, combining theories of gravity, thermodynamics, and quantum field theory with theories of structure formation, nuclear physics, and condensed matter physics. Its observational tools include the most intricate and expensive scientific experiments ever devised, from large-scale interferometry, high-energy particle accelerators, and deep-sea neutrino detectors, to space-based observatories such as Hubble, Wilkinson microwave anisotropy probe (WMAP) and Planck. The recent and stunning detection of massive gravitational encounters between black holes by means of gravitational waves is only one of several windows that have recently been opened into the study of distant and exotic objects, and to ever-earlier epochs of the universe.

Cosmology is in a golden age of discovery, the likes of which have rarely been seen in the physical sciences. Theory has hardly kept up, but its bringing together of the fundamental theories of physics is also historic in its vitality. It draws them together in ways that put pressure on each: whether because, as in quantum mechanics, cosmology is an application in which there is no 'external observer'; or because, as with the standard model of particle physics and general relativity, there is tension between their basic principles; or because, as in statistical mechanics, it highlights the extraordinary importance of the initial conditions of the universe for local physics. Add to these components the existing foundational problems of each discipline even in non-cosmological settings: the measurement problem in quantum mechanics, the 'naturalness' problem of the Higgs mass and the cosmological constant or 'dark energy' (so-called 'fine-tuning' problems), and the information-loss paradox of black-hole physics. The result is a heady brew – and this is not even to mention the enigma that is dark matter, making up the bulk of the gravitating matter of the universe, its nature still unknown.

What place, in this perfect storm, for philosophy? Some see none: 'philosophy is dead', according to Stephen Hawking, and needs no engagement from scientists. And indeed, where philosophers of physics have made inroads on conceptual questions in physics, they have tended to focus on cleanly defined theories treated in isolation. Synthetic theories, in complex applications, are messy and ill suited to rigorous analysis, axiomatisation, or regimentation by other means – the standard tools of philosophy. And yet, time and again scientists ignorant of philosophy go on to do it anyway, badly. Philosophical questions are natural to the growing child and beset anyone with an enquiring mind; they are suppressed at best by ring-fencing, at worst by decree. And cosmology has long been a testing ground of philosophy. According to some, its central domain is 'the problem of cosmology'; that is the problem of understanding the world, including ourselves, and our knowledge, as part of the world, to echo Karl Popper.

For all of that, 'philosophy of cosmology' as a body of philosophical literature engaged with contemporary cosmology does not yet exist. This book marks a beginning. In it we have gathered essays edging out from cosmological problems to philosophy, and from philosophical problems to cosmology; it is at the points at which they meet that we hope for the greatest synergies. Both disciplines are at a turning point: cosmology, in virtue of the greatly accelerated rate of data acquisition that brings with it a new set of fundamental problems with few or no precursors in any of the empirical sciences; philosophy of physics, in virtue of significant progress in the last two decades on the foundations of quantum mechanics, in particular an understanding of quantum theory as a realistic theory applicable to the universe as a whole. Add to this new perspectives on the meaning of physical probability, the role of probability in the foundations of statistical mechanics, and the significance of the initial state of the universe for local physics and the arrow of time.

In their contributions to the book, philosophers explore still wider concerns in epistemology, metaphysics, and philosophy of mathematics. We have urged them, just as contributors working in cosmology proper, to push in the least comfortable directions and to expose rather than conceal what is conceptually obscure in their undertakings. As editors, we are united in the view that the hard problems of cosmology should be thrown into as sharp a relief as possible, and in the simplest terms possible, if there is to be any hope of overcoming them. It is to that end that this book, and the research programmes from which it was created, were conceived.

We thank the John Templeton Foundation for a grant that enabled the preparation of this volume by supporting a series of workshops and a conference.

Acknowledgments

The editors acknowledge that this project was made possible through the support of a grant from the John Templeton Foundation. The opinions expressed in this publication are those of the authors and do not necessarily reflect the views of the John Templeton Foundation. Our gratitude goes especially to the late Dr Jack Templeton for his personal support of this project.

We would like to thank Barry Loewer for his valuable input and Jeremy Butterfield for his insightful advice throughout the conception, planning and execution of the project. Our special thanks go to Christopher Doogue for his administrative assistance, Cambridge University Press for supporting this book project and, especially, Vince Higgs, Charlotte Thomas, Lucy Edwards, Helen Flitton, Viv Lillywhite and Timothy Hyland for their editorial management.

Last, but not least, we acknowledge and value the help of all our colleagues whose contributions made this project a reality.

Part I
Issues in the Philosophy of Cosmology

1

The Domain of Cosmology and the Testing of Cosmological Theories

GEORGE F. R. ELLIS

This chapter is about foundational themes underlying the scientific study of cosmology:

- What issues will a theory of cosmology deal with?
- What kinds of causation will be taken into account as we consider the relation between chance and necessity in cosmology?
- What kinds of data and arguments will we use to test theories, when they stretch beyond the bounds of observational probing?
- Should we weaken the need for testing and move to a post-empirical phase, as some have suggested?

These are philosophical issues at the foundation of the cosmological enterprise. The answer may be obvious or taken for granted by scientists in many cases, and so seem hardly worth mentioning; but that has demonstrably led to some questionable statements about what is reliably known about cosmology, particularly in popular books and public statements. The premise of this chapter is that it is better to carefully think these issues through and make them explicit, rather than having unexamined assumed views about them shaping cosmological theories and their public presentation. Thus, as in other subjects, being philosophical about what is being undertaken will help clarify practice in the area.

The basic enterprise of cosmology is to use tested physical theories to understand major aspects of the universe in which we live, as observed by telescopes of all kinds. The foundational issue arising is the uniqueness of the universe [66, 27, 28]. Standard methods of scientific theory testing rely on comparing similar objects to determine regularities, so they cannot easily be applied in the cosmological context, where there is no other similar object to use in any comparison. We have to extrapolate from aspects of the universe to hypotheses about the seen and unseen universe as a whole. Furthermore, physical explanations of developments in the very early universe depend on extrapolating physical theories beyond the bounds where they can be tested in a laboratory or particle collider. Hence philosophical issues of necessity arise as we push the theory beyond its testable limits. Making them explicit clarifies what is being done and illuminates issues that need careful attention.

The nature of any proposed cosmological theory is characterized by a set of features which each raise philosophical issues:

3

1. Scope and goals of the theory
2. Nature of the theory: kinds of causality/explanation envisioned
3. Priors of the theory: the range of alternatives set at the outset
4. Outcomes of the theory: what is claimed to be established or explained
5. Data for the theory: what is measured
6. Limits to testing of outcomes
7. Theory, data, and the limits of science

The following sections will look at each in turn.

1.1 Scope and Goals of the Theory

I distinguish here between the physical subject of cosmology, and the wider concept of cosmologia, which also entertains philosophical issues.

1.1.1 Cosmology

I define "cosmology" as theory dealing with physical cosmology and related mathematical and physical issues. Specifically, it proposes and tests mathematical theories for the physical universe on a large scale, and for structure formation in that context. It is a purely scientific exercise, relating to the hot big bang theory of the background model of cosmology, probably preceded by an earlier phase of exponential expansion, to theories of structure formation in this context, and to the many observational tests of these proposals (see e.g. Silk [88], Dodelson [24], Mukhanov [69], Durrer [25], Ellis, Maartens and MacCallum [31], Peter and Uzan [81]). It is a very successful application of general relativity theory to the large-scale structure of spacetime in a suitable geometrical and physical context. However, because of the exceptional nature of cosmology as a science, it pushes science to the limits, particularly when considering what can be said scientifically about the origins of the universe, or existence of a multiverse.

1.1.2 Cosmologia

By contrast, "cosmologia" considers all this, but in addition deals in one way or another with the major themes of the origin of life and the nature of existence that are raised because the physical universe (characterised by "cosmology") is the context for our physical being. It does so in a way compatible with what we know about physical cosmology and physics. Cosmologia necessarily considers major themes in philosophy and metaphysics, perhaps relating them to issues of meaning and purpose in our lives, and is of great public interest and concern.

1.1.3 Which are we Studying?

The major point to make here is that theorists tackling cosmological issues need to make clear which topic they are dealing with: cosmology, or cosmologia. Either is a legitimate topic for investigation, and one can choose which to tackle, but it should be very clear which is the topic of discussion. What is not legitimate is to use only the methods of cosmology, and claim to solve problems of cosmologia. If one wants to tackle cosmologia, one must use adequate methods to do so, involving an adequate scope of enquiry, methods, and data, as discussed further below.

Now one may think the latter ("cosmologia") is not what academic cosmologists should be dealing with; and indeed most groups dedicated to cosmology in physics or astronomy departments restrict themselves to cosmology as defined here. However, some scientists studying cosmology in such departments are proclaiming confidently about issues of cosmologia in highly publicized books and lectures (e.g. Susskind [92], Hawking and Mlodinow [52], Krauss [57]). Thus in these cases the drive to consider these broader issues comes from cosmologists themselves.

The theme of this section can be stated as:

Issue 1: *We can legitimately consider either the enterprise of cosmology as defined here, or extend our investigations to the further issues referred to in cosmologia; but we must make a clear decision as to which stance we are taking, and then adhere to this clearly and consistently in our work. The scope of the theory proposed should be made very clear at the outset.*

In this way, workers in the field can make quite clear if they are engaged in a purely scientific enterprise of cosmology, or are also entering the philosophical and metaphysical waters embodied in cosmologia, commenting on issues of meaning and purpose as well as discussing issues in physical cosmology as evidenced by astronomical observations.

1.2 Kinds of Causality/Explanation Envisioned

Given a statement of scope of the theory, the next issue is what kind of causality will be taken into account. What kinds of causal influences are assumed to be possible in the universe? How do they relate to the great issues of chance, necessity, and purpose?

Assuming we are dealing with cosmology rather than cosmologia, the basic underlying assumption is that what happens at the cosmic scale is the outcome of the interplay of chance and necessity alone. The fundamental problem is that in the cosmological context of a unique universe, the difference between them is blurred.

1.2.1 Necessity

The inexorable nature of necessity is taken to be embodied in fixed and immutable physical laws, expressed in mathematical form [76]. They are taken to be eternal and unchanging, being the same everywhere in the universe at all times and places. The nature of the causal laws proposed (variational principles, symmetry principles, equations of state, and so on) characterizes the causal factors in action in the physical universe that underlie necessity.

Sometimes it is suggested that the laws of physics change with time or place, for example through variation of constants such as the gravitational constant G [96], or choice of a string theory vacuum; but if so, a higher set of laws will be proposed that determine how this happens, for example laws determining how G changes with time, or the laws of string theory that show how string vacua affect local physics. If this is not done, we have no ability to predict physical outcomes scientifically. If this is done, then it is this higher-level set of laws that are the fundamental unchanging ones that govern what happens: in essence one's first stab at finding the unchanging laws was wrong, but that does not mean they do not exist: the higher level set are of this nature.

The Nature of the Laws of Physics

The laws of physics are not physical entities; they are abstract relations characterizing the behavior of matter or fields. It is not clear if the laws are prescriptive, controlling what happens, or descriptive, describing what happens. If they are prescriptive, where or in what way do they exist? How do they get their causal power? If they are descriptive, something else controls how matter behaves: what is it, and how does it get its causal power? And then, why does matter everywhere behave in the same way, so that it is described by mathematical laws?

Possibility Spaces

It is not clear how to obtain traction in considering these issues. A proposal that to some degree sidesteps them is the idea that possibility spaces are the best way to describe the effects of physical laws. Deterministic laws $\mathcal{L}$ are associated with possibility spaces Ω_P that delimit their possible outcomes, and express the nature of necessity by characterizing what is and what is not possible. In effect, we characterize causality not by the nature of the laws themselves, but by the nature of their solution spaces, which strictly constrain what is possible in the physical world. These include phase spaces in classical theory, Hilbert spaces in quantum theory, the landscape of string theory, and so on. Possibility spaces are hierarchically structured, with multiple layers of description and effective behavior depending on the level of averaging used. Constraints such as conservation laws characterize allowable trajectories in these spaces, and so largely define the dynamics (e.g. in Hamiltonian systems).

1.2.2 Ontology and Epistemology

A distinction is crucial here. Possibility spaces themselves exist as unchanging abstract (Platonic) spaces Ω_P limiting all possible structures and motions of physical systems. They are the same at all places and times. Our knowledge of them however is a representation of that space that is changing with time. That is, we represent Ω_P by some projection:

$$\mathcal{P}(t) : \Omega_P \rightarrow E_P \qquad (1.1)$$

into a representation space E_P where $\mathcal{P}(t)$ depends on the representation we use, and changes with time. This does not mean that physics itself, or the possibilities it allows, are changing: it is just that our knowledge of it is changing with time. Ontology (what possibilities exist, as a matter of fact) is entailed by the nature of Ω_P. Epistemology (what we know about it) is determined by the projection $\mathcal{P}(t)$. The representation space E_P will be represented via some coordinate system and set of units, which can be altered without changing the nature of the entities being represented. Such coordinate changes therefore represent symmetries in the space of possible representations.

The Nature of the Laws

The fundamental issue for cosmology is two-fold. Firstly,

- What kinds of causal laws $\mathcal{L}$ and associated possibility spaces Ω_P hold in the universe? What are their properties?

That is a scientific issue, determined by experiment and observation. According to our current understanding, the laws of physics are locally describable in terms of suitable mathematical equations. They will involve:

- a description of the variables involved and their interactions, governed by specific charges and masses,
- associated variational principles, leading to dynamic equations,
- symmetry principles and associated conservation laws and constraints,
- a specification of the geometry and number of dimensions of the space in which this all happens.

This leads to appropriate partial differential equations, such as Maxwell's equations, Yang–Mills equations, the Einstein field equations, the Dirac equation, and so on, that can be used to calculate the outcome of the application of the laws, given suitable initial data. There will be alternative ways of expressing these laws, for example it may be possible to express them in Hamiltonian or Lagrangian form, or as path integrals, or as partial differential equations, or as integral equations. Given suitable constraints on the physical situation considered, the partial differential equations may reduce to families of ordinary differential equations for suitable variables, which can be expressed in dynamical systems terms.

Why do they have their Specific Nature?

Secondly,

- What underlies the existence of the specific laws $\mathcal{L}$ that hold and associated physical possibility spaces Ω_P? Why do they have the nature they have?

This is a philosophical issue, because there is no experiment we can do to test this feature: we can test what characteristics they have, indeed this is what we do when we determine the effective laws of physics, but we cannot test why they have this nature or what underpins their existence. As far as physical cosmology is concerned, the standard assumption is that their nature, having been determined in some unknown way before the universe began, cannot be different. They just are what they are.

1.2.3 Contingency

Given the laws of physics, the outcome depends on a set of boundary conditions $\mathcal{B}$ for the relevant equations. These are contingent in that, unlike the laws of physics, they could have been otherwise. This is the essential difference between laws and boundary conditions.

Then one has to ask what fixes the specific boundary conditions that actually occurred. In the past, the assumption was that it was fixed by a symmetry principle (the "Cosmological Principle") [11]. The current tendency is to assume they are fixed by chance, that is, some random process determines them [46]. The outcome (the universe that actually comes into being) is then fixed by the deterministic laws $\mathcal{L}$ that map Ω_P into itself, giving a unique outcome from the initial data:

$$\mathcal{L}(\mathcal{B}) : \Omega_P \rightarrow \Omega_P \qquad (1.2)$$

with different outcomes for different initial data $\mathcal{B}$.

1.2.4 The Relation Between Them

But what does contingency mean in the context of the universe as a whole? Is it a meaningful concept? If so how do we cash it out? The difference is clear in the case of systems situated in the universe, but problematic for the universe itself.

The problem is that as we have only one universe to observe, so we have only one set of initial conditions to relate to; and no way to test if they could indeed have been different. All we experience is that one specific set of initial conditions has indeed occurred. It could be that only one set of initial conditions is possible: in which case their value is fixed by a law, not by a choice among a number of possible different initial states, which is the essence of the idea of contingency.

Are there laws for the cosmos as a whole? This is highly disputed territory. In the 1960s–1970s the idea of a Cosmological Principle was proposed [11], which in effect is a law of initial conditions for cosmology, determining what spacetimes actually occur out of all those that are possible. It determines the subspace $\mathcal{C}$ of realizable cosmologies out of the set of all possibilities Ω_P. Note that this is still an abstract set of possibilities, as it will be a family of space-times rather than only a specific one with all parameters determined, assumed to lead somehow to one or other actually existing universe out of these possibilities. More recently the proposal that there is a law of initial conditions for the universe as a whole has been made by some authors on the basis of various ideas about quantum cosmology (see Hartle [48]). But the concept of a law that applies to only one object is in conflict with the essential nature of a law, namely that it applies to a class of similar objects. To give it meaning one in effect has to introduce an actually existing ensemble $\mathcal{E}$ of universes out of those in $\mathcal{C}$, thus denying the uniqueness of the universe, together with an explicit or implicit probability measure $\mathcal{M}_E$ on this ensemble, allowing one to talk about chance in this context.

Three problems occur. First, these proposals are plagued by infinities that tend to occur in such ensembles. Thus the outcome of using a proposed measure may be ill defined. Second, while we can argue for specific such measures $\mathcal{M}_E$ on various grounds, we cannot test any proposed measure by observation or experiment, because we cannot check the outcomes as regards numbers of universes that occur with the predicted frequencies. It is therefore an untestable choice, with the outcome determined by *a priori* philosophical assumptions (such as a "principle of mediocrity" [99]). In practice, measures on the spaces $\mathcal{E}$ or $\mathcal{C}$ of cosmological models are chosen to give desired results: specifically, trying to make the one universe we can see appear to be probable. Finally, one has to clarify if this measure is proposed as relating to a physically existing ensemble of universes, or a conceptual one. If the former, how do we show it exists? If the latter, how does it influence what actually happens?

Issue 2: *How can there be a meaningful difference between chance (i.e. contingent boundary conditions for cosmology) and necessity (the deterministic physical laws that act in an inevitable way, as characterized by possibility spaces) in the context of the existence of the unique single universe we probe by astronomical observations and physical experiments?*

The distinction seems rather arbitrary. We cannot establish the difference observationally or experimentally. Nevertheless we usually assume there is such a difference, as that is how ordinary physics works.

1.2.5 Creation of the Universe

These issues come to a head in terms of theories of creation of the universe (e.g. Hartle and Hawking [49]). The problem with theories of creation of the universe is that we can only proceed by applying physical ideas that we determine within the universe, and extrapolate them to applying to the universe itself. It is not clear that makes sense, *inter alia* because there is no concept of space or time before the universe exists, and certainly there is no way to test their validity (we cannot rerun the project and try with varied starting conditions, for example).

To additionally propose that one has a theory of creation of the universe "from nothing" (e.g. Krauss [57]) does not make scientific sense, for one can only develop a creation theory by assuming the nature of the laws to be applied: quantum field theory, for example [80], perhaps extended to the standard model of particle physics, or something like it. These are assumed to causally precede the coming into being of the physical universe. Calling that "nothing" is sleight of hand [4]: it is a very complex structure that is assumed to in some sense pre-exist the coming into being of the universe (for they are assumed to cause its existence). So where or in what sense do all these items exist prior to the existence of the universe itself? Presumably they are meant to inhabit some kind of Platonic space that pre-exists the universe, which somehow gains causal power over material entities even though space and time do not exist then. That is a very powerful form of existence.

There is a lot that needs clarification here. Supposing this was indeed clarified into a consistent theory, it would not be a testable theory of how things happened, even if some proposed outcomes might be testable.

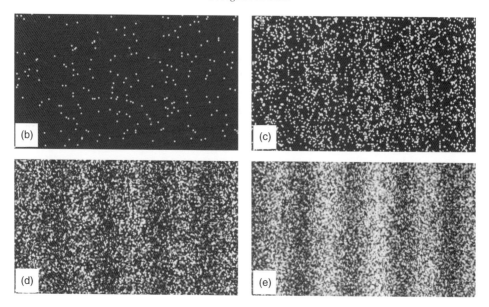

Figure 1.1 **Irreducible randomness occurs in experimental quantum physics**. Double-slit-experiment performed by Dr. Tonomura showing the build-up of an interference pattern of single electrons. The numbers of electrons are (b) 200, (c) 6,000, (d) 40,000, (e) 140,000 [from Wikimedia commons]. Quantum theory predicts the statistics of outcomes that eventually builds up, but cannot even in principle predict where each individual electron will arrive.

1.2.6 Chance: Randomness

Chance occurs within the universe, due to statistical interaction between emergent levels of structure. However, attractors in possibility spaces determine basins of attraction, and so relate chance and necessity: substantial variation of data can lead to the same outcomes, reducing the effect of randomness. In the case of strange attractors, the effect is the opposite: outcomes are unpredictable in practice, even though they are predictable in principle.

Irreducible Quantum Uncertainty

However, in addition, irreducible quantum uncertainty occurs in the universe due to quantum physics effects: the initial data do not determine the outcome uniquely, even in principle, because quantum physics underlies all physics. This is illustrated in the double-slit experiment outcomes shown in Figure 1.1.

Assuming that standard quantum theory as tested by many experiments is right, there is no way even in principle of determining where each individual photon will end up on the screen after passing through a very narrow double slit, although the statistics of the outcomes are fully determined by the Schrödinger equation [37, 55, 108].

Now it is rather strange that many working physicists talk about quantum physics as if it were a deterministic theory. This is because the Schrödinger equation is a deterministic equation, uniquely determining the evolution of the wave function Ψ from initial data. But the value of Ψ does not determine specific physical outcomes: it determines only the statistics of these outcomes [37, 38, 55]. When a measurement takes place, the unitary Schrödinger equation does not apply: a superposition of eigenstates collapses to a single eigenstate (see e.g. Zettli [108]: 158), which is a non-unitary change. The philosophy seems to be that as the behavior of an ensemble of events, and in particular the statistics of associated energy changes and scattering angles, is uniquely determined, quantum mechanics is a deterministic theory. But there is no ensemble to consider without individual events! (as is seen in Figure 1.1): and irreducible uncertainty applies to the individual events. Physics is in fact unable to predict unique specific outcomes from initial data: it does not (at the micro level) determine the particular individual things that happen.

Various attempts have been made to show this is not the case, e.g. the Many Worlds (Everett) interpretation of quantum physics [55], and the Bohm Pilot Wave Theory [8]. In these cases one has a deterministic (unitary) outcome taking place behind the scenes of what is actually experienced by physicists working in a laboratory. None of these proposals alters the fact that experiments definitively demonstrate that we actually experience irreducible quantum uncertainty in a laboratory as we try to predict the outcomes for specific individual events from measurable initial data [37, 108].

Unpredictability in Cosmology

Now one might think that this is irrelevant to physics at a cosmological scale, because this only happens at microscales. However, if the current standard model of cosmology (see Section 1.4) is correct, this is not the case, because in this theory, quantum fluctuations during the inflationary era are amplified to macroscopic scales by the exponential expansion that occurs, and then result in the seed perturbations on the last scattering surface that lead to the formation of clusters of galaxies and galaxies by a bottom-up process (small-scale structures form first, and then assemble to form larger scale structures later).

This means that if we knew everything measurable about the universe at the start of inflation, this would not determine the specific individual astronomical structures that we can observe in the universe at later times. The statistics of these structures is predictable and indeed is used as a test of cosmological theory [24, 69, 25, 31, 81]. The existence of specific individual structures such as our own galaxy is not caused by the quantum processes in operation, for this is the nature of quantum physics as has been conclusively determined by laboratory experiment, as shown in Figure 1.1: it is subject to irreducible uncertainty. Given complete data at the start of the universe, cosmological theory cannot predict the specific structures that will later come into being. The conclusion is that necessity does not always apply in the universe, even at cosmological scales. However, the statistics of what happens is uniquely determined.

1.2.7 The Origin of Life in the Universe

It is a moot issue as to whether cosmology should be considered as extending to discussion of the origins of life in the universe. Cosmologia must certainly do so – it is central to that enterprise. The viewpoint suggested here will be that the origin of life lies outside the domain of cosmology proper, as it raises so many new kinds of issues to do with the nature of causation in biological systems that are unrelated to the gravitational and astrophysical issues attended to by cosmology.

In particular, a key feature of life is that all biological structures have a function or purpose (Hartwell *et al.* [50]), and that must be taken into account in an adequate way if one enters this territory. Life also crucially involves the idea of adaptive selection to the social and ecological environment. It is dealing with other kinds of causal processes than are considered by physical cosmology. They certainly do take place in the universe as we know it, but physics *per se* does not consider them.

1.3 Priors of the Theory

The priors of the theory are the range of alternatives set at the outset. Given an understanding of the relation between chance and necessity, the next issue is the range of causal and contingent priors: what is presumed to exist at the start of the evolution of the universe? What range of alternatives will be considered for necessity, and for contingent elements? Only if we consider sufficient alternatives can we consider how robust our theory is in the space of theories.

1.3.1 Necessity: the Prior Alternatives Considered

The possibility spaces specified in a theory depend on the range of alternatives we are prepared to consider, which are defined by the priors of that theory.

What range of physical laws or regularities is assumed to hold to embody necessity? For example:

- What range of theories of gravitation will we consider? – general relativity, scalar-tensor theories, f(R) theories, bimetric theories, conformal theories, fourth-order theories, theories with torsion, and so on (see e.g. Gair *et al.* [42]).
- What kinds of theories of dark energy will be included, as well as a cosmological constant? Will they for example include quintessence, phantom energy, or chameleon theories?
- What possibilities will we take into account for dark matter? MACHOs, WIMPS, axions, supersymmetric partners of known particles?
- What inflaton possibilities are included? Single field, multiple field, torsion? Will they for example include a non-minimally coupled Higgs field?
- Will we look at alternatives to inflation? – will we consider pre-big bang models, string gas cosmology, any of the various cyclic models?

- Will we look at versions of inflationary theory without an initial singularity [30, 23] (which are possible, because they do not obey the energy conditions needed for the singularity theorems [51])?
- Are fundamental constants to be taken as priors, or to be determined by the theory? Will they be allowed to vary in time [96]?
- If one is to attempt a theory of creation of the universe, what will be the prior assumptions? What quantum gravity options will we explore? (string theory, loop quantum gravity, causal set theory, something else [67])? Do we assume physical laws preceded the existence of the universe (in causal terms), or came into being with the universe?
- Will we propose a model starting from "nothing"? If so, what is "nothing"? Can "nothing" exist? Can it cause anything?

Overall, there are many possibilities. To check the uniqueness and robustness of the standard model in the space of all models, we should check alternatives; we cannot test alternatives unless we consider them.

1.3.2 Geometry and Topology: the Priors Considered

Similarly, what range of initial or boundary conditions for the theory will be taken into account in order to establish its robustness?

There is a larger range of contingent possibilities (initial conditions) than often considered [31]:

- Will we consider alternative spatial topologies [59], or only universes with simply connected 3-spaces? There is an infinite number of topologies for the case $k = -1$.
- In particular, will we consider small universes, that we could have seen around since decoupling [32, 21]? These are the only universes where we can see all that exists in the universe, and perhaps even see our own galaxy as it was in the past.
- Will we consider models with positive and negative spatial curvature? Many papers do not, but positive curvature makes a huge difference as regards possibilities in the future and the past [84], as well as suggesting a finite universe and simplifying boundary value problems.
- Will we consider anisotropic modes (as in Bianchi spatially homogeneous models [103])? If one believes in generic conditions, they will be there too.
- Will we consider inhomogeneous (Lemaître–Tolman) models [10], where the Copernican principle does not hold? These have the potential to explain the supernova magnitude–redshift relations without the need for any dark energy or cosmological constant [16]. One can only see how viable they are by examining their observational implications.
- Will we consider models based on voids, such as the Lindquist/Wheeler models, where the large-scale homogeneity is not assumed from the outset but rather is derived as an averaged outcome of masses imbedded in vacuole regions [20]?

We need some foundation to proceed as the basis for adequately studying the relation of chance and necessity. As an example, it is somewhat paradoxical that while inflation is claimed to solve the smoothness problem, almost all studies only consider almost-Friedmann–Lemaître–Robertson-Walker (FLRW) models – which start off extremely smooth.

1.3.3 Priors for Studying Cosmologia

Extending the scope of consideration to cosmologia will imply consideration of priors concerning a range of further issues, for example the existence of possibility spaces to do with life and thought. Thus these may include considering as priors.

Possibility spaces Ω_L for the existence of life, for example those discussed by Wagner [101]. If our studies extend to the existence of life, we must necessarily be concerned about the prior conditions that allow life to exist. The possibility of life existing is an eternal unchanging feature of the universe, depending on the nature of the laws of physics and their relation to biological possibilities; whether it actually comes into existence is a separate issue, based on contingent features related to the nature of the particular universe or universes that happen to come into existence.

Abstract Platonic spaces related to the logical operation of the mind, for example a Platonic world Ω_M of mathematical abstractions (Changeux and Connes [17], Penrose [75]), learnt about by the human mind through its neural network structure (Churchland [18]), and *inter alia* underlying the practice of theoretical physics. The point is that key parts of mathematics are discovered rather than invented, e.g. the fact that the square root of 2 is an irrational number, the value of π, and Pythagoras' theorem. These are abstract truths that are independent of the existence and culture of human beings, and pre-exist the existence of life, perhaps even of the universe itself. While they need not be considered by cosmology, cosmologia must consider them too, for they are part of what exists and helps shape human thoughts and culture.

In each case the distinction between what actually exists (the Platonic reality) and what we know about them (our representations of this reality), as discussed in Section 2.1, will hold. Why they exist with the form they have, and the nature of their relation to possibility spaces for physical laws, is a deep philosophical mystery.

1.4 Theoretical Outcomes

The standard theory gives a set of outcomes of explanatory and descriptive models that are generally agreed on, plus some further claims that are more controversial.

1.4.1 Outcomes of the Theory: models

Outcomes include models of what exists at a cosmological scale, claims about physical causes of their existence in terms of effects of forces acting on matter, claims about having tested these statements through astronomical observations as well as laboratory or collider experiments where relevant, and claims of unifications of previously separate domains and results to create such models.

The basic model [24, 25, 31, 81] is a model of the smoothed-out geometry of the universe – usually a Robertson–Walker spacetime – and its time evolution, driven by matter, radiation, and possibly a cosmological constant. It extends to including astrophysical and physical aspects of the early universe associated with the hot big bang era such as baryosynthesis, nucleosynthesis, a radiation-dominated to matter-dominated transition, and decoupling of matter and radiation leading to relic blackbody cosmic radiation. It will usually include an even earlier inflationary era driven by one or more effective scalar fields, and a later dark energy-dominated era.

Outcome 1: Background Model

The standard major outcome is the standard expanding universe first background FLRW model governed by gravity [51]. Its metric in comoving coordinates is:

$$ds^2 = -dt^2 + a^2(t)\{dr^2 + f^2(r)d\Omega^2\} \tag{1.3}$$

where $a(t)$ is the scale factor, $f(r)$ is $\{\sin r, r, \sinh r\}$ if the normalised spatial curvature k is $\{+1, 0, -1\}$, and $d\Omega^2$ is a the metric of a unit 2-sphere [84]. The dynamics is very early inflation followed by a hot big bang radiation-dominated era, including nucleosynthesis, followed by decoupling and a matter-dominated era. At late times, dark energy, possibly a cosmological constant, dominates and causes an acceleration of the rate of expansion. Alternative models may allow for anisotropy [103] or inhomogeneity [10, 16, 20, 31].

Predictions: cosmological relics The standard model supplemented by physics and astrophysics firstly predicts what can be called geological data: that is, data about present-day entities that relates to events on our world line in the far distant past [26]. These include for example element abundances due to primordial nucleosynthesis [71, 89] followed by stellar nucleosynthesis. These all put constraints on the background cosmology. Then there is the relic background radiation; but that is detected by telescopes, so is dealt with in the next paragraph.

Predictions: astronomical data Secondly, there are astronomical data gathered from light emitted on our past light cone [58, 26]. The standard model predicts area distance versus redshift relations which are observable via standard candles or standard-size objects [87], together with number count versus redshift relations for objects whose observable numbers are related to the density of matter in the universe by a standard bias factor. It also predicts cosmic blackbody radiation as a relic of the hot big bang era, with a very accurate blackbody spectrum [39] and complex angular power spectrum [25].

These predictions entail a number of model parameters [81, 1, 60] such as the Hubble constant, the density of baryonic matter, the density of cold dark matter, and the value of the cosmological constant, that are not fixed by the theory but rather are observationally

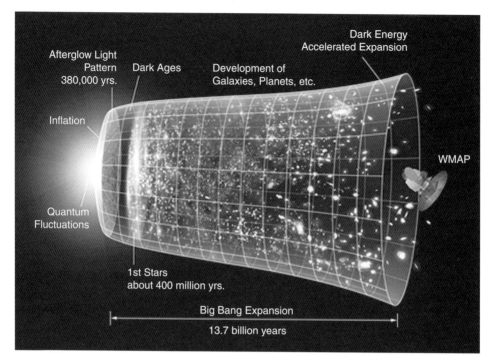

Figure 1.2 **The expansion history of the universe** (from: WMAP website http://map.gsfc.nasa.gov/.).

determined. Extended models predict anisotropies that can occur in the model [58] or geometrical effects such as transverse versus tangential Hubble parameter and redshift space distortions [64].

Outcome 2: Perturbed Models

The further outcome is statistical predictions for perturbations about the background spacetime, giving a model of structure growth [24] and its effects on CBR anisotropies [86, 25].

The standard model proposes creation of seed fluctuations by quantum effects during inflation [69], followed by baryon-radiation acoustic oscillations in the hot big bang era, all leading to seed perturbations on the last scattering surface. These then lead to growth of inhomogeneities by gravitational attraction acting on cold dark matter and baryons, modified by dark energy, providing a physical model for the emergence of large scale structures [81].

Predictions: The model predicts the statistics of matter (power spectra and correlation functions) together with a background radiation anisotropy power spectrum exhibiting a Sachs–Wolfe plateau at large angular scales and baryon acoustic oscillations (BAO) peaks at smaller angular scales [24, 25, 81]. It has a basic parameter (the amplitude of perturbations on the last scattering surface) that is not predicted by the theory but rather is observationally determined. Together with other parameters and observations such as the spectral tilt and the B-Mode polarization spectrum, it can be used to choose between various inflationary proposals.

This is all summarized in the iconic Figure 1.2, provided by the WMAP team. From the inflationary era onwards, this model is tested by many kinds of data that fit together in a solid way to support it (see Section 5.5). However major issues remain.

Outcome Limits

There are substantial limits to the outcomes provided by the standard model of cosmology, within the domain of what it aims to explain. We do not have a unique model for inflation (we do not know what inflaton is), nor do we have a good understanding of what dark matter is, nor do we know the nature of dark energy. We also do not have a good understanding of the quantum-to-classical transition that changes quantum fluctuations in the inflationary era into classical fluctuations at the surface of last scattering [79, 91], although there are some tentative models of this process [12].

Despite these lacuna, it gives a good model of the geometry and evolution of the universe and of structures in it, confirmed by the data (see Section 1.5 below).

1.4.2 Proposed Outcomes: Extended Models

The theory is extended by some workers in two different ways, that are not observationally testable in the same way as the standard model.

Extension 1: The Start of the Universe

A variety of models has been proposed for the start of the universe in terms of its origins from some kind of pre-existing state (perhaps a collapsing phase, or a state with positive definite signature, or a non-singular quantum gravity era), through some kind of laws of creation of the universe, or through some specific assumed boundary conditions (including the Hawking–Hartle no-boundary proposal [49]). These then provide conditions for a start to the inflationary era. There is some observational constraint on such proposals through their influence on the inflationary era, but the assumed geometry at that time cannot be observed and the physics or pre-physics presumed to hold at this time cannot be experimentally tested.

However, whatever happens then is required to produce quite special initial conditions for inflation in relation to the space of all possible geometries and their associated gravitational entropy (Penrose [74]). As mentioned above, you cannot study this by only considering perturbed FLRW models: you have to take into account highly inhomogeneous and anisotropic geometries, which greatly outnumber almost all FLRW models; and in most of them, inflation will never start, because anisotropy terms dominate the scalar field terms [85].

Extension 2: The Universe Beyond the Visual Horizon

Some workers make strong claims about what exists beyond the visual horizon; specifically there are claims of the existence of a multiverse of one kind or another [62, 107, 45, 94], for example eternal new inflation, claimed to lead to an infinite number of pocket universes, and eternal chaotic inflation can follow from specific choices of the inflaton potential [46]. Sometimes these may lead to bubble collisions which are in principle observable [40]; however many would not. It is also sometimes claimed that physical parameters such as the cosmological constant will be different in different bubbles; however this is an additional assumption, requiring a hypothesized mechanism for creating this difference. It does not follow from the geometry *per se*. And if it is assumed to follow from the landscape of string theory/M theory, this is of course an unconfirmed speculation about the nature of fundamental physics [7].

The physics presumed to lead to these results (i.e. the specific assumed inflaton potentials that underlie them) cannot be directly tested; however one can get some limits on them from their influence on structure formation and its effects on cosmic microwave background (CMB) anisotropies and polarization [65]. The associated measures that are claimed to make such results probable or improbable are not testable.

1.4.3 Outcomes of the Theory: Unifications

A major goal of science is to unify previously disparate areas, for example unifying the force causing apples to fall with that causing the Moon to orbit the Earth (Newton), or unifying in a single set of equations electricity and magnetism (Maxwell).

Outcome 3: Unifications of Physics and Cosmology

Thus a key further outcome of cosmological theory is unification of a series of physics theories with cosmological theory. The standard model unifies:

- Gravitation on Earth and in the Solar System with the dynamics of the universe, leading to prediction of the time evolution of the universe [41, 84];
- Atomic physics in a laboratory with tight coupling in the hot big bang era and the process of decoupling, leading to prediction of the existence of cosmic blackbody background radiation [5, 71, 25];
- Nuclear physics in a reactor with the process of cosmological nucleosynthesis, leading to prediction of primordial element abundances [71, 102].

Limits on Unifications

The intended unification of particle physics with early universe processes has however not come to fruition because the process of baryosynthesis is not yet linked to laboratory experiments, and the inflaton has not been identified with any known physical field, and hence is not actually linked to established particle physics. The latter unification would however be achieved if the inflaton were the Higgs particle [35], as originally supposed [46].

1.5 Data for the Theory: Testing of Models

A key issue is: what data are to be used to test the theory? Basic physics data are assumed as the foundation for the theory. The major data specifically for cosmology come from telescopes at all wavelengths, ground based and satellite based, and will eventually extend to gravitational wave detectors and perhaps neutrino telescopes.

1.5.1 Laboratory Data and Solar System Tests

These give data in two ways.

The Nature of Particles and Forces

Firstly, they explore the nature of the gravitational force that underlies cosmological dynamics [51]. They confirm that Einstein's General Theory of Relativity is an excellent classical theory of gravitation, correctly predicting all solar system gravitational effects to high accuracy and they place limits on deviation from General Relativity Theory.

They also explore the nature of particles that interact via those forces. In particular, dark matter is not ordinary baryonic matter, and its nature is unknown. Laboratory experiments and collisions at particle colliders constrain theories regarding dark matter in important ways, and might even lead to its discovery.

Cosmic Relics

Secondly, they confirm the nature of what presently exists as a result of the physics operating in the early history of the universe: the existence of matter (leptons, baryons), photons, neutrinos, and of the elements that make up our existence (particularly carbon, hydrogen, nitrogen, oxygen, phosphorus) arising first from primordial nucleosynthesis and then from processing in stellar interiors. Relics from the past are evidence about the past, and so constrain our past thermal and physical history. Measurements of solar system and stellar abundances of elements confirm the standard picture of nucleosynthesis in a satisfactory way, up to some queries about the abundance of lithium that still need clarification [89].

The theory predicts some relics that are probably not directly observable: namely cosmological neutrino and gravitational wave backgrounds. However, their indirect effects may be observable.

1.5.2 Astronomical Data

The specific astronomical data testing cosmological theories are [58, 31, 81]:

- Galaxy and supernovae magnitude–redshift curves,
- Source number counts as a function of redshift or luminosity,

- Matter power spectra, including a BAO peak,
- Matter 2-point and 3-point angular correlation functions,
- The CMB energy spectrum, particularly deviations from a blackbody form,
- CMB angular power spectra, with a Sachs–Wolfe plateau and acoustic peaks,
- CMB polarization power spectra, also with peaks,
- Weak gravitational lensing observations, constraining dark matter.

The analysis of data is made complex by peculiar velocities and associated redshift space distortions [78], as well as weak lensing of distant sources [104]. The CMB bispectrum is a probe of the non-Gaussianity of primordial perturbations. The astronomical data test both the background model and structure formation in that model.

1.5.3 The Basic Model: Isotropy and Spatial Homogeneity

The first question is whether we can show that the averaged geometry of the universe is indeed spatially homogeneous and isotropic, as represented by the FLRW family of models.

Spatial Isotropy

Isotropy of observations on large averaging scales is well established. In particular the matter distribution is isotropic on large enough averaging scales, and the CMB temperature is isotropic to one part in 10^5 once the dipole due to our motion relative to the universal rest frame is removed.

This establishes that in comoving coordinates the metric can be written as:

$$ds^2 = -f^2(r,t)\, dt^2 + a^2(r,t)\{dr^2 + R^2(t,r)d\Omega^2\} \tag{1.4}$$

with $f(r,t) = a(r,t) = 1$ on the central world line $r = 0$. We are supposed to be close to that worldline.

Spatial Homogeneity

However, testing spatial homogeneity, that is, showing the metric in Eq. (1.4) can be reduced to Eq. (1.3), is more difficult, because we can only see down one past light cone. Nevertheless it is indeed testable in various ways [64]:

- Galaxy number counts suggest spatial homogeneity, as pointed out by Hubble and Peebles [72]. However, we observe distant galaxies with large look-back times, so their intrinsic luminosity may have been different at that time, and this compromises the test [68]. Indeed radio source number counts are incompatible with a Robertson–Walker geometry unless there is substantial evolution of either source numbers or luminosity.
- Good standard candles such as type Ia supernovae can be used to test spatial homogeneity through a null test independent of the gravitational theory and also of the equation of state of matter [19].

- Time drift of cosmological redshifts can also be used to test spatial homogeneity [97], but the measurements are a long term project.
- The most sensitive test however is via the kinematic Sunyaev–Zeldovich effect [110], which implies that deviations from spatial homogeneity cannot be large enough to explain the apparent acceleration of the universe at late times [64].[1]

The latter data in particular give us good grounds for adopting an FLRW model for the visible universe on large scales. The Copernican principle does not have to be adopted as an *a priori* philosophical postulate, as was assumed in previous decades (see e.g. Bondi [11]): it is an observationally testable hypothesis.

1.5.4 Characterizing Geometry: Perturbed FLRW Models

Given that the universe is well described by an FLRW model, one wants the specific parameters describing the best-fit such model.

The Parameters

The key set of parameters describing these models is [1, 60]:

1. The Hubble constant, $H_0 = 100h$ km/s/Mpc.
2. The physical density of baryons $\Omega_b h^2$.
3. The physical density of cold dark matter $\Omega_{dm} h^2$.
4. The energy density of dark energy in units of the critical density, Ω_Λ.
5. The equation of state of dark energy w ($w = -1$ for a cosmological constant).
6. The reionization optical depth, τ.
7. The power-law spectral index n_s of primordial density (scalar) perturbations ($n_s = 1$ for scale-free primordial perturbations).
8. The amplitude of primordial scalar curvature perturbations, A_s.
9. The perturbation scalar to tensor ratio r.

Items 1–6 are properties of the background model, while items 7, 8, and 9 are to do with perturbations. However, tests of the background model, determining all these parameters, are best done via studying the matter power spectrum and CMB power spectrum and anisotropy power spectrum, which are properties of perturbations of the background model. One uses statistical fitting that determines all these parameters together (background and perturbations) from a combination of observations via many telescopes [24, 81, 1].

We normally assume the initial fluctuations are adiabatic and Gaussian, as predicted by simple inflationary theory [69]. One can add variables that test this assumption. Given these data, it puts limits on the nature of dark matter, of dark energy, and on the nature of the inflationary era. We can also test if we live in a small universe in various ways [32], but particularly by looking for identical circles in the CMB sky [21]. This adds up to a detailed model of physical cosmology that is well tested by observations.

---

[1] This is in effect an implementation of the Almost Ehlers–Geren–Sachs theorem [90].

Outcome: *The standard model of cosmology is well tested within the observational horizons. Its basic parameters are well constrained by observations, as is the existence of dark matter and dark energy, and observations also constrain the kinds of inflationary epoch that might have occurred. However, it has problems to do with the nature of the inflaton, dark matter, and dark energy. None of these has been identified.*

Thus the standard model may be regarded as observationally well established, but with significant unknown aspects of important dynamical elements.

1.5.5 Predictions Made

The gold standard of scientific theory is predicting the results of observations before they are made. Cosmology can indeed claim to have done that, albeit in a slightly tortuous way, in the following cases:

1. The expansion of the universe was predicted by Friedmann in 1922 before it had been observationally tested [41]. It was independently predicted by Lemaître in 1927 [61], with prediction of a linear redshift–distance relation, as well as presentation of some observational data supporting this idea, before Hubble's famous paper in 1929 [54].
2. The existence of cosmic blackbody radiation was predicted by Alpher and Hermann in 1948 [5], well before it was detected in 1965. The COBE and WMAP satellites verified the blackbody nature of the CBR spectrum with great precision; it has a temperature of 2.72548 ± 0.00057 K [39].
3. Acoustic peaks in the CMB power spectrum were predicted by Peebles and Yu [73] with details of the relation filled in by Bond and Efstathiou [9] and then many others. Recent satellite observations, e.g. by the Planck satellite [1], confirm these predictions in a very satisfactory way.

The latter is particularly impressive, as the structure of the curves is determined by just a few parameters [81].

1.5.6 Relation to Cosmologia

Clearly none of the above data or models have anything direct to do with understandings of purpose, meaning, and the existence of life, except weakly in terms of providing conditions for galaxies and stars to come into existence. While these are necessary for the existence of life, they are far from sufficient, and it says nothing about the nature of or concerns of any intelligence that may emerge. It is theory and data suited for the study of cosmology, not cosmologia.

Issue 3: Adequate data *The data considered must be adequate to support the theory developed, whether cosmology or cosmologia. If it is to do just with cosmologia, the data discussed in Section 1.5.5 will suffice. If it has to do with the existence of life and intelligence, it must include priors and data relevant to these areas too.*

Extending the scope of consideration to cosmologia will imply consideration of data about a range of further issues, for example the existence of possibility spaces to do with life and thought mentioned in Section 1.3.3. However they are handled, it will look at data to do with intention and meaning and purpose as well as chance, at data to do with good and evil as well as impersonal laws of physics. It is a very different project from cosmology.

1.6 Limits to Testing of Outcomes

We can test outcomes by astronomical, physical, and geological data; how uniquely does this constrain the model? There are substantial limits to the testing of cosmological models firstly because of the uniqueness of the universe, and secondly because we observe it from just one spacetime event.

1.6.1 The Basic Limitations

The basic limitations result from the uniqueness of the Universe on the one hand and its vast size and age on the other.

Uniqueness

Firstly, the uniqueness of the universe [66, 27] prevents us from comparing it with other similar objects or re-running it again with other initial conditions. Hence we cannot do the kind of experimental testing that is possible for theories about other physical objects. However, we are able to test its unique history that in fact happened through cosmological relics, that is, "geological" kinds of data.

This is quite different from experimental science, where we can set up experiments with different initial conditions to see the outcomes. It is also different from other historical sciences, such as the study of evolution of life and of continental drift. In those cases we can look at many items of data that are relevant to huge numbers of plants and animals affected by evolutionary theory in the first case, and the various continents and sea beds affected by continental drift in the second. Theory can thus be tested in different contexts, unlike the case of cosmology.

Only One Viewpoint

Secondly, because of the vast scale and age of the universe, on a cosmological scale we are unable to move from a single event in spacetime [26].

The Earth's distance from the Sun is about 8 light minutes, a galaxy is about 50,000 light years in diameter, and the Hubble scale is about 10^{10} light years. The recorded history of the human race is about 10^{-7} of the age of the universe. Consequently we only see a two-dimensional projection of a four-dimensional spacetime, because (at a cosmological scale) we can only observe the universe from just one spacetime event "here and now" [58, 86, 26] by means of light travelling towards us at the speed of light. We can easily test isotropy, but we cannot easily determine how far away distant objects are, and so cannot so easily test spatial homogeneity. That is why distance determination is so important in cosmology. Nevertheless we are able to observationally check for spatial homogeneity, as mentioned above, provided we observe objects whose time evolution is well constrained, or indirectly use the properties of CMB propagation in an inhomogeneous universe.

Despite these limitations, we can reasonably justify the Standard Model of cosmology [24, 69, 31, 81] discussed above:

- inflation occurs with generation of fluctuations, followed by reheating,
- a hot big bang era with baryosynthesis, nucleosynthesis, and BAO, followed by decoupling,
- a matter-dominated era and structure formation, eventually succeeded by a dark energy dominated era.

This can be regarded as well-established scientific theory. However, two limitations are major hurdles to extending tests of the standard model further, as discussed in Sections 1.6.2 and 1.6.3.

Additionally, as implied in Section 1.2.4, we face the problem of *cosmic variance*: we have theories that predict statistics of models but we see only one spacetime, so we may live in a universe which is an outlier in terms of the statistics of the models. On scales significantly smaller than the Hubble scale we can do the same measurement in different patches in the sky and average over the patches. On the Hubble scale we cannot do such an averaging: we have just one bubble we can view at that scale.

1.6.2 The Physics Horizon

The first issue is major limits on testing the relevant physics at early times due to energy limits on what can be done via particle colliders. The highest collision energies currently attainable at the large hadron collider (LHC) are 14 TeV $= 14 \times 10^{12}$ eV. Suppose we could go 1,000 times higher to 10^{16} eV; we still could not experimentally examine the energy scale of inflation predicted by the simplest models ($10^{15} - 10^{16}$ GeV $= 10^{24} - 10^{25}$ eV), which is close to that expected for baryosynthesis (10^{15} GeV $= 10^{24}$ eV), much less determine the nature of what happened in the quantum gravity era, expected at the Planck scale of around 10^{19} GeV $= 10^{28}$ eV.

This limit of about 10^{16} eV may be dubbed the *physics horizon* [28]: one cannot experimentally determine the relevant physics at higher energies: it is not possible to build a particle collider on Earth that could do it. Thus unlike the case of nucleosynthesis, where the theory can be related to data from nuclear reactors on Earth, one has no way of experimentally determining the nature of what happened in the inflationary era – unless the inflaton is the Higgs particle, in which case this is indeed possible [35]. Otherwise the relevant physics is out of experimental reach and hence theories about what happened then are destined to remain untested speculation. The same applies even more so in the case of the quantum gravity era.

1.6.3 Causal and Visual Horizons

Secondly, there is a causal horizon, the *particle horizon* [83, 51], and an observational horizon, the *visual horizon* [31].

Particle Horizon

Anything we can detect must have arrived here at the speed of light, or a lesser speed. We can therefore only probe as far as our past light cone allows since the start of the universe. This limit – the particle horizon – is given at a time t by the comoving matter coordinate value $r = u_{ph}(t)$ where:

$$u_{ph}(t) = \int_0^t \frac{dt}{a(t)} \;\Rightarrow\; d_{ph}(t) = a(t)u_{ph}(t) \tag{1.5}$$

This value depends on what happens all the way back to the start of the universe, and is much larger when inflation takes place than in a non-inflationary universe. In an Einstein–de Sitter universe, Eq. (1.5) gives $d_{ph}(t_0) = 3/H_0$.

This implies that experimental or observational tests made here and now cannot probe causal influences arriving from distances greater than $d_{ph}(t_0)$. Structure formation studies that refer to scales greater than the "horizon scale" are referring to the Hubble scale at early times (when it was much smaller), not to the particle horizon at the present time.

Visual Horizon

Because we can only observe the universe by photons arriving at the speed of light, we can only see as far as our past light cone allows since the time of decoupling of matter and radiation (the universe was opaque before then), and we cannot see to earlier times. Thus what we can see is limited in both space and time. This visual horizon at a time t [31] is given by the comoving matter coordinate value $r = u_{ph}(t)$:

$$u_{vh}(t) = \int_{t_{dec}}^t \frac{dt}{a(t)} \;\Rightarrow\; d_{vh}(t) = a(t)u_{vh}(t) \tag{1.6}$$

where t_{dec} is the time of decoupling of electromagnetic radiation. It is independent of what happened at earlier times, and so in particular is unaffected by whether inflation took place or not. It places major limits on the data obtainable from distant cosmological domains. This is indicated in Figure 1.3, where light cones are at $\pm 45^o$ and the lower line is the surface of last scattering. The only matter in the universe that is observable is that whose world lines intersect our past light cone between now and the surface of last scattering. In an Einstein–de Sitter universe, it is almost the same as the particle horizon, but in an inflationary universe they are very different. The occurrence of inflation does not affect the visual horizon but affects the particle horizon.

This limit applies to all forms of electromagnetic radiation. There are similar horizons for neutrino and gravitational wave observations, where t_{dec} in Eq. (1.6) is replaced by the times of neutrino and graviton decoupling, respectively. No developments in technology will alter these limits.

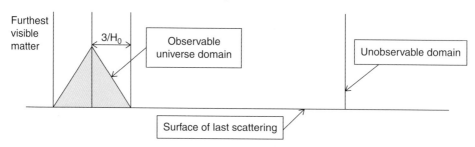

Figure 1.3 Observable limits indicated in a Penrose diagram, where comoving coordinates are used and light rays are at ±45°. The line at the bottom is the surface of last scattering; visual horizons are unaffected by what happens earlier.

The implication is that anything we say about exterior regions, or matter in them, is unverifiable. The universe out there may be spatially homogeneous, or it may not. The matter density may stay constant or rise or fall or oscillate. We could possibly be surrounded by a singular sphere at finite radius, or by an asymptotically flat empty domain. We can claim what we like, and no observation will ever be able to either confirm or disprove what we say [28].

1.6.4 Claimed Multiverses

There are at present many claims for physical existence of various types of multiverse: vast numbers of universe domains, something like ours, but perhaps with varying parameters or even varying physics [82, 46, 107, 93, 45]. The assumption is that we can extrapolate from the observable domain to 100 Hubble radii, 10^{1000} Hubble radii, or much much more, and predict at least statistically what exists out there. Indeed the claim is often made that there is an infinite number of such other universe domains ("pocket universes") out there, usually assumed to have different physics in each [13, 45, 95].

Now no one disputes that for a reasonable distance beyond the visual horizon, things are likely to look like what we see: in that sense other universe domains exist beyond what is visible. But the further out we make claims about what exists, the less justifiable they are (see Figure 1.3). No observational data whatever are available for claims about what happens $10^{10000000}$ Hubble radii away, as implied by these theories. To be sure, most of these theories only predict statistics of what is there, not what actually exists: but there is no way to test these statistical predictions. Thus multiverse claims are not directly observationally testable.

Implied by Known Physics that Leads to Chaotic Inflation

One way of justifying the multiverse claim is to propose it is the outcome of physics that can be regarded as well founded, and makes our observed universe domain probable. The two dominant ways of justifying this are the chaotic inflationary proposal where quantum fluctuations in conjunction with a ϕ^2 potential lead to continual bubble formation [62], and the proposal of Coleman–de Luccia tunnelling between quantum vacua [2].

Now not all inflation is chaotic [63], so evidence for inflation is not necessarily evidence for chaotic inflation. The proposal of a ϕ^2 potential can be tested by its effect on CBR anisotropies, and the Planck data are against it [65]. The key physics for the tunnelling proposal (Coleman–de Luccia tunnelling) in false vacuum-driven inflation is extrapolated from known and tested physics to new contexts; the extrapolation is unverified and indeed is unverifiable; it may or may not be true. The physics is hypothetical rather than tested, as one might assume from some presentations. The situation is not

$$\{\text{Known Physics}\} \Rightarrow \text{Multiverse} \tag{1.7}$$

but rather

$$\{\text{Known Physics}\} \Rightarrow ?? \Rightarrow \{\text{Hypothetical Physics}\} \Rightarrow \text{Multiverse} \tag{1.8}$$

It is a great extrapolation from known physics. This extrapolation is untestable: it may or may not be correct.

If we propose some mechanism for creating a multiverse, we can test it to a limited degree by observing what is in our particular domain, and claim that the multiverse makes probable what we see. But we cannot prove that what we see is indeed probable, because there is no unique measure that applies to such universes [44, 47]. Proposed measures suffer from divergence problems because of the infinities of universes they are supposed to apply to [63, 100]. Further, you cannot test a measure, precisely because we see only one universe domain. In the end, there is no proof the universe we see is probable. That is a philosophical assumption that may or may not be true. It is not a necessity that the universe is probable.

The Anthropic Multiverse

The problem of the fine tuning of the laws of physics that allows life to exist (the Anthropic question [15, 13, 82, 92]) has often been cited in favour of chaotic inflation with different physics or constants of physics in each bubble.

In particular, this is often argued as regards the problem of vacuum energy [106, 82]. The Quantum Field Theory vacuum energy gives a huge value, discrepant with the cosmologically observed value for Λ by 120 orders of magnitude if this vacuum energy gravitates, as is the obvious assumption. This is solved if Λ takes different values in each bubble universe in a multiverse, as might be suggested by string theory for example, for then life will only exist in bubbles with small values of Λ because these will be the only ones with galaxies and stars. Hence the small value of Λ is claimed to be the result of an anthropic selection effect in a physically existing multiverse. Weinberg [105] used this argument to predict a small positive value for Λ, which is an impressive achievement.

This argument however assumes not just that different bubbles exist but also physics that allows different values of Λ, perhaps string theory with many vacua, together with a mechanism that ensures different vacua are realised in different bubbles. The first is problematic [7]; it is particularly this last step that is an untestable assumption about physics. Furthermore, the argument from Λ to the multiverse depends on there being no other way of solving the vacuum energy problem; otherwise it is one of several such proposals and one has to choose between them using philosophical principles, which cannot themselves be tested by a scientific process.

A way round the vacuum energy dilemma used to motivate the multiverse is as follows: the vacuum does not gravitate if we use trace-free Einstein equations plus separate conservation equations (an approach closely related to "unimodular gravity"). This works as follows [36]: the Einstein equations

$$R_{ab} - \tfrac{1}{2}Rg_{ab} + \Lambda g_{ab} = \kappa T_{ab} \tag{1.9}$$

(ten equations) imply energy conservation:

$$T^{ab}_{\;;b} = 0 \tag{1.10}$$

It is plausible to assume Λ would include a vacuum energy term. If instead of Eq. (1.9), we take its trace-free part, we get

$$R_{ab} - \tfrac{1}{4}Rg_{ab} = \kappa(T_{ab} - \tfrac{1}{4}Tg_{ab}) \tag{1.11}$$

(9 equations). Assume Eq. (1.11) is the real gravitational equation, rather than Eq. (1.9); then the trace T of T_{ab} does not gravitate and neither does Λ, so the vacuum energy has no handle on spacetime curvature. Now assume Eq. (1.10) holds separately, and using this, integrate the divergence of Eq. (1.11). This formally recovers Eq. (1.9), but now Λ is a constant of integration that has nothing to do with vacuum energy [106, 36]. Inflation works fine in this modified theory [29] (basically because the Klein Gordon equation for ϕ involves only the gradient of $V(\phi)$, not its absolute value), and variation principles can be found for it [6]. This solves the major discrepancy between the quantum vacuum energy and the observed value of Λ that is otherwise a great problem for gravitational physics, without invoking a multiverse.

Bubble Collisions

In multiverse models based on tunnelling, there is a competition between the rate of nucleation of bubbles and the rate of expansion of the inflating region. If they are delicately balanced, there will be some bubble collisions but the bubble will largely remain separated, and inflaton will continue forever. If the expansion rate is rather higher, there will be no bubble collisions. If it is rather lower, all the bubbles will merge into one and inflation will come to an end.

The one way of confirming the physical model that underlies chaotic inflation predictions is if we could observe the effects of such bubble collisions [43, 3]. It is proposed that such collisions would lead to anomalous circles in CBR anisotropy observations. This would be pretty convincing of the anthropic multiverse proposal if some physical constant such as the fine structure constant or particularly the cosmological constant were different within such circles [70]. A further interesting proposal is to look for collisions via the kinetic Sunyaev Zel'dovich effect [109]. So looking for such traces is certainly worthwhile. However, only some multiverse models make this prediction, so not seeing them will not exclude multiverse models, just that subclass of models. If any of these effects were found that would be positive, but one would have to rigorously exclude other effects that could lead to them, which may be difficult.

Cyclic/Bouncing Universe

A variety of cyclic universes have been proposed that are in effect eternally existing multiverses, with different universe bubbles being created again and again in time rather than occurring at the same time in space. This potentially may lead to the issue of eternal return.

The problem here is that inflation effectively wipes out traces of previous eras [47], so again there is no easy proof of such claims. It has been claimed that they too might lead to circular patterns in the CMB sky, in the case of Penrose' conformal cyclic cosmology [77], or specific patterns of perturbations resulting from previous collapsing phases, in a variety of bouncing models. However, these assume either some kind of generalization of general relativity, where a regular bounce can take place despite the probable growth of inhomogeneity and anisotropy during the collapse, or some form of hypothetical higher-dimensional "brane" structure. It is not clear that this physics can be tested.

Implication of all the Above

The multiverse proposal is not provable either by observation, or as an implication of well-established physics. It is under-determined by the data; in particular theories proposing how different physical parameters will necessarily occur in different bubbles, should those occur, are untested and indeed untestable. It may be true, but cannot be shown to be true by observation or experiment, although a subset of multiverse theories (those with bubble collisions) could possibly receive some limited observational support, by showing that a few bubble collisions have taken place.

Thus it is not a part of testable scientific theory in the usual sense. However, it does have great explanatory power: it does provide an empirically based rationalization for fine tuning, developing from known physical principles. But even if we develop a multiverse theory which can explain in an anthropic way why these parameters have the values they have, this theoretical argument does not amount to a unique prediction, because multiverses allow anything to occur.

Here one must distinguish between explanation and prediction. Successful scientific theories make specific predictions which can be tested. The multiverse theory cannot make any unique predictions because, unless one adds extra untestable philosophical assumptions as an add-on, it can explain anything at all. Any theory that is so flexible is not testable because almost any observation can be accommodated. It is plausibly scientifically informed philosophy rather than science in the usual sense.

Note: we are concerned here with real physically existing multiverses, not with potential or hypothetical multiverses. Those exist as theories that we can consider and develop as we wish. It is the claim that they are physically realised that is the issue at stake.

Multiverses and Cosmologia

Because the multiverse is most often proposed in relation to the anthropic issue, it often is implicitly [82] or explicitly [92, 52, 57] phrased in that context; hence it is related to cosmologia rather than just cosmology. It is then stated to solve the philosophical problems of existence in a purely scientific way.

However, the multiverse does not solve any of the fundamental problems of existence – they just recur in this new context:

- Why the multiverse?
- Why its laws? Why the specific constants associated with those laws?
- Why is it a suitable home for life?

To suggest it solves them is naïve – it just postpones them. It is not a solution to the problems of cosmologia.

1.6.5 Infinities

A particular problem with multiverse theories is the often claimed existence of physically existing infinities of galaxies in them [47, 63]. This raises a more general issue: will physically existing infinities occur in the theory? If so will they play an essential or inessential role?

The key factor here is that infinity is not a very large number: it is an unattainable state, rather than an attainable state. It is not a very large number, it is an amount greater than any number that can exist. It is needed in mathematics, but cannot occur in physical reality. David Hilbert stated its relation to physics very clearly ([53]: 151):

 The infinite is nowhere to be found in reality, no matter what experiences, observations, and knowledge are appealed to. It neither exists in nature nor provides a legitimate basis for rational thought ... The role that remains for the infinite to play is solely that of an idea ... which transcends all experience and which completes the concrete as a totality ..."

The interesting cosmological claim in this regard is the idea that a spatial infinity necessarily results instantaneously from the tunnelling processes that are supposed to lead to new universe bubbles [40]. However, this does not work in a finite time if we remember that the origin of such a bubble cannot be exactly a spacetime point [34]. The potential spatial infinity arising from the tunnelling takes an eternity to attain completion, no matter how small the bubble nucleus is, and so is never attained in any finite time. That is the true nature of infinities in cosmology. They are always out of reach and remain potential rather than actualised. One can therefore propose as a basic principle:

Finiteness *Infinities, which are essentially unattainable by their very definition, should not occur as essential elements in any proposed physical theory, including cosmology.*

This contrasts with their frequent invocation by some cosmological theorists. It is a huge act of hubris to extrapolate from the observable universe domain to a supposed physically existing infinity, never encountering any limit (remember the conformal diagram in Figure 1.3). Hilbert gives a more supportable position.

In any event, claims of physically existing infinities of universes are not scientific statements – if science involves testability by either observation or experiment. No matter how long you count entities, you cannot prove there is an infinite number of them – if indeed you can see them, which you can not in this case (horizons!).

1.6.6 Creation

Various theories claim to explain the creation or start of the universe on a scientific basis, essentially in terms of some or other of the known laws of physics [98, 49, 57]. But as mentioned in Section 1.2.5, this is a very problematic claim.

Classical general relativity theory predicts that if the usual energy conditions are satisfied, there should be a start to the universe [51]. Now this is a very dramatic event: if it occurs, it is a spacetime singularity, representing a start to matter and fields, to space and time, to physics itself. Can this be considered in a causal way? The whole concept of causality is called into question in this context, because it depends on the existence of the categories of space and time – which did not exist before the universe came into being. If so how, and in terms of what pre-existing entities or powers did this happen?[2]

The attempts made to bring this within the powers of physics, however, assume some or other of these elements – specifically, the laws of physics, involving *inter alia* the entire structure of quantum field theory, Hilbert spaces, variational principles, symmetry principles, and so on [80] – preceded the existence of the universe and so were able to bring it into being. In what way can these exist before the universe exists? What alternative ways are there of considering this?

These issues were mentioned in Section 1.2.5. But the further issue here is, suppose we have a theory for creation of the universe (out of nothing, or out of something), can we scientifically test such proposals? How do we test what occurred before physical events existed? How do we test if laws based in spacetime features somehow had an existence before space or time had come into being?

It is clear we can do no laboratory or collider experiments that are relevant to such theories. If we have such a theory it may make some claims about what cosmological features will be visible today. But there will be no way to tie any such prediction uniquely to such a theory, as there is already a number of alternatives, usually with adjustable parameters that can be used to fit any predicted cosmological effects to the data.

---

[2] I am very aware that the term "happen" hardly makes sense in this context; but we have no other way to think about it than to use this or a similar term.

1.6.7 Limits of Testing

The burden of this section is to recognise the limits of testing theories in the cosmological context.

Issue 4: Limits on testing and observations *There are strict limits on observations in cosmology both in terms of space and time. There are also limits on testing the physics relevant to cosmological theories. These limits strongly constrain the degree to which we can observationally determine the nature of cosmological models on the largest scales and at the earliest times.*

Despite these limits, many strong claims are being made about what can be determined in cosmology. By great ingenuity we have in fact a well-tested model of what exists in the observable universe domain (Figure 1.3). Outside that domain, our models are not constrained by observations in the same way.

1.7 Theory, Data, and the Limits of Science

The very nature of the scientific enterprise is at stake in the string theory and multiverse debates: their proponents are proposing weakening the nature of scientific proof in order to claim that multiverses together with string theory provide a scientific explanation for what we see. This is a dangerous tactic. Leonard Susskind explicitly states the criteria for scientific theories should be weakened [56], as do Richard Dawid [22] and Sean Carroll [14]. The cosmological ideas and physical theories put forward are seen as being so compelling that we can loosen the need for empirical testing.

On the face of it, this proposal for "post empirical science" (Dawid) is a highly questionable proposal [33]. This is specifically where careful philosophical attention is needed to throw light on the argument put forward.

1.7.1 Criteria for a Scientific Theory

The usual criteria for a theory being scientific are [27, 28]:

1. Satisfactory structure: (a) internal consistency, (b) simplicity (Ockham's razor), (c) "beauty" or "elegance", (d) usually, a mathematical formulation.
2. Intrinsic explanatory power: (a) logical tightness, (b) scope of the theory – unifying otherwise separate phenomena.
3. Extrinsic explanatory power: (a) connectedness to the rest of science, (b) extendability – a basis for further development.
4. Observational and experimental support: (a) the ability to make quantitative predictions that can be tested; (b) confirmation: the extent to which the theory is supported by such tests.

The problem is that in the context of cosmology, these will conflict with each other. You have to choose between them. It is particularly the last that characterizes a theory as scientific, and it is the one that is being brought into question.

In order to support multiverses and string theory as scientific theories, of course the first prize would be to show that these theories are indeed after all testable, in some way not recognized in the discussion above. Failing that, one has to consider in what way one can justify them as scientific when such tests are not available.

1.7.2 Arguments

Various arguments have been given by Dawid and Carroll, related to the underdetermination of the theories by the possible observations. There is not space to do a proper response here: there is of course a large literature on the nature of scientific theories and their development and philosophical status. The purpose of this chapter however is to point out that there is serious philosophical work to be done in developing a rigorous framework for selecting what should be included in the scientific fold, and what is rather regarded as philosophy. I will just make a few comments.

The main arguments for weakening the requirement for experimental testing seem to be the following:

- The theory proposed is the only game in town. In the case of cosmology, it is the only explanation of the very small, observationally determined value of the cosmological constant, in contrast to quantum field theory predictions. In the case of string theory, it is the only well-developed proposal for unification of all four fundamental forces.
  But then firstly, it must indeed be the only explanation. As indicated above, trace-free gravity is a viable way to solve the vacuum energy problem: a multiverse is not the only possibility. Secondly, there may be no such game at all. In the case of string theory, it is assumed one can unify the strong, electroweak, and gravitational forces in a unified framework. But from a general relativity viewpoint, gravity is fundamentally not a force at all: it is an effect of spacetime curvature. It is therefore perfectly possible there is no such unification.
- If the theory provides unexpected unifications of different areas of physics, it must be true.
  But this is a statement about mathematical models, it does not always imply physical realization of these mathematical theories; that has to be decided by experimental test. Thus for example anti-de Sitter/conformal field theory (CFT) dualities are proposed to give results about solid state physics, an astounding such unification. But multiverses do not provide any specific physics unification.
- Extension of mathematical models in this way to develop elegant new theories has worked well in the past, so it will work again in the future.
  This ignores the cases where it did not work, for example the SU(5) Grand Unified Theory that was thought to be right because it was so simple and beautiful a unification of the electroweak and strong forces, but was eventually shown to be wrong by proton decay experiments.

As regards theories about the coming into existence of the universe itself, and what makes specific laws fly, there are many games in town. The one thing that is quite clear is that they cannot all be true, because they contradict each other.

1.7.3 What Can we Know?

This chapter has explored the limits to what can be tested in cosmology, and set the framework for considering its relation to the nature of science.

Main Thesis 1: *The standard model of cosmology is well tested within the observational horizons. However, it has problems to do with dark matter, dark energy, and the transition of quantum fluctuations to classical fluctuations. Additionally the nature of the inflaton is unknown.*

Main Thesis 2: *Because of the limits on testing and observation, unless we live in a small universe there will always be uncertainty about what exists on the largest scales. In particular this applies to the supposed nature of any multiverse that is claimed to exist. If theories insist on claiming the physical existence of infinities of any kinds of entities, they are making claims that are well beyond the bounds of testable science.*

Main Thesis 3: *Statements about the nature of/causes of the origin of the universe will of necessity always be speculative rather than proved science. Attempts to claim one has solved the issue of the origin of the universe and of Cosmologia on the basis of scientifically testable theories are unfounded. They are attempts to pass off philosophical predilections as established science. They mislead the public about what science can say about important issues.*

Philosophers of science should team up with scientists to clarify the boundaries of science in this case where testability is pushed beyond its limits.

References

[1] P. R. Ade *et al.*, (Planck Collaboration) (2015) Planck 2015 results. XIII. Cosmological parameters. arXiv:1502.01589v2.

[2] A. Aguirre, (2007) Eternal Inflation, past and future. arXiv:0712.0571.

[3] A. Aguirre and M. C. Johnson, A status report on the observability of cosmic bubble collisions. *Reports on Progress in Physics*, **74** (2011), 074901.

[4] D. Albert, *On the Origin of Everything: A Universe From Nothing, by Lawrence M. Krauss*, New York Times book review, March 23, 2012. [www.nytimes.com /2012/03/25/books/review/a-universe-from-nothing-by-lawrence-m-krauss.html].

[5] R. A. Alpher and R. C. Herman, On the Relative Abundance of the Elements. *Physical Review*, **74** (1948), 1737–42.

[6] E. Alvarez, (2012) The weight of matter. arXiv:1204.6162.

[7] T. Banks, (2012) The Top 10^{500} Reasons Not to Believe in the Landscape. arXiv:1208.5715.

[8] D. J. Bohm and B. J. Hiley, The de Broglie pilot wave theory and the further development of new insights arising out of it. *Foundations of Physics*, **12** (1982), 1001–16.

[9] R. Bond and G. Efstathiou, Cosmic background radiation anisotropies in universes dominated by nonbaryonic dark matter. *Astrophysical Journal*, **285** (1984), L45–8.

[10] H. Bondi, Spherically symmetrical models in general relativity. *Monthly Notices of the Royal Astronomical Society*, **107** (1947), 410–25.

[11] H. Bondi, *Cosmology*. (Cambridge: Cambridge University Press, 1961).

[12] P. Canate, P. Pearle and D. Sudarsky, (2012) CSL Wave Function Collapse Model as a Mechanism for the Emergence of Cosmological Asymmetries in Inflation. arXiv:1211.3463.

[13] B. J. Carr, Ed. *Universe or Multiverse?* (Cambridge: Cambridge University Press, 2007).

[14] S. Carroll, (2014) "Falsifiability" *Edge Essay 2014: What Scientific Idea Is Ready For Retirement*? [https://edge.org/response-detail/25322].

[15] B. Carter, Large Number Coincidences and the Anthropic Principle in Cosmology. *IAU Symposium 63: Confrontation of Cosmological Theories with Observational Data* (1974), pp. 291–8 (Dordrecht: Reidel, 1974).

[16] M.-N. Celerier, The Accelerated Expansion of the Universe Challenged by an Effect of the Inhomogeneities. A Review. *New Advances in Physics*, **1** (2007), 29. [arXiv:astro-ph/0702416].

[17] J.-P. Changeux and A. Connes, *Conversations on Mind, Matter, and Mathematics*, (Princeton: Princeton University Press, 1998).

[18] P. Churchland, *Plato's camera: How the physical brain captures a landscape of abstract universals*, (Cambridge, MA: MIT Press, 2012).

[19] C. Clarkson, B. Bassett and T. H. C. Lu, A general test of the Copernican Principle. *Physical Review Letters*, **101** (2008), 011301 [arXiv: 0712.3457].

[20] T. Clifton and P. G. Ferreira, Archipelagian cosmology: Dynamics and observables in a universe with discretized matter content *Physics Review*, **D 80** (2009), 103503; Erratum *Physical Review*, **D 84** (2011), 109902 [arXiv:0907.4109].

[21] N. J. Cornish, D. N. Spergel and G. D. Starkmann, Circles in the sky: Finding topology with the microwave background radiation. *Classical and Quantum Gravity*, **15** (1998), 2657–70 [arXiv:gr-qc/9602039].

[22] R. Dawid, *String Theory and the Scientific Method*, (Cambridge: Cambridge University Press, 2013).

[23] S. del Campo, E. I. Guendelman, R. Herrera and P. Labrana, (2015) Classically and Quantum stable Emergent Universe from Conservation Laws. arXiv:1508.03330.

[24] S. Dodelson, *Modern Cosmology*, (Elsevier, 2003).

[25] R. Durrer, *The Cosmic Microwave Background*, (Cambridge: Cambridge University Press, 2008).

[26] G. F. R. Ellis, General relativity and cosmology. In *General Relativity and Cosmology*, Varenna Course No. XLVII, ed. R. K. Sachs (Academic, New York, 1971). Reprinted as Golden Oldie, *General Relativity and Gravitation*, **41** 2009, 581–660.

[27] G. F. R. Ellis, Issues in the Philosophy of Cosmology. In *Handbook in Philosophy of Physics*, ed. J Butterfield and J Earman (Elsevier, 2006). [arXiv:astro-ph/0602280].

[28] G. F. R. Ellis, On the Philosophy of Cosmology. *Studies in History and Philosophy of Modern Physics*, **46** (2013), 5–23.

[29] G. F. R. Ellis, The Trace-Free Einstein Equations and inflation. *General Relativity and Gravitation*, **46** (2014), 1619. [arXiv:1306.3021].

[30] G. F. R. Ellis and R. Maartens, The Emergent Universe: inflationary cosmology with no singularity? *Classical and Quantum Gravity*, **21** (2004), 223–32. [arXiv:gr-qc/0211082].

[31] G. F. R. Ellis, R Maartens and M. A. H. MacCallum, *Relativistic Cosmology*, (Cambridge: Cambridge University Press, 2012).

[32] G. F. R. Ellis and G. Schreiber, Observational and dynamical properties of small universes. *Physics Letters*, **A 115** (1986), 97–107.

[33] G. F. R. Ellis and J. Silk, Scientific Method: Defend the Integrity of Physics. *Nature*, **516** (2014), 321–3.

[34] G. F. R. Ellis and W. R. Stoeger, A Note on Infinities in Eternal Inflation. *General Relativity and Gravitation*, **41** (2009), 1475–84. [arXiv:1001.4590].

[35] G. Ellis and J. P. Uzan, Inflation and the Higgs particle. *Astronomy and Geophysics*, **55** (2014), 1.19–1.20.1.

[36] G. F. R. Ellis, H. van Elst, J. Murugan and J.-P. Uzan, On the trace-free Einstein equations as a viable alternative to general relativity. *Classical and Quantum Gravity*, **28** (2011), 225007. [arXiv:1008.1196].

[37] R. P. Feynman, *The Feynman Lectures on Physics, Vol. 1*. (Reading, MA: Addison-Wesley, 1963), Chapter 46.

[38] R. P. Feynman and A. R. Hibbs, *Quantum Mechanics and Path Integrals*, ed. D F Styer, (Mineola, New York: Dover, 1965).

[39] D. J. Fixsen, The Temperature of the Cosmic Microwave Background. *The Astrophysical Journal*, **707** (2009), 916–20. [arXiv:0911.1955].

[40] B. Freivogel, M. Kleban, M. R. Martinez and L. Susskind, Observational consequences of a landscape. *Journal of High Energy Physics*, **0603** (2006), 039. [arXiv:hep-th/0505232].

[41] A. Friedmann, Über die Krümung des Raumes. *Zs f Phys* **10** (1922), 377–86. Reprinted as a Golden Oldie: *General Relativity and Gravitation*, **31** (1999), 1991.

[42] J. R. Gair, M. Vallisner, S. L. Larson and J. G. Baker, Testing General Relativity with Low-Frequency, Space-Based Gravitational-Wave Detectors. *Living Reviews in Relativity*, **16** (2013), 7 [www.livingreviews.org/lrr-2013-7].

[43] J. Garriga, A. H. Guth and A. Vilenkin, Eternal inflation, bubble collisions, and the persistence of memory. *Physical Review*, **D76** (2007), 123512. [arXiv:hep-th/0612242].

[44] G. W. Gibbons and N. Turok, Measure problem in cosmology. *Physical Review*, **D77** (2008), 063516.

[45] B. Greene, *The Hidden Reality: Parallel Universes and the Deep Laws of the Cosmos*, (New York: Knopf, 2007).

[46] A. Guth, Eternal inflation and its implications. *Journal of Physics*, **A 40** (2007), 6811–26. [arXiv:hep-th/0702178].

[47] A. H. Guth and V. Vanchurin, (2012) Eternal inflation, global time cutoff measures, and a probability paradox. [arXiv1211.1347].

[48] J. B. Hartle, Theories of Everything and Hawking's Wave Function of the Universe. In *The Future of Theoretical Physics and Cosmology*, ed. G. W. Gibons, E. P. S. Shellard and S. J. Rankin (Cambridge: Cambridge University Press, 2003). [arXiv:gr-qc/0209047].

[49] J. B. Hartle and S. W. Hawking, Wave function of the Universe. *Physics Review*, **D 28** (1983), 2960.

[50] L. H. Hartwell, J. J. Hopfield, S. Leibler and A. W. Murray, From molecular to modular cell biology. *Nature*, **402** (1999), C47.

[51] S. W. Hawking and G. F. R. Ellis, *The Large Scale Structure of Space-time*. (Cambridge: Cambridge University Press, 1973).

[52] S. Hawking and L. Mlodinow, *The Grand Design*, (London: Bantam, 2011).

[53] D. Hilbert, On the Infinite. In *Philosophy of Mathematics*, ed. P Benacerraf and H Putnam, (Englewood Cliffs, NJ: Prentice Hall, 1964) pp. 134-51.

[54] E. Hubble, A relation between distance and radial velocity among extra-galactic nebulae. *Proceedings of the National Academy of Sciences*, **15** (1929), 168–73.

[55] C. J. Isham, *Lectures on Quantum Theory: Mathematical and Structural Foundations*, (London: Imperial College Press, 1929).

[56] H. Kragh, (2012) Criteria of Science, Cosmology, and Lessons of History. arXiv:1208.5215.

[57] L. Krauss, *A Universe From Nothing: Why There Is Something Rather Than Nothing*, (New York: Free Press, 2012).

[58] J. Kristian and R. K. Sachs, Observations in Cosmology *Astrophysical Journal* **143** (1966), 379–99. Reprinted as GRG Golden Oldie: *General Relativity and Gravitation*, **43** (2011), 337–58.

[59] M. Lachieze-Rey and J. P. Luminet, Cosmic Topology. *Physics Reports*, **254** (1995), 135–214. [arXiv:gr-qc/9605010].

[60] D. Larson, J. L. Weiland, G. Hinshaw and C. L. Bennett, Comparing Planck and WMAP: Maps, Spectra, and Parameters. *The Astrophysical Journal*, **801** (2015), 9. [arXiv:1409.7718].

[61] G. Lemaitre, A Homogeneous Universe of Constant Mass and Increasing Radius Accounting for the Radial Velocity of Extra-Galactic Nebulae. *Annales de la Société Scientifique de Bruxelles*, **A47** (1927), 49–59. Published in translation in MNRAS **91** (1931), 483–90, and republished as GRG Golden Oldie: *General Relativity and Gravitation*, **45** (2013), 1635–46.

[62] A. D. Linde, Eternal chaotic inflation. *Modern Physics Letters*, **A01** (1986), 81.

[63] A. Linde and M. Noorbala, Measure problem for eternal and non-eternal inflation. *Journal of Cosmology and Astroparticle Physics*, **2010** (2010), 008.

[64] R. Maartens, Is the Universe homogeneous? *Philosophical Transactions of the Royal Society*, **A 369** (2011), 5115–37.

[65] J. Martin, C. Ringeval and V. Vennin, Encyclopaedia Inflationaris. arXiv:1303.3787.s.

[66] W. H. McCrea, Cosmology. *Reports on Progress in Physics*, **16** (1953), 321.

[67] J. Murugan, A. Weltman and G. F. R. Ellis, *Foundations of Space and Time: Reflections on Quantum Gravity*, (Cambridge: Cambridge University Press, 2012).

[68] N. Mustapha, C. W. Hellaby and G. F. R. Ellis, Large Scale Inhomogeneity Versus Source Evolution – Can We Distinguish Them Observationally? *Monthly Notices of the Royal Astronomical Society*, **292** (1999), 817-830 [gr-qc/9808079].

[69] V. Mukhanov, *Physical Foundations of Cosmology*, (Cambridge: Cambridge University Press, 2005).

[70] K. A. Olive, M. Peloso and J.-P. Uzan, The wall of fundamental constants. *Physical Review*, **D83** (2011), 043509. [arXiv:1011.1504].

[71] P. J. E. Peebles, Primordial Helium Abundance and the Primordial Fireball. II. *Astrophysical Journal*, **146** (1966), 542–52.

[72] P. J. E. Peebles, *The Large-Scale Structure of the Universe*, (Princeton: Princeton University Press, 1980).

[73] P. J. E. Peebles and J. T. Yu, Primeval Adiabatic Perturbation in an Expanding Universe. *Astrophysical Journal*, **162** (1970), 815–36.

[74] R. Penrose, Difficulties with inflationary cosmology. In E. J. Fergus (ed.). 14th Texas Symposium Relativistic Astrophysics, *Proceedings of the New York Academy of Science*, New York, 1989.

[75] R. Penrose, *The Large, the Small, and the Human Mind*, (Cambridge: Cambridge University Press, 1997).

[76] R. Penrose, *The Road to Reality: A complete guide to the Laws of the Universe*, (London: Jonathan Cape, 2004).

[77] R. Penrose, *Cycles of time: An extraordinary new view of the universe*, (New York: Knopf, 2011).

[78] W. J. Percival, L. Samushia, A. J. Ross, C. Shapiro and A. Raccanelli, Redshift-space distortions. *Philosophical Transactions of the Royal Society A*, **369** (2011), 5058–67.

[79] A. Perez, H. Sahlmann and D. Sudarsky, On the quantum origin of the seeds of cosmic structure. *Classical and Quantum Gravity*, **23** (2006), 2317–54. [arXiv:gr-qc/0508100].

[80] M. E. Peskin and D. V. Schroeder, *An Introduction To Quantum Field Theory*, (New York: Westview Press, 1995).

[81] P. Peter and J.-P. Uzan, *Primordial Cosmology*, (Oxford: Oxford Graduate Texts, 2013).

[82] M. J. Rees, *Our cosmic habitat*, (Princeton: Princeton University Press, 2001).

[83] W. Rindler, Visual Horizons in World Models. *Monthly Notices of the Royal Astronomical Society*, **116** (1956), 662–77. Reprinted: *General Relativity and Gravitation*, **34** (2002), 133.

[84] H. P. Robertson, Relativistic Cosmology. *Review of Modern Physics*, **5** (1933), 62–90. Reprinted as: Golden Oldie: *General Relativity and Gravitation*, **44** (2012), 2115–44.

[85] A. Rothman and G. F. R. Ellis, Can inflation occur in anisotropic cosmologies? *Physics Letters*, **B180** (1986), 19–24.

[86] R. K. Sachs and A. M. Wolfe, Perturbations of a Cosmological Model and Angular Variations of the Microwave Background. *Astrophysical Journal*, **147** (1967), 73–89. [Reprinted as Golden Oldie: *General Relativity and Gravitation*, **39** (2007), 1944].

[87] A. Sandage, The Ability of the 200-Inch Telescope to Discriminate Between Selected World Models. *Astrophysical Journal*, **133** (1961), 355–92.

[88] J. Silk, *A Short History of the Universe*, (New York: Scientific American Library, 1997).

[89] G. Steigman, Primordial Nucleosynthesis: Successes And Challenges. *International Journal of Modern Physics*, **E15** (2005), 1–36. [arXiv:astro-ph/0511534].

[90] W. Stoeger, R. Maartens and G. F. R. Ellis, Proving almost-homogeneity of the universe: an almost-Ehlers, Geren and Sachs theorem. *Astrophysical Journal*, **443** (1995), 1–5.

[91] D. Sudarsky, Shortcomings in the understanding of why cosmological perturbations look classical. *International Journal of Modern Physics*, **D20** (2011), 509–52. [arXiv:0906.0315v3].

[92] L. Susskind, *The Cosmic Landscape: String Theory and the Illusion of Intelligent Design*, (New York: Little, Brown, and Co., 2006).

[93] M. Tegmark, Is 'the theory of everything' merely the ultimate ensemble theory? *Annals of Physics*, **270** (2006), 1–51.

[94] M. Tegmark, Parallel universes. In J. D. Barrow (ed.), *Science and ultimatereality: From quantum to cosmos*. (Cambridge: Cambridge University Press, 2004). [astro-ph/0302131].

[95] M. Tegmark, *Our Mathematical Universe: My Quest for the Ultimate Nature of Reality*, (Knopf: Doubleday Publishing Group, 2014).

[96] J.-P. Uzan, Varying constants, gravitation and cosmology. *Living Reviews in Relativity*, **14** (2010), 2. (arXiv:10095514).

[97] J.-P. Uzan, C. Clarkson and G. F. R. Ellis, Time drift of cosmological redshifts as a test of the Copernican principle. *Physical Review Letters*, **100** (2008), 191303. [arXiv 0801.0068].

[98] A. Vilenkin, Creation of universes from nothing. *Physics Letters*, **B117** (1982), 25.

[99] A. Vilenkin, The Principle of Mediocrity. arXiv:1108.4990.

[100] A. Vilenkin, Global structure of the multiverse and the measure problem, *AIP Conference Proceedings*, **1514** (2013), 7.

[101] A. Wagner, *Arrival of the fittest*, (London: Penguin, Random House, 2014).

[102] R. V. Wagoner, W. A. Fowler and F. Hoyle, On the Synthesis of Elements at Very High Temperatures. *Astrophysical Journal*, **148** (1967), 3–49.

[103] J. Wainwright and G. F. R. Ellis, (Ed) *Dynamical systems in cosmology*, (Cambridge: Cambridge University Press, 1996).

[104] J. Wambsganss, Gravitational Lensing in Astronomy. *Living Reviews in Relativity* **1** (1998), 12. [www.livingreviews.org/lrr-1998-12].

[105] S. Weinberg, Anthropic Bound on the Cosmological Constant. *Physics Review Letters*, **59** (1987), 2607–10.

[106] S. Weinberg, The cosmological constant problem. *Reviews in Modern Physics*, **61** (1989), 1–23.

[107] S. Weinberg, Living in the multiverse. In B. Carr (ed.) *Universe or Multiverse?* (Cambridge: Cambridge University Press, 2007). [arXiv:hep-th/0511037].

[108] N. Zettili, *Quantum Mechanics: Concepts and Applications*, (Chichester: John Wiley, 2001).

[109] P. Zhang and M. C. Johnson, Testing eternal inflation with the kinetic Sunyaev Zel'dovich effect. arXiv:1501.00511.

[110] P. Zhang and A. Stebbins, Confirmation of the Copernican principle through the anisotropic kinetic Sunyaev Zel'dovich effect. *Philosophical Transactions of the Royal Society*, **A 369** (2011), 5138–45.

2

Black Holes, Cosmology and the Passage of Time: Three Problems at the Limits of Science

BERNARD CARR

2.1 Introduction

The boundary between science and philosophy is often blurred at the frontiers of knowledge. This is because one is dealing with proposals which are not amenable to the usual type of scientific tests, at least initially. Some scientists have an antipathy to philosophy and therefore regard such proposals disparagingly. However, that may be short-sighted because historically science had its origin in natural philosophy and the science/philosophy boundary has continuously shifted as fresh data accumulate. The criteria for science itself have also changed. So ideas on the science/philosophy boundary may eventually become proper science. Sometimes the progress of science may even be powered from this boundary, with new paradigms emerging from there.

A particularly interesting example of this in the context of the physical sciences is cosmology. This is because the history of physics involves the extension of knowledge *outwards* to progressively larger scales and *inwards* to progressively smaller ones, and the scientific status of ideas at the smallest and largest scales has always been controversial. Cosmology involves both extremes and so is doubly vulnerable to anti-philosophical criticisms. While *cosmography* concerns the structure of the Universe on the largest scales, these being dominated by gravity, *cosmogeny* studies the origin of the Universe and involves arbitrarily small scales, where the other forces of nature prevail. Indeed, there is a sense in which the largest and smallest scales merge at the Big Bang. So cosmology has often had to struggle to maintain its scientific respectability and more conservative physicists still regard some cosmological speculations as going beyond proper science. One example concerns the current debate over the multiverse. The issue is not just whether other Universes exist but whether such speculations can be classified as science even if they do, since they may never be seen.

While most of this chapter focuses on cosmology, two other problems straddling the boundary between physics and philosophy are also discussed. The first concerns black holes. Although these objects were predicted by general relativity a century ago, Albert Einstein thought they were just mathematical artefacts and it was 50 years before observational evidence emerged for their physical reality. The first type to be established were stellar black holes but subsequently evidence has emerged for increasingly massive ones, with 'intermediate mass black holes' being associated with gamma-ray bursts and 'super-massive black holes' powering quasars and perhaps even forming seeds for galaxies. Theorists also speculate about 'primordial black holes' which formed in the early stages of the Big Bang and could be much smaller than a solar mass. Like other Universes, we cannot be certain that primordial black holes existed – not because they are too far away but because they formed too early. Nevertheless, thinking about them has led to important physical insights and placed interesting constraints on the early Universe.

Studies of cosmology and black holes have much in common. Historically, both have involved considering ever larger and smaller scales, and both become very speculative at the largest and smallest scales. Although Einstein himself was slow to appreciate their significance, they both derive from general relativity and yet may play a role in some final theory which goes beyond this.

The final problem is even more speculative and concerns the flow of time. Unlike the other two problems, this one clearly goes beyond relativity theory and many would argue that it goes beyond physics altogether. However, I take a contrary view and propose an interpretation which involves higher dimensions and a very speculative link between cosmology and black holes. So this third problem brings the other two together, although I should caution that my proposal trespasses even further into the domain of philosophy.

The plan of this chapter is as follows. Section 2.2 expands on the notion of the outward and inward journeys of physics. Section 2.3 discusses the multiverse and M-theory as examples of ideas at the limits of science. Section 2.4 introduces the concept of 'meta-cosmology' to describe ideas on the border of cosmology and philosophy. Section 2.5 reviews the evolving relationship between cosmology and metacosmology from a historical perspective. Section 2.6 focuses on black holes and shows that similar considerations apply in this context. Section 2.7 proposes a very tentative model for the passage of time. Section 2.8 draws some general conclusions about the nature of science and the possible relevance of mind.

2.2 The Macro-Micro Connection and the Triumph of Physics

The outward journey into the *macroscopic* domain and the inward journey into the *microscopic* domain – mentioned in the Introduction – have revealed ever larger and smaller levels of structure in the Universe: planets, stars, galaxies, clusters of galaxies and the entire observable Universe in the macroscopic domain; cells, DNA, atoms, nuclei, sub-atomic particles and the Planck scale in the microscopic domain. These scales of structure are summarised in the image of the *Cosmic Uroborus* (the snake eating its own tail) shown in Figure 2.1. The numbers at the edge indicate the scale of these structures in centimetres. As one moves anticlockwise from the tail to the head, the scale increases through 60 decades: from the Planck scale (10^{-33}cm) – the smallest meaningful distance allowed by quantum gravity – to the scale of the observable Universe (10^{27}cm). So one can regard the Cosmic Uroborus as a clock, in which each minute corresponds to a decade in scale.

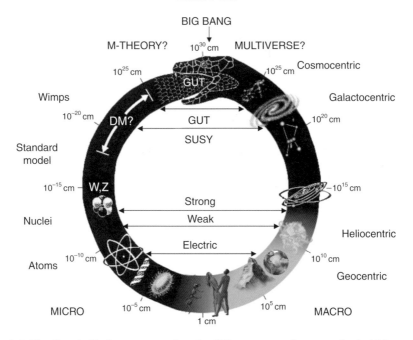

Figure 2.1 The Cosmic Uroborus summarises the different types of structure in the Universe. Also shown are the cross-links associated with various forces. The macro and micro scales merge at the Big Bang, where new physics may arise.

The inward and outward journeys have also led to the discovery of the forces which determine the forms of these structures and the associated laws of nature. These forces are indicated by the horizontal lines in Figure 2.1 and link the macroscopic and microscopic domains. For example, the 'electric' line connects an atom to a planet because the force which binds the electron to a nucleus in an atom and the intermolecular force which binds solid objects are both electrical in origin. The 'strong' and 'weak' lines connect a nucleus to a star because the strong force which holds nuclei together also provides the energy released in the nuclear reactions which power a star and the weak force which causes nuclei to decay also prevents stars from burning out too soon. The 'SUSY' (supersymmetry) line connects 'weakly interacting massive particles' to galaxies because these may provide the 'cold dark matter' (CDM) halos revealed by galactic rotation curves. The 'GUT' (grand unified theory) line connects with large-scale structure because the density fluctuations which led to this originated when the temperature of the Universe was high enough for GUT interactions to be important.

This demonstrates that the macroscopic and microscopic domains are intimately linked, so that the outward and inward journeys are not disconnected but constantly throw light on each other. Indeed physics has revealed a unity about the Universe which makes it clear that everything is connected in a way which would have seemed inconceivable a few decades ago. The Big Bang might be regarded as the ultimate micro-macro link since it implies that the entire Universe was once compressed to a tiny region of huge density. This is why the head of the snake meets the tail. Since light travels at a finite speed, we can never see further than the distance light has travelled since Big Bang; this is about 40 billion light-years, three times the age of the Universe times the speed of light because the cosmic expansion helps light travel further. Near this horizon, more powerful telescopes probe to earlier times rather than larger distances, so early Universe studies have led to an exciting collaboration between particle physicists and cosmologists. We now have a fairly complete picture of the history of the Universe after the first microsecond (Carr, 2013a).

The inward and outward journeys have also led to new conceptual ideas and changes in our worldview. The outward one has led to the shifts from geocentric to heliocentric to galactocentric to cosmocentric worldviews and to the radical change of view of space and time entailed in relativity theory. The inward one has led to atomic theory, quantum theory and a progressively unified view of the forces of nature. Indeed, these developments have shattered our perspective of the microworld just as much as relativity theory has shattered our perspective of the macroworld. We do not fully understand what happens as one approaches the top of the Uroborus – one encounters the multiverse on the macroscopic side and M-theory on the microscopic side – but the possibility of incorporating gravity into the unification of forces has led some physicists to proclaim that we are on the verge of obtaining a 'Theory of Everything' (TOE).

An intriguing feature of the top of the Uroborus is the possibility of extra dimensions. Unifying all the subatomic interactions requires extra wrapped-up dimensions of the kind proposed by Theodor Kaluza (1921) and Oskar Klein (1926) to explain electromagnetism. For example, 'superstring' theory suggests there could be six and the way they are compactified is described by the Calabi-Yau group. There were originally five different superstring theories but it was later realised that these are all parts of a single more embracing model called 'M-theory', which has seven extra dimensions (Witten, 1995). In one variant of this, proposed by Lisa Randall and Raman Sundrum (1999), the eleventh dimension is extended, so that the physical world is viewed as a four-dimensional (4D) 'brane' in a higher-dimensional 'bulk'. The development of these ideas is summarised in Figure 2.2. We do not experience these extra dimensions directly because their effects only become important on the smallest and largest scales (i.e. at the top of the Uroborus).

2.3 The Multiverse and M-Theory

In contemplating the top of the Cosmic Uroborus, where we encounter the multiverse and M-theory, it is clear that we are stretching physics to its limits. This is because we cannot see further than the distance light has travelled since the Big Bang (10^{25}m) or smaller than the distance (10^{-19}m) probed by the Large Hadron Collider (LHC), both these regimes being relevant to cosmology. However, whatever the practical problems of acquiring data in these regimes, it would be perverse to claim that there is no interesting physics there. In this section, I will briefly discuss the multiverse and M-theory; see Carr (2007) for a more complete treatment.

In the standard Big Bang theory, there should be unobservable expanding domains beyond the horizon which are still part of our Big Bang. This is what Max Tegmark (2003) calls the 'Level I' multiverse and it is relatively uncontroversial. If taken to extremes, it leads to some bizarre possibilities (like our having identical clones at some stupendous distance) and George Ellis highlights the hubris involved in such an extrapolation (Carr and Ellis, 2008). However, it would be hard to deny its existence at some level.

Recent developments in cosmology and particle physics have led to the more radical proposal that there could also be other Big Bangs which are distinct from ours. These might be regarded as the inevitable outcome of the physical processes that generated our own Universe and form what Tegmark classifies as the 'Level II' multiverse. Some of the proposals come from cosmologists and others from particle physicists, so the Level II multiverse might be regarded as the culmination of scientific attempts to understand the largest and smallest scales in Figure 2.1. In a way, Ellis's hubris argument *supports* the Level II multiverse, because it requires even more hubris to assume that the Level I multiverse extends forever. Indeed, the density fluctuations seen in the cosmic microwave background (CMB) would be of order unity if extrapolated to a scale of 10^{100} times the current horizon, so this might be taken to indicate the scale of Level II structure. Although this is large, it is gratifying that it is much less than the scale at which Tegmark's clones appear!

Let us first examine the cosmological proposals. Some invoke 'oscillatory' models in which a single Universe undergoes cycles of expansion and recollapse (Tolman, 1934). In this case, the different Universes are strung out in time. Others invoke the inflationary scenario, in which our observable domain is a tiny part of a bubble, which underwent an accelerated expansion phase at some early time as a result of false vacuum effects (Guth, 1981). In this case, there could be many other bubbles, corresponding to other Universes spread out in space. A variant of this idea is 'eternal' inflation, in which the Universe is

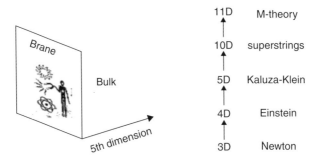

Figure 2.2 The sequence of extra dimensions invoked in modern physics (right), one of which is extended in brane cosmology (left).

continually self-reproducing, so that there are an infinite number of bubbles extending in both space and time (Vilenkin, 1983; Linde, 1986).

Turning to multiverse proposals inspired by particle physics, we have seen that in one version of M-theory our Universe is a 4D brane in a five-dimensional (5D) bulk. In this case, there might be many other branes, corrresponding to a multiverse extending in the fifth dimension, and collisions between neighbouring ones might even generate Big Bangs (Steinhardt and Turok, 2006). In the standard version of M-theory, with no extended dimension, the vacuum state is determined by the Calabi-Yau compactification. Recent developments suggest that the number of these compactifications could be enormous (e.g. 10^{500}), each one corresponding to a Universe with a different value of the cosmological constant (Λ) and a different set of physical constants (Bousso and Polchinski, 2000). So in this 'string landscape' scenario, the constants would be contingent on which Universe we happen to occupy (Susskind, 2005).

One problem with the last scenario is that the observed cosmological constant seems implausibly small. In principle, its value could lie anyway between plus and minus the Planck density, which is 120 orders of magnitude larger than the observed value. There is also the puzzling feature that the observed vacuum density is currently very similar to the mean matter density, a coincidence which would only apply at a particular cosmological epoch. One cannot predict the distribution of Λ across the different Universes precisely in this picture but it would be very unlikely to have a peak in the observed range. However, as first pointed out by Steven Weinberg (1987) and later discussed by George Efstathiou (1995) and Alex Vilenkin (1995), the value of Λ is constrained because galaxies could not form if it were much larger than observed. Since the growth of density perturbations is quenched once Λ dominates the density, if bound systems have not formed by then, they never will. Thus, if one were to contemplate an ensemble of Universes with a wide spread of values of Λ, our own existence would require that we occupy one of the tiny fraction in which the value is sufficiently small. This is an example of an 'anthropic' fine-tuning argument – indeed, in terms of the precision required, it may be the most impressive tuning of all.

The oldest form of the multiverse is the 'many worlds' interpretation of quantum mechanics (Everett, 1957), in which there is no quantum collapse but the Universe branches every time an observation is made. Tegmark classifies this as the 'Level III' multiverse and it is the most natural framework in which to describe quantum cosmology, which applies when the classical spacetime description of general relativity breaks down at the Planck time (10^{-43} s). In this approach, one has a superposition of different histories for the Universe and uses the path integral formalism to calculate the probability of each history (Hartle and Hawking, 1983). In some models, this replaces the Big Bang with a bounce and leads to a form of the cyclic model.

The most speculative version of the multiverse, described by Tegmark as 'Level IV', postulates disconnected Universes governed by different laws based on different mathematical structures. It derives from an underlying philosophical stance that everything that can happen in physics does happen, so any mathematically possible Universe must exist 'somewhere' (Tegmark, 1998). This is certainly the most philosophical of the multiverse proposals.

Despite the popularity of the multiverse proposal, some physicists are deeply uncomfortable with it. The ideas involved are highly speculative and they are currently – and may always remain – untestable in the sense that astronomers may never be able to observe the other Universes directly. For these reasons, some physicists do not regard these ideas as coming under the purview of science at all. Since our confidence in them is based on faith rather than experimental data, they seem to have more in common with religion. On the other hand, evidence for other Universes might eventually be forthcoming – for example, from the scars of collisions with other Universes in the CMB (Garriga *et al.*, 2007; McEwen *et al.*, 2012) or from dark flows (Kashlinsky *et al.*, 2009) or from cosmic wakes (Mersini-Houghton and Holman, 2009).

Another argument might be that an idea can be regarded as scientific if it is implied by a theory which is testable. So if a theory *predicts* the multiverse, then verifying that theory would at least provide partial evidence. For example, this would apply if the extra dimensions of M-theory were detected with the LHC. If they are not detected, one might argue that extra dimensions are mathematical rather than physical (Woit, 2006; Smolin, 2007). But many ideas in modern physics might be regarded as mathematical and the picture of ultimate reality provided by theorists deviated from the common-sense reality provided by our physical senses long before we reached the stage of the multiverse and M-theory.

If no direct evidence for other Universes or M-theory is forthcoming, one could still argue that the anthropic fine-tunings provide *indirect* evidence for other Universes (Carter, 1974; Carr and Rees, 1979; Barrow and Tipler, 1986; Hogan, 2000). Although the multiverse proposal was not originally motivated by an attempt to explain these tunings, it seems clear that the two concepts are interlinked. For if there are many Universes, the question arises as to why we inhabit this particular one and (at the very least) one would have to concede that our own existence is a relevant selection effect. A huge number of Universes will allow all possibilities and combinations to occur, so somewhere – just by chance – things will be right for life. Many physicists therefore regard the multiverse as providing the most natural explanation of the anthropic fine-tunings. If one wins the lottery, it is natural to infer that one is not the only person to have bought a ticket.

In assessing the multiverse interpretation of the fine-tunings, a key issue is whether some of the physical constants are contingent on accidental features of symmetry-breaking and the initial conditions of our Universe or whether some fundamental theory will determine all of them uniquely (Rees, 2001). The two cases correspond to the multiverse and single Universe options, respectively. This relates to Einstein's famous question: 'Did God have any choice when he created the Universe?' If the answer is no, there would be no room for the Anthropic Principle. If the answer is yes, then trying to predict the values of the constants would be as forlorn as Kepler's attempts to predict the spacing of the planets based on the Platonic solids.

Even if one accepts that the anthropic fine-tunings derive from a multiverse selection effect, what determines the selection? If it relates to the presence of observers, is some minimum threshold of intelligence required or does the mere existence of consciousness suffice? Or does it just reflect some form of 'life principle', as advocated by Paul Davies (2006)? I have argued elsewhere (Carr, 2007) that the Anthropic Principle should really be called the Complexity Principle, in which case the connection with consciousness or life may be incidental.

Another important question is whether our Universe is typical or atypical within the ensemble. Advocates of the Anthropic Principle usually assume that life-forms similar to our own will be possible in only a tiny subset of Universes. More general life-forms may be possible in a somewhat larger subset but life will not be possible everywhere. On the other hand, by invoking a Copernican perspective, Lee Smolin (1997) has argued that *most* of the Universes should have properties like our own, so that ours is typical. His own favoured model is a particular example of this; I will describe this in some detail because it links the topics of this chapter.

Smolin argues that the physical constants have *evolved* to their present values through a process akin to mutation and natural selection. The underlying physical assumption is that whenever matter gets sufficiently compressed to undergo gravitational collapse into a black hole, it gives birth to another expanding Universe in which the fundamental constants are slightly mutated. Since our own Universe began in a state of great density (i.e. with a Big Bang), it may itself have been generated in this way (i.e. via gravitational collapse in some parent Universe). Cosmological models with constants permitting the formation of black holes will therefore produce progeny (which may in turn produce further black holes since the constants are nearly the same), whereas those with the wrong constants will be infertile. Through successive generations of Universes, the physical constants will then naturally evolve to have the values for which black hole (and hence baby Universe) production is maximised. Smolin's proposal involves very speculative physics, since we have no understanding of how the baby Universes are born, but it has the virtue of being testable, since one can calculate how many black holes would form if the parameters were different. There is no need for the constants to be determined by a TOE and no need for the Anthropic Principle, since observers are just an incidental consequence of the Universe being complex enough to give rise to black holes.

2.4 Cosmology and Metacosmology

Although cosmology is now recognised as part of mainstream physics, it is different from most other branches of science: one cannot experiment with the Universe or observe other ones, and speculations about processes at very early and very late times depend upon theories of physics which may never be directly testable. This is why some cosmological speculations are dismissed as being too philosophical. I use the term 'metacosmology' to describe aspects of cosmology which might be regarded as bordering on philosophy (Carr, 2014), although we will see that one cannot delineate the cosmology/metacosmology border precisely.

(a)

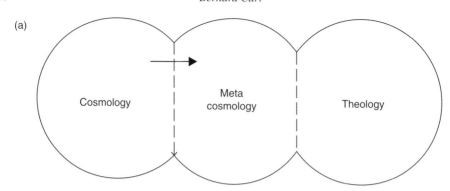

Figure 2.3(a) This illustrates the sequence from cosmology to metacosmology to theology. The arrow indicates that the cosmology/metacosmology boundary evolves, so that today's metacosmology becomes tomorrow's cosmology.

(b)

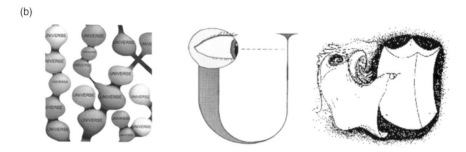

Figure 2.3(b) Three possible explanations of the anthropic fine-tunings, these becoming increasingly 'wild' from a physics perspective as one moves to the right. The sequence is connected with Figure 2.3(a) but the precise correspondence is fuzzy.

Cosmology also relates to theology. All cultures have their creation myths (a sceptic might claim that the Big Bang theory is just the most recent one) and issues about the origin and future of the Universe also arise in religion. Of course, the remit of religion goes well beyond the materialistic issues which are the focus of cosmology. However, in so much as religious and cosmological truths overlap, they must be compatible. This has been stressed by George Ellis (this book), who distinguishes between 'cosmologia', which takes into account 'the magnificent gestures of humanity', and cosmology, which just focuses on physical aspects of the Universe. Most scientists are even more uncomfortable straying into the domain of theology than philosophy. So it may be appropriate to regard cosmology, metacosmology and theology as forming a sequence, with metacosmology occupying a middle ground, as illustrated in Figure 2.3(a).

As an example of a problem which may impinge on all three areas, consider the issue of the anthropic fine-tunings. One might consider three interpretations of these tunings, as illustrated in Figure 2.3(b). The first possibility is that there is not just one Universe but lots of them, all with different coupling constants, and that we necessarily reside in one of the

small fraction which satisfies the anthropic constraints (Carr, 2007). At least some physicists would regard this explanation as scientific and therefore put it in the cosmology box
of Figure 2.3(a). The second possibility, based on the notion that the Universe is described
by a quantum mechanical wave function and that consciousness is required to collapse
this, is that the Universe does not exist until consciousness has arisen. Once it has done so,
one might think of it as reflecting back on the Big Bang, thereby forming a closed circuit
which brings the world into existence (Wheeler, 1977). Even if consciousness really does
collapse the wave function (which is far from certain), this explanation might be regarded
as metaphysical and so deemed to be in the metacosmology box. The third possibility is
that the fine-tunings reflect the existence of a 'Creator' or 'God' who tailor-made the Universe for our benefit by placing a 'pin' carefully in the space of coupling constants (Leslie,
1989; Holder, 2004). This explanation clearly belongs to the theology box.

These possibilities represent a sliding scale of increasing unpalatability for most physicists, with the A word (Anthropic), the C word (Consciousness) and the G word (God)
being increasingly taboo. The hard-line physicist may not be entirely comfortable invoking a multiverse but it is better than appealing to consciousness and definitely preferable
to God. This does not mean any of the three explanations is less logical; it just reflects
the physicist's bias towards an explanation which is scientific. In any case, the dichotomy
between God and multiverse is simplistic (Collins, 2007). While the fine-tunings certainly
do not provide unequivocal evidence for God, nor would the existence of a multiverse preclude God. On the other hand, if there's only one Universe, the argument for a fine-tuner
may become more compelling, I will not discuss such theological issues any further here.

It should be stressed that opinions differ as to the location of the cosmol-
ogy/metacosmology boundary in Figure 2.3(a). Some physicists would regard the multiverse as philosophy, with only a single self-created Universe of the kind envisaged by
Stephen Hawking (2001) qualifying as physics. Davies (2006) even relegates it to theology, regarding the concepts of a multiverse and a Creator as equally metaphysical. The
boundary between cosmology and metacosmology is therefore fuzzy. More importantly, I
argue below that it is constantly evolving (hence the arrow in Figure 2.3(a), so that today's
metacosmology becomes tomorrow's cosmology. The controversy over whether the multiverse should be classified as cosmology or metacosmology was the focus of my dialogue
with Ellis, in which I defended and he opposed its scientific status (Carr and Ellis, 2008).
My present view is that the multiverse should currently be regarded as metacosmology but
I anticipate that the cosmology/metacosmology boundary will eventually shift sufficiently
for it to be reclassified as cosmology. So the arrow in Figure 2.1 represents my attempt to
compromise with Ellis.

Similar issues arise at the microscopic frontier, which I have argued is also part of
cosmology. Although the advent of atomic theory in the eighteenth century yielded crucial insights into thermodynamics and chemistry, many scientists were sceptical of atoms
which could not be seen. Later it was realised that an atom is mainly empty space and that
its constituents are not solid particles but described by a quantum wave function which is
smeared out in space. But the long debate about the interpretation of quantum mechanics

Table 2.1. *Lessons of history*

Lesson 1	Theoretical prejudice should not blind one to the evidence
Lesson 2	New observational developments are hard to anticipate
Lesson 3	Don't reject a theory because it has no observational support
Lesson 4	Don't be deterred by the opposition of great scientists
Lesson 5	Be prepared to apply known physics in new domains
Lesson 6	Majority opinion and expectation are often wrong
Lesson 7	Metacosmology becomes cosmology because of new data
Lesson 8	The tide of history may be against the cosmocentric view
Lesson 9	The nature of legitimate science changes with time

is still dismissed as philosophy by many physicists. We have seen that proposed TOEs are sometimes dismissed as mathematics but whether that should be regarded as physics or philosophy is itself contentious.

2.5 Historical perspective of Cosmology/Metacosmology

In this section I will argue that the boundary between metacosmology and cosmology is continuously shifting as new ideas evolve from being pre-scientific to scientific. The domain of legitimate science is thus always expanding. Some historical examples will be used to illustrate this point, each of which conveys a particular lesson. These lessons are discussed in more detail elsewhere (Carr, 2014) and are summarised in Table 2.1.

To the ancient Greeks, the heavenly spheres were the unchanging domain of the divine and therefore outside science. It required Tycho Brahe's observation of a supernova in 1572 and the realisation that its apparent position did not change as the Earth moved around the Sun to dash that view. Because this contradicted the Aristotelian view that the heavens cannot change, the claim was at first received sceptically. Frustrated by those who had eyes but would not see, Brahe wrote: 'O crassa ingenia. O coecos coeli spectators'. [Oh thick wits. Oh blind watchers of the sky.] *Lesson 1: Theoretical prejudice should not blind one to the evidence.* I would claim that the analogue of Tycho's supernova in the multiverse debate is the fine-tunings, even though we are literally 'blind' in that sense that we cannot see the other Universes.

Long after Galileo had realised that the Milky Way is nothing more than an assemblage of stars and Newton had shown that the laws of nature could be extended beyond the Solar System, there was still a prejudice that the investigation of this region was beyond the domain of science. In 1835 August Comte commented on the study of stars: 'We will never be able by any means to study their chemical compositions. The field of positive philosophy lies entirely within the Solar System, the study of the Universe being inaccessible in any possible science.' Comte had not foreseen the advent of spectroscopy, which identified absorption features in stellar spectra with chemical elements. *Lesson 2: New observational developments are hard to anticipate.* Perhaps one day we will find extra dimensions at the LHC or create baby Universes in the laboratory or visit other Universes through wormholes.

Cosmology attained the status of a proper science in 1915, when the advent of general relativity gave it a secure mathematical basis. Nevertheless, for a further decade there was resistance to the idea that science could be extended beyond the Galaxy. Indeed, many astronomers refused to believe that there *was* anything beyond. Although Immanuel Kant had speculated, as early as 1755, that some nebulae are 'island Universes', similar to the Milky Way, most astronomers continued to adopt a Galactocentric view until the 1920s. Indeed, the term 'Universe' became taboo in some quarters. Ernest Rutherford once remarked 'Don't let me hear anyone use the word "Universe" in my Department!' a comment some might echo today about the multiverse. The controversy came to a head in 1920 when Heber Curtis defended the island Universe theory in a famous debate with Harlow Shapley. The issue was finally resolved in 1924, when Edwin Hubble measured the distance to M31 using Cepheid variable stars.

A few years later Hubble – using radial velocities for several dozen nearby galaxies obtained by Vesto Slipher – discovered that all galaxies are moving away from us with a speed proportional to their distance. Alexander Friedmann had predicted this in 1922 on the basis of general relativity but Einstein rejected this model at the time because he believed the Universe (i.e. the Milky Way) was static and he invoked the cosmological constant to allow this possibility. In fact, Einstein continued to uphold the static model even after the evidence was against it and he only accepted the expanding model – admitting his 'biggest blunder' – several years after Hubble published his data. *Lesson 3: Don't reject a theory because it has no observational support.* Knowing how much weight to attach to theory and observation can be tricky.

Georges Lemaître – who derived Friedmann's equations independently – was the first person to consider the implications of the Universe having started in a state of great compression and is known as the father of the Big Bang theory. This is now almost universally accepted but the reaction of some of his contemporaries is revealing. Einstein remarked to Lemaître in 1927: 'Your maths is correct but your physics is abominable'. Arthur Eddington was equally loathe to accept the implications of the cosmic expansion: 'Philosophically, the notion of a beginning of the present order of nature is repugnant to me.' In contrast to Lemaître, he regarded the Big Bang as an unfortunate fusion of physics and theology. *Lesson 4: Don't be deterred by the opposition of great scientists.* What is regarded as respectable in cosmology is determined by a handful of influential people and this works both ways. One reason the Anthropic Principle has become more respectable is that several physicists of great stature have now embraced it.

Even after Hubble's discovery gave cosmology a firm empirical foundation, it was many decades before it gained full scientific recognition. When Ralph Alpher and Robert Herman were working on cosmological nucleosynthesis in the 1940s, they recall: 'Cosmology was then a sceptically regarded discipline, not worked in by sensible scientists' (Alpher and Hermann, 1988). Although they were only using known physics, people were sceptical of applying this in unfamiliar contexts. Only with the detection of the CMB in 1964 was the hot Big Bang theory established as a branch of mainstream physics and subsequent studies of this radiation have established cosmology as a precision science. *Lesson 5: Be prepared to apply known physics in new domains.* Admittedly, some cosmological speculations depend upon unknown physics but that relates to Lesson 4.

The last few decades have seen even more dramatic developments. First, although one expects the expansion of the Universe to slow down because of gravity, observations of distant supernovae suggest that it is accelerating (Riess *et al.*, 1998; Perlmutter *et al.*, 1999). We do not know for sure what is causing this but it must be some exotic form of 'dark energy', most probably related to the cosmological constant introduced by Einstein to make the Universe static. These discoveries have led to the concordance 'ΛCDM' model. For almost 90 years (ever since the demise of Einstein's static model) it had been assumed that the cosmological constant is zero, even though theorists had no clear understanding of why. *Lesson 6: Majority opinion and expectation is often wrong.* However, the nature of the dark energy is still uncertain. We cannot be sure that it is a cosmological constant and some people would advocate a more radical departure from the standard model (Lahav and Massini, 2014).

According to the inflationary scenario (Guth, 1981), the early Universe would also have undergone an accelerating phase as a result of the vacuum energy. For many years, such speculations were dismissed by some cosmologists as being too remote from observations. However, this changed when anisotropies in the CMB were discovered by COBE (Smoot *et al.*, 1992) and then probed to ever greater precision with WMAP (Spergel *et al.*, 2003) and the Planck Collaboration (2014). The fluctuations have exactly the form predicted by the inflationary scenario and this allows us to determine cosmological parameters very precisely. This illustrates a crucial point about the evolution of the cosmology/metacosmology boundary. *Lesson 7: Metacosmology becomes cosmology because of new data.* The following quote from Efstathiou (2013) is pertinent here: 'Such ideas may sound wacky now, just like the Big Bang did three generations ago. But then we got evidence and it changed the way we think about the Universe.' However, it should be cautioned that many claimed effects go away (e.g. magnetic monopoles, various dark matter candidates, primordial gravitational waves), so one needs to wait for observational claims to be confirmed.

We have seen that another idea which has become popular is that of the multiverse. More conservative cosmologists would prefer to maintain the *cosmocentric* view that ours is the only Universe but one cannot exclude the possibility that our observable Universe is just a miniscule part of a much larger physical reality. In some ways, this parallels the debate about extragalactic nebulae a century ago. The evidence for other Universes can never be as decisive as that for extragalactic nebulae but the transformation of worldview required may be just as necessary. *Lesson 8: The tide of history may be against the cosmocentric view.* Helge Kragh (2013) has criticised me for using historical considerations as an argument for the multiverse itself but my argument is merely intended as a sociological comment and I would never claim that historical considerations carry the same weight as empirical evidence.

The final lesson of history is nicely reflected in a comment by Weinberg (2007): 'We usually mark advances in the history of science by what we learn about nature, but at certain critical moments the most important thing is what we discover about science itself. These discoveries lead to changes in how we score our work, in what we consider to be an acceptable theory.' Weinberg is referring specifically to M-theory and the multiverse but this insightful remark may be of more general application. *Lesson 9: The nature of legitimate science changes with time.* I return to this issue at the end of this chapter.

2.6 Black Holes and the Limits of Science

General relativity predicts that, if the matter in a region is sufficiently compressed, gravity becomes so strong that it forms a black hole. In the simplest case, where space has no hidden dimensions, the size of the black hole (the Schwarzschild radius) is proportional to its mass, $R_s = 2GM/c^2$. Thus the density to which matter must be squeezed scales as the inverse square of the mass. The Sun would need to be compressed to a radius of about 3 km to become a black hole, corresponding to a density of about 10^{19} kg m^{-3} or 100 times nuclear density. The Sun itself is not expected to evolve to a black hole but there is a wide range of masses above 1 $M_\odot$ in which black holes could form at the present epoch and they might form below 1 $M_\odot$ at much earlier epochs. [See Chapter 1 of Calmet *et al.* (2014) for a detailed discussion.]

The most plausible mechanism for black hole formation is the collapse of stars which have completed their nuclear burning. However, this only happens for sufficiently massive stars. Those smaller than 4 $M_\odot$ evolve into white dwarfs because the collapse of their remnants can be halted by electron degeneracy pressure. Stars larger than 4 $M_\odot$ but smaller than about 100 $M_\odot$ burn stably until they form an iron/nickel core, at which point no more energy can be released by nuclear reactions, so the core collapses. If the collapse can be halted by neutron degeneracy pressure, a neutron star will form and a reflected hydrodynamic shock then ejects the envelope of the star, giving rise to a type II supernova. If the core is too large, however, it necessarily collapses to a black hole. Above 40 $M_\odot$ the core collapses directly but for 20–40 $M_\odot$ collapse is delayed and occurs due to fallback of ejected material (MacFadyen *et al.*, 2001).

The first stars to form in the Universe may have been larger than 100 $M_\odot$. Such 'Very Massive Objects' (VMOs) are radiation-dominated and unstable to nuclear-energised pulsations during their hydrogen- and helium-burning phases. It used to be thought that the resulting mass loss would be so rapid as to preclude the existence of VMOs. However, the pulsations are now expected to be dissipated as a result of shock formation and this could quench the mass loss enough for them to survive for their main-sequence time (which is just a few million years). VMOs encounter an instability when they commence oxygen-core burning because the temperature in this phase is high enough to generate electron-positron pairs (Fowler and Hoyle, 1964). This has the consequence that smaller cores explode, while larger ones collapse to 'Intermediate Mass Black Holes' (IMBHs). Both numerical (Woosley and Weaver, 1982) and analytical (Bond *et al.*, 1984) calculations indicate that this happens for VMOs above about 200 $M_\odot$. When IMBHs were suggested as dark matter candidates 30 years ago (Carr *et al.*, 1984), their existence was regarded sceptically. However, it is now thought that IMBHs may power ultra-luminous X-ray sources or be associated with Gamma-Ray Bursts. They may also exist in the nuclei of some Globular Clusters, formed perhaps from the coalescence of smaller mass black holes.

Stars in the mass range above 10^5 $M_\odot$ are unstable to general relativistic instabilities and may collapse directly to 'Supermassive Black Holes' (SMBHs) without any nuclear burning at all (Fowler, 1966). One could plausibly envisage their formation at the centres of dense star clusters through dynamical relaxation: the stars would be disrupted through collisions and a single supermassive star could then form from the newly released gas. SMBHs might also form from the coalescence of smaller holes or from accretion onto a single hole of more modest mass. SMBHs are known to reside in galactic nuclei (Kormendy and Richstone, 1995). Our own galaxy harbours a 4×10^6 $M_\odot$ black hole and quasars – which represent an earlier evolutionary phase of galaxies – are thought to be powered by 10^8 $M_\odot$ black holes. The largest black hole to date has a mass of 2×10^{10} $M_\odot$ (Thomas *et al.*, 2016). SMBH formation does not entail extreme physical conditions, since an object of 10^9 $M_\odot$ would only have the density of water on falling inside its event horizon. Nevertheless, it has taken much longer for their existence to be accepted than it has for stellar black holes.

In the early 1970s it was realised that primordial black holes (PBHs) could have formed in the early Universe and be much smaller than a solar mass. This is because the cosmological density was very high at early times, exceeding nuclear density within the first microsecond of the Big Bang and rising indefinitely at earlier times. A comparison of cosmological density and Schwarzschild density implies that a PBH forming at time t must have a mass $c^3 t/G \sim 10^5$ (t/s) $M_\odot$. This is just the mass within the cosmological horizon, so they could span an enormous mass range: from 10^{-5} g for those forming at the Planck time to 1 $M_\odot$ for those forming at the quark-hadron phase transition (10^{-5}s) to 10^5 $M_\odot$ for those forming at 1 s (Hawking, 1971). Such PBHs could have formed either from initial inhomogeneities or spontaneously at some sort of cosmological phase transition. [See Chapter 4 of Calmet *et al.* (2014) for a detailed discussion.]

PBHs smaller than 10^{21} kg (about the mass of the Moon) might be regarded as 'microscopic', in the sense that they are smaller than a micron, but they could still have interesting astrophysical consequences. For example, they could collide with the Earth or have detectable lensing and dynamical effects or provide the dark matter (Carr *et al.*, 2010). Those smaller than 10^{12} kg (about the mass of a mountain) would be smaller than a proton and have dramatically different consequences. This is because in 1974 Hawking discovered that a black hole radiates thermally with a temperature inversely proportional to its mass (Hawking, 1974). For a solar-mass black hole, the temperature is around 10^{-6} K, which is negligible. But for a black hole of mass 10^{12} kg, it is 10^{12} K, corresponding to the emission of 100 MeV gamma-rays. Because the emission carries off energy, the mass of the black hole decreases. As it shrinks, it gets hotter, emitting increasingly energetic particles and shrinking ever faster. The time for a black hole to evaporate completely is proportional to the cube of its initial mass. For a solar-mass hole, this is an unobservably long 10^{64} y but it is the present age of the Universe (10^{10} y) for 10^{12} kg. Thus PBHs with this initial mass would be completing their evaporation today and smaller ones would have evaporated at an earlier cosmological epoch. Since microscopic PBHs (i.e. those smaller than 10^{21} kg) are hotter than the CMB, they can also be regarded as 'quantum' in the sense that their emission is not suppressed by accretion of radiation.

When the black hole mass gets down to about 10^6 kg, the evaporation becomes explosive, the remaining energy being released in just a second. Such explosions are unlikely to be detectable in the simplest picture but David Cline and collaborators (e.g. Cline and Otwinowski, 2009) have suggested that the PBH explosions might explain some short-timescale gamma-ray bursts. Future observations will settle this issue but even if PBH explosions are not detected, or even if PBHs never formed, Hawking's work was still a tremendous conceptual advance. This is because it unified three previously disparate areas of physics: general relativity, quantum theory and thermodynamics. So it has been useful to study PBHs even if they never formed!

Hawking's theory was only a first step towards a full quantum theory of gravity, since his analysis breaks down when the density reaches the Planck value of about 10^{97} kg m^{-3} because of quantum gravitational fluctuations in the spacetime metric. An evaporating black hole reaches this density when it gets down to a radius of 10^{-35} m and a mass of 10^{-8} kg. Such a 'Planckian black hole' is much smaller in size but much bigger in mass than an elementary particle. A theory of quantum gravity would be required to understand the formation and evaporation of such an object. This might even allow black holes to leave stable Planck-mass relics rather than evaporating completely, in which case these relics could be dark matter candidates (MacGibbon, 1978).

Another factor may come into play as a black hole shrinks towards the Planck scale: the existence of extra dimensions. As indicated in Figure 2.2, the unification of the forces of nature may require the existence of extra 'internal' dimensions beyond the four dimensions of spacetime. These are usually assumed to be compactified on the Planck scale, in which case their effects are unimportant for black holes heavier than the Planck mass. However, they are much larger than the Planck length in some models and this has the consequence that gravity should grow much stronger at short distances than implied by the Newtonian inverse-square law (Arkani-Ahmed *et al.*, 2000). In other models, the extra dimensions are 'warped' so that matter is confined to a 4D hypersurface, but this has the same gravity-magnifying effect (Randall and Sundrum, 1999). In either case, the standard estimate of the Planck energy (and hence the minimum mass of a black hole) could be too high.

This has the important implication that black holes could be made in accelerators. For example, protons at the LHC reach an energy of roughly 10 TeV, which is equivalent to a mass of 10^{-23} kg. When two such particles collide, one might wonder whether they can get close enough to form a black hole. In the standard picture, the likelihood of this is very small because 10^{-23} kg is much less than the Planck mass of 10^{-8} kg. However, if there

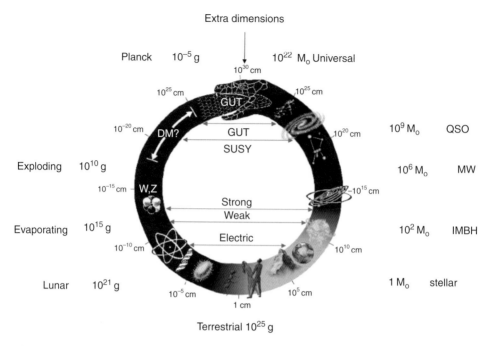

Figure 2.4 The Cosmic Uroboros applied to black holes as a link between macro and micro physics. QSO stands for 'Quasi-Stellar Object', MW for 'Milky Way', IMBH for 'Intermediate Mass Black Hole' and LHC for 'Large Hadron Collider'.

are large extra dimensions, the Planck scale is lowered and the energy required to create black holes could lie within the LHC range. They would evaporate almost immediately, lighting up the particle detectors like Christmas trees (Carr and Giddings, 2005). Although there is still no evidence for this, it opens up the exciting prospect of probing black hole evaporation, higher dimensions and quantum gravity itself.

The crucial role of black holes in linking macrophysics and microphysics is summarised in Figure 2.4. The various types of black holes are labelled by their mass, this being proportional to their size if there are three spatial dimensions. On the right are the well-established astrophysical black holes. On the left – and possibly extending somewhat to the right – are the more speculative PBHs. If the extra dimensions at the top of the Uroborus are much larger than the Planck length, then we have seen that black holes might be produced in accelerators. These are not themselves primordial but this would have important implications for PBH formation.

The history of cosmology and black holes reveals interesting similarities. Both have involved an expansion to larger and smaller scales – often against scepticism – and the ideas involved may become untestable at the largest and smallest scales. At the micro end, we have the possibility of extra dimensions and the Black Hole Uncertainty Principle correspondence (Carr, 2013b). At the macro end, we have the birth of Universe through black

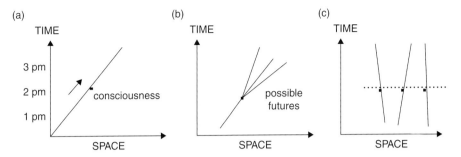

Figure 2.5 Illustrating three problems of consciousness from a relativistic perspective: (a) the passage of time; (b) the selection of possible futures; (c) the coordination of time for spatially separated observers.

hole formation (Smolin, 1997) and the persistence of black holes through a cosmological bounce (Carr and Coley, 2011). All these proposals might be regarded as ultra-speculative. So the macro and micro frontiers correspond to physics/philosophy boundaries for black holes as well as cosmology and similar issues arise (e.g. the detection of PBH explosions in the black hole scenario versus bubble collisions in the multiverse scenario).

2.7 The Flow of Time and Higher Dimensions

A long-standing problem on the interface of physics and philosophy concerns the passage of time. In the 'block Universe' of special relativity, the three-dimensional object is just the 'constant-time' cross-section of an immobile 4D world-tube and we come across events as our field of consciousness sweeps through the block. However, nothing within the spacetime picture describes this sweeping or identifies the particular moment at which we make our observations. Past and present and future coexist, so if one regards consciousness as crawling along the world-line of the brain, like a bead on a wire, as illustrated in Figure 2.5(a), that motion itself cannot be described by relativity theory. Thus there is a fundamental distinction between physical time (associated with special relativity and the outer world) and mental time (associated with the experience of 'now' and the inner world). Many people have made this point (Weyl, 1949; Brain, 1960; Davies, 1985; Lockwood, 1989; Smythies, 2003). Indeed, there is a huge philosophical literature on this topic and an ongoing controversy between the presentists and eternalists (Price, 1996; Savitt, 2006; Earman, 2008).

This also relates to the problem of free will. In a mechanistic Universe, a physical object (such as an observer's body) is usually assumed to have a well-defined future world-line. However, one intuitively imagines that at any particular experiential time there are a number of possible future world-lines, as illustrated in Figure 2.5(b), with the intervention of consciousness allowing the selection of one of these. Admittedly, this choice may be illusory but that is how it feels. The middle line in the figure shows the mechanistic (unchanged) future, while the other lines show two alternative (changed) futures. This implies that the past is fixed but that the future is undetermined. The failure of relativity to describe the process of future becoming past and different possible future world-lines may also relate to quantum theory. This is because the collapse of the wave function to one of a number of possible states entails a basic irreversibility. One way of resolving this is to invoke the 'many worlds' picture (Everett, 1957), which is reminiscent of Figure 2.5(b).

Another relevant question is how the 'beads' of different observers are correlated. If two observers interact (i.e. if their world-lines cross), they must presumably be conscious at the same time (i.e. their 'beads' must traverse the intersection point together). However, what about observers whose world-lines do not intersect? Naively identifying contemporaneous beads by taking a constant time slice, as illustrated by the broken line in Figure 2.5(c), might appear to be inconsistent with special relativity, since this rejects the notion of simultaneity at different points in space. However, the notion of a preferred time is restored in general relativity because the large-scale isotropy and homogeneity of the Universe single out a special 'cosmic time', measured by clocks comoving with the expanding cosmic background. Even for an inhomogeneous cosmological model, preferred spatial hypersurfaces can be specified as having constant proper time since the Big Bang (Ellis, 2014). One also needs some concept of simultaneity at different points in space in quantum mechanics in order to describe the Einstein–Podolsky–Rosen (1935) paradox. The problem of reconciling relativity theory and quantum mechanics may therefore connect to the problem of understanding consciousness.

One model for describing the flow of time and collapse of the wave function invokes a 'growing block universe'. This is illustrated in Figure 2.6, which represents potential futures by dotted lines, some of which solidify as decisions or observations are made (cf. Ellis, 2014). Note that it is the transition between the four diagrams which corresponds to the *flow* of time. However, even this does not describe the *flow* of time because all the information is displayed in the last picture. It is just a block Universe with dotted lines, so we need some extra ingredient.

One possible extra ingredient, proposed by the philosopher C. D. Broad (1953), is a *second* type of time (t_2), or at least a higher dimension, with respect to which our motion through physical time (t_1) is measured. This is illustrated in Figure 2.7(a), which represents the progression of consciousness in a 4D space as a path in a 5D space. At any moment in t_2 a physical object will have either a unique future world-line (in a mechanistic model) or a number of possible world-lines (in a quantum model). The intervention of consciousness or quantum collapse allows the future world-line to change or be selected from, respectively, as indicated by the solid lines in Figure 2.6. Since the future is not absolutely predetermined in this model, there is still a difference between the past and the future. As illustrated in Figure 2.7(b), at any point in t_2 the past in t_1 is uniquely prescribed but the future is fuzzy.

This interpretation of the flow of time is also suggested by the Randall–Sundrum proposal, illustrated in Figure 2.2, in which spacetime is regarded as a 4D brane embedded in a

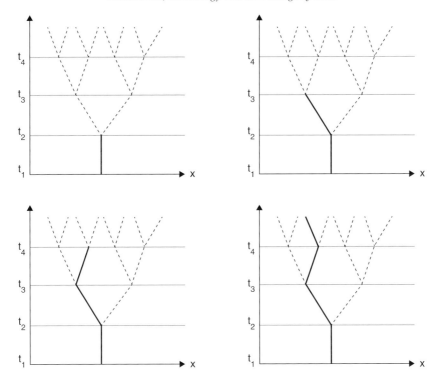

Figure 2.6 Sequence of decisions in growing black universe, with past shown solid and possible futures shown dotted.

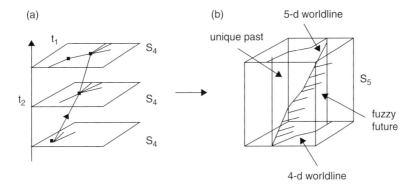

Figure 2.7 Describing flow of conscious time in a 4D structure with second time (a) and 5D structure (b).

5D bulk. In the simplest case, the brane corresponds to the flat spacetime of special relativity. However, there is a cosmological version of this picture – called 'brane cosmology' –

in which the brane is curved and space is expanding (Maartens, 2004). The cosmic expansion can then be interpreted as being generated by the brane's motion through the fifth dimension. My proposal identifies this fifth dimension with the extra dimension invoked to explain the flow of time. Although models with multiple times have also been proposed by physicists (Weinstein, 2008), it is not generally appreciated that the Randall–Sundrum picture may resolve a long-standing philosophical problem.

The relativistic solution which describes the motion of the brane through the bulk is the 5D Schwarzschild–Anti de Sitter model (Bowcock *et al.*, 2000; Mukohyama *et al.*, 2000). The coordinates within the brane are (r, θ, φ, T) with T being cosmic time. The fifth coordinate R corresponds to the embedding dimension, with the brane being located at $R = a(T)$ where $a(T)$ is the cosmic scale factor. So the extra dimension is spacelike but implicitly related to T. This is why the brane can be regarded as moving through the bulk and why the extra dimension is naturally identified with hypersurfaces of homogeneity $(T = const.)$. The model has two parameters: the mass of the 5D black hole (m) and the 5D cosmological constant (Λ). The Randall–Sundrum brane has $m = 0$ and fixed a, so it is static and the background is de Sitter. If $m \neq 0$, the metric takes the 5D Schwarzschild–Anti de Sitter form, with the fifth coordinate playing a role analogous to the 'radial' distance in the 4D Schwarzschild solution. The solution contains a black hole event horizon. This means that the Universe effectively emerges out of a 5D black hole and that the fifth coordinate becomes timelike at sufficiently early times. This might be contrasted with the proposal of Smolin (1997), in which the Universe emerges from a 4D black hole.

Clearly, invoking a second time dimension only generates a *global* flow of time. It does not explain the sense of individual identity (i.e. the first person perspective). As discussed elsewhere, for this one needs to introduce the notion of a 'specious present' and I relate this to the presence of other compactified dimensions (Carr, 2015). However, this proposal obviously does not resolve all the philosophical problems associated with the passage of time (Price, 1996).

2.8 Concluding Remarks on the Limits of Science

In this chapter I have argued that the domain of legitimate science grows with time, partly because new data become accessible and partly because the nature of science changes. In concluding, I will focus on the latter possibility and try to identify the crucial features of science with particular reference to astronomy. I will end with some remarks about the relevance of mind to physics.

It used to be assumed that science depends on *experiments* but astronomers cannot experiment with stars and galaxies. They can only let nature do this for them by observing billions of them in different states of evolution. Cosmology is in a worse state because we can only observe our own Universe and it is usually assumed that *observability* is essential to science. However, there are many entities posited by physics which cannot be seen (e.g. quarks and the interiors of a black hole); one just needs some aspects of a theory to be observable for it to be regarded as scientific. Others have emphasised the importance of *testability* or *falsifiability* in science. But on what timescale should one demand this? Is it reasonable to deny that M-theory qualifies as science because it has not been vindicated experimentally after 30 years? It might take a hundred years but the definition of science should not depend upon how long a problem takes to solve.

The question of whether M-theory and the multiverse are part of legitimate science is clearly unresolved at present. This is why I relegate the multiverse to the domain of meta-cosmology, leaving open the possibility that it will eventually be promoted to cosmology. Even if they are, there is one sense in which the current situation is very special. This is because for the first time the macro and micro physics/philosophy boundaries in Figure 2.1 have merged, the very large and very small being unified through quantum gravity. The question then becomes: does this merging represent the *completion* of science or merely a *transformation* in its nature – of the kind to which Weinberg refers and which tends to happen with every paradigm shift (Kuhn, 1970)?

Personally I favour the latter view and I will end by suggesting some possible features of the new type of science. Firstly, since we are in the domain of quantum gravity, one feature is likely to be a transcendence of the usual spacetime description. In particular, the new paradigm may involve extra dimensions, this being a feature of both M-theory and some multiverse proposals. Another feature of the new science may be a more explicit reference to *consciousness*. The mainstream view is that consciousness has a purely passive role in the Universe. In fact, most physicists assume that it is beyond their remit altogether because physics is concerned with a 'third person' account of the world (*experiment*) rather than a 'first person' account (*experience*). They infer that their focus should be the objective world, with the subjective element being banished as much as possible.

On the other hand, other physicists are sceptical of claims to be close to a TOE, when such a conspicuous aspect of the world is neglected. Thus Noam Chomsky (1975) declares 'Physics must expand to explain mental experiences'. Roger Penrose (1994) predicts 'We need a revolution in physics on the scale of quantum theory and relativity before we can understand mind'. Andrei Linde (2004) suggests that 'consciousness is as fundamental to the cosmos as space-time and mass-energy'. Many physicists disagree with this and regard the C word as just as taboo as the A word. However, in the last few decades there have been several hints from physics itself (e.g. the Anthropic Principle) that mind may be *fundamental* rather than *incidental* features of the Universe.

I will end on a provocative note by returning to the image of the Cosmic Uroborus in Figure 2.1. Although we have known since Copernicus that we are not at the centre of the Universe geographically, Figure 2.1 suggests we are at the centre of the scales of structure. This is because simple physics shows that the size of a human – or any living being – is roughly the geometric mean of the Planck length and the size of the observable Universe. Figure 2.1 also encapsulates the triumph of physics in explaining the dazzling array of increasingly complex structures which have evolved in the 14 billion years since the Big Bang. Since the culmination of this complexity – at least on Earth – is the human brain, whose remarkable attributes include consciousness, it is curious that this attribute is almost completely neglected by physics. The Uroborus also represents the way in which we have systematically expanded our outermost and innermost limits of awareness through scientific progress (i.e. it represents a blossoming of consciousness). Thus the *physical* evolution of the universe from the Big Bang (at the top of the Uroborus) through various stages of complexity to humans (at the bottom) is just the start of a phase of *intellectual* evolution, in which mind works its way up both sides to the top again. Perhaps it is not inconceivable that the marriage of relativity theory and quantum theory at the top of the Uroborus will involve mind in some fundamental way.

References

Alpher, R. A. and Hermann, R. (1988) Reflections on early work on big bang cosmology. *Physics Today* **41**, 24–34.

Arkani-Hamed, N., Dimopoulos, S. and Dvali, G. (2000) The universe's unseen dimensions. *Scientific American* **283**, 2–10.

Barrow, J. D. and Tipler, F. (1986) *The Anthropic Cosmological Principle*. Oxford: Oxford University Press.

Bond, J. R., Arnett, W.D. and Carr, B. J. (1984) The evolution and fate of Very Massive Objects. *Astrophys. J.* **280**, 825–47.

Bousso, R. and Polchinksi, J. (2000) Quantization of four-form fluxes and dynamical neutralization of the cosmological constant. *J. High Energy Phys.* 06, 006.

Bowcock, P., Charmousis, C. and Gregory, R. (2000) General brane cosmologies and their global spacetime structure. arXiv:hep-th/0007177.

Brain, Lord (1960) Space and sense-data. *British Journal for the Philosophy of Science* **11**, 177–91.

Broad, C. D. (1953) *Religion, Philosophy and Psychical Research*. New York: Harcourt, Brace.

Calmet, X., Carr, B. J. and Winstanley, E. (2014) *Quantum Black Holes*. Springer.

Carr, B. J. (2007) *Universe or Multiverse?* Cambridge: Cambridge University Press.

Carr, B. J. (2013a) Lemaître's prescience: the beginning and end of the cosmos. In R. D. Holder and S. Mitton, eds. *Georges Lemaître: Life, Science and Legacy*, Vol. 395 of Astrophysics and Space Science Library. Springer, pp. 145–72.

Carr, B. J. (2013b) Black holes, the generalized uncertainity principle and higher dimensions. *Mod. Phys. Lett.* **A 28**, 1340011.

Carr, B. J. (2014) Metacosmology and the limits of science. In M. Eckstein, M. Heller and S. Szybka, eds. *Mathematical Structures of the Universe*, Copernicus Center Press, pp. 407–32.

Carr, B. J. (2016) The flow of time, self-identity and higher dimensions. Preprint.

Carr, B. J. and Coley, A. A. (2011) Persistence of black holes through a cosmological bounce. *Inter. J. Math. Phys. D.* **29**, 2733–8.

Carr, B. J. and Ellis, G. F. R. (2008) Universe or multiverse? *Astron. Geophys.* **49**, 2.29–2.37.

Carr, B. J. and Giddings, S. (2005) Quantum Black Holes. *Scientific American* **232**, 48–55.

Carr, B. J. and Rees, M. J. (1979) The anthropic principle and the structure of the physical world. *Nature* **278**, 605.

Carr, B. J., Bond, J. R. and Arnett, W. (1984) Cosmological consequences of Population III stars. *Astrophys. J.* **277**, 445–69.

Carr, B. J., Kohri, K., Sendouda, Y. and Yokoyama, J. (2010) New cosmological constraints on primordial black holes. *Phys. Rev. D.* **81**, 104019.

Carter, B. (1974) Large number coincidences and the anthropic principle in cosmology. In M. S. Longair, ed. *Confrontation of Cosmological Models with Observations,* Dordrecht: Reidel. p. 291.

Chomsky, N. (1975) *Reflections of Language.* New York: Pantheon Books.

Cline, D. and Otwinowski, S. (2009) Evidence for primordial black hole final evaporation: Swift, BATSE and KONUS and comparisons of VSGRBs and observations of VSB that have PBH time signatures. ArXiv: 0908.1352.

Collins, R. (2007) The multiverse hypothesis: A theistic perspective. In B. J. Carr, ed. *Universe or Multiverse?* Cambridge: Cambridge University Press, pp. 459–480.

Davies, P. C. W. (1985) *God and the New Physics.* Harmondsworth: Penguin.

Davies, P. C. W. (2006) *The Goldilocks Enigma: Why is the Universe Just Right for Life?* London: Allen Lane.

Earman, J. (2008) Reassessing the prospects of a growing block model of the universe. *Intern. Stud. Phil. Sci.* **22**, 135–64.

Efstathiou, G. (1995) An anthropic argument for a cosmological constant. *Mon. Not. R. Astr. Soc.* **274**, L73.

Efstathiou, G. (2013) "The Daily Galaxy." www.dailygalaxy.com/my_weblog/2013/06/the-dark-flow-the-existence-of-other-universes-new-claims-of-hard-evidence.html. Quoted on June 3.

Einstein, A., Podolsky, B. and Rosen, N. (1935) Can a quantum mechanical description of reality be considered complete? *Physical Review* **47**, 777–80.

Ellis, G. F. R. (2014) The evolving block universe and the meshing together of times. *Ann. New York Acad. Sci.* **1326**, 26–41.

Everett, H. (1957) Relative state formulation of quantum mechanics. *Rev. Mod. Phys.* **29**, 454.

Fowler, W. (1966) The stability of supermassive stars. *Astrophys. J.* **144**, 180–200.

Fowler, W. and Hoyle, F. (1964) Neutrino processes and pair formation in massive stars and supernovae. *Astrophys. J. Supp.* **9**, 201–320.

Garriga, J., Guth, A. and Vilenkin, A. (2007) Eternal inflation, bubble collisions, and the persistence of memory. *Phys. Rev. D.* **76**, 123512.

Guth, A. H. (1981) Inflationary universe: a possible solution to the horizon and flatness problems. *Phys. Rev. D.* **23**, 347.

Hartle, J. B. and Hawking, S. W. (1983) Wave function of the Universe. *Phys. Rev. D.* **28**, 2960–75.

Hawking, S. W (1971) Gravitationally collapsed objects of very low mass. *Mon. Not. Roy. Astron. Soc.* **152**, 75–8.

Hawking, S. W. (1974) Black hole explosions? *Nature* **248**, 30–1.

Hawking, S. W. (2001) *The Universe in a Nutshell.* London: Bantam Press.

Hogan, C. J. (2000) Why the Universe is just so. *Rev. Mod. Phys.* **72**, 1149.

Holder, R. D. (2004) *God, the Multiverse and Everything: Modern Cosmology and the Argument from Design.* Aldershot: Ashgate.

Kaluza, T. (1921) Zum Unitätsproblem in der Physik. *Sitzungsber. Preuss. Akad. Wiss. Berlin* (Math. Phys.) 966–72.

Kashlinsky, A., Atrio-Barandela, F., Kocevski, D. and Ebeling, H. (2009) A measurement of large-scale peculiar velocities of clusters of galaxies: technical details. *Astrophys. J.* **691**(2), 1479.

Klein, O. (1926) Quantentheorie und Funfdimensionale Relativitätstheorie. *Zeit. Phys.* A **37**, 895–906.

Kormendy, J. and Richstone, D. (1995) Inward bound – the search for supermassive black holes on galactic nuclei. *Ann. Rev. Astron. Astrophys.* **33**, 581–624.

Kragh, H. (2013) The criteria of science, cosmology and the lessons of history. In M. Heller, B. Brozek and L. Kurek, eds. *Between Philosophy and Science*, Krakow: Copernicus Center Press, pp. 149–72.

Kuhn, T. S. (1970) *The Structure of Scientific Revolutions*. Chicago: Chicago University Press.

Lahav, O. and Massimi, M. (2014) Dark energy, paradigm shifts, and the role of evidence. *Astron. Geophys.* **55**(3), 3.12–3.15.

Leslie, J. (1989) *Universes*. London: Routledge.

Linde, A. D. (1986) Eternally existing self-reproducing chaotic inflationary universe. *Phys. Lett. B*, **175**, 395.

Linde, A. D. (2004) Inflation, Quantum Cosmology and the Anthropic Principle. In J. D. Barrow *et al.*, eds. *Science and Ultimate Reality,* Cambridge: Cambridge University Press, pp. 426–58.

Lockwood, M. (1989) *Mind, Brain and Quantum: The Compound I.* Oxford: Blackwell Press.

Maartens, M. (2004) Brane-World Cosmology. http://relativity.livingreviews.org/lrr-2004-7.

McEwen, J. D., Feeney, S. M., Johnson, M. C. and Peiris, H. V. (2012) Optimal filters for detecting cosmic bubble collisions. *Physical Review D* **85**:103502.

MacFadyen, A. J., Woosley, S. E. and Heger, A. (2001) Supernovae, jets, and collapsars. *Astrophys. J.* **550**, 410–25.

MacGibbon, J. (1978) Can Planck-mass relics of evaporating black holes close the universe? *Nature* **329**, 308–9.

Mersini-Houghton, L. and Holman, R. (2009) 'Tilting' the Universe with the landscape multiverse: The dark flow. *J. Cosm. Astropart. Phys.* 02,006.

Mukhoyama, S., Shiromizu, T. and Maeda, K. (2000) Global structure of exact cosmological solutions in the brane world. *Phys. Rev D* **62**, 024028.

Penrose, R. (1994) *Shadows of the Mind: A Search for the Missing Science of Consciousness.* Oxford: Oxford University Press.

Perlmutter, S., *et al.* (1999) Measurements of Omega and Lambda from 42 High-Redshift Supernovae. *Astrophys. J.* **517**, 565.

Planck Collaboration (2014) Planck 2013 results. XVI. Cosmological parameters. *Astron. Astrophys.* **566**(A54).

Price, H. (1996) *Time's Arrow and Archimedes' Point*. New York: Oxford University Press.

Randall, L. and Sundrum, R. (1999) An alternative to compactification. *Phys. Rev. Lett.* **83**, 4690.

Rees, M. J. (2001) *Just Six Numbers: The Deep Forces That Shape the Universe*. London: Weidenfeld and Nicholson.

Riess, A. G. *et al.* (1998) Observational Evidence from Supernovae for an Accelerating Universe and a Cosmological Constant, *Astron. J.* **116**, 1009.

Savitt, S. F. (2006) Being and becoming in modern physics. In *Stanford Encyclopedia of Philosophy.* http://plato.stanford.edu/entries/spacetime-bebecome/.

Smolin, L. (1997) *The Life of the Cosmos.* Oxford: Oxford University Press.

Smolin, L. (2007) *The Trouble with Physics.* New York: Allen Lane.

Smoot, G., *et al.* (1992) Structure in the COBE differential microwave radiometer first-year maps. *Astrophys. J. Lett.* **396**, L1–5.

Smythies, J. R. (2003) Space, time and consciousness. *Journal of Consciousness Studies* **10**, 47–56.

Spergel, D. N., *et al.* (2003) First-year Wilkinson Microwave Anisotropy Probe (WMAP) observations. *Astrophys. J. Supp.* **148**, 175.

Steinhardt, P. J. and Turok, N. (2006) Why the cosmological constant is small and positive. *Science* **312**, 1180–3.

Susskind, L. (2005) *The Cosmic Landscape: String Theory and the Illusion of Intelligent Design.* New York: Little Brown & Co.

Tegmark, M. (1998) Is the theory of everything merely the ultimate ensemble theory? *Ann. Phys.* **270**, 1.

Thomas, J. *et al.* (2016) A 17 billion solar mass black hole in a group galaxy with a diffuse core. *Nature* **532**, 340.

Tegmark, M. (2003) Parallel universes. *Scientific American* **288**, 41.

Tolman, R. (1934) *Relativity, Thermodynamics and Cosmology.* Oxford: Clarendon Press.

Vilenkin, A. (1983) Birth of inflationary universes. *Phys. Rev. D.* **27**, 2848.

Vilenkin, A. (1995) Predictions from quantum cosmology. *Phys. Rev. Lett.* **74**, 846.

Weinberg, S. W. (1987) Anthropic bound on the cosmological constant. *Phys. Rev. Lett.* **59**, 2607.

Weinberg, S. W. (2007) Living in the multiverse. In B. J. Carr, ed. *Universe or Multiverse?* Cambridge: Cambridge University Press, p. 29.

Weinstein, S. (2008) Multiple time dimensions. arXiv:0812.3869 [physics.gen.ph].

Weyl, H. (1949) *Philosophy of Mathematics and Natural Science.* New Jersey: Princeton University Press, p 116.

Wheeler, J. (1977) Genesis and Observership. In R. Butts and J. Hintikka, eds. *Foundational Problems in the Special Sciences.* Dordrecht: Reidel, p. 3.

Witten, E. (1995) String theory dynamics in various dimensions. *Nuc. Phys. B* **443**, 85–126.

Woit, P. (2006) *Not Even Wrong: The Failure of String Theory and the Continuing Challenge to Unify the Laws of Physics.* New York: Basic Books.

Woosley, S. E. and Weaver, T. A (1982) Theoretical models for supernovae. In M. J. Rees and R. J. Stoneham, eds. *Supernovae: A survey of current research.* Dordrecht, pp. 79.

3

Moving Boundaries? – Comments on the Relationship Between Philosophy and Cosmology

CLAUS BEISBART

3.1 Introduction

A popular account of the development of science, in particular of cosmology, and of the relation between science and philosophy goes like this: Science is inherently progressive and ever extends our knowledge. Step by step, phenomena at smaller and larger scales become known to us. Philosophy, by contrast, is concerned with those questions that we cannot (yet) answer in a scientific way due to a lack of empirical evidence. Accordingly, as science progresses, the boundary between the sciences and philosophy shifts. As a result, philosophy continuously loses ground, thus becoming more and more marginal – or so the view is. The aim of this comment is to discuss this account of science and philosophy with a special emphasis on cosmology.

3.2 A Popular Story About the Development of Science and Philosophy

Let us first unfold the popular story in more detail. There is quite a lot to support it. Early attempts to account for the natural world and to explain the observed phenomena, e.g. in ancient Greece, were fraught with speculation. There was simply no alternative, since, in modern terms, background knowledge was small and data sparse. Accordingly, to the extent that there was natural science, it was speculative and in this sense philosophical. Philosophy and the natural sciences were not, and could not be, properly distinguished, as is evident e.g. from the fact that a renowned philosopher, viz. Plato, wrote a dialogue concerned with what we now call cosmology.

Given the lack of knowledge and data, it is no surprise that there were rival views on how the world is composed and how far it extends. As to the small scales, there was controversy as to whether the natural world is built out of tiny, indivisible bodies, as the atomists claimed. There were competing views about the largest scales too: Whereas the atomists favoured an infinite world, the school that gained predominance for quite some time claimed that the Universe was finite and consisted largely of what is now considered to be the solar system (consult Kragh (2007) for a history of cosmology).

After the so-called Scientific Revolution, and particularly during the twentieth century, our knowledge of the natural world has rapidly increased. The hypothesis that bodies are made up of a fairly small number of elements gained momentum after the works by John

Dalton. Around 1900, the hypothesis that matter is made of atoms that were invisible at the time gained support because it could provide detailed accounts of various experiments. In 1911, Ernest Rutherford proposed a model according to which atoms in turn consist of nuclei and electrons. Soon after, it was found that the nucleus is composed of protons and neutrons. According to the standard model, which is the basis of contemporary research in elementary particle physics, protons and neutrons consist of quarks. We are now talking about scales well below 10^{-15} m.

Cosmology is intimately related to progress in the opposite direction. To recap the history briefly: During the Scientific Revolution, geocentrism was sacrificed in favour of heliocentrism. Galileo Galilei discovered that the Milky Way is composed of stars too, giving rise to the idea that we live in a huge, but flattened system of stars. Immanuel Kant and others suspected that there are other galaxies of this type; a hypothesis that was confirmed when the Great Debate about nebulae was decided. In our days, large-scale structure surveys such as the Two-degree Field project and the Sloan Digital Sky Survey (SDSS) redshift surveys map millions of galaxies. Objects have been observed that are of the order of 10^{25} m away from us. A recently very fashionable hypothesis concerns scales even larger than the observable Universe; it has it that the Universe is only part of a huge Multiverse of Universes that differ in their values of important parameters.[1]

No doubt then that the Universe of scientific knowledge has been expanding in both directions and that we can now zoom in on the microcosm and the macrocosm in some detail. And it is not just the case that our knowledge extends to spatial scales that are ever more remote from our own; rather, our knowledge regarding temporal scales has also been expanding; and we can now tell a detailed story about how the observable Universe evolved from something like 13.8 billion years ago.

Philosophy, by contrast, has not been progressive in this way. There is no realm of things with respect to which philosophy has attained more and more knowledge in the way physics did. Moreover, scientific findings have step by step cleared up what seemed to be philosophical grounds; a lot of claims that philosophers have issued about the structure of the material world, e.g. about the orbits of planets, have been proved false. Other speculations such as the views of the atomists may seem visionary and pioneering in the light of modern findings. Nevertheless, the extent to which previous philosophical views about the structure of the material world hold true does not turn on the philosophical merits the views were supposed to have; it seems rather a lucky coincidence if philosophical views accord with the findings of modern science. And today, no philosopher would dare to develop views about the macro or the micro-world without heavily leaning on science; every other approach would seem crazy.

So much for the popular story. A variety of it is given in this book by Bernard Carr (see Chapter 2). Other physicists subscribe to something like this too. For instance, Hawking and Mlodinow (2010, p. 13) admit that questions about the Universe were once discussed

---

[1] See e.g. North (1965), Kanitscheider (1991) and Kragh (2007) for the history of cosmology; see Mason (1962) for a general history of the sciences.

by philosophers. As to our own time, their diagnosis is: 'Philosophy has not kept up with the modern developments of science, particularly physics. Science has become the bearer of the torch of discovery in our quest of knowledge'.[2]

3.3 How May the Story Continue?

If the development of physics is correctly pictured in this account, then a natural question to ask is: Will the story go on in the way it has been evolving thus far? Will physicists continue to extend our knowledge to ever smaller and larger scales? Will physicists find out that the particles that they take now to be elementary are composed of even smaller particles? Likewise, concerning large scales, will the Multiverse hypothesis stabilise to scientific knowledge, and, if so, will the Multiverse later be found to be part of an even larger I-don't-know-what-verse?

There are two ways in which the story may come to an end, concerning small or large scales, respectively. The first alternative is that physicists hit the limits of being. Maybe, certain particles or entities of somehow different types are simply elementary and are not composed of anything else, and the laws that hold about these entities are the foundations of the whole of physics. Likewise, concerning large scales, maybe, everything that exists is indeed confined in certain scales. If the respective scales have become known to physicists, there is no further progress to be made. The only question then is whether we can ever know we have reached the limits of being? Can we know that there is nothing above or below the scales that we have learned about up to this point? If the answer is yes concerning both scales, the exploration of ever smaller and larger scales will have come to its definite end. This will not be the end of physics though, because a huge part of physics has always been concerned with discovering, describing and explaining yet unknown phenomena at scales between the smallest and largest known scales. There is no reason to expect that a theory of everything exonerates physicists of this task.

The other alternative is that physicists reach the limits of what humans can know before they reach the limits of being. Maybe, certain scales are simply beyond what we can ever grasp, describe and explain. Under this alternative, the question is whether we can ever know that we have reached the limits of what we can know.

The question of whether the scientific endeavour to study ever smaller and larger scales may come to a halt is at the centre of an important part of the 'Critique of Pure Reason' by Immanuel Kant (see particularly A405–A591, B432–B619). For Kant, progress in all sorts of directions is demanded by reason: It is the task of reason to find the conditions of those things already established (A508 f./B536 f.). The problem though is, according to Kant, that reason does not content itself with stepwise progress; rather it develops views and arguments that try to anticipate whether or not progress will go on for ever and whether there are limits of being (e.g. A VII f. and A408–420/B435–448). As a result, reason is caught in a contradiction because there seem to be decisive arguments both in favour, and against,

---

[2] See below for a related quotation by Hawking.

the view that there is an infinite chain of conditions. The deeper reason for the problems is that reason leaves the limits of any possible experience if it tries to infer whether the chain of conditions is infinite or not (Kant, 1781). The search for ever larger scales is covered by what Kant calls the first antinomy (A426–433/B454–461). In Kant's words the question is whether the world has an infinite spatial extension. The search for ever smaller particles is discussed in what Kant calls the second antinomy (A434–437/B462–B465). In Kant's terms the crucial question is whether substances consist of simple parts.

Kant's main target is a metaphysical cosmology, i.e. an attempt to answer questions about the basic structure of reality in terms of *a priori* reasoning. But his conclusions extend to the empirical sciences because he thinks that the questions cannot be answered by the sciences, either. He argues that, as far as the world of experiences goes, there are no definite answers as to whether substances are composed of simple parts and whether the world is infinite (A502–507/B530–535). This implies that physics will never come to a point that is known to be the end concerning either large or small scales. Kant's discussion in the 'Critique of Pure Reason' is thus challenging in view of the natural sciences too.

Kant's terminology may strike us as old-fashioned metaphysical parlance, and, even worse, the way he has formulated some of his questions presupposes a physics that is now outdated. For instance, Kant assumes that space is infinite, which need not be the case according to modern physics. Kant's discussion thus needs a thorough analysis. It may nevertheless turn out that he has a point. In my view, he ultimately poses important questions: What would it mean to say that we arrive at a point at which our knowledge covers every scale? And how can there ever be scientific evidence that there is no more to come?

3.4 Have We Already Reached the Limits of What We Can Know?

However, we may think of these questions and of limits of being and possible knowledge, already up to now the exploration of smaller and larger scales has become increasingly difficult. The reason is of course that we move away from those scales that we can directly observe. A possible worry thus is that the story we are discussing is not exactly right. Maybe, the alleged findings about very small and large scales do not really qualify as knowledge. This certainly seems to be the case concerning the Multiverse at the largest or concerning String Theory at the smallest scales; we are here talking about no more than hypotheses, which may be attractive on scientific grounds but which do not yet qualify as knowledge. But the point may even extend to scales that are closer to us. Maybe, we do not really know that there are quarks or electrons or that there are billions of other galaxies. Protons and quarks are only detected using instruments the operation of which is built upon a great number of assumptions. The identification and description of very distant objects such as high-redshift galaxies and quasars draws on theories and assumptions too. The Multiverse hypothesis even goes beyond what we may observe according to the General Theory of Relativity because it postulates a plurality of Universes from which no signal can yet have reached us. If the hypothesis is accepted, then that is only because it is assumed to explain some features of our observable Universe.

Philosophy of science features an interesting debate as to how seriously we should take the findings that are now accepted by scientists. A particular focus is on the question to what extent science has brought us knowledge about unobservables. Whereas realists take it that physics has given us knowledge about scales much beyond what we may observe, some non-realist positions such as Bas van Fraassen's constructive empiricism deny this (van Fraassen, 1980). Interestingly, the debate is almost exclusively focused on small scales, thus neglecting the realm of cosmology. One reason is that an influential notion of observability from the philosophical literature (van Fraassen, 1980, p. 16) abstracts from our spatio-temporal position and thus does not take into account limitations that derive from so-called horizons (see below). But some of the worries of non-realists refer to large scales too. One worry is particularly important: The available data or even the data of which we may avail ourselves may underdetermine the choice between theories and hypotheses, simply because several hypotheses are compatible with the data. If this is so, it seems that we cannot know which of the hypotheses is true.

Whether this type of argument is sound depends on two questions: First, are there alternatives to what we believe about small or large scales, or do we at least have reasons to assume that there are such alternatives that comply with all data? Second, if there are several hypotheses that are compatible with the same available data, are there nevertheless good reasons to embrace one hypothesis as the true one, e.g. because it has much more explanatory power than its rivals?

Let us briefly discuss the questions for both small and large scales. At small scales, a lot of findings are uncontested, e.g. that there are six types of quark, as the Standard Model of particle physics has it. There is of course the question of whether quarks are composed of other things or whether they are certain strings, but we may say that these are simply questions about the composition and properties of quarks, and that the openness of these questions does not put into doubt what we do already know about quarks. But there is in fact a problem here because the particles from the standard model are described in terms of quantum mechanics. This theory has puzzled philosophers and physicists for quite a while, a main question being what the theory really tells us about the world. There are various so-called interpretations of quantum mechanics. According to Bohmian mechanics, for instance, the ultimate quantum objects are particles that are not so different from classical particles, except that they follow a dynamics that is different from classical mechanics (see Goldstein (2013) for an introduction). Other interpretations have it that there is simply the wave function in configuration space (Ney and Albert, 2013) or that quantum mechanics describes a multitude of classical branches, as the so-called many worlds interpretation has it (for an introduction see Vaidman (2015))[3]. Since quantum mechanics provides the conceptual framework of the standard model, the problem is that we do not really know what any quantum-mechanical particle, be it a quark or an electron, is: Are we talking about

---

[3] Most interpretations of quantum mechanics are still focused on non-relativistic quantum mechanics.

classical particles, are the so-called particles really just an aspect of the wave function, etc.? Note that we are here not just concerned with some features of electrons, quarks, etc., but rather with what they really are. For the most part, the different interpretations of quantum mechanics cannot be distinguished in terms of possible observations. This does not mean though that there are equal grounds to accept each of the hypotheses. There are now intensive debates as to which interpretation is to be preferred; the debates are much, but not exclusively, advanced by philosophers. That philosophers engage in these debates, which cannot be decided in terms of observations, is of course no coincidence; it accords well with the account of scientific progress that we are discussing. And it need not be the case that the debates can never be decided in a reasonable way; it may turn out that most reasons speak in favour of one interpretation. Note in particular that quantum mechanics itself may be further developed by physicists, thus facilitating the interpretation. Nevertheless, right now, the choice between the different interpretations of quantum mechanics is undetermined by data and very likely also by other reasons, to choose a particular interpretation. If this situation continues, then the picture we are discussing is over-optimistic because we have hit the limits of what we can know.

For large scales, there is underdetermination too. General results show that our observations are compatible with various markedly distinct space-time structures (Manchak, 2009). Very roughly, the problem is that, according to the theories of relativity, transmission of physical information is restricted to the speed of light. This means that every observer in a space-time has a horizon and can only have obtained signals from some part of the space-time. Since the observable realm fits many possible space-times, various space-time models are compatible with the observations that an observer could have made.

The crucial question then is whether there are other reasons to argue in favour of one or the other space-time. No physicist would dare to commit herself to a particular space-time model, but e.g. Multiverse theorists make at least some claims that stretch beyond the scales we can observe. One idea might be that explanatory concerns provide justification for doing so: A particular account of the way of explaining what we observe may be so convincing that we take it to represent the Universe even at larger scales. There are problems though with this strategy, e.g. because at least some of them are very expensive in terms of ontology (see Beisbart (2009) and Ellis (2014)). If the problems cannot be solved, then a permanent underdetermination prevents us from knowing scales well beyond our horizon.

3.5 What Significance Do Shifting Standards of Science Have for the Story?

One possible way to get around this problem seems to be to say that science has been evolving and continues to evolve and that the standards of science change during this evolution too. One then can admit that e.g. the Multiverse hypothesis may not qualify as scientifically established by standards that were once held to be scientific, but this would not matter too much if the Multiverse hypothesis complies with those standards that are postulated in some future.

This move is based upon correct observations about the history of science. To begin with the obvious, science has certainly changed in that the content of what was counted as scientific knowledge has altered. The point applies to cosmology in particular, because our conception of the Universe has changed a lot. Of course, in a purely formal sense, the Universe has always counted as everything there is in terms of matter. But the views about what this Universe is have changed tremendously during centuries.

Moreover, as e.g. historian and philosopher of science Thomas S. Kuhn has stressed (Kuhn, 1962), along with the scientific findings, the standards with which the sciences are required to comply have changed. The idea is that basic hypotheses about a field and the standards come in a package. For instance, if you think that all material bodies are nothing but atoms, you will particularly value a special type of explanation, viz. micro-explanation, which accounts for some phenomena by drawing on the underlying atoms, their properties and mutual forces. If you disagree with the atomistic picture, your standards on good explanations are likely to differ.

But the assumption that the standards of scientific work can change creates its own problems. Suppose that we have to choose between two rival theories that each suggests its own standards. By what standards can we decide which theory to choose? If there are no standards that are not theory-relative, then each theory may seem valuable by its own standards, without there being any rational way to choose among the theories. Kuhn is famous for drawing such a conclusion, but this conclusion was to a large part received with hostility. Resistance against Kuhn's conclusion was not just based upon wishful thinking, but also on arguments that attempted to show that certain theory changes are well supported by quite uncontroversial reasons.

The lessons for our purposes are clear enough: If some people doubt that scientific progress has indeed yielded knowledge about one or the other part of the Universe, it will not do to insist that the standards of scientific research are subject to change and that the findings that are doubted by some qualify as scientific knowledge according to current or, maybe, future standards. Rather, a case should be made that the changes concerning the standards which are appealed to are superior to rivals e.g. because they reflect a better understanding of some parts of the world or because they have proved fruitful. It is doubtful whether such a case has yet been made concerning String Theory or the Multiverse hypothesis (but see Dawid (2013) for an interesting discussion about the epistemological underpinnings of String Theory).

3.6 Is No Task Left for Philosophy?

Things are different if we turn to results that have withstood critical scrutiny for quite some while. It is plausible to say that results about quarks and electrons or about the evolution of the observable Universe have much improved our knowledge about the micro and the macro-cosmos, respectively (this point holds true even if we do not know what quantum particles really are). Let us thus assume that, to a significant extent, the story under investigation has it right for the evolution of physics. What are the consequences for philosophy? Is the implication that it has lost significant grounds and that nothing is left over for it? Hawking (1992, pp. 174–5) writes:

In the eighteenth century, philosophers considered the whole of human knowledge, including science, to be their field and discussed questions such as: Did the Universe have a beginning? However, in the nineteenth and twentieth centuries, science became too technical and mathematical for the philosophers, or anyone else except a few specialists. Philosophers reduced the scope of their inquiries so much that Wittgenstein, the most famous philosopher of this century, said, 'The sole remaining task for philosophy is the analysis of language.' What a comedown from the great tradition of philosophy from Aristotle to Kant!

Is Hawking right? The answer is clearly No. There is clear historical and systematic evidence for this answer.

Turn first to the historical evidence. Hawking is of course right that the origin of the Universe was an important theme for Kant, but is not any more for most twentieth century philosophers. But to use Kant as an example, he was thinking about a lot of other questions too. In his 'Critique of Pure Reason' he assumes that metaphysics is concerned with other questions such as: Is the human soul a simple substance? Are we free to choose? Does God exist (e.g. B7)? But in the Critique, his project was not to answer these questions. Rather, Kant urged a critical philosophy that first examines to what extent human beings can answer the questions of metaphysics (B22 f.). And, as a matter of fact, he argued that human beings would never be able to answer the questions because they transcend any possible experience. His own philosophy in the Critique thus is a thorough investigation of human reason and human knowledge; the same is true for many other classic works in philosophy, e.g. the writings of John Locke and David Hume. This type of work is not in any way affected by the progress of physics (even though some of this work is now related to research carried out in cognitive science). More generally, questions about the micro and the macrocosm or about the origin of the world define one, but not an exclusive or predominant occupation of philosophers during the history of philosophy.

Turn now to systematic reasons for why there are still lots of grounds for philosophy. It is sometimes said that cosmology is about everything there is (e.g. Peacock, 1999, p. xi), but this statement is at best premature, if not misleading or false. Cosmology as a physical science is concerned with the spatial and temporal distribution of material stuff. As such, it does not deal with e.g. normative issues, with mental phenomena, e.g. consciousness, or with the existence of universals. Cosmology and physics, as they are currently done, do not answer the question of what we should do and what are good reasons to believe something. They are not concerned with the understanding of consciousness. And the question of whether there are mind-independent universals, e.g. natural kinds or properties, is orthogonal to cosmology too.

It may eventually turn out that everything that we might want to say about normative questions and about consciousness can ultimately be reduced to physical theories. It may further turn out that there are simply no universals or that they are part of physics. If this is so and if a couple of other things we are talking about do not have an existence that can be fully described in terms of physics, then cosmology is about everything. But whether this condition is fulfilled, is up for philosophical debate.

Questions about consciousness or the existence of universals may seem a bit far-fetched; but note that there are more mundane questions that are currently discussed in philosophy. These questions are actually posed by physics. One has already been mentioned above, viz. the question of what quantum mechanics implies about the basic structure of the material world. Another question is how we should think of space and time. Time in particular is hard to understand: It is a basic experience that time passes, but it is notoriously difficult to make sense of this in terms of physical theories. Physics does not fully deal with questions like these; it is rather philosophers of physics who discuss them.

3.7 What Sort of Knowledge Do Philosophers Gain?

It is a fact that there is now a division of labour between physicists and philosophers of physics. Philosophers do not investigate the redshifts of galaxies, and at least most physicists do not address the question of how time passes. But this division of labour raises a question: If philosophers think about space and time e.g., how are they entitled to their claims? What are the sources of knowledge upon which they rely? How can they ever argue for their conclusions without drawing on experience? And if they only draw on experience, what is the point of claiming that there is a legitimate role for a philosophy of physics quite apart from physics itself?

This question raises deeper issues, e.g. whether there are brands of knowledge that are not empirical. There are a number of respectable suggestions for such knowledge, e.g. the Kantian proposal that some knowledge is not empirical because it names the necessary conditions of experience. But this is not the place to discuss this and other proposals. I will only raise a weaker point: It is not so clear how far our experience extends. Can we still say that we know by experience that there are atoms? And are there empirical reasons to think that the Universe is only part of a larger Multiverse? If it is not clear how far our experience goes, then why require that philosophy be based upon something other than experience?

In his passage quoted above on p. 72, Hawking complains that some philosophers have contented themselves with the clarification of concepts. His remarks echo proposals by the logical positivists and others. The positivists claimed that there are only two varieties of knowledge, viz. empirical and purely conceptual knowledge. Quine, in his 'Two Dogmas of Empiricism', famously claimed that the distinction cannot be drawn in the way suggested. He proposed a picture in which there is a web of beliefs that somehow latch onto experience. Some of our beliefs are very close to what we experience, others are more remote. The former are part of the natural sciences; the latter may become more philosophical (Quine, 1951). If this is correct, then there is not a sharp distinction, but rather a continuous transition between the natural sciences and philosophy. Whatever we may think of this picture in general, it seems quite appropriate for a huge part of modern philosophy of physics. The philosophy of physics does not aim at a variety of knowledge that is completely independent of experience. It does draw on experience, but it is concerned with more general issues than physics is. The distinction between physics and its philosophy is mainly pragmatic, deriving from the fact that physics has become so complicated that a division of labour seems appropriate.

3.8 Conclusion

To summarise: In this commentary, I have raised a number of questions about a common picture of how physics, particularly cosmology, on the one hand, and philosophy, on the other hand, have evolved and how they are related to each other. I have raised a number of questions about this picture. Some of the questions point to difficulties about the picture and reveal that it is too optimistic. I have not answered all my questions; but the discussion should at least have pointed to the presuppositions of the picture. A lot of questions are left open then – many for the philosophy of physics, in particular of cosmology.

References

Beisbart, C. (2009) Can We Justifiably Assume the Cosmological Principle in Order to Break Model Underdetermination in Cosmology? *Journal for General Philosophy of Science* **40**, 175–205.

Dawid, R. (2013) *String Theory and the Scientific Method.* Cambridge: Cambridge University Press.

Ellis, G. F. R. (2014) On the Philosophy of Cosmology. *Studies in History and Philosophy of Modern Physics* **46**, 5–23.

Goldstein, S. (2013) Bohmian Mechanics. In E. N. Zalta, ed. *The Stanford Encyclopedia of Philosophy* (Spring 2013 Edition). http://plato.stanford.edu/archives/spr2013/entries/qm-bohm/.

Hawking, S. (1992) *A Brief History of Time.* London: Transworld Publishers Ltd.

Hawking, S. and Mlodinow, L. (2010) *The Grand Design. New Answers to the Ultimate Questions of Life.* London: Transworld Publishers.

Kanitscheider, B. (1991) *Kosmologie*, 2nd edn. Stuttgart: Philipp Reclam jun.

Kant, I. (1781) *Kritik der reinen Vernunft.* Riga: Johann Friedrich Hartknoch (2nd edn 1787), English translation: Kant, I. (1998) *Critique of Pure Reason*, translated by P. Guyer and A. W. Wood; Cambridge Edition of the Works of Kant. Cambridge: Cambridge University (page numbers refer to the 1781 and 1787 editions, abbreviated as A and B, respectively).

Kragh, H. (2007) *Conceptions of Cosmos. From Myths to the Accelerating Universe: A History of Cosmology.* Oxford: Oxford University Press.

Kuhn, T. S. (1962) *The Structure of Scientific Revolutions.* Chicago: University of Chicago Press.

Manchak, J. B. (2009) Can We Know the Global Structure of Spacetime? *Studies in History and Philosophy of Modern Physics* **40**, 53–6.

Mason, S. F. (1962) *A History of the Sciences.* New York: Macmillan.

Ney, A. and Albert, D. Z. (eds.) (2013) *The Wave Function: Essays on the Metaphysics of Quantum Mechanics.* New York: Oxford University Press.

North, J. D. (1965) *The Measure of the Universe. A History of Modern Cosmology.* Oxford: Clarendon Press.

Peacock, J. A. (1999) *Cosmological Physics.* Cambridge: Cambridge University Press.

Quine, W. V. O. (1951) *Two Dogmas of Empiricism. Philosophical Review* **60**, 20–43.

Vaidman, L. (2015) Many-Worlds Interpretation of Quantum Mechanics. In E. N. Zalta, ed. *The Stanford Encyclopedia of Philosophy* (Spring 2015 Edition). http://plato.stanford.edu/archives/spr2015/entries/qm-manyworlds/.

van Fraassen, B. (1980) *The Scientific Image.* Oxford: Clarendon Press.

4

On the Question Why There Exists Something Rather Than Nothing

RODERICH TUMULKA

4.1 Introduction

In my opinion, nothing useful has ever been written on the question in the title, and small is the contribution that I have to offer. I outline an explanation for why there is something rather than nothing, an explanation which, however, I believe is incorrect because it makes a certain empirical prediction (absence of qualia) that is incorrect. Nevertheless, it may be interesting to discuss this reasoning. It allows, in principle though not in practice, to derive the laws of nature and all physical facts about the universe. Then I elucidate which objections to this explanation are, in my opinion, valid and which are not.

Explanation in physics usually works this way: observable phenomena get explained by physical theories. A physical theory is the hypothesis that the physical world consists of certain kinds of physical objects governed by certain laws. From this hypothesis we derive, or we make it plausible that it can be derived, that the phenomenon in question (typically) occurs; then we say that the theory explains the phenomenon. The physical theory does not explain why these kinds of physical objects exist, why others do not, and why these laws hold; instead of explaining them, the theory merely posits them. At best, some theories are simpler and more elegant than others (e.g. Einstein's general relativity more than Newton's theory of gravity). But no physical theory comes close to explaining why these laws hold, or why there exist any physical objects at all. Thus, no physical theory contributes to the question in the title.

Specifically, some physical theories allow for the possibility of a vacuum state, i.e. that at a certain time there is no matter in space, and for the possibility of a transition from a vacuum state to a non-vacuum state, i.e. that at some other time there is some matter in space. While such a theory has some explanatory value, it does not touch upon the question in the title, as it does not explain the physical laws, nor why space-time exists,[1] nor why certain kinds of matter (described by certain kinds of mathematical variables) exist and others do not. For related discussion, see Holt's overview [4] and Albert's critique [1] of Krauss's book [5].

---

[1] I also call space-time, not only pieces of matter, a physical object.

In this chapter I outline a novel type of explanation of why anything exists. Obviously, the reasoning is very different from usual physical theories. The explanation also aims to explain why the world is the way it is. In particular, it allows in principle to derive the laws of physics and other testable consequences. I will answer objections that I think are not valid, and I will argue that the explanation I describe, although it is a coherent reasoning, ultimately fails for a reason connected to the mind–body problem (see Chalmers, 1996 [3] and references therein): it predicts the non-existence of qualia.

Even if the explanation fails it may be worth considering because there are, as far as I am aware, no other explanations for why anything exists. Perhaps, the closest that any reasoning before came to such an explanation was Anselm of Canterbury's (1033–1109) ontological argument for the existence of god [2], endorsed by Leibniz (see Loewer, 1978 [6] for a critique). I believe that Anselm's argument is not coherent and mine is; but I will not discuss Anselm's argument here.

I have the sense that many scientists and philosophers think that the question in the title is meaningless and cannot be answered. On the contrary, I think that the question is meaningful, as I understand its meaning, and the explanation I describe suggests to me that a rational answer might be conceivable.

4.2 The Explanation

The reasoning begins with the facts of mathematics. There is no mystery about why they are true. Their truth lies in the nature of mathematics and is explained by their content. Even if the physical world had been created by a god, the facts of mathematics would have been facts prior to that act, that is, it would not have been within the god's power to change these facts. Along with the mathematical facts come mathematical objects, such as the empty set, the set containing only the empty set, the ordinal numbers (an ordinal number is a set α of sets that is well-ordered by the $\in$ relation and $\in$-transitive, i.e. contains each of its elements as a subset), the natural numbers regarded as the finite ordinal numbers, and the things that can be constructed from the natural numbers. I take it that mathematical objects exist in some sense. Not in the same way as physical objects, but in the way appropriate for mathematical objects.

Now suppose, for the sake of the reasoning, that mathematical objects could naturally be arranged in a way that looks like a discrete space-time, say a discrete version of a four-dimensional manifold. For example, it can be argued that all mathematical objects are ultimately sets; and all sets naturally form a directed graph with the $\in$ relation as the edges; and all sets naturally form a hierarchy according to their type.[2] A graph is somewhat similar to a discrete space-time; indeed some proposed definitions of discrete space-time say that such a thing is a partially ordered set – which is more or less a directed graph. The type has certain similarities with time (see Remark 6. in Section 4.3). The example breaks down at some point, however, since I do not see any similarity between the sets of a given

---

[2] The type of a set [7] is an ordinal number such that a set of type α contains only sets of type smaller than α.

type and three-dimensional space. But suppose that the mathematical objects did naturally form a discrete space-time $\mathcal{M}$.

Suppose further that there is a property M that some mathematical objects have and others do not, and which bears a natural similarity to the property of being matter (I will come back to this in Remark 6. below). Then some space-time points in $\mathcal{M}$ will be 'occupied by matter' (in the sense of having the property M) and others will not. They may form world-lines or other subsets of $\mathcal{M}$. Thus, in a certain way of looking at the world of mathematics, it appears like a space-time with matter moving in there. For example, think of all mathematical objects as being sets, take $\mathcal{M}$ again to be the graph of all sets with the $\in$ relation, and let M be the property of being an ordinal number. That is a simple property that some objects have and others do not. If we think of type as time then, since there is exactly one ordinal for every type, at every time there is exactly one space-time point containing matter – a particle world-line. The example is of limited use since it represents a space-time with a single particle. But suppose that for some natural property M, the pattern of matter in space-time is more complicated, more like patterns in the physical world.

Suppose further that in $\mathcal{M}$ and in this pattern of matter, there are some sub-patterns which, if they actually were the patterns of physical matter in a physical space-time, would constitute intelligent beings. Then I feel that we are justified to say that in some sense these beings actually exist, since they are made of the elements of $\mathcal{M}$, which do exist. Let me call these beings the mathians. To the mathians, $\mathcal{M}$ must appear like a physical space-time, as it is the space-time in which they live, and the objects with property M must appear like matter, as it is the matter of which they are made.

If the mathians ask what the physical laws of their space-time are, their question can be answered by pure thinking, by pure mathematics. In fact, every historical fact of their universe is determined by pure mathematics, although the mathians may not be able to carry out the thinking needed to determine these facts.

If the mathians ask why their universe exists, then their question can indeed be answered. The correct answer is that $\mathcal{M}$, and the pattern of matter in $\mathcal{M}$ and the mathians, exist because they *are* mathematical objects, which exist by their nature. If the mathians ask why the matter in their universe is distributed in a certain way, then their question can be answered. The correct answer is that the distribution of matter in their universe follows logically from the mathematical facts, which are necessarily true. It seems plausible to me that the mathians may or may not be aware that they are the mathians, and that their universe is $\mathcal{M}$, and that there is a reason for why their universe exists and looks the way it does. Or some mathians may be aware and others may not. I see no reason why all mathians would have to be aware that they are mathians.

Now suppose that *we* are mathians, and that our space-time is $\mathcal{M}$. Then there *is* an explanation to why our universe exists, and to why matter is distributed in a certain way in it. That is the explanation I promised. Let me call it the i-explanation. The i-explanation is that the matter of which we consist ultimately consists of mathematical objects, and the facts of our universe are ultimately mathematical facts; and there is no mystery to why mathematical objects exist or to why mathematical facts are true.

4.3 Remarks

1. The statement that the matter of which we consist ultimately consists of mathematical objects may sound similar to statements such as that (i) a particle is a unitary representation of the Lorentz group, or that (ii) reality according to quantum field theory consists of field operators, or that (iii) reality according to quantum physics consists of pure information ('it from bit'). But it is not similar.

   Statement (i) really means, I take it, that the possible types of physical particles in our universe correspond to the unitary representations of the Lorentz group; not that an individual particle literally *is* one of these representations. According to the *i*-explanation, in contrast, an individual particle (if our universe contains point particles) at a particular point in time actually *is* a certain mathematical object. This object, by the way, is not as simple and beautiful as a unitary representation of the Lorentz group; rather, it must be an exorbitantly complicated mathematical object that is unlikely ever to be individually considered by a mathematician because there are so many other objects with similar properties.

   Let me turn to statement (ii). Its meaning is unclear to me, as I do not understand how operators can be a mathematical description of matter. I do understand, in contrast, how a subset of space-time can be a mathematical description of matter, and that is what the *i*-explanation provides.

   Statement (iii) means, I take it, that physical facts are not objective but exist only if observed by intelligent beings. According to the *i*-explanation, in contrast, there are objective physical facts, whether observed or not, namely that space-time is given by $\mathscr{M}$ and that matter is located at those space-time points with property M.

2. There are two ways in which mathematical objects and facts are relevant to the physical world according to the *i*-explanation. First, they may form part of the physical world, and second, they may apply to the physical world. For example, suppose $\mathscr{M}$ was the graph of all sets with the $\in$ relation. The number 5, understood as an ordinal number, is a set and therefore a space-time point in $\mathscr{M}$. However, the number 5 may also apply to other space-time regions, for example because there are 5 particles in a certain space-time region, i.e. there are 5 sets with property M in a certain family of sets. We are familiar with the second role of mathematical objects but not with the first.

3. Needless to say, the *i*-explanation does *not* claim that every physical theory (say, Newtonian mechanics) is realised in some (part of a) physically existing universe, in contrast to an idea proposed by Tegmark [8, 9]. For example, any particular conceivable universe governed by Newtonian mechanics is a mathematical object, and thus corresponds to a space-time point rather than to the whole universe or a region therein; in particular, the universe $\mathscr{M}$ is not governed by Newtonian mechanics.

4. It is useful to distinguish between a specific *i*-explanation and the *i*-framework. To specify an *i*-explanation, it is necessary to specify how to arrange the totality of mathematical objects as a discrete space-time, and to specify the property M. Put differently,

it is necessary to specify an *isomorphism*, or *identification*, i between the mathematical world and the physical world. I have not specified i, so I have actually not provided an i-explanation. I have only outlined how such an explanation would work if i could be provided. That is, I have described the *i-framework*.

As long as no candidate for i has been specified, it does not perhaps seem particularly likely that such an isomorphism i exists. The failure of the example involving the graph of all sets with the $\in$ relation, and my failure to come up with a better example, make it seem even less likely. Yet, it seems conceivable that i exists, and it seems that if i exists then the i-explanation does explain why there exists something rather than nothing.

5. If i exists and we know what it is then it is possible, at least in principle, to derive all laws of physics. It is also possible in principle, in this case, to determine all historical facts, past and future, unless there are limitations to the mathematical facts we can find out about. However, it is conceivable that the computational cost of finding out about interesting historical facts by analysing the corresponding mathematical facts is prohibitively expensive, and in particular that finding out facts about our future, such as next week's stock prices, by means of a computation of the corresponding mathematical facts necessarily takes longer than it takes those facts to occur (in the example, longer than one week).

6. It follows in particular that any specific candidate for i can be, at least in principle, tested empirically, as it entails a particular distribution of matter in space-time that can be compared to the one in our universe. But even without such test results, some candidates for i may be more persuasive, or more attractive, than others on purely theoretical grounds. For example, we may judge candidates for i by their elegance and would require, in particular, that M be a very simple property: If its definition were enormously complicated, then that would cost the explanation much of its explanatory value. Furthermore, an i-explanation would seem more compelling if M could be argued to be not any old mathematical property but a very special one, one that plays a crucial mathematical role. For example, the property of a set to be both well ordered by $\in$ and $\in$-transitive is a very special property, as testified to by the fact that mathematicians have a name (ordinal number) for such sets; basically, this property is special because well-orderings play a crucial mathematical role in set theory.

    Furthermore, an i-explanation would seem more compelling if analogies can be pointed out between M and the physical property (of a space-time point) of being occupied with matter. In the same vein, an i-explanation would seem more compelling if analogies can be pointed out between $\mathcal{M}$ and space-time. For example, if $\mathcal{M}$ is the graph of all sets directed by the $\in$ relation, I can point out that the concept of *type* of a set [7] bears some analogies with the concept of *time*: Both are linearly ordered, and, in fact, since the type governs the order in which sets must be defined so as to allow only well-defined objects as elements when forming a set, it seems natural to think of the type as a kind of 'logical time of set theory'.

4.4 Objections

Objection: To the question of why anything exists, most of the argument is irrelevant. Already in the first paragraph of Section 4.2, it was claimed that mathematical objects exist. Therefore, something exists, and the remainder of the argument did not contribute to answering the question in the title. On the other hand, the first paragraph of Section 4.2 did more or less take for granted that mathematical objects exist, but did not explain why they exist, and did not answer the question in the title.

Answer: There are two relevant senses of existence: mathematical and physical. The title refers to physical existence. The first paragraph of Section 4.2 refers to mathematical existence. Why something exists mathematically is easy to explain: Mathematical objects do, as soon as they are conceivable. Why something exists physically is hard to explain. In particular, physical objects do *not* exist physically as soon as they are conceivable. The argument takes the mathematical existence of mathematical objects for granted and aims at explaining the physical existence of physical objects. That is why it is not over after the first paragraph.

* * *

Objection: No statement about physical existence can follow from statements exclusively concerned with mathematics. That is a matter of elementary logic, like 'ought' cannot follow from 'is'.

Answer: That is true only as long as nothing is known about the meaning and the nature of physical existence. The hypothesis of our reasoning is that the physical existence of a space-time point x ultimately means the mathematical existence of a certain mathematical object O corresponding to x, and that the physical existence of matter at x ultimately means that O has the property M. If this is the case then statements about physical existence can clearly follow from mathematical statements.

* * *

Objection: The i-explanation aims at explaining the existence of a physical world, of space-time, matter in motion, and all facts that supervene on matter in motion. But it does not lead to qualia (i.e. conscious experiences, such as experience of the colour red [3]). The mathians do not have qualia (i.e. they are not conscious, they do not experience the colour red). If they claim they do then they are mistaken. I know that I have qualia, so I cannot be a mathian. If the i-explanation were correct then qualia would not exist. Thus, the i-framework makes a prediction, the absence of qualia, that can be regarded as an empirical prediction and that our findings disagree with.

Answer: I think that this objection is a good argument against the i-explanation. It may be tempting to hypothesise that qualia ultimately *are* mathematical properties, but that does not fit with the nature of qualia: no mathematical structure would explain the way the colour red looks (see, e.g. Chalmers, 1996 [3]), so the nature of an experience of red cannot be mathematical. So I agree that the *i*-framework makes a prediction, the absence of qualia, that is empirically false. I conclude that the *i*-explanation, while it is a possible explanation of the existence of the physical world, is not the correct explanation.

Acknowledgements

I thank Ovidiu Costin, Robert Geroch, and Barry Loewer for comments on an earlier version of this chapter. I acknowledge funding from the John Templeton Foundation (grant no. 37433).

References

[1] Albert, D. Z. 2012. On the Origin of Everything. Review of Krauss, L. (2012) *A Universe from Nothing: Why There is Something Rather than Nothing*. Free Press. *The New York Times* (25 March 2012).
[2] Anselm of Canterbury, 1078. *Proslogion*. English translation in Deane, S. N. *St. Anselm: Basic Writings*. Chicago: Open Court, 1962.
[3] Chalmers, D. 1996. *The Conscious Mind*. Oxford: Oxford University Press.
[4] Holt, J. 2017. Why the Universe Exists. To appear in Ijjas, A. and Loewer, B. eds. *A Guide to the Philosophy of Cosmology*. Oxford: Oxford University Press.
[5] Krauss, L. 2012. *A Universe from Nothing: Why There is Something Rather than Nothing*. New York: Free Press.
[6] Loewer, B. 1978. Leibniz and the Ontological Argument. *Philosophical Studies*, **34**, 105–9.
[7] Russell, B. 1908. Mathematical logic as based on the theory of types. *American Journal of Mathematics*, **30**, 222–62.
[8] Tegmark, M. 1998. Is "the Theory of Everything" Merely the Ultimate Ensemble Theory? *Annals of Physics*, **270**, 1–51. http://arxiv.org/abs/gr-qc/9704009.
[9] Tegmark, M. 2008. The Mathematical Universe. *Foundations of Physics*, **38**, 101–50. http://arxiv.org/abs/0704.0646.

Part II

Structures in the Universe and the Structure of Modern Cosmology

5

Some Generalities About Generality

JOHN D. BARROW

5.1 Introduction

The equations of general relativity and its extensions are mathematically complicated and their general coordinate covariance offers special challenges to anyone seeking exact solutions or conducting numerical simulations. They are non-linear in a self-interacting (non-Abelian) way because the mediator of the gravitational interaction (the graviton) also feels the gravitational force. By contrast in an Abelian theory, like electromagnetism, the photon does not possess the electric charge that it mediates. As a result of this formidable complexity and non-linearity, the known exact solutions of general relativity have always possessed special properties. High symmetry, or some other simplifying mathematical property, is required if Einstein's equations are to be solved exactly. General solutions are out of reach.

This 'generality' problem has been a recurrent one in relativistic cosmology from the outset in 1916 when Einstein [1] first proposed a static spatially homogeneous and isotropic cosmological model with non-Euclidean spatial geometry in which gravitationally attractive matter is counter-balanced by a positive cosmological constant. This solution turned out to be unstable [2–6]. Subsequently, the appearance of an apparent 'beginning' and 'end' to simple expanding-universe solutions led to a long debate over whether these features were also unstable artefacts of high symmetry or special choices of matter in the known cosmological solutions, as Einstein thought possible. The quest to decide this issue culminated in a new definition of such 'singularities' which allowed precise theorems to be proved without the use of special symmetry assumptions. In fact, by using the geodesic equations, their proofs made no use of the Einstein equations [7, 8]. Special solutions of Einstein's equations, like the famous Gödel metric [9] with its closed timelike curves, also provoked a series of technical studies of whether its time-travelling paths are a general feature of solutions to Einstein's equations, or just isolated unstable examples. In the period 1967–1980 there was considerable interest in determining whether the observed isotropy of the microwave background radiation could be explained because it appeared to be an unstable property of expanding universes [10, 11]. The mechanism of 'inflation', first proposed in 1981 by Guth [12], provided a scenario in which this conclusion could be reversed, and isotropy could be a stable (or asymptotically stable) property of expanding-universe solutions, by widening the range of conditions on the allowed forms of matter that could dominate the expansion dynamics of the very early universe [13–18]. Just to show how knowledge, fashion, and belief change, the requirements on the density, ρ, and pressure, p, of matter content needed for inflation to occur ($\rho + 3p < 0$) are exactly the opposite of those assumed ($\rho + 3p > 0$) in the principal singularity theorems of Penrose and Hawking [7, 8, 19] in order to establish sufficient conditions for a singularity (at least one incomplete geodesic) to have occurred in our past.

In the study of differential equations, an exact solution is called *stable* if small perturbations remain bounded as time increases; it is called *asymptotically stable* if the perturbations die away to zero with increasing time. Our solar system is dynamically stable but not asymptotically stable. Another useful pair of definitions are those introduced by Hawking [20] in 1971, who uses the same word in a technically different way. He defines a 'stable' or 'open' property of a dynamical system to be one that occurs from an open set (rather than merely a single point) in initial data space. However, it is possible for a property of a cosmological model to be stable but be of no physical interest: it is a necessary but not a sufficient property for physical relevance because the property in question could be stable only in open neighbourhoods of initial data space describing universes with other highly unrealistic properties (contraction or extreme anisotropy, for example). A 'generic' or 'open dense' property will be one that occurs near almost every initial data set (that is, it is open dense on the space of all initial data). A sufficient condition for a stable property to be of physical interest is that it is generic in this sense [21].

In this chapter we will discuss approaches to the problem of assessing generality and some of the results that arise in typical and topical cosmological problems. We will try to avoid significant technicalities. There is a deliberate emphasis upon fundamental questions of interest to philosophers of science rather than upon the astrophysical complexities of the best-fit cosmological models or the galaxy of inflationary universe models. Attention will be focussed on classical general relativity; aspects of quantum cosmology will be treated in other chapters.

5.2 General Relativistic and Newtonian Cosmology

General relativity is a much larger theory than Newtonian gravity. It has ten symmetric metric potentials, g_{ab}, instead of one Newtonian gravitational potential, Φ, and ten field equations (Einstein's equations) instead of a single one (Poisson's equation) to determine them from material content of space and time. Newtonian gravity has a fixed time and a fixed space geometry which is usually taken to be a monotonous linear time plus a three-dimensional (3D) Euclidean space (although another fixed curved space could be assumed simply by using the appropriate ∇^2 operator in Poisson's equation).

Despite appearances, Newton's theory is not really complete and Newtonian cosmology is not a well-posed theory [22, 23]. Unlike general relativity, it contains no propagation equations for the shear distortion and the formulation of anisotropic cosmological models requires these to be put in by hand. As a result the Newtonian description of an isotropic and homogeneous cosmology looks exactly like general relativity [24] because these shear degrees of freedom are necessarily absent. This feature manifests itself in results for the general asymptotic behaviour of the Newtonian n-body problem in the unbound (expanding) case. Rigorous results can be obtained for the moment of inertia (or radius of gyration), or rotation of the total finite mass of n-bodies, but not for its shape [22, 25].

All solutions of Einstein's equations describe entire universes. The relative sizes of the two theories mean that infinitely many of these general relativistic solutions possess no Newtonian counterpart. However, there are also Newtonian 'universes' which have no counterpart in general relativity. For example, there are shear-free Newtonian solutions with expansion and rotation: these cannot exist in general relativity [26]. More striking, there exist solutions of the Newtonian n-body problem (for $n > 4$) in which a system of point particles expands to infinite size in finite time, undergoing an infinite number of oscillations in the process [27]. For example, two counter-rotating binary pairs, all of equal mass, with a lighter particle oscillating between their centres along a line perpendicular to their orbital planes, can expand to infinite size as a result of an infinite number of recoils in a finite time! This is only possible because Newtonian point particles can get arbitrarily close to one another and so the $1/r^2$ forces between them can become arbitrarily large. In general relativity, this cannot happen. When two point particles of mass M approach closer than $4GM/c^2$ an event horizon forms around them. This is a simple example of a form of 'cosmic censorship' that saves us from the occurrence of an actual observable infinity, in Aristotle's sense [28], locally. In general relativity there is evidence that under broad conditions there is a maximum force, equal to c^4/G [29], as well as the more fundamental maximum velocity for information transfer, c: neither of these relativistic limits on velocity and force strength exist in Newtonian theory.

5.3 Generality – Some Historic Cases

There has been a succession of cosmological problems where particular solutions were found with striking properties that required further analysis to determine whether those properties were general features of cosmological solutions to the Einstein equations.

5.3.1 *Static Universes*

The first isotropic and homogeneous cosmological model found by Einstein [1] was a static universe with zero-pressure matter, positive cosmological constant and a positive curvature of space. Subsequently, this solution was shown to be unstable when it was perturbed within the family of possible isotropic and homogeneous solutions of Einstein's equations by Eddington and implicitly by Lemaître [30], who found the general solutions of which Einstein's universe was a particular, and clearly unstable, case. These demonstrations led to the immediate abandonment of the static universe and, in Einstein's case, of the cosmological constant as well [31].[1] It turns out that this stability problem is more complicated than it appears and has only been completely explored, when other forms of matter are present, quite recently. The static universe is only unstable against small inhomogeneous perturbations on scales exceeding the Jeans length when $p/\rho < 1/5$. When $1 \geq p/\rho > 1/5$, the Jeans length for the inhomogeneities exceeds the size of the universe and so the instability does not become Jeans unstable and amplify in time [2–6].

5.3.2 Singularities

There is a long and interesting history of attempts to interpret and avoid 'singularities' in the cosmological solutions of Einstein's equations. In the first expanding solutions with zero-pressure matter found by Friedmann [32], it appeared that there was a necessary beginning to the expansion with infinite density at a finite time in the past and there could (in spatially closed cases) also be an apparent end to the universe at a finite time in the future. One response to these infinities, particularly by Einstein, was to question whether they would remain if the family of solutions was widened. First, Einstein asked whether the addition of pressure would resist the compression to infinite density. Lemaître showed that adding pressure actually made the problem worse by hastening the appearance of the infinite density [33]. The reason is a relativistic one. Whereas in Newtonian physics any pressure resists gravitational compression, in relativity pressure also gravitates because it is a form of energy (and so has an equivalent mass via '$E = mc^2$') and increases the compression (see also Chandrasekhar and Miller, 1974 [34]). Next, Einstein wondered whether it was the perfect isotropy of the expanding-universe solutions that was responsible. If anisotropy was allowed then perhaps the compression would be defocused and the singularity avoided. Again, Lemaître was easily able to show that simple anisotropic universes have the same types of singularity and they are approached quicker than in the isotropic case [33].

These investigations by Lemaître amounted to tests of the stability of the singularity's occurrence within different wider sets of initial data. Many cosmologists were convinced by these examples that singularities were ubiquitous in these types of cosmology unless new forms of matter could be found which resisted compression to infinite density. One such material was the C-field of the steady state theory, introduced by Hoyle to describe 'continuous creation' of matter in a de Sitter universe that avoided the high-density singularities of the Friedmann–Lemaître models. However, all its null geodesics are past incomplete and so it is technically singular [35] (a feature that has been recently rediscovered in another context [36]).

---

[1] Einstein supposedly said that introducing the cosmological constant was the biggest blunder of his life, presumably because it meant that he failed to predict the expansion of the universe, but I can find no primary trace of such a remark. Einstein did make a similar remark about his signing of the letter to Roosevelt urging the construction of an atomic bomb (although he didn't work on the Manhattan Project because he was judged to be a potential security risk). In 1954, he called his decision to sign 'the one great mistake in my life' [31] which suggests he never made such a remark about the cosmological constant.

Later, in the period 1957–1966, a different approach was pursued for a while by members of Landau's school in Moscow, notably by Khalatnikov and Lifshitz [37]. Initially, they set out to show that singularities did not occur in the general solution of the Einstein equations. Their argument, which was incorrect, was that because singularities arose in solutions that were not general (like the Friedmann solutions or the anisotropic Kasner universes) they would appear in the general solution. The circumstantial evidence for this conclusion was the belief that the singularity arising in these cosmological solutions was just a singularity in the coordinate system used to describe the dynamics and so was unphysical (as is the 'singularity' that arises at the North Pole of the Earth where the meridians intersect in standard mapping coordinates). When it occurred you could change to a new set of coordinates until they too became singular (as they always did) and so on *ad infinitum*. Unfortunately, it is important to investigate what happens in the limit of this process: a true physical singularity remains – as became increasingly clear when the problem was subjected to a different sort of analysis. The singularity theorems of Hawking and Penrose [7, 8] were able to define sufficient conditions for the formation of a singularity by adopting a definition of a singularity as an inextensible path of a particle or light-ray in spacetime. These theorems made no reference to special symmetries or the subtleties of coordinate choices. Singularities were where time ran out: part of the edge of spacetime [38]. It remained to be shown that these endpoints were caused by infinities in physical quantities. To some extent this can be done but the full story is by no means complete, even now. These theorems ended the argument about whether cosmological singularities were physically real and general and later work by Belinskii, Khalatnikov and Lifshitz refocused upon finding the general behaviour near a physical singularity [39].

It is important to stress that the singularities are *theorems*, not *theories*. They give *sufficient* conditions for singularities so if their assumptions are not all met this does not mean that there is no inevitable singularity, merely that no conclusions can be drawn. The interesting historical aspect which we signalled in the introduction is that the sufficient conditions generally included the requirement that the matter content of the universe obeys $\rho + 3p > 0$. We no longer believe this inequality holds for all matter sources. Indeed, the observations that the universe is accelerating could be claimed to show that the assumption that $\rho + 3p > 0$ is false.

5.3.3 Isotropisation

After the discovery in 1967 of the high level of isotropy in the CMB temperature distribution [40] there was a long effort to explain why this was the case. Up until then, cosmologists had assumed an isotropic and homogeneous background universe and regarded the presence of small inhomogeneities (like galaxies) as the major mystery requiring a simple explanation. The discovery of the CMB isotropy placed created a new perspective in which it was the high isotropy and uniformity of the assumed background universe that was the major mystery. A new approach, proposed by Misner and dubbed 'chaotic cosmology' sought to show that general cosmological initial conditions would end up leading to an isotropically expanding universe after more than about 10 billion years [10, 11]. This programme had an interesting methodological aspect. If it could be shown that almost all initial conditions (subject to some weak conditions of physical reasonableness) would lead to an isotropic universe, then observations of the isotropy level could not tell us anything about the initial conditions: memory of them would have been erased by the expansion.

In studying whether this idea could work it was again the issue of generality that was crucial. It was asking whether isotropically expanding universes were stable or even asymptotically stable attractors at late times. Two types of analysis were performed. The first just asked whether anisotropic cosmologies would approach isotropy at late times if they just contain zero-pressure matter and radiation. The second, which Misner proposed, was to ask what happened if dissipative stresses could arise because of the presence of collisionless particles, like neutrinos and gravitons, at particular epochs in the very early universe. Perhaps large initial anisotropies could be damped out by these dissipative processes, leaving an isotropically expanding universe?

Several interesting approaches to these questions were developed. On the physical level Barrow and Matzner [41] showed that the chaotic cosmology philosophy could not work in general because the dissipation of anisotropies and inhomogeneities must produce heat radiation, in accord with the second law of thermodynamics. The earlier dissipation occurred the larger the entropy per baryon produced. So, the observed entropy per baryon today (of about 10^9) placed a prohibitively strong bound on how much anisotropy could have been damped out during the history of the universe. The presence of particle horizons with proper radii proportional to t in the early universe also placed a major constraint on the damping of any large scale inhomogeneities by causal processes. Misner [42] attempted to circumvent this by discovering the remarkable possibility that spatially homogeneous universes of Bianchi type IX (dubbed the 'Mixmaster' universe because of this property) could potentially allow light to travel around the universe arbitrarily, often on approach to $t = 0$ as a result of their chaotic dynamics. Unfortunately, this horizon removal mechanism was ineffective in practice because of the improbability of the horizonless dynamical configurations and the fact that only about 20 chaotic oscillations of the scale factor could have occurred between the Planck time (10^{-43}s) and the present [43]. There was also the concern that if general relativistic cosmology was a well-behaved initial-value problem then one could always concoct anomalously anisotropic universes today that could be evolved back to their initial conditions at any arbitrary early time [44]. These would provide counter examples to the chaotic cosmology scheme, although one has to be careful with this argument because the counter examples could all have physically impossible initial conditions, and in fact they often do [45].

In 1973 Collins and Hawking [46] carried out some interesting stability analyses of isotropic universes to discover if isotropy was a stable property of homogeneous initial data. The results found were widely discussed and reported at a semi-popular level but needed to be treated cautiously because of the fine detail in the theorems. They reported that for ever-expanding universes isotropic expansion was 'unstable' but if attention was narrowed to spatially flat initial data with zero-pressure matter then isotropy was 'stable'. The cosmological constant was assumed to be zero. The definition of stability used was in fact *asymptotic stability* and so the proof that isotropy was unstable just meant that anisotropies did not tend to zero as $t \to \infty$. In fact, closer analysis showed that in general $\sigma/H \to$ constant in open universes (and it is impossible for σ/H to grow asymptotically) with $\rho + 3p > 0$. This means that isotropy is *stable,* although not *asymptotically stable* [47, 48]. The other technicality is that this result is a consequence of the fact that these open universes become vacuum (or spatial curvature dominated) at late times. The behaviour of the anisotropy is therefore an asymptotic property of vacuum cosmologies and does not tell us anything about the past history of the universe at redshifts $z > z_c$, where $z_c < O(1)$ is the redshift where the expansion becomes curvature dominated. These stability results therefore did not help us understand the sort of initial data that could give rise to high isotropy after about 10 billion years of expansion.

5.3.4 Cosmic No-Hair Theorems

Amid all this interest in explaining the isotropy of the universe there was one prescient approach by Hoyle and Narlikar [49] that predated the discovery of the high isotropy of the CMB. In 1963 they pointed out that in the standard big bang model the isotropy and average uniformity of the universe was a mystery but in the steady state universe it would be naturally explained. To support this claim they showed that the de Sitter universe is stable against scalar, vector and tensor perturbations. Thus, a steady state universe, described by the de Sitter metric of general relativity would always display high isotropy and uniformity. In fact, had they only known it, they could have predicted that the only spectrum of perturbations consistent with the steady state universe is the constant curvature spectrum with constant (small) metric perturbations on all scales that is observed with high accuracy today – although ironically via perturbations in the CMB whose existence the steady state model could not explain. Any departure from this spectrum, with $\delta\rho/\rho \propto L^{-2}$ on length scale L, would either create divergent metric potential perturbations as $L \to \infty$ or $L \to 0$.

The idea of the inflationary universe provided a new type of explanation for the observed isotropy and uniformity of the universe from general initial conditions, but with one important difference from past expectations – the isotropy and homogeneity was predicted to be *local*. The inflationary universe theory proposed that there was a finite interval of time, soon after the apparent beginning of the expansion (typically at $\sim 10^{-35}$s in the original conception of the theory), when the expansion of the universe would accelerate due to the presence of very slowly evolving scalar fields of the sort that appeared in new theories of high-energy physics. These would contribute stresses with $\rho + 3p < 0$ and cause the expansion scale factor to accelerate. When this occurs the expansion rapidly approaches isotropy. Anisotropies fall off very rapidly and isotropic expansion is asymptotically stable. In the most likely scenario, where $p = -\rho$, the expansion behaves temporarily like

Hoyle's steady state universe and grows exponentially in time. However, the inflation needs
to end, and this only happens if the scalar fields responsible decay into ordinary particles
and radiation with $\rho + 3p > 0$. When this happens the usual decelerating expansion is
resumed but with anisotropies so diminished in amplitude that they remain imperceptibly
small at late times [13–18].

Inflation works by taking a small patch of the universe that is small enough for light
signals to cross it at an early time t_I and expanding it so dramatically (exponentially in
time) that it grows larger than the entire visible universe today in the short period during which inflation occurs. Thus the isotropy and high uniformity of the visible universe
today are a reflection of the fact that it is the expanded image of a region that was small
enough to be coordinated by light-like transport processes and damping when inflation
occurred. If inflation had not occurred, and the expansion had merely continued along its
standard decelerating trajectory then the initially smooth and isotropic region would not
have expanded significantly by the present time, t_0. Here is a simple calculation of how
this happens.

Suppose the preset temperature of the CMB is $T_0 = 3K$ and when inflation occurred it
was $T_I = 3 \times 10^{28}K$. Then, since $T \propto a^{-1}$, the scale factor has increased by a factor of
$T_I/T_0 = 10^{28}$. At time t_I, the horizon size is equal to $d(t_I) = 2ct_I$ where $t_I \simeq 10^{-35}$s, so
a horizon-sized region of size $2ct_I$ at t_I would only have expanded to a size $2 \times 3 \times 10^{10}$
cm s^{-1} $\times 10^{-35}$s $\times 10^{28} = 6 \times 10^3$ cm by the present day. This is not of any relevance
for explaining isotropy and uniformity over scales of order $ct_0 \sim 10^{28}$ cm today. However,
suppose inflation occurs at t_I and inflates the expansion scale factor by a factor of e^N. We
will now be able to enlarge the causally connected region of size $2ct_I$ up to a scale of
$e^N \times 6 \times 10^3$ cm. This will exceed the size of the visible universe today if $e^N \times 6 \times 10^3 \gtrsim$
10^{28}. This is easily possible with $N > 60$. If the expansion is exponential with $a(t) \propto e^{Ht}$
then we only need inflation to last from about 10^{-35} s until 10^{-33} s in order to effect
this. The regularity of the universe is therefore explained without any dissipation taking
place. A very tiny smooth patch is simply expanded to such an extent that its smooth and
isotropic character is reflected on the scale of the entire universe today. It is very likely
(just as it is more likely that a randomly chosen positive integer will be a very large one)
that the amount of inflation that occurred will be much larger than 60 e-folds. Yet, the
result is to predict that the universe will be uniform on the average out to the inflated scale
$e^N \times 6 \times 10^3$ but may be rather non-uniform if we could see further. In some variants of
the theory many other fundamental features of the universe (values of constants of Nature,
space dimensions, laws of physics) are different beyond the inflated scale as well. While
there have always been overly positivistic philosophers who have cautioned against simply
assuming that the unobserved part of the (possibly infinite) universe is the same on average
as the observed part, this is the first time there has been a positive prediction that we should
not expect them to be the same.

The result of a sufficiently long period of accelerated expansion is to drive the local
expansion dynamics of the universe towards the isotropic de Sitter expansion, with

asymptotic form of the metric at large time, t, of the form ($\alpha, \beta = 1, 2, 3$) [18]

$$ds^2 = dt^2 - g_{\alpha\beta}dx^\alpha dx^\beta,$$

$$g_{\alpha\beta} = \exp[2Ht]a_{\alpha\beta}(\mathbf{x}) + b_{\alpha\beta}(\mathbf{x}) + \exp[-Ht]c_{\alpha\beta}(\mathbf{x}) + \cdots,$$

where H is the constant Hubble rate, with $3H^2 = \Lambda$, and $a_{\alpha\beta}(\mathbf{x}), b_{\alpha\beta}(\mathbf{x})$ and $c_{\alpha\beta}(\mathbf{x})$ are arbitrary symmetric spatial functions. The Einstein equations allow only two of the $a_{\alpha\beta}$ and two of the $c_{\alpha\beta}$ to be freely specifiable and all the $b_{\alpha\beta}$ are determined by them. Thus there are four independently arbitrary spatial functions specifying the solution on a spacelike surface of constant t in vacuum. Notice the spatial functions $a_{\alpha\beta}(\mathbf{x})$ at leading order in the metric (H is a constant though). This is why the metric only approaches de Sitter locally, exponentially rapidly inside the event horizon of a geodesically moving observer. If black body radiation ($p = \rho/3$) is added then a further four arbitrary spatial functions are required (three for the normalised four-velocity components and one for the density) and

$$\rho \propto \exp[-4Ht],$$

$$u_0 \to 1, u_\alpha \propto \exp[Ht]c^\beta_{\alpha;\beta},$$

$$c^\alpha_\alpha = 0.$$

Hence, we see that the three-velocity $V^2 = u_\alpha u^\alpha$ tends to a constant as $t \to \infty$. The asymptotic state is therefore de Sitter plus a constant (or 'tilted') velocity field which affects the metric at third order (via $c_{\alpha\beta}(\mathbf{x})$). This is easy to understand physically if we consider a large rotating eddy that expands with the universe and has angular velocity $\omega = Va^{-1}$. Its angular momentum is $Ma^2\omega \propto (\rho a^3)a^2(Va^{-1})$ and this is conserved as the universe expands. Since the radiation density falls as $\rho \propto a^{-4}$, we have V constant as $a \to \infty$.

The number of free spatial functions specifying this asymptotic solution is eight in the case with radiation (and the same holds when any other perfect fluid matter is present). In the next section we will show that this is characteristic of a part of the general solution of Einstein's equations.

In conclusion we see that a finite period of accelerated expansion is able to drive the expansion towards isotropy from a very large class of initial conditions (not all initial conditions, since the universe must not, for example, recollapse before a period of accelerated expansion begins). We have just discussed the most extreme form of accelerated expansion with constant Hubble expansion rate, H, and $a \propto \exp[Ht]$ but similar conclusions hold for power-law inflation, with $a \propto t^n, n > 1$, and intermediate inflation, with $a \propto \exp[At^n]$, with $A > 0$, and $0 < n < 1$ constants. The key conceptual point is that inflation explains the present isotropy without dissipating initial anisotropies in the way that the chaotic cosmology programme imagined and so it evades the Barrow–Matzner entropy per baryon constraint [41]. Instead, it drives the initial inhomogeneities far beyond the visible horizon today and the stress driving the acceleration dominates over all forms of anisotropy at large expansion volumes and times. The earlier analyses of the stability of isotropic expansion by Collins and Hawking, and others [46] had restricted attention to forms of matter in the universe with $\rho + 3p > 0$ and always assumed $\Lambda = 0$ because there was no reason to think otherwise at that time. As a result, they had excluded the possibility of accelerated expansion which can solve the isotropy problem without any dissipation occurring if it can arise for a finite period of time in the early universe.

5.3.5 The Initial Value Problem

The attempts to explain the isotropy of the universe from arbitrary initial conditions gave rise to another interesting perspective that is worth highlighting. General relativity is an initial value problem and so for 'well-behaved' cosmological solutions this means that the present state of the universe described by any solution of Einstein's equations is a continuous function of some 'initial data' at any past time. In a technical sense it might appear that given any state of the universe today – highly anisotropic, for example – then there exists some initial data set that evolves to give that state regardless of the action of any damping effects. Hence, there could never be a theory that could explain the actual state of the universe today as the result of evolution from any (or almost any) initial conditions. The problem with this argument is that the initial conditions that do evolve to counter-factual cosmological states at late times may arise only from initial data states that are completely unphysical in some respect [45]. Take a simple example of a Bianchi type I anisotropic universe. The anisotropy energy density and radiation energy density fall as

$$\sigma^2 = \sigma_0^2 (1 + z)^6,$$

$$\rho_\gamma = \rho_{\gamma 0} (1 + z)^4.$$

We can choose values of the constant σ_0^2 so that the universe's expansion is dominated by anisotropy today – just pick $\sigma_0^2 = \rho_{\gamma 0} \simeq 10^{-34}$ gm cm^{-3} to specify the present data. However, if we run this apparent counter-example back to the time when the radiation temperature is $T_{pl} \simeq 10^{32} K = 3(1 + z_{pl})$ when its energy density equals the Planck density, 10^{94} gm cm^{-3}, we require the anisotropy energy density to be 10^{64} times larger than the Planck energy density at that time – a completely unphysical situation. Alternatively, if we had taken the anisotropy energy density to be the Planck density at z_{pl} then we have the strange initial condition that the radiation density is 10^{64} times smaller despite all forms of energy being in quantum gravitational interaction at that time.

  This is a (deliberately) dramatic example but the basic problem with the argument is one that we can find with other arguments regarding the generality of more complicated outcomes in cosmology. For example, there have been claims (and claims to the contrary) that inflation is not generic for Friedmann universes containing scalar fields with a quadratic self-interaction potential [50–52]. The claim is based on using the Hamiltonian measure in the phase space for the dynamics to show that the bulk of the initial data measure is for solutions which do not inflate. This type of initial data corresponds to solutions with a huge initial kinetic term $(\dot{\phi}^2)$ which dominate the potential $V(\phi) = m^2\phi^2$ by a large factor so that the potential never comes to dominate the dynamics by any pre-specified epoch. However, this does not look very natural because it requires the two forms of energy density to differ by an enormous factor when one of them equals the Planck energy density (above which we know nothing about what happens since general relativity, quantum mechanics and statistical mechanics all break down). The better course is not to let any energy density exceed the Planck value but (surprisingly) this appears to be controversial.

5.4 Naive Function Counting

There have been several attempts to reduce the description of the astronomical universe to the determination of a small number of measurable parameters. Typically, these will be the free parameters of a well defined cosmological model that uses the smallest number of constants that can provide a best fit to the available observational evidence. Specific examples are the popular characterisations of cosmology as a search for 'nine numbers' [53], 'six numbers' [54], or the six-parameter minimal ΛCDM model used to fit the WMAP [55] and Planck data sets [56]. In all these, and other, cases of simple parameter counting there are usually many simplifying assumptions that amount to ignoring other parameters or setting them to zero; for example, by assuming a flat Friedmann background universe or a power-law variation of density inhomogeneity in order to reduce the parameter count and any associated degeneracies. The assumption of a power-law spectrum for inhomogeneities will reduce a spatial function to two constants, while the assumption that the universe is described by a Friedmann metric plus small inhomogeneous perturbations both reduces the number of metric unknowns and converts functions into constant parameters. In what follows we are going to provide some context for the common minimal parameter counts cited above by determining the total number of spatial functions that are needed to prescribe the structure of the universe if it is assumed to contain a finite number of simple matter fields. We are not counting fundamental constants of physics, like the Newtonian gravitation constant, the coupling constants defining quadratic lagrangian extensions of general relativistic gravity, or the 19 free parameters that define the behaviours of the 61 elementary particles in the standard three-generation $U(1) \times SU(2) \times SU(3)$ model of particle physics. However, there is some ambiguity in the status in some quantities. For example, as to whether the dark energy is equivalent to a true cosmological constant (a fundamental constant), or to some effective fluid or scalar field, or some other emergent effect [57]. Some fundamental physics parameters, like neutrino masses, particle lifetimes, or axion phases, can also play a part in determining cosmological densities but that is a secondary use of the cosmological observable. Here we will take an elementary approach that counts the number of arbitrary functions needed to specify the general solution of the Einstein equations (and its generalisations). This will give a minimalist characterisation that can be augmented by adding any number of additional fields in a straightforward way. We will also consider the count in higher-order gravity theories as well as for general relativistic cosmologies. We enumerate the situation in spatially homogeneous universes in detail so as to highlight the significant impact of their spatial topology on evaluations of their relative generality.

Let us move on to a more formal discussion of how to specify the generality of solutions to Einstein's equations by counting the number of free functions (or constants) that a given solution (or approximate solution) contains. In view of the constraint equations and coordinate covariances of the theory this requires a careful accounting.

The cosmological problem can be formulated in general relativity using a metric in a general synchronous reference system [58]. Assume that there are F matter fields which are non-interacting and each behaves as a perfect fluid with some equation of state $p_i(\rho_i)$, $i = 1, \ldots F$. They will each have a normalised four-velocity field, $(u_a)_i$, $a = 0, 1, 2, 3$. These will in general be different and non-comoving. Thus each matter field is defined on a spacelike surface of constant time by four arbitrary functions of three spatial variables, x^α since the u_0 components are determined by the normalisations $(u_a u^a)_i = 1$. This means that the initial data for the F non-interacting fluids are specified by $4F$ functions of three spatial variables. If we were in an N-dimensional space then each fluid would require $N + 1$ functions of N spatial variables and F fluids would require $(N + 1)F$ such functions to describe them in general.

The 3D metric requires the specification of 6 $g_{\alpha\beta}$ and 6 $\dot{g}_{\alpha\beta}$ for the symmetric spatial 3×3 metric in the synchronous system but these may be reduced by using the four coordinate covariances of the theory and a further four can be eliminated by using the four constraint equations of general relativity. This leaves four independently arbitrary functions of three spatial variables [58] which is just twice the number of degrees of freedom of the gravitational spin-2 field. The general transformation between synchronous coordinate systems maintains this number of functions [58]. This is the number required to specify the general vacuum solution of the Einstein equations in a three-dimensional (3D) space. In an N-dimensional space we would require $N(N + 1)$ functions of N spatial variables to specify the present data for $g_{\alpha\beta}$ and $\dot{g}_{\alpha\beta}$. This could be reduced by $N + 1$ coordinate covariances and $N + 1$ constraints to leave $(N - 2)(N + 1)$ independent arbitrary functions of N variables [59]. This even number is equal to twice the number of degrees of freedom of the gravitational spin-2 field in $N + 1$ dimensional spacetime.

When we combine these counts we see that the general solution in the synchronous system for a general relativistic cosmological model containing F fluids requires the specification of $(N - 2)(N + 1) + F(N + 1) = (N + 1)(N + F - 2)$ independent functions of N spatial variables. If there are also S non-interacting scalar fields, ϕ_j, $j = 1, \ldots, S$, present with self-interaction potentials $V(\phi_j)$ then two further spatial functions are required (ϕ_j and $\dot{\phi}_j$) to specify each scalar field and the total becomes $(N + 1)(N + F - 2) + 2S$. For the observationally relevant case of $N = 3$, this reduces to $4(F + 1) + 2S$ spatial functions.

For example, if we assume a simple realistic scenario in which the universe contains separate baryonic, cold dark matter, photon, neutrino and dark energy fluids, all with separate non-comoving velocity fields, but no scalar fields, then $F = 5$ and our cosmology needs 24 spatial functions in the general case. If the dark energy is not a fluid, but a cosmological constant with constant density and $u_i = \delta_i^0$, then the dark energy 'fluid' description reduces to the specification of a single constant, $\rho_{DE} = \Lambda/8\pi G$, rather than four functions and reduces the total to 21 independent spatial functions. However, if the cosmological constant is an evolving scalar field then we would have $F = 4$ and $S = 1$, and now 22 spatial functions are required. Examples of full-function asymptotic solutions were found for perturbations around de Sitter space-time by Starobinsky [18], the approach to 'sudden' finite-time singularities [60, 61] by Barrow, Cotsakis and Tsokaros [62], and near quasi-isotropic singularities with $p > \rho$ 'fluids' by Heinzle and Sandin [63].

These function counts of 21–24 should be regarded as lower bounds. They do not include the possibility of a cosmological magnetic field or some other unknown matter fields. They also treat all light ($<< 1MeV$) neutrinos as if they are identical (heavy neutrinos can be regarded as CDM if they provide the largest contribution to the matter density but if they are not responsible for the dominant dark matter then they should be counted as a further contribution to F). If there are matter fields which are not simple fluids with $p(\rho)$ – for example an imperfect fluid possessing a bulk viscosity or a gas of free particles with anisotropic pressures – then additional parameters are required to specify them. There can still be overall constraints – a trace-free energy-momentum tensor, for example, in the cases of electric and magnetic fields or Yang–Mills fields – and we would just count the number of independent terms in the total energy-momentum tensor [64].

In the case of the Planck or WMAP mission data analyses, six constants are chosen to define the standard (minimal) ΛCDM model. For WMAP [55], these are the present-day Hubble expansion rate, H_0, the densities of baryons and cold dark matter, the optical depth, τ, at a fixed redshift, and the amplitude and slope of an assumed power-law spectrum of curvature inhomogeneities on a specified reference length scale. This is equivalent to including three matter fields (radiation, baryons, cold dark matter) but the standard ΛCDM assumes zero spatial curvature, k, *ab initio*, so a relaxation of this would add a curvature term or a dark energy field, because when $k \neq 0$ the latter could no longer be deduced from the other densities and the critical density (defined by H_0). The light neutrino densities are assumed to be calculable from the radiation density using the standard cosmological thermal history, so there are effectively $F = 5$ matter fields (with k set to zero in the base model and a metric time derivative determined by H). All deviations from isotropy and homogeneity enter only at the level of perturbation theory and are characterised by the spectral amplitude and slope on large scales; the amplitude on small scales ('acoustic peaks' in the power spectrum) is determined from that on large scales by an $e^{-2\tau}$ damping factor determined by the optical depth parameter τ. The Planck mission's parameter choice is equivalent to this [56].

Although a general solution of the Einstein equations requires the full complement of arbitrary functions, different parts of the general solution space can have behaviours of quite different complexity. For example, when $N \leq 9$ there are homogeneous vacuum universes which are dynamically chaotic [39, 65, 66] but the chaotic behaviour disappears when $N \geq 10$ even though the number of arbitrary constants remains maximal for each N [67]. Hence, the dynamical complexity can fail to be captured by the function-counting approach.

5.4.1 Einstein's 'strength'

As an interesting historical aside, we should mention Einstein's attempt to study the power of mathematical formalisms to describe physical theories by ascribing to them a numerical measure of their predictive power, which he called the 'strength' of a system of differential equations. It was to be measured by the number of free pieces of initial data needed to determine the general solutions of the equations. Einstein believed that 'The smaller the number of free data consistent with the system of field equations, the 'stronger' is the system. It is clear that in the absence of any other viewpoint from which to select the equations, one will prefer a "stronger" system to a less strong one' [68]. This was the method Einstein proposed to follow in his quest for a unified field theory (how different to the methodology that led to all his past great successes). The enumeration of the strength of a system of equations for d variables began by expanding an analytic function of these variables in a Taylor series about a point and noting that at n^{th} order the total number of terms in the expansion is

$$\binom{n+d-1}{n} \equiv \frac{(n+d-1)!}{n!(d-1)!}.$$

If there are field equations which ensure that when the function is specified arbitrarily on a $d-1$ dimensional (spatial) surface then those in the remaining (temporal) dimension are determined by them, and only $\binom{d+1}{n}$ of the Taylor series coefficients remain arbitrary. The fraction of coefficients that remain free is therefore

$$\frac{\binom{d-1}{n}}{\binom{d}{n}} = \frac{d-1}{n+d-1}.$$

In the case of Einstein's equations we have coordinate covariances and constraint equations to use to reduce the count of free functions. The resulting strength turns out to be identical to the count of independent pieces of initial data for the metric and its first derivative that we have just described, giving a strength of four in vacuum. A similar count can be done for Maxwell's equations (which have the same strength), or other equations of mathematical physics. A fuller discussion is given in Schutz (1975) [69] and Mariwalla (1974) [70].

5.5 More General Gravity Theories

There has been considerable interest in trying to explain the dark energy as a feature of a higher-order gravitational theory that extends the lagrangian of general relativity in a non-linear fashion [71–74]. This offers the possibility of introducing a lagrangian that is a function $L = f(R, R_{ab}R^{ab})$ of the scalar curvature R and/or the Ricci scalar $R_{ab}R^{ab}$ in anisotropic models, with the property that it contributes a slowly varying dark energy-like behaviour at late times without the need to specify an explicit cosmological constant. How-ever, these higher-order lagrangian theories (excluding the Lovelock lagrangians in which the variation of the higher-order terms contribute pure divergences [75] and so the field equations are always second order in any spatial dimension) all have fourth-order field equations in 3D space when $f \neq A + BR$, with A, B constants. This means that the initial data set for such theories is considerably enlarged because we must specify $\ddot{g}_{\alpha\beta}$ and $\dddot{g}_{\alpha\beta}$ in addition to $g_{\alpha\beta}$ and $\dot{g}_{\alpha\beta}$. In N space dimensions, this results in a further $N(N+1)$ function of N variables and so a general cosmological model with F fluids and S scalar fields requires a specification of $2(N^2 - 1) + F(N+1) + 2S = (N+1)(F + 2N - 2) + 2S$ independent arbitrary spatial functions. For $N = 3$, this is $16 + 4F + 2S$. General relativity with four matter fields plus a cosmological constant requires 20 spatial functions plus one constant, in general, whereas a higher-order gravity theory with four matter fields and no scalar fields (and no cosmological constant because it should presumably emerge from the metric behaviour) requires the specification of 32 spatial functions.

5.6 Reducing Functions to Constants

The commonest simplification used to reduce the size of the cosmological characterisation problem is to turn the spatial functions into constants. This simplification will be exact if the universe is assumed to be spatially homogeneous. The set of possible spatially homogeneous and isotropic universes with natural topology is based upon the classification of homogeneous three-spaces created by Bianchi [76–79] (together with the exceptional case of Kantowski–Sachs–Kompanyeets–Chernov with $S^1 \times S^2$ topology [80, 81], which we will ignore here as it displays non-generic behaviour).

The most general Bianchi-type universes are those of types $VI_h, VII_h, VIII$ and IX. Of these, only types VII_h and IX, respectively, contain open and closed isotropic Friedmann subcases. These most general Bianchi types are all defined by four arbitrary constants in vacuum plus a further four for each non-interacting perfect fluid source. Therefore, in 3D spaces, the most general spatially homogeneous universes containing F fluids are defined by $4(1 + F)$ arbitrary *constants*. This suggests that they might be the leading order term in a linearisation of the general inhomogeneous solution in the homogeneous limit. However, things might not be so simple. The four-function space of solutions to Einstein's models like type IX with compact spaces has a conical structure at points with Killing vectors and so linearisation about the points must control an infinite number of spurious linearisations (associated with all the tangents that can be drawn through the point of the cone but do not run down the side of the cone) that are not the leading-order terms in any convergent series expansion of a true solution [21, 82].

The Bianchi classification of spatially homogeneous universes derives from the classification of the group of isometries with 3D subgroups that act simply transitively on the manifold. Intuitively, these give cosmological histories that look the same to observers in different places on the same hypersurface of constant time.

The Bianchi types are subdivided into two classes [83]: Class A contains types $I(1 + F), II(2 + 3F), VI_0(3 + 4F), VII_0(3 + 4F), VIII(4 + 4F)$ and $IX(4 + 4F)$, while Class B contains types $V(1 + 4F), IV(3 + 4F), III(3 + 4F), VI_{-1/9}(4 + 3F), VI_h(4 + 4F)$ and $VII_h(4 + 4F)$. The brackets following each Roman numeral of the Bianchi-type geometry contain the number of constants defining the general solution when F non-interacting perfect fluids, each with $p > -\rho$, are present, so $F = 0$ defines the vacuum case. For example, Bianchi type I denoted by $I(1+F)$ is defined by one constant in vacuum (when it is the Kasner metric) and one additional constant for the value of the density when each matter field is added. For simplicity, we have ignored scalar fields here, but to include them simply add $2S$ inside each pair of brackets. The Euclidean metric geometry in the type I case requires $R_{0\alpha} = 0$, identically, and so the three non-comoving velocities (and hence any possible vorticity) must be identically zero. This contains the zero-curvature Friedmann model as the isotropic (zero parameter) special case. In the next simplest case, of type V, the general vacuum solution was found by Saunders [84] and contains one parameter, but each additional perfect-fluid adds four parameters because it requires specification of a density and three non-zero u_α components. The spatial geometry is a Lobachevsky space of constant negative isotropic curvature. The isotropic subcases of type V are the zero-parameter Milne universe in vacuum and the F-parameter open Friedmann universe containing F fluids.

In practice, one cannot find exact homogeneous general solutions containing the maximal number of arbitrary constants because they are too complicated mathematically, although the qualitative behaviours are fairly well understood, and many explorations of the observational effects use the simplest Bianchi I or V models (usually without including non-comoving velocities) because they possess isotropic three-curvature and add only a simple fast-decaying anisotropy term (requiring one new constant parameter) to the Friedmann equation. The most general anisotropic metrics which contain isotropic special cases, of types VII and IX, possess both expansion anisotropy (shear) and anisotropic three-curvature. Their shear falls off more slowly (logarithmically in time during the radiation era) and the observational bounds on it are much weaker [35, 46, 85–88].

5.7 Some Effects of the Topology of the Universe

So far, we have assumed that the cosmological models in question have the 'natural' topology, that is R^3 for the 3D flat and negatively curved spaces and S^3 for the closed spaces. However, compact topologies can also be imposed upon flat and open universes to make their spatial volumes finite and there has been considerable interest in this possibility and its observational consequences for optical images of galaxies and the CMB [89–91].

The classification of compact negatively curved spaces is a challenging mathematical problem. When compact spatial topologies are imposed on spatially flat and open homogeneous cosmologies, this produces a major change in their relative generalities and the numbers of constants needed to specify them in general.

The most notable consequences of a compact topology on 3D homogeneous spaces is that the Bianchi universes of types IV and VI_h no longer exist at all and open universes of Bianchi types V and VII_h must be isotropic with spaces that are quotients of a space of constant negative curvature, as required by Mostow's Rigidity theorem [92–96]. The only universes with non-trivial structure that differs from that of their universal covering spaces are those of Bianchi types I, II, III, VI_0, VII_0 and VIII. The numbers of parameters needed to determine their general cosmological solutions when F non-interacting fluids are present and the spatial geometry is compact are now given by $I(10 + F), II(6 + 3F), III(2 + N'_m + F), VI_0(4+4F), VII_0(8+4F)$ and $VIII(4+N_m+4F)$, again with $F = 0$ giving the vacuum case, as before, and an addition of $2S$ to each prescription if S scalar fields are included. Here, N_m is the number of moduli degrees of freedom which measures the complexity of the allowed topology, with $N_m \equiv 6g + 2k - 6 \equiv N'_m - 2g$, where g is the genus and k is the number of conical singularities of the underlying orbifold [94, 95]. It can be arbitrarily large.

The rigidity restriction that compact types V and VII_h must be isotropic means that compactness creates general parameter dependencies of $V(F)$ and $VII_h(F)$ which are the same as those for the open isotropic Friedmann universe, or the Milne universe in vacuum when $F = 0$.

The resulting classification is shown in Table 5.1 [97]. We see that the introduction of compact topology for the simplest Bianchi type I spaces produces a dramatic increase in relative generality. Indeed, they become the most general vacuum models by the parameter-counting criterion. An additional nine parameters are required to describe the compact type I universe compared to the case with non-compact Euclidean R^3 topology. The reason for this increase is that at any time the compact three-torus topology requires three identification scales in orthogonal directions to define the torus and three angles to specify the directions of the vectors generating this lattice plus all their time-derivatives. This gives 12 parameters, of which two can be removed using a time translation and the single non-trivial Einstein constraint equation, leaving ten in vacuum compared to the one required in the non-compact Kasner vacuum case.

The following general points are worth noting:

1. The imposition of a compact topology changes the relative generalities of homogeneous cosmologies.
2. The compact flat universes are more general in the parameter-counting sense than the open or closed ones.
3. Type VIII universes, which do not contain Friedmann special cases but can in principle become arbitrarily close to isotropy are the most general compact universes.

The most general case that contains an isotropic special case is that of type VII_0 – recall that the VII_h metrics are forced to be isotropic so open Friedmann universes now become asymptotically stable [94] and approach the Milne metric, whereas in the non-compact case they are merely stable and approach a family of anisotropic vacuum plane waves

Table 5.1. *The number of independent arbitrary constants required to
prescribe the general 3D spatially homogeneous Bianchi-type universes
containing F perfect fluid matter sources in cases with non-compact and
compact spatial topologies. The vacuum cases arise when $F = 0$. If S
scalar fields are also present then each parameter count increases by 2S.
The type IX universe does not admit a non-compact geometry and
compact universes of Bianchi types IV and VI_h do not exist. Types III and
VIII have potentially unlimited topological complexity and arbitrarily
large numbers of defining constants parameters through the unbounded
topological parameters $N_m \equiv 6g + 2k - 6$ and $N'_m = N_m + 2g$, where g is
the genus and k is the number of conical singularities of the underlying
orbifold [97].*

Cosmological	No. of defining parameters with F non-interacting fluids	
Bianchi type	Non-compact topology	Compact topology
I	$1 + F$	$10 + F$
II	$2 + 3F$	$6 + 3F$
VI_0	$3 + 4F$	$4 + 4F$
VII_0	$3 + 4F$	$8 + 4F$
VIII	$4 + 4F$	$4 + N_m + 4F$
IX	$-$	$4 + 4F$
III	$3 + 4F$	$2 + N'_m + F$
IV	$3 + 4F$	$-$
V	$1 + 4F$	F
VI_h	$4 + 4F$	$-$
VII_h	$4 + 4F$	F

[47]. This peculiar hierarchy of generality should be seen as a reflection of how difficult it is to create compact homogeneous spaces supporting these homogeneous groups of
motions.

5.8 Inhomogeneity

The addition of inhomogeneity turns the constants defining the cosmological problem into
functions of three space variables. For example, we are familiar with the linearised solutions for small density perturbations of a Friedmann universe with natural topology which
produces two functions of space that control temporally growing and decaying modes.
The function of space in front of the growing mode is typically written as a power-law
in length scale (or wave number) and so has arbitrary amplitude and power index (both
usually assumed to be scale-independent constants to first or second order) which can be

fitted to observations. Clearly there is no limit to the number of parameters that could be introduced to characterise the density inhomogeneity function by means of a series expansion around the homogeneous model (and the same could be done for any vortical or gravitational-wave perturbation modes) but the field equations would leave only eight independent functions. Further analysis of the function characterising the radiation density is seen in the attempts to measure and calculate the deviation of its statistics from Gaussianity [98] and to reconstruct the past light-cone structure of the universe [99]. Any different choice of specific spatial functions to characterise inhomogeneity in densities or gravitational waves requires some theoretical motivation. What happens in the inhomogeneous case if open or flat universes are given compact spatial topologies is not known. As we have just seen, the effects of topology on the spatially homogeneous anisotropic models was considerable whereas the effects on the overall evolution of isotropic models (as opposed to the effects on image optics) is insignificant. It is generally just assumed that realistically inhomogeneous universes with non-positive curvature (or curvature of varying sign) can be endowed with a compact topology and, if so, this places no constraints on their dynamics. However, both assumptions would be untrue for homogeneous universes and would necessarily fail for inhomogeneous ones in the homogeneous limit. It remains to be determined what topological constraints arise in the inhomogeneous cases. They could be weaker because inhomogeneous anisotropies can be local (far smaller in scale than the topological identifications) or they could be globally constrained like homogeneous anisotropies. Newtonian intuitions can be dangerous because compactification of a Newtonian Euclidean cosmological space seems simple but if we integrate Poisson's equation over the compact spatial volume we see that the total mass of matter must be zero. This follows from Poisson's equation since

$$0 = \int_V \nabla^2 \Phi dV = 4\pi G \int_V \rho dV = 4\pi GM,$$

where V is the compact spatial volume, M the total mass, and Φ is the Newtonian gravitational potential.

In practice, there is a divide between the complexity of inhomogeneity in the universe on small and large scales. On large scales there has been effectively no processing of the primordial spectrum of inhomogeneity by damping or non-linear evolution. Its description is well approximated by replacing a smooth function by a power-law defined by two constants, as for the microwave background temperature fluctuation spectrum or the two-point correlation function of galaxy clustering. Here, the defining functions may be replaced by statistical distributions for specific features, like peaks or voids in the density distribution. On small scales, inhomogeneities that entered the horizon during the radiation era can be damped out by photon viscosity or diffusion and may leave distortions in the background radiation spectrum as witness to their earlier existence. The baryon distribution may provide baryon acoustic oscillations which yield potentially sensitive information about the baryon density [55, 56]. On smaller scales that enter the horizon later, where damping and non-linear self-interaction have occurred, the resulting distributions of luminous and dark matter are more complicated. However, they are correspondingly more difficult to predict in detail and numerical simulations of ensembles of models are used to make predictions down to the limit of reliable resolution. Predicting their forms also requires a significant extension of the simple, purely cosmological enumeration of free functions that we have discussed so far. Detailed physical interactions, 3D hydrodynamics, turbulence, shocks, protogalaxy shapes, magnetic fields, and collision orientations, all introduce additional factors that may increase the parameters on which observable outcomes depend. The so-called bias parameter, equal to the ratio of luminous matter density to the total density, is in reality a spatial function that is being used to follow the ratio of two densities because one (the dark matter) is expected to be far more smoothly distributed than the other. All these small scale factors combine to determine the output distribution of the baryonic and non-baryonic density distributions and their associated velocities.

5.8.1 *Links to Observables*

The free spatial functions (or constants) specifying inhomogeneous (homogeneous) metrics have simple physical interpretations. In the most general cases the four vacuum parameters can be thought of as giving two shear modes (i.e. time-derivatives of metric anisotropies) and two parts of the anisotropic spatial curvature (composed of ratios and products of metric functions). In the simplest vacuum models of type I and V the three-curvature is isotropic and there is only one shear parameter. It describes the allowed metric shear and in the type V model a second parameter is the isotropic three-curvature (which is zero in type I) – just like k in the Friedmann universe models. When matter is added there is always a single ρ (or p) for each perfect fluid and up to three non-comoving fluid velocity components. If the fluid is comoving, as in type I only the density parameter is required for each fluid; in type V there can also be three non-comoving velocities. The additional parameters control the expansion shear anisotropy, anisotropic three-curvature. They may all contribute to temperature anisotropy in the CMB radiation but the observed anisotropy is determined by an integral down the past null cone over the shear (effectively the shear to Hubble rate ratio at last scattering of the CMB), rather than the Weyl curvature modes driven by the curvature anisotropy (which can be oscillatory [100], and so can be periodically very small even though the envelope is large), while the velocities contribute dipole variations. Thus, it is difficult to extract complete information about all the anisotropies from observations of the lower multipoles of the CMB alone in the most general cases [99, 101, 102].

   At present, the observational focus is upon testing the simplest possible ΛCDM model, defined by the smallest number (six) of parameters. As observational sensitivity increases it will become possible to place specific bounds or make determinations of the full spectrum of defining functions (or constants), or at least to confirm that they remain undetectably small as inflation would lead us to expect. In an inflationary model they can be identified with the spatial functions defining the asymptotic expansion around the de Sitter metric [18].

There have also been interesting studies of the observational information needed to determine the structure of our past null cone rather than constant-time hypersurfaces in the Universe [103], extending earlier investigations of the links between observables and general metric expansions by McCrea [104] and by Kristian and Sachs [105].

The high level of isotropy in the visible universe, possibly present as a consequence of a period of inflation in the early universe [12], or special initial conditions [45, 63, 107–110], is what allows several of the defining functions of a generic cosmological model to be ignored on the grounds that they are too small to be detected with current technology. An inflationary theory of the chaotic or eternal variety, in which inflation only ends locally, will lead to some complicated set of defining functions that exhibit large smooth isotropic regions within a complicated global structure which is beyond our visual horizon and unobservable (although not necessarily falsifiable within a particular cosmological model). However, despite the success of simple cosmological theories in explaining almost all that we see in the universe, it is clear that there is an under-determination problem: we cannot make enough observations to specify the structure of space-time and its contents, even on our past light cone, let alone beyond it. It is not a satisfactory methodology to use observations to construct a description of space-time. Rather, we proceed by creating parametrised descriptions that follow from solutions of Einstein's equations (or some other theory) and then constrain the free parameters by using he observational data. Despite the widespread lip service paid to Popper's doctrine of falsification as a scientific methodology, its weaknesses are especially clear in cosmology. It assumes that all observations and experimental results are correct and unbiased – that what you see is what you get. In practice, they are not and you never know whether observational data are falsifying a theory, or are based on wrong measurements, or subject to some unsuspected selection effect [111]. All that observational science can ever do is change the likelihood of a particular theory being true or false. Sometimes the likelihood can build up (or down) to such an extent that we regard a theory being tested (like the expansion of the universe) as 'true' or (like cold fusion) as 'false'.

Acknowledgements

Support from the STFC (UK) and the JTF Oxford-Cambridge Philosophy of Cosmology programme grant is acknowledged.

References

[1] A. Einstein, S. -B. Preuss. *Akad. Wiss.* **142** (1917).
[2] A. S. Eddington, *Mon. Not Roy. Astron. Soc.* **90**, 668 (1930).
[3] E. R. Harrison, *Rev. Mod. Phys.* **39**, 862 (1967).
[4] G. W. Gibbons, *Nucl. Phys.* **B 292**, 784 (1987).
[5] *Ibid.*, **310**, 636 (1988).
[6] J. D. Barrow, G. F. R. Ellis, R. Maartens and C. Tsagas, *Class. Quant. Grav.* **20**, L155 (2003).

[7] S. W. Hawking and G. F. R. Ellis, *The Large Scale Structure of Space-Time,* (Cambridge University Press, Cambridge, 1973).

[8] Hawking S. W. and Penrose R., *Proc. R. Soc. London A* **314**, 529 (1970).

[9] K. Gödel, *Rev. Mod. Phys.* **21**, 447 (1949).

[10] C. W. Misner, *Phys. Rev. Letts.* **19**, 533 (1967).

[11] C. W. Misner, *Ap. J.* **151**, 431 (1968).

[12] A. H. Guth, *Phys. Rev. D* **23**, 347 (1980).

[13] R. M. Wald, *Phys. Rev. D* **28**, 2118 (1982).

[14] J. D. Barrow, *The Very Early Universe*, ed. G. W. Gibbons, S. W. Hawking & S. T. C. Siklos, (Cambridge U. P., Cambridge,1983) pp. 267–72.

[15] W. Boucher and G. W. Gibbons, *ibid.* pp. *273–8.*

[16] L. Jensen and J. Stein Schabes, *Phys. Rev. D* **35**, 1146 (1987).

[17] J. D. Barrow, *Phys. Lett. B* **187**, 12 (1987).

[18] A. A. Starobinsky, *Sov. Phys. JETP Lett.* **37**, 66 (1983).

[19] R. Earman, *Bangs, Crunches, Whimpers and Shrieks: Singularities and acausalities in relativistic spacetimes,* (Oxford U.P., Oxford, 1995).

[20] S. W. Hawking, *Gen. Rel. Gravitation* **1**, 393 (1971).

[21] J. D. Barrow and F. J. Tipler, *Phys. Reports* **56**, 371 (1979).

[22] J. D. Barrow and G. Götz, *Class. Quantum Gravity*, **6**, 1253 (1989).

[23] J. D. Norton, in *The Expanding Worlds of General Relativity*, eds. H. Goenner, J. Renn, J. Ritter, and T. Sauer, *Einstein Studies*, **7**, 271–323 (1999).

[24] E. A. Milne and W. H. McCrea, *Quart. J. Math. Oxford Ser.* **5**, 73 (1934).

[25] D. Saari and Z. Xia, *Hamiltonian Dynamics and Celestial Mechanics*, (Amer. Math. Soc., Providence RI, 1996).

[26] G. F. R. Ellis, gr-qc/1107.3669.

[27] Z. Xia, *Ann. Math.,* **135**, 411 (1992).

[28] J. D. Barrow, *The Infinite Book*, (Jonathan Cape, London, 2005).

[29] J. D. Barrow and G. W. Gibbons, *Mon. Not. Roy. Astron. Soc.* **446**, 3874 (2014).

[30] G. Lemaître, *Mon. Not. Roy. Astron. Soc.* **91**, 483 (1931).

[31] R. Clark, *Einstein: the life and times*, (Avon Books, New York, 1972), p. 752.

[32] A. Friedmann, *Zeit f. Phys.* **10**, 377 (1922).

[33] G. Lemaître, *Ann. Soc. Sc. Bruxelles*, **53A**, 51 (1933).

[34] S. Chandrasekhar and J. C. Miller, *Mon. Not. Roy. Astron. Soc.* **167**, 63 (1974).

[35] J. D. Barrow and F. J. Tipler, *The Anthropic Cosmological Principle*, (Oxford U.P., Oxford, 1986), p. 612.

[36] A. Borde and A. Vilenkin, *Phys. Rev. Lett.* 72, 3305 (1994).

[37] E. M. Lifshitz and I. M. Khalatnikov, *Adv. Phys.* **12**, 185 (1963).

[38] J. D. Barrow, *The Origin of the Universe*, (Basic Books, New York, 1993), chapter 3.

[39] V. A. Belinskii, I. M. Khalatnikov and E. M. Lifshitz, *Adv. Phys.* **19**, 525 (1970).

[40] R. B. Partridge and D. T. Wilkinson, *Phys. Rev. Lett.* **18**, 557 (1967).

[41] J. D. Barrow and R. M. Matzner, *Mon. Not. Roy. Astron. Soc.* **181**, 719 (1977).

[42] C. W. Misner, *Phys. Rev. Lett.* **22**, 1071 (1969).

[43] A. G. Doroshkevich and I. D. Novikov, *Sov. Astron.* **14**, 763 (1971).

[44] C. B. Collins and J.M. Stewart, *Mon. Not. Roy. Astron. Soc.* **153**, 419 (1971).

[45] J. D. Barrow, *Phys. Rev. D* **51**, 3113 (1995).

[46] C. B. Collins and S.W. Hawking, *Mon. Not. R. Astron. Soc.* **162**, 307 (1972).

[47] J. D. Barrow and D. H. Sonoda, *Phys. Reports* **139**, 1 (1986).

[48] J. D. Barrow, *Quart. J. Roy. Astron. Soc.* **23**, 344 (1982).

[49] F. Hoyle and J. V. Narlikar, *Proc. Roy. Soc.* **A 273**, 1 (1963).
[50] L. A. Kofman, A. D. Linde and V. F. Mukhanov, *JHEP* 0210, 057 (2002).
[51] G. W. Gibbons and N. Turok, *Phys. Rev. D* **77**, 063516 (2008).
[52] A. Corichi and D. Sloan, arXiv:1310.6399.
[53] M. Rowan Robinson, *The Nine Numbers of the Cosmos*, (Oxford U.P., Oxford, 1999).
[54] M. J. Rees, *Just Six Numbers: The deep forces that shape the universe*, (Phoenix, London, 2001).
[55] D. N. Spergel *et al.,* (WMAP), *Ap. J. Supplt.* **148**, 175, (2003).
[56] P. A. R. Ade *et al.,* (Planck Collaboration Paper XVI), arXiv:1303.5076.
[57] J. D. Barrow and D. Shaw, *Phys. Rev. Lett.* **106**, 101302 (2011).
[58] L. Landau and E. M. Lifshitz, *The Classical Theory of Fields*, 4th rev. edn. (Pergamon, Oxford, 1975).
[59] J. D. Barrow, Gravitation and Hot Big Bang Cosmology, In *The Physical Universe: The Interface Between Cosmology, Astrophysics and Particle Physics*, eds. J. D. Barrow, A. Henriques, M. Lago & M. Longair, pp. 1–20, (Springer-Verlag, Berlin, 1991).
[60] J. D. Barrow, *Class. Quant. Grav.* **21**, L79 (2004).
[61] J. D. Barrow and A. A. H. Graham, arXiv: 1501.04090.
[62] J. D. Barrow, S. Cotsakis and A. Tsokaros, *Class. Quant. Grav.* **27**, 165017 (2010).
[63] M. Heinzle and P. Sandin, *Comm. Math. Phys.* **313**, 385 (2012).
[64] J. D. Barrow, *Phys. Rev. D* **89**, 064022 (2014).
[65] J. D. Barrow, *Phys. Rev. Lett.* **46**, 963 (1981) and Phys. Reports **85**, 1 (1982).
[66] D. Chernoff and J. D. Barrow, *Phys. Rev. Lett.* **50**, 134 (1983).
[67] J. Demaret, M. Henneaux and P. Spindel, *Phys. Lett.* **164**, 27 (1985).
[68] A. Einstein, *The Meaning of Relativity*, Appendix II, p. 136, 6th rev. edn. (Methuen, London and NY, 1956).
[69] B. Schutz, *J. Math. Phys.* **16**, 855 (1975).
[70] K. H. Mariwalla, *J. Math. Phys.* **15**, 468 (1974).
[71] J. D. Barrow and A. C. Ottewill, *J. Phys. A* **16**, 2757 (1983).
[72] V. Faraoni and T. Sotiriou, *Rev. Mod. Phys.* **82**, 451 (2010).
[73] T. Clifton, P. G. Ferreira, A. Padilla, C. Skordis, *Phys. Reports* **513**, 1 (2012).
[74] T. Clifton and J. D. Barrow, *Phys. Rev. D* **72**, 123003 (2005).
[75] D. Lovelock, *J. Math. Phys.* **12**, 498 (1971).
[76] L. Bianchi, *Mem. Matematica Fis. d. Soc. Ital. delle Scienza,* Ser. Terza **11**, 267 (1898) reprinted in *Gen. Rel. Grav.* **33**, 2171 (2001).
[77] A. H. Taub, *Ann. Math.* **53**, 472 (1951).
[78] M. A. H. MacCallum, in *General Relativity: An Einstein Centenary Survey*, eds. S. W. Hawking and W. Israel, (Cambridge UP, Cambridge, 1979), pp. 533–76.
[79] G. F. R. Ellis, S. T. C. Siklos and J. Wainwright, in *Dynamical Systems in Cosmology*, eds. J. Wainwright and G. F. R. Ellis (Cambridge UP, Cambridge, 1997), pp. 11–42.
[80] R. Kantowski and R. K. Sachs, *J. Math. Phys.* **7**, 443 (1966).
[81] A. S. Kompaneets and A. S. Chernov, *Sov. Phys. JETP* **20**, 1303 (1964).
[82] A. E. Fischer, J. E. Marsden and V. Moncrief, *Ann. Inst. H. Poincaré* **33**, 147 (1980).
[83] G. F. R. Ellis and M. A. H. MacCallum, *Comm. Math. Phys.* **12**, 108 (1969).
[84] P. T. Saunders, *Mon. Not. R. Astron. Soc.* **142**, 213 (1969).
[85] A. G. Doroshkevich, V. Lukash and I. D. Novikov, *Sov. Phys. JETP* **37**, 739 (1973).
[86] J. D. Barrow, *Mon. Not. R. Astron. Soc.* **175**, 359 (1976).

[87] J. D. Barrow, R. Juszkiewicz and D. N. Sonoda, *Mon. Not. R. Astron. Soc.* **213**, 917 (1985).

[88] J. D. Barrow, *Phys. Rev. D* **55**, 7451 (1997).

[89] G. F. R. Ellis, *Gen. Rel. Gravn.* **2**, 7 (1971).

[90] R. Aurich, S. Lustig, F. Steiner and H. Then, *Class. Quant. Grav.* **21**, 4901 (2004).

[91] P. A. R. Ade *et al.,* (Planck Collaboration Paper XXVI), arXiv:1303.5086.

[92] A. Ashtekar and J. Samuel, *Class. Quant. Grav.* **8**, 2191 (1991).

[93] H. V. Fagundes, *Gen. Rel. Gravn.* **24**, 199 (1992).

[94] J. D. Barrow and H. Kodama, *Int. J. Mod. Phys. D* **10**, 785 (2001).

[95] J. D. Barrow and H. Kodama, *Class. Quant. Grav.* **18**, 1753 (2001).

[96] H. Kodama, *Prog. Theor. Phys.* **107**, 305 (2002).

[97] J. D. Barrow, *Phys. Rev. D* **89**, 064022 (2014).

[98] P. A. R. Ade *et al.,* (Planck Collaboration Paper XXIV), arXiv:1303.5084.

[99] R. Maartens, *Phil. Trans. R. Soc. A* **369**, 5115 (2011).

[100] W. C. Lim, U. S. Nilsson and J. Wainwright, *Class. Quant. Grav.* **18**, 5583 (2001).

[101] W. Stoeger, M. Araujo, T. Gebbie, *Ap. J.* **476**, 435 (1997).

[102] U. S. Nilsson, C. Uggla, J. Wainwright and W. C. Lim, *Ap. J.* **522**, L1 (1999).

[103] G. F. R. Ellis, S. D. Nell, R. Maartens, W. R. Stoeger and A. P. Whitman, *Phys., Rep.* **124**, 315 (1985).

[104] W. H. McCrea, *Zeit. Astrophys.* **9**, 290 (1934). W. H. McCrea, *Zeit. Astrophys.* **18**, 98 (1939), reprinted in *Gen. Rel. Grav.* **30**, 315 (1998).

[105] J. Kristian and R. K. Sachs, *Ap. J.* **143**, 379 (1966).

[106] A. Guth, *Phys. Rev. D* **23**, 347 (1981).

[107] J. D. Barrow, Nature **272**, 211 (1978).

[108] R. Penrose, *Cycles of Time*, (Bodley Head, London, 2010).

[109] P. Tod, arXiv:1309.7248.

[110] J. Hartle and S. W. Hawking, *Phys. Rev. D* **28**, 2960 (1983).

[111] J. D. Barrow, in *Seeing Further: The Story of Science and the Royal Society*, Royal Society 350th Anniversary Volume, ed. B. Bryson, (Harper Collins, London, 2010), pp. 361–84.

6

Emergent Structures of Effective Field Theories

6.1 Introduction

Science and philosophy have a strong interaction and an unquestionable complementarity but one needs to draw a clear line between what they tell on the world, on their scope and implications.

6.1.1 Role of Cosmology

Cosmology does indeed play an important role in this debate. It has long been part of mythology and philosophy and, during the twentieth century, adopted a scientific method, thanks to the observational possibilities offered by astrophysics. We can safely state that scientific cosmology was born with Einstein's general relativity, a theory of gravitation that made space and time dynamical, a physical field that needs to be determined by solving dynamical equations known as Einstein field equations. The concept of spacetime was born with this theory, and in its early years cosmology was a space of freedom to think general relativity (Eisenstaedt, 1989). It led to the formulation of a 'standard' model, often referred to as the big bang model, that allows us to offer a history of the matter in the universe, in particular by providing an explanation for the synthesis of each of the elements of the Mendeleev table, and a history of the formation of the large scale structure. Cosmology has also excluded many possible models and hypotheses, such as the steady state model or topological defects as the seeds of the large scale structure.

The current standard cosmological model can be summarised by Figure 6.1 from which one can easily conclude that all the phenomena that can be observed in our local universe, from primordial big bang nucleosynthesis (BBN) to today, rely mostly on general relativity, electromagnetism and nuclear physics, that is on physics below 100 MeV and well under control experimentally (see Ellis *et al.* (2012); Mukhanov (2005); Peter and Uzan (2009) for textbooks on modern cosmology). This is an important property since the fact that we can understand our observable universe does not rely on our ability to construct e.g. a theory of quantum gravity. This is the positive aspect, the negative one being that it may be very difficult to find systems in which gravity, quantum physics and observations appear together, inflationary perturbations and black holes being probably the only ones. At later

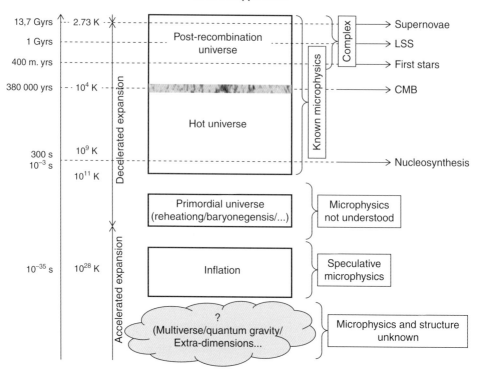

Figure 6.1 Summary of the history of our local universe in the standard cosmological model. It indicates the cosmic time elapsed after the big bang, as derived from the Friedmann equations, the temperature of the photon bath filling the universe. The local universe provides observations on phenomena from big bang nucleosynthesis to today spanning a range between 10^{-3} s to 13.7 Gyrs. One major transition is the equality which separates the universe in two eras: a matter-dominated era during which large scale structure can grow and a radiation-dominated era during which the radiation pressure plays a central role, in particular in allowing for acoustic waves. Equality is followed by recombination, which can be observed through the CMB anisotropies. For temperatures larger than 10^{11} K, the microphysics is less understood and more speculative. Many phenomena such as baryogenesis and reheating still need to be understood. Earlier, the microphysics relies on assumptions that may seem out of reach today. The whole history of our universe appears as a parenthesis of decelerated expansion, during which complex structures can form, in between two periods of accelerated expansion, which do not allow for this complex structure to either appear or even survive. From Uzan (2013).

time, the dynamics of the large scale structure becomes non-linear and requires the use of numerical simulations to be fully understood. At early time, the physics becomes more speculative. Extensions of the standard model of fields and interactions are then needed to understand baryogenesis, to construct the physics of the inflationary phase, etc.

Both the existence of a structured and complex universe and the fact that we can understand it call for an explanation: our universe from 1 s after the big bang to today, 13.7 Gyr after the big bang, can be described by a parenthesis between two inflationary phases during which the laws of physics that have been determined locally at low energy and linear regime are sufficient, making them easy to solve.

6.1.2 Cosmological Models

A cosmological model is a mathematical representation of our universe that is based on the laws of nature that have been derived and validated locally in our Solar system. It thus stands at the cross-road between theoretical physics and astronomy. Its construction depends on our knowledge of microphysics but also on *a priori* hypothesis on the geometry of the spacetime describing our universe. It follows that such a construction relies on four main hypotheses (see Uzan (2010) for a detailed description): (H1) a theory of gravity, (H2) a description of the matter contained in the universe and their non-gravitational interactions, (H3) symmetry hypothesis, and (H4) a hypothesis on the global structure, i.e. the topology, of the universe. These hypotheses are not on the same footing since (H1) and (H2) refer to the physical theories. These two hypotheses are however not sufficient to solve the field equations and we must make an assumption on the symmetries (H3) of the solutions describing our universe on large scales while (H4) is an assumption on some global properties of these cosmological solutions, with the same local geometry.

6.1.3 Cosmic Extrapolations

In such an approach, we need to push the theories and models that have been validated in our local neighbourhood beyond their domain of validity, in energy, space and/or time. In doing so, we usually adopt a 'radical conservatist' attitude, following the words by Wheeler (Deutsch, 1997), in which we continue to adopt our models and theories until we reach either an inconsistency with experiment or an internal theoretical inconsistency. The former clearly provides an experimental insight that our extrapolation is not correct while the second calls for new concepts to be forged in order to resolve inconsistencies that may appear due to the 'collision' of two theories (I detailed the way fundamental constants played a major role in this evolution in Uzan and Lehoucq (2005), which is unfortunately not translated in English). Indeed in such an approach, we tend to attribute a high credence to principles and mathematical structures that have proved to be very efficient to formulate the laws of nature locally. This high credence is not a proof of validity of the extrapolation but only reflects the fact that no catastrophe arises from using these theories far outside their domain of validity, and there is no need to invoke new structures or theories, hence the conservatism.

As an example, any of the four hypotheses can be extrapolated independently, giving different images of the nature of the universe, all compatible with our knowledge of the local observable universe. For instance, it can be shown that as soon as topological structure is invoked (hypothesis H4), hence leading to a universe of finite spatial volume, it cannot be distinguished from an infinite space as soon as its typical size is larger than 1.15

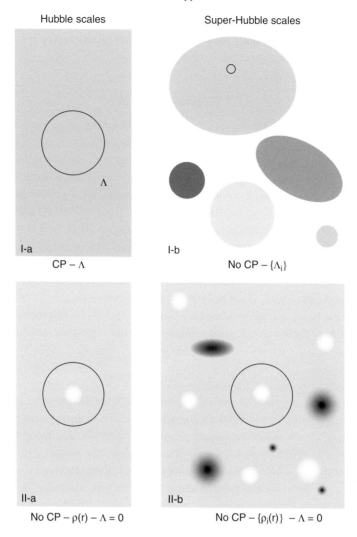

Figure 6.2 On the scales of the observable universe (circle), the acceleration of the universe can be explained by a cosmological constant in which case the construction of the cosmological model relies on the Copernican principle (upper left). To make sense of a cosmological constant, one argues one will introduce a large structure known as the multiverse (upper right) which can be seen as a collection of universes of all sizes and in which the values of the cosmological constant, as well as other constants, are randomised. The anthropic principle then states that we observe only those universes where the value of these constants are such that observers can exist. In this sense we have to abandon the Copernican principle on the scales of the multiverse. An alternative is to assume that there is no need for a cosmological constant or new physics, in which case we have to abandon the Copernican principle and assume e.g. that we are living in an under-dense region (lower left). However, we may recover the Copernican principle on larger scales if there exists a distribution of over- and under-dense regions of all sizes and densities on super-Hubble scales, without the need for a multiverse. In such a view, the Copernican principle will be violated on Hubble scale, just because we live in such a structure which happens to have a size comparable to the one of the observable universe. This latter view is now not favoured by observations. From Uzan (2010).

times the radius of the last scattering surface (Fabre *et al.*, 2013). Above such a size, the extrapolation of an infinite universe is no more under control and cannot be distinguished observationally from a finite universe. Another example concerns the implications of the existence of a cosmological constant driving the late time acceleration of the expansion of our universe (see Figure 6.2). According to our credence on the validity of the Coperni-can principle (H3) and the scale on which it applies, one can argue for the existence of a multiverse, large-scale inhomogeneities or new physics (see e.g. Carr (2007) for a discus-sion of the different possibilities). The Copernican principle can be tested on the size of the observable universe (Caldwell and Stebbins, 2008; Goodman, 1995; Uzan *et al.*, 2008) making the model of II.a of Figure 6.2 very constrained, hence decreasing significantly the credence of extrapolation II.b. But it does not increase the one of models I.b even though I.a gives a more accurate description of our local universe. Similarly, modifications of general relativity on large scale can be tested (Uzan and Bernardeau, 2001).

In this process, we use different kinds of proofs. First, the historical proof consists of starting from our local observations, to organize them consistently in order to draw con-clusions on the conditions in the past. In this abductive inference, we can determine the scenario that maximises the consistency of the facts and its explicative power (in a class of scenarios that needs to be mathematically well defined). It follows that our conclusions are only probable. Second, the experimental or observational proofs aim at understanding the local laws and causes. They allow one to exclude hypotheses and can modify the inference of the historical abduction (see e.g. the discussion on constants below). Hence cosmology is a constant interplay between local and glocal features. It is important to explain to what extent our conclusions are robust.

In cosmology, when confronted with an inconsistency, one always has to consider three options. One can either invoke the need for new physics or a modification of the laws of physics we have extrapolated (e.g. on large cosmological distance, low curvature regime, etc.) or have a more conservative attitude concerning fundamental physics and modify the cosmologogical hypothesis. The first option is indeed to be very conservative and doubt the observations (astrophysical solution).

At this stage, we need to recall that it is central to distinguish the conclusions that have been validated by experiments or observations and extrapolations with high credence.

Theoretical extrapolation is a valuable enterprise in order to understand the cartography of the space of possible worlds that a set of theories allows one to construct (this is indeed our favorite game). Usually, we are aware of the limitations of these speculations, because we know they are usually built on many speculative layers. As an internal debate, it causes no problem. But we need to be careful as to what is delivered to a larger audience, col-leagues from other fields and the larger public. In particular, drawn by extrapolations and theoretical stories on the different ways the universe may or may not be, public debates tend to slide to some relativism (sometimes even leading some extreme social constructivism to qualify the scientific method of myth).

To that purpose, I think we need to reaffirm that science is first characterised by a methodological contract. It represents the minimum we have to follow in order to consider our results as scientific. It can indeed be improved and is tacit between all of us. Our goal is to deliver to the public objective knowledge about the world. This implies that we need to adopt a materialistic method. For instance, we assume the existence of some reality, independent of our own existence and on us studying it; we approach it in a neutral way with the goal of revealing the way it appears to us, in its phenomena that are reproducible and testable. This materialism is a methodology and indeed not a philosophy nor the product of any philosophy. It is simply a methodology that protects us from relativism or spirituality that can alter or instrumentalise its productions.

Indeed, science and philosophy have a strong relation. Scientific results set a passive constraint on any philosophy and philosophy helps us in coining our concepts. Needless to say there is porosity and that the boundary has been fluctuating over centuries. To the outside world, we need to make sure that science is first judged by its procedures rather than by its results, that our attitude does not consist in finding an explanation for 'everything' now but in classifying the problems and answering some well-defined questions, hence the notion of a good model associated to a domain of validity. If we are not able to formulate clear rules that dictate the construction of knowledge, then we will have to stick to facts and results rather than to methods and we may slowly slide into a debate between two beliefs. This knowledge belongs to the whole society (which is actually the reason I believe we have to be public servants!), so is freely available.

In our debate, in a closed room (even though there was a camera, sic!) we can indeed draw personal metaphysical reflections based on our scientific constructions and what we have understood from the world. Interpretation of physical theories is a matter of debate and we shall not avoid it, the typical example being quantum mechanics. Interestingly, some theories give their own interpretation, as is the case of general relativity, while others still allow a huge number of possible interpretations leading to radically different metaphysical constructions. But we have to avoid confusion between the legitimacy of a personal metaphysical position and the consensus of an entire community. Cosmology (similarly to Darwinian evolution) because it entangles historical and experimental proofs, because people wrongly project hopes (and fears) on the fact that it can resolve some philosophical issues such as origin, is more exposed to relativism.

6.1.4 Outline

This text proposes a reflexion on these themes. To that purpose, I shall first, in Section 6.2, discuss the structures of physical theories, the implication of the existence of complexity and the notion of effective theories. Section 6.3 describes a recent example of how a mathematical structure now taken for granted in theoretical physics, namely the existence of a Lorentzian metric, can be circumvented and how it leads to new possibilities concerning the notion of time. Section 6.4 considers the role of fundamental constants in the theories of physics, discussing their importance, the fact that they may be only effective quantities and then not constant as well as their implication for the fine-tuning argument.

6.2 On the Construction of Physical Theories

The history of physics suggests that there are no mathematical structures that can be safely supposed to survive the evolution of our knowledge.

Part of the art of theoretical physics is to find the mathematical structures that allow us to formalise and 'decomplexify' the laws of nature. These structures include the description of space-time (dimension, topology,...), of the matter and interactions (fields, symmetries,...). While there is a large freedom in the choice of these mathematical structures, the developments of theoretical physics taught us that some of them are better suited to describe some classes of phenomena. But these choices are only validated by the mathematical consistency of the theory and, in the end, by the agreement of their consequences with experiments. Mathematics has also the power to reveal aspects of nature that are not accessible directly. This is the case of the Dirac equation that revealed the existence of anti-matter, from a pure mathematical consistency popping out of the equations. Indeed, experiments are then needed to check this insight. The magic here is that most of the time, such structures have turned out to exist. This power of mathematics has led some (wrongly) to think that it may be sufficient to stick to the mathematical consistency of the construction.

Some structures, considered as fundamental in the domain of validity of a theory, can be replaced by other structures in a new description, e.g. the spacetime of Newtonian physics can be formally seen as the limit of the Minkowski structure in regimes where $v \ll c$. In this respect, the role of fundamental constants as concept synthesisers or as limiting quantities is central to signal the need for new mathematical structures outside a domain of validity (Uzan, 2003). At each step, some properties, such as the topology of space, the number of spatial dimensions or the numerical values of the free parameters that are the fundamental constants (Uzan, 2003), may remain *a priori* free in one framework, or imposed in another framework (e.g. the number of space dimensions is fixed in string theory (Polchinsky, 1998)). It may even be that different structures can reproduce what we know about physics and one has to rely on less-defined criteria, such as simplicity and economy, to choose between them.

A natural question is to determine when a mathematical structure of a physical theory, that we know is limited in its validity, reveals one of the underlying structures of nature and can be given a high credence when we extrapolate the theory or can be generalised in the formulation of a more fundamental theory.

6.2.1 Effective Theories

The fact that we can understand the universe and its laws has a strong implication on the structure of the physical theories. At each step in their construction, we have been dealing with phenomena below a typical energy scale, for technological constraints, and it turned out (experimentally) that we have always been able to design a consistent theory valid in such a restricted regime. This is not expected in general and is deeply rooted in the mathematical structure of the theories that describe nature.

We can call such a property a *scale decoupling principle* and it refers to the fact that there exist energy scales below which effective theories are sufficient to understand a set of physical phenomena. *Effective theories* are the most fundamental concepts in the scientific approach to the understanding of nature and they always come with a domain of validity inside which they are efficient to describe all related phenomena. They are a successful explanation at a given level of complexity based on concepts of that particular level. It also means that they can only answer a limited set of questions and indeed cannot be blamed for this. For instance, we do not need to have understood and formulated string theory to formulate nuclear physics and we do not need to know anything about nuclear physics to develop atomic physics or chemistry, needless to say biology. Similarly electromagnetism describes how protons and electrons attract each other independently of the fact that we have understood, or not, that protons are made of quarks and gluons. This implies that the structures of the theories are such that there is a kind of stability and independence of higher levels with respect to more fundamental ones. This property is also important from an experimental point of view, since we try to restrict to a system that is decoupled as much as one can from its environment so that it can be assumed isolated. Truly isolated systems never exist and the effect of the environment may play an important role in cases such as quantum entanglement or decoherence. It follows that various disciplines have developed independently in almost quasi-autonomous domains, each of them having its own ontology and dynamics that are independent of our ability to formulate a theory explaining these concepts. Sometimes two such effective theories will collide and show inconsistency that will need, in order to be resolved, to introduce new concepts, more fundamental, from which the concepts of each of the theories can be derived in a limiting behaviour, at least in principle (e.g. while the wave function of quantum mechanics has properties of waves and particles, one cannot state *a priori* when a photon will behave as a particle or a wave as in e.g. a Young experiment photon by photon, so that in general one simply needs to abandon the old and often intuitive concepts at the price of otherwise entering in endless debates). For example, Maxwell electromagnetism and Galilean kinematics are incompatible at high velocity, which is at the origin of special relativity with the new concept of spacetime, or the concept of wave function from which the preexisting concepts of particle and wave are limiting behaviours. Note that it implies that concepts that were thought to be incommensurable (such as space and time, or momentum and wave number) need to be unified, which is usually achieved by the introduction of new fundamental constants (speed of light, or Planck constant, in the two examples at hand) that were not considered as fundamental (or even to exist) in the previous theories (see e.g. Uzan and Lehoucq (2005) for a full description of the role of constants in that context).

It follows that the set of theories we are using to describe the world around us can be split into a hierarchy of levels of reality (see Figure 6.3) as characterised by their corresponding academic subjects (Ellis, 2005, 2012). We emphasise that this hierarchy is based on a hierarchy of explanations (space science is needed to do astronomy). It is however different to what happens dynamically since the formation of planets requires first stars to have existed. Note also that the right branch is related, from a spacetime point of view, to the

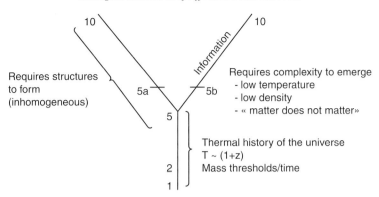

Figure 6.3 Hierarchy of theories in terms of their level of complexity as proposed by Ellis (2005, 2012) rephrased to make explicit the bifurcation that appears at the level 5 of complexity at which one needs to distinguish mineral chemistry (5a) and organic chemistry (5b). At this transition a central emergent quantity, *information*, appears on the left branch leading to new phenomena such as reproduction-selection, evolution, life and consciousness. The development of the complexity levels of the right branch are however conditional to the one of the left branch.

left since one needs stars/planets for life to develop (there is hierarchy in the formation of structure).

Higher level behaviours are constrained from lower level laws from which they emerge. The relation between the different levels has the following properties (the way the different levels of complexity interact has been studied in depth by Ellis (2005, 2012)):

- Higher level behaviours are constrained by the lower level laws from which they emerge. This is the usual *bottom-up causation* in which microscopic forces determine what happens at the higher levels. The more fundamental gives the space of possibilities in which a higher level can develop, by constraining e.g. causality, the type of interactions or structures that can exist. As an example in nuclear physics, a free neutron is unstable and will decay into a proton. While a fundamental process of nuclear physics, it can be understood from the quark composition of the neutron and protons why neutrons decay into protons and not the reverse.
- *Scale separation* implies that at each level of complexity, one can define fundamental concepts that are not affected from the fact that they may not be fundamental at a lower level (e.g. we can consider protons and neutrons as fundamental particles to describe many nuclear properties and forget that they are made of quarks). In that sense, much of the higher level phenomena remain quite independent of the microscopic structures, fields and interactions.
- At least for the lowest levels, the fact that physical theories are renormalisable implies that they influence the higher levels mostly through some numbers. This is in particular the case of the fundamental constants of a given effective theory. They cannot be explained within the framework of this particular theory since there exists nothing

more fundamental. They can however be replaced by more fundamental constants of an underlying level. For instance, the mass or the gyromagnetic factor of the proton are fundamental constants of nuclear physics. They can however be 'explained', even if the actual computation may be difficult (see Luo *et al.* (2011)), in terms of constants of the lower level (such as the quark masses, binding energies). This explanation of the constants of an effective theory may reveal new phenomena that cannot be dealt with before (e.g. the fundamental parameters of the effective theory may now be varying) but these phenomena have to be at the margin (or below the error bars) of the experiences that have validated the effective theory, since otherwise it would not have been a good theory (with reproducible predictions)!

- Not all the concepts of a higher level can be explained in terms of lower level concepts. Each level may require its own concepts that do not exist, or even are not related, to lower level concepts. These are *emergent properties* so that the whole may not be understood in terms of its parts. A typical example is the notion of information that cannot be reduced to chemistry or physical concepts. The concept of temperature however can be explained in terms of lower level concepts such as mean kinetic energy.
- The fact that there exists a lower level of complexity and thus microscopic degrees of freedom implies that these degrees of freedom can be heated up so that we expect to have apparitions of *entropy* and of *dissipation*. Black-hole entropy and Hawking temperature are one argument for the fact that gravity may not be a fundamental interaction.
- The higher levels of complexity can backreact on the lower level. This is the notion of *top-down causation*. It can take different forms such as contextuality, selection effect, control loop, etc. (see Ellis (2005, 2012)). Note that the understanding of phenomena of the whole, i.e. of a higher level of complexity, contributes to the understanding of the properties and dynamics of the parts as they function in, and also allow for the existence of, the global structure.

6.2.2 *Emergence of Complexity in our Universe*

The universe is, at the moment, the largest embedding structure in which all phenomena of lower complexity levels do develop. It is thus important to describe how the structures of the universe and of its history offer the possibility for the emergence of other levels. The cosmological model describes an expanding universe cooling down and developing nonlinear structures through the action of gravity, from an initial state at thermal equilibrium. We can thus summarise the evolution of our universe (see Figure 6.1) as

$$(\text{hot, homogenous, simple}) \rightarrow (\text{cold, inhomogeneous, complex}).$$

In this particular case, most of the history of the universe relies on the properties of general relativity. The fact that it is universal, and the equivalence principle, tell us that the microscopic nature of the matter is (mostly) irrelevant to determine the large scale structure of spacetime since all that matters is the averaged stress-energy tensor on the scales of interest.

In the early universe, higher levels of complexity cannot emerge, simply because the photon thermal bath is too energetic for complex structures to be stable. For instance, there cannot be any light nuclei before BBN so that at temperatures above 100 MeV there exist only free protons, neutrons, neutrinos, photons, electrons and positrons at thermal equilibrium under the action of the weak and electromagnetism interactions (see chapter 4 of Peter and Uzan (2009)). Above 100 MeV, nuclear physics is not needed to describe the universe. In the earlier phases, prior to preheating the matter content of the universe was constituted of a scalar field, the inflaton, so that even particle physics was irrelevant. At preheating, one needs to have a description of particle physics to describe the universe (in particular, one needs to know the number of relativistic degrees of freedom, the mass thresholds, their coupling to the inflaton itself, etc...) Before recombination, the matter in the universe remains ionised so that there is atomic physics going on (even though atomic transitions are important to describe the process of recombination) and there cannot be any molecules at that stage since they would instantaneously be destroyed. To start any chemistry (level 5), one needs to wait for the first stars to synthetise elements heavier than the existing hydrogen, helium, beryllium and lithium. The transition to level 5 thus requires structures such as galaxies and first stars to start forming (in fact one even needs to wait longer since these heavier nuclei have to be released back to the instellar medium). This is summarised in Figure 6.4.

It follows that no real complexity can emerge in the universe before it has cooled enough and before structures start forming. We can estimate that before a redshift of ~ 30 no level of complexity above 5b is actually reached in our universe. The complexity levels of the left branch start developing with the formation of the large scale structure. As shown in Figure 6.4 not only do the higher levels of complexity appear only at late time, but also in a very inhomogeneous way since they require the density of matter to be large enough for e.g. molecular interactions to be non-negligible.

There is a feature that comes with the appearance of the different levels of complexity. At each step, only a tiny part of the matter content is involved in the next complexity level. This could be referred to as a *decimation principle*. For instance, before baryogenesis, matter and anti-matter are in equilibrium. When they annihilate, only a tiny part of the matter content survives and most of it is turned into radiation hence increasing the entropy of the universe. Then, only a small part of the nuclei is able to participate to chemistry since hydrogen and helium constitute 75 per cent and 24 per cent, respectively. Then, only a tiny part of the molecules is organic and only a small part of it contributes to self-reproducing molecules. Whether this is of some relevance and a generic feature of the emergence of higher levels of complexity still needs to be investigated.

6.2.3 Causality

This structure of layers of theories of increasing complexity and of nested Russian babouchka dolls of effective theories has also implications for the notion of causaliy.

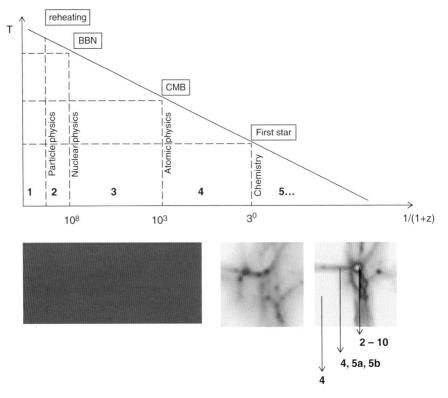

Figure 6.4 The different levels of complexity presented in Figure 6.3 can appear only in regions (and times) where the physical conditions allow for these complex structures, usually more fragile, to be stable. Before decoupling, i.e. 300,000 years after the big bang, temperature is too high for atoms to exist and only very light nuclei have been formed so that no molecule can exist. The universe remains mostly homogeneous, hot, with a very simple chemical composition and no level above 4 can emerge. Only after decoupling do the large scale structures start growing under the effect of gravity (since then radiation pressure has become negligible). The universe becomes inhomogeneous and locally galaxies, stars and then planets can form, allowing (1) for spacetime regions where complexity levels above 5b can be reached and (2) for a more variegated chemical composition with heavier nuclei and then molecules. From Uzan (2013).

George Ellis (2005, 2012) has described how higher levels can backreact on the properties of lower levels by top-down action. There are however limitations on how it can proceed. Understanding this will teach us what to expect from an explanatory point of view from different effective physical theories.

Below level 5, the causality of all theories is inherited from special relativity. Locally they all have a Lorentzian structure so that causality is dictated by the lightcone structure of the embedding spacetime. The physical conditions at a spacetime point M are dictated by initial conditions within the past lightcone and the equations of evolution. These theories are often asked to have a well-posed Cauchy problem, which ensures that any future field configuration is completely determined by initial conditions and that the notion of 'future configuration' is globally well defined, which is thought as encompassing the notion of predictability (this can be actually an argument to exclude some theories; see e.g. Esposito-Farèse *et al.* (2010); Fleury *et al.* (2014)). This property is related to the existence of a fundamental constant which characterises the maximum speed of propagation of any interaction identified to the speed of light. While this is the case with our current theories, it may not be the case for quantum theories of gravity in which Lorentz invariance may be broken, in some bimetric theories of gravity in the more speculative theories we describe in the next section.

Above level 5, top-down action becomes efficient. In general, it implies that the conditions at a given spacetime position become mostly independent of the initial conditions (Ellis, 2005, 2012). As argued above, the appearance of this level of complexity is localised both in time and space, so that the region of the universe in which the top-down causation is efficient is limited. Indeed, it has to be limited by the usual relativistic causality but it is expected that the typical propagation speed of the action of the higher levels on the lower ones is much smaller than the speed of light, so that the future lightcone of influence is expected to be very narrow. For example, in principle humanity can backreact on stellar physics to increase the lifetime of the Sun once it is understood that it can be performed by homogenising the Solar fuel. It is however obvious that the time to achieve the required technological evolution is much larger than 8 minutes. In the same way, the simple fact that the first artificial probes have been exiting the Solar system only recently demonstrates that the domain of influence of (human) life remains restricted to the Solar system. It is thus safe to neglect it for the evolution of our galaxy. Note however that it may well modify planetary science since terraformation of Mars is not completely out of reach.

6.2.4 Mathematics

Among the features of our theories is the constant use and efficiency of mathematics.

The lowest levels of complexity enjoy the property that all elements are undistinguishable, in the sense that any electron is strictly similar to any other electron. Moreover, there is a limited (and actually small) number of different building blocks. This implies that there exists a one-to-one mapping between the physical entities and their mathematical descriptions. There is actually no need to distinguish them (even though caution is in order since the mathematical structure can be changed in case of the discovery of a new property). Once these structures are fixed, the way they can interact is actually fixed (by symmetries and consistency between the different structures), and causality is mainly fixed by the Lorentzian structure, as we discussed. We can thus conclude that, at these levels, *mathematics are prescriptive*.

In higher levels of complexity, the situation is different because of combinatorics. One can construct about a hundred stable nuclei with many hundreds of isotopes, which leads to a number of molecules that cannot be estimated, up to macromolecules such as DNA. In terms of causality, it is impossible to get rid of randomness, even classically, since as it appears the fortuitous conjunction at a spacetime point of two independent causal chains. It follows that we cannot associate a single mathematical structure to each physical entity. We can indeed model a class of them in terms of a mean individual and fluctuations. This is indeed a powerful procedure but one in which the *mathematics are only descriptive*. They become a powerful (or not!) tool. But the true mathematical structure of the underlying level is slowly hidden and diluted by complexity.

6.3 Lorentzian Structure

Among all the mathematical structures used in theoretical physics, and in the framework of metric theories of gravitation, the signature of the metric is in principle arbitrary.

6.3.1 Lorentz Signature

Indeed, it seems that on the scales that have been probed so far there is the need for only one time dimension and three spatial dimensions. It is also now universally accepted that the relativistic structure, and in particular as the cleanest way to implement the notion of causality, is a central ingredient of the construction of any realistic field theory. Spacetime enjoys a locally Minkowski structure. When gravity is included, the equivalence principle implies (this is not a theoretical requirement, but just an empirical fact, required at a given accuracy) that all the fields are universally coupled to the same Lorentzian metric. Thus, we now take it for granted that the spacetime is a four-dimensional (4D) manifold endowed with a metric of signature $(-,+,+,+)$.

While the existence of two time directions may lead to confusions (Bars, 2001; Halsted, 1982), several models for the birth of the universe (Friedman, 1998; Gibbons and Hartle, 1990; Gott and Li, 1998; Hartle and Hawking, 1983) are based on a change of signature via an instanton in which a Riemannian and a Lorentzian manifold are joined across a hypersurface which may be thought of as the origin of time. While there is no time in the Euclidean region, where the signature is $(+,+,+,+)$, it flips to $(-,+,+,+)$. Eddington (1923) even suggested that it can flip across some surface to $(-,-,+,+)$ and signature flips also arise in brane (Gibbons and Ishibashi, 2004; Mars *et al.*, 2007) or loop quantum cosmology (Mielczarek, 2014).

It is legitimate to investigate whether the signature of this metric is only a convenient way to implement causality or whether it is just a property of an effective description of a microscopic theory in which there is no such notion. What does it take to construct a realistic field theory in a positively definite metric?

6.3.2 A Clock Field as a New Ingredient

Let us start by physics in flat spacetime and consider a 4D Riemannian manifold $\mathcal{M}$ with a positive definite Euclidean metric $g^{\mathrm{E}}_{\mu\nu} = \delta_{\mu\nu}$ in Cartesian coordinates. The theories on this manifold have thus no natural concept of time.

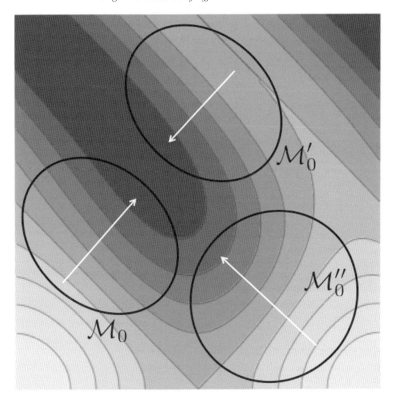

Figure 6.5 Example of a spatial configuration of the clock field. Locally, one can define regions such as $\mathcal{M}_0$, $\mathcal{M}_0'$ and $\mathcal{M}_0''$, in each of which a time direction emerges. Indeed this direction does not preexist at the microscopic level and can be different from patches to patches. From Mukohyama and Uzan (2013).

If one considers a scalar field χ and assumes that its Lagrangian is a scalar and led to second-order equations, the only terms that can be included are a 'kinetic term', $\frac{1}{2}\delta^{\mu\nu}\partial_\mu\chi\partial_\nu\chi$, and a potential term $V(\chi)$. As a consequence its field equation is elliptic, determining a statical configuration.

In order to make dynamics emerge locally, we introduce a scalar field ϕ such that: (1) its derivative has a non-vanishing vacuum expectation value in a region $\mathcal{M}_0$ of the Riemannian space (see Figure 6.5) and (2) it couples to all other fields.

The first condition is implemented by assuming that $\partial_\mu\phi = $ const. $\neq 0$ in $\mathcal{M}_0$. We can thus set $\partial_\mu\phi = M^2 n_\mu$ on $\mathcal{M}_0$ with n_μ a unit constant vector. Its norm $X_E \equiv \delta^{\mu\nu}\partial_\mu\phi\partial_\nu\phi = M^4$ is constant and satisfies $X_E > 0$ on $\mathcal{M}_0$. Now, under this assumption, one of the coordinates of the Euclidean space can be chosen as $dt = n_\mu dx^\mu$, that is

$$t \equiv \frac{\phi}{M^2} \tag{6.1}$$

up to a constant that can be set to zero without loss of generality and we introduce three independent coordinates x^i ($i = 1 \ldots 3$) on the hypersurfaces Σ_t normal to n^μ.

The second condition is implemented by considering that χ couples only to $\partial_\mu \phi$ so that one can consider the action

$$S_\chi = \int d^4x \left[-\frac{1}{2} \delta^{\mu\nu} \partial_\mu \chi \partial_\nu \chi - V(\chi) + \frac{1}{M^4} \left(\delta^{\mu\nu} \partial_\mu \phi \partial_\nu \chi \right)^2 \right]. \tag{6.2}$$

On $\mathcal{M}_0$ where the condition (6.1) holds, it is straightforward to deduce that

$$S_\chi \rightarrow \int dt d^3x \left[-\frac{1}{2} \eta^{\mu\nu} \partial_\mu \chi \partial_\nu \chi - V \right] \tag{6.3}$$

$$\equiv \int_{\mathcal{M}_0} dt d^3x \left[-\frac{1}{2} \eta^{\mu\nu} \partial_\mu \chi \partial_\nu \chi - V \right]. \tag{6.4}$$

Hence, on $\mathcal{M}_0$, the action (6.2) describes the dynamics of a scalar field propagating in an effective 4D Minkowski spacetime with metric $\eta_{\mu\nu} = \text{diag}(-1,+1,+1,+1)$. The apparent Lorentzian dynamics, with a preferred time direction, results from the coupling to ϕ. As a consequence, ϕ is related to what we usually call 'time', so that we shall call it a *clock field*.

6.3.3 Classical Field Theory in Flat Spacetime

In the previous example, the clock field allows for the emergence of an effective Lorentzian dynamics because the scalar field is actually coupled to the effective metric $\hat{g}^{\mu\nu} = \delta^{\mu\nu} - \frac{2}{M^4} \delta^{\mu\alpha} \delta^{\nu\beta} \partial_\alpha \phi \partial_\beta \phi$ that reduces on $\mathcal{M}_0$ to $\eta_{\mu\nu}$.

This construction can actually be generalised easily to vector fields and Dirac spinors (Mukohyama and Uzan, 2013), and to Majora and Weyl spinors (Kehayias et al., 2014), which allows us (Mukoyama and Uzan, 2013) to construct the full action of the standard model of particle physics in flat spacetime at the classical level.

6.3.4 Gravity and Physics in Curved Spacetime

It is even possible to extend this construction to gravity (Mukohyama and Uzan, 2013).

To this purpose, we shall consider a general 4D Riemannian manifold $\mathcal{M}$ with a positive definite metric $g^E_{\mu\nu}$. To minimise the number of degrees of freedom, we demand that the equation of motion for ϕ is second order. Hence, its action is restricted to the Riemannian version of the Horndeski theory (Horndeski, 1974) with shift symmetry. Equivalently, it is given by the shift-symmetric generalised Galileon (Deffayet et al., 2011).

For the effective equations, i.e. once the Lorentzian structure has emerged, we would like to ensure that the system is invariant not only under time translation but also under CPT. For this reason, we require that besides the shift symmetry ($\phi \rightarrow \phi + \text{const.}$) the theory also enjoys a Z_2 symmetry ($\phi \to -\phi$) for the clock field action. With these symmetries, the Riemannian action reduces to

$$S_g = \int d^4x \sqrt{g_E} \left\{ G_4(X_E)R_E + \mathcal{K}(X_E) \right.$$
$$\left. - 2G_4'(X_E) \left[(\nabla_E^2 \phi)^2 - (\nabla_\mu^E \nabla_\nu^E \phi)^2 \right] \right\}, \tag{6.5}$$

where $X_E \equiv g_E^{\mu\nu} \partial_\mu \phi \partial_\nu \phi$.

As demonstrated in Mukohyama and Uzan (2013), it reduces on $\mathcal{M}_0$ to

$$S_g = \int d^4x \sqrt{-g} \left\{ f(X)R + 2f'(X) \left[(\nabla^2 \phi)^2 - (\nabla^\mu \nabla^\nu \phi)(\nabla_\mu \nabla_\nu \phi) \right] \right.$$
$$\left. + P(X) \right\}. \tag{6.6}$$

in terms of $X = -g^{\mu\nu} \partial_\mu \phi \partial_\nu \phi$ where $g_{\mu\nu}$ is the emergent Lorentzian metric and where f and P are two functions related to G_4 and $\mathcal{K}$. This action (6.6) is a special case of the covariant Galileon (Deffayet *et al.*, 2011) and the equations of motion are second order. It can also be shown (Mukohyama and Uzan, 2013) that the action for scalar and vector fields can easily be generalised to curved spacetime, while the case of spinor is still an open question.

6.3.5 Discussion

This shows that by introducing a coupling of the standard fields to a clock field, we have shown that a Lorentzian dynamics can emerge on a patch $\mathcal{M}_0$ of a Riemannian space, including gravity. This goes far beyond earlier attempts (Girelli *et al.*, 2009).

It is important to emphasise that: (1) this construction is, for now, limited to classical field theories; (2) when restricted to $\mathcal{M}_0$ all fields propagate in the same effective Minkowski metric so that the equivalence principle is safe in first approximation; (3) indeed the couplings to the clock field have been tuned for that purpose. Indeed the action (6.2) could have been chosen as

$$S_\chi = \int d^4x \left[-\frac{\kappa_\chi}{2} \delta^{\mu\nu} \partial_\mu \chi \partial_\nu \chi - V(\chi) + \frac{\alpha_\chi}{2M^4} \left(\delta^{\mu\nu} \partial_\mu \phi \partial_\nu \chi \right)^2 \right],$$

and a Lorentzian signature is recovered only if $\alpha_\chi > \kappa_\chi > 0$. In the case where these constants are not tuned, different fields can have different lightcones. (4) In the bosonic sector, since the theory is invariant under the Euclidean parity ($x^\mu \to -x^\mu$) and field parity ($\phi \to -\phi$), both P and T invariances in the Lorentzian theory are ensured. This explains why we have included only quadratic terms in $\partial_\mu \phi$ in the actions for scalars and vectors. (5) In the fermionic sector, one of the coupling terms is not CPT invariant after the clock field has a vacuum expectation value. (6) The configuration of the clock field is not arbitrary but determined by solving its equation of motion. Since its action enjoys a shift symmetry, it will take the form of a current conservation.

In curved spacetime, gravity takes the form of a covariant Galilean theory that depends on two free functions constrained by the stability of a Friedmann–Lemaître background with respect to linear scalar and tensor perturbations. For the matter sector, the actions for scalar and vector fields are easily generalised and each depends on two free parameters (κ, α) that are allowed to be functions of X_E in general but may as well be assumed constant. Besides, there is an environmental parameter which characterises the clock field configuration on the patch $\mathcal{M}_0$. The emergent model has the following properties: (1) it induces two components which respectively behave as dark matter and dark energy when considering the dynamics of a homogeneous cosmology. This sets two constraints for the cosmology to be consistent with standard cosmology, at least at the background level. (2) In general, scalars and vectors propagate in two different effective metrics. In order for the weak equivalence principle to hold, we have to impose that these two metrics coincide. In the simplest situation in which the coeficients (κ, α) are assumed to be constant, one only requires a tuning on the parameters of the Lagrangians, but then it is satisfied whatever the configuration of the clock field. In this sense the tuning is not worse than the usual assumption that all the fields propagate in the same metric. (3) In general, the parameters entering our effective Lorentzian actions are environmentally determined. This means that if X_E is not strictly constant on $\mathcal{M}_0$, fundamental constants may be spacetime dependent, which is strongly constrained (Uzan, 2003, 2005, 2009, 2011). (4) The speed of light and of graviton may not coincide, which is constrained by the observations of cosmic rays (Moore and Nelson, 2001) because particles propagating faster than the gravity waves emit gravi-Cerenkov radiation.

To conclude, from a theoretical point of view, such a construction gives a new insight into the need for Lorenzian metric as a fundamental structure. As we have shown, this is not a mandatory requirement and a decent field theory, at least at the classical level, can be constructed from a Riemannian metric. Such a formalism may be fruitful in the debate on the emergence of time and, speculating, for the development of quantum gravity.

It also opens up a series of questions and possibilities. We can list (1) the development of a quantum theory, (2) the possibility from the classical viewpoint that singularities in our local Lorentzian region may be related to singularities in the clock field (e.g. similar to topological defects) and not in the metric of the Euclidean theory. These are, for now, speculations but they illustrate that this framework may be fruitful for extending our current field theories, including general relativity.

6.4 Fundamenal Constants

Fundamental constants play an important role in physics. In particular, they set the order of magnitude of phenomena; they allow one to forge new concepts; they characterise the domain of validity of the theory in which they appear.

In gravitation, their constancy is related to the validity of the Einstein equivalence principle and, in cosmology, they play a central role in multiverse construction since they are used to quantify the level of fine-tuning.

6.4.1 Role in Physical Theories

We shall start by defining the fundamental constants of a theory as any parameter that is not determined by the theories we are using.

These parameters have to be assumed constant from a theoretical point of view; since they enjoy no equation of evolution they cannot be expressed in terms of more fundamental quantities so that they can only be measured in the laboratory. From an experimental point of view, the reproducibility of experiments that have been used to validate the theories in which they appear ensures that this is a good approximation at the level of accuracy of these experiments on the time scales they span. This also means that testing for the constancy of these parameters is a test of the theories in which they appear and allows one to extend the knowledge of their domain of validity, at the limit of what metrology can achieve.

The number of fundamental constants depends on what we consider as being the fundamental theory of physics. Today, gravitation is described by general relativity, and the three other interactions and whole fundamental fields are described by the standard model of particle physics. In such a framework, one has 22 unknown constants (the Newton constant, six Yukawa couplings for the quarks and three for the leptons, the mass and vacuum expectation value of the Higgs field, four parameters for the Cabibbo–Kobayashi–Maskawa (CKM) matrix, three coupling constants, a UV cut-off to which one must add the speed of light and the Planck constant (Ellis and Uzan, 2005; Hogan, 2000; Uzan, 2003)).

Indeed, when introducing new, more unified or more fundamental theories the number of constants may change so that the list of what we call fundamental constants is a time-dependent concept and reflects both our knowledge and ignorance (Weinberg, 1983). For instance, we know today that neutrinos have to be somewhat massive. This implies that the standard model of particle physics has to be extended and that it will involve at least seven more parameters (three Yukawa couplings and four CKM parameters). On the other hand, this number can decrease, e.g. if the non-gravitational interactions are unified. In such a case, the coupling constants may be related to a unique coupling constant α_U and a mass scale of unification M_U through $\alpha_i^{-1}(E) = \alpha_U^{-1} + (b_i/2\pi) \ln(M_U/E)$, where the b_i are numbers which depend on the explicit model of unification. This would also imply that the variations, if any, of various constants will be correlated.

The tests of the constancy of fundamental constants take all their importance in the realm of the tests of the equivalence principle (Will, 1993). This principle, which states the universality of free fall, the local position invariance and the local Lorentz invariance, is at the basis of all metric theories of gravity and implies that all matter fields are universally coupled to a unique metric $g_{\mu\nu}$ which we shall call the physical metric, $S_{\text{matter}}(\psi, g_{\mu\nu})$. The dynamics of the gravitational sector is dictated by the Einstein–Hilbert action $S_{\text{grav}} =$

$\frac{c^3}{16\pi G}\int\sqrt{-g_*}R_*\mathrm{d}^4x$. General relativity assumes that both metrics coincide, $g_{\mu\nu} = g^*_{\mu\nu}$, which implements the equivalence principle in its strong form.

The test of the constancy of constants is obviously a test of the local position invariance hypothesis and thus of the equivalence principle. Let us also emphasise that it is deeply related to the universality of free fall (Dicke, 1964) since if any constant c_i is a spacetime dependent quantity so will be the mass of any test particle. It follows that the action for a point particle of mass m_A is given by

$$S_{p.p.} = -\int m_A[c_j]c\sqrt{-g_{\mu\nu}(x)v^\mu v^\nu}\mathrm{d}t$$

with $v^\mu \equiv \mathrm{d}x^\mu/\mathrm{d}t$ so that its equation of motion is

$$u^\nu\nabla_\nu u^\mu = -\left(\frac{\partial\ln m_A}{\partial c_i}\nabla_\beta c_i\right)(g^{\beta\mu} + u^\beta u^\mu). \tag{6.7}$$

It follows that a test body does not enjoy a geodesic motion and experience an anomalous acceleration which depends on the sensitivity $f_{A,i} \equiv \partial\ln m_A/\partial c_i$ of the mass m_A to a variation of the fundamental constants c_i. In the Newtonian limit, $g_{00} = -1 + 2\Phi_N/c^2$ so that $\mathbf{a} = \mathbf{g}_N + \delta\mathbf{a}_A$ with the anomalous acceleration $\delta\mathbf{a}_A = -c^2\sum_i f_{A,i}\left(\nabla c_i + \frac{\mathbf{v}_A}{c^2}\dot{c}_i\right)$. Such deviations are strongly constrained in the Solar system and also allow us to bound the variation of the constants (Dent, 2007, 2008).

There are many ways to design a theory with dynamical constant. As long as one sticks to field theories, the recipe is simple: one needs to promote a constant of the theory to the status of a dynamical field, allowing e.g. for a kinetic term and a potential. The coupling to the 'standard fields' can be guessed from the theory in which these quantities are constants but, indeed the functional forms of the couplings remain arbitrary. This has two consequences. First, the equations derived with this parameter constant will be modified and one cannot just let it vary in the equations derived by assuming it is constant. Second, the variation of the action with respect to this new field provides a new equation describing the evolution for this new parameter (i.e. of the constant). The field responsible for the time variation of the ï£¡ï£¡constantï£¡ï£¡ is also responsible for a long-range (composition-dependent) interaction, i.e. at the origin of the deviation from General Relativity, because of the coupling of the standard matter fields.

6.4.2 Constraints on Their Time Variation

Since any physical measurement reduces to the comparison of two physical systems, one of them often used to realise a system of units, it only gives access to dimensionless numbers. This implies that only the variation of dimensionless combinations of the fundamental constants can be measured and would actually also correspond to a modification of the physical laws (see e.g. Uzan (2003, 2011)). Changing the value of some constants while letting all dimensionless numbers remain unchanged would correspond to a change of

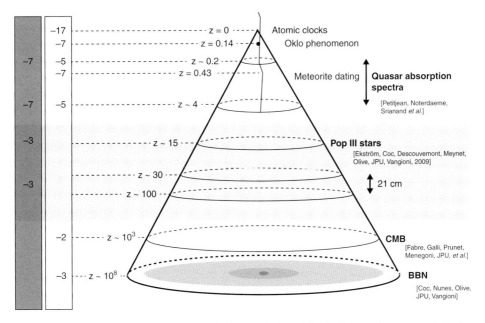

Figure 6.6 Summary of the constraints on the time variation of the fundamental constants. It depicts the various systems in the spacetime diagram with our past lightcone. The left bar gives the typical magnitude of the constraint on the relative variation of the fine structure constant -5 meaning $<$ 10^{-5} etc.) and the red bar a forecast of the expected improvment in the coming years. From Uzan (2011).

units. It follows that from the 22 constants of our reference model, we can pick three of them to define a system of units (such as e.g. c, G and h to define the Planck units) so that we are left with 19 unexplained dimensionless parameters, characterising the mass hierarchy, the relative magnitude of the various interactions, etc. Indeed, sometimes one refers to a variation of a constant with dimension, typically G. This is dangerous but usually one has been setting the units in such a way that this constant is dimensionless. For instance a varying G theory is meant as a theory in which $Gm_e^2/\hbar c$ is varying and all mass ratios are kept constant (see the recent criticism by Duff (2014) showing it may not be clear for everyone).

The various physical systems that have been considered can be classified in many ways (Uzan, 2003, 2011).

First, we can classify them according to their look-back time and more precisely their spacetime position relative to our actual position. This is summarised in Figure 6.6 which represents our past lightcone, the location of the various systems (in terms of their redshift z) and the typical level at which they constrain the time variation of the fine structure constant. These systems include atomic clock comparisons ($z = 0$, Blatt *et al.* (2008); Bize *et al.* (2003); Cingöz *et al.* (2008)), the Oklo phenomenon ($z \sim 0.14$, Damour and

Table 6.1. *Summary of the systems considered to set constraints on the variation of the fundamental constants. We summarise the observable quantities (see text for details), the primary constants used to interpret the data and the other hypotheses required for this interpretation. [α: fine structure constant; μ: electron-to-proton mass ratio; g_I: gyromagnetic factor; E_r: resonance energy of the samarium-149; λ: lifetime; B_D: deuterium binding energy; Q_{np}: neutron-proton mass difference; τ: neutron lifetime; m_e: mass of the electron; m_N: mass of the nucleon].*

System	Observable	Primary constraints	Other hypothesis
Atomic clock	$\delta \ln \nu$	g_I, α, μ	-
Oklo phenomenon	isotopic ratio	E_r	geophysical model
Meteorite dating	isotopic ratio	λ	-
Quasar spectra	atomic spectra	g_p, μ, α	cloud properties
21 cm	T_b	g_p, μ, α	cosmological model
CMB	T	μ, α	cosmological model
BBN	light element abundances	$Q_{np}, \tau, m_e, m_N, \alpha, B_D$	cosmological model

Dyson (1996); Kuroda (1956)), meteorite dating ($z \sim 0.43$, Olive *et al.* (2004); Peebles and Dicke (1962)), both having a spacetime position along the world line of our system and not on our past lightcone, quasar absorption spectra ($z = 0.2 - 4$, Chand *et al.* (2004, 2005); Petitjean *et al.* (2009); Webb *et al.* (1999, 2001)), population III stars (Ekström *et al.*, 2010) cosmic microwave background anisotropy ($z \sim 10^3$; Ade *et al.* (2014)) and primordial nucleosynthesis ($z \sim 10^8$, Coc *et al.* (2006, 2007)). Indeed higher redshift systems offer the possibility to set constraints on a larger time scale, but at the price of usually involving other parameters such as the cosmological parameters. This is particularly the case of the cosmic microwave background and primordial nucleosynthesis, the interpretation of which requires a cosmological model.

The systems can also be classified in terms of the physics they involve in order to be interpreted (see Table 6.1). For instance, atomic clocks, quasar absorption spectra and the cosmic microwave background require only the use of quantum electrodynamics to draw the primary constraints, so that these constraints will only involve the fine structure constant α, the proton-to-electron mass ratio μ and the various gyromagnetic factors g_I. On the other hand, the Oklo phenomenon, meteorite dating and nucleosynthesis require nuclear physics and quantum chromodynamics to be interpreted.

For any system, setting constraints goes through several steps that we sketch here. First, any system allows us to derive an observational or experimental constraint on an observable quantity $O(G_k, X)$, which depends on a set of primary physical parameters G_k and a set of external parameters X, that usually are physical parameters that need to be measured or constrained (e.g. temperature,...). These external parameters are related to our knowledge of the physical system and the lack of their knowledge is usually referred to as systematic uncertainty.

From a physical model of the system one can deduce the sensitivities of the observables to an independent variation of the primary physical parameters

$$\kappa_{G_k} = \frac{\partial \ln O}{\partial \ln G_k}. \tag{6.8}$$

As an example, the ratio between various atomic transitions can be computed from quantum electrodynamics to deduce that the ratio of two hyperfine-structure transitions depends only on g_I and α while the comparison of fine-structure and hyperfine-structure transitions depends on g_I, α and μ. For instance (Karshenboim, 2006) $\nu_{Cs}/\nu_{Rb} \propto \frac{g_{Cs}}{g_{Rb}} \alpha^{0.49}$ and $\nu_{Cs}/\nu_H \propto g_{Cs} \mu \alpha^{2.83}$.

The primary parameters are usually not fundamental constants (e.g. the resonance energy of the samarium E_r for the Oklo phenomenon, the deuterium binding energy B_D for nucleosynthesis, etc.) The second step is thus to relate the primary parameters to (a choice of) fundamental constants c_i. This would give a series of relations

$$\Delta \ln G_k = \sum_i d_{ki} \Delta \ln c_i. \tag{6.9}$$

The determination of the parameters d_{ki} requires first to choose the set of constants c_i (do we stop at the masses of the proton and neutron, or do we try to determine the dependencies on the quark masses, or on the Yukawa couplings and Higgs vacuum expectation value, etc.? See e.g. Dent (2007) for various choices) and also requires us to deal with nuclear physics and the intricate structure of quantum chromodynamics (QCD). In particular, the energy scale of QCD, Λ_{QCD}, is so dominant that at lowest order all parameters scale as Λ_{QCD}^n so that the variation of the strong interaction would not affect dimensionless parameters and one has to take the effect of the quark masses.

6.4.3 Fine Tuning

Fundamental constants are the centre of interest of the fine-tuning argument, according to which the laws of nature have to be adjusted in order for complexity to exist.

As we have seen in the previous section, the previous study on the variation of the fundamental constants allows us to quantify the fine tuning. For example, the study of population III stars (Ekström *et al.*, 2010) shows that the strength of the nuclear interaction cannot vary by more that 1/1000 for carbon to be produced.

The question which I am not going to address here is when do we start worrying about such a fine tuning? I refer to the extensive discussion by Barnes (2012) and its references.

Among the solutions that cannot be denied, the multiverse scenario is usually proposed. It is based on the idea of a meta-theory (often thought of as string theory, but not necessarily) that leads to a distribution of universes with different low energy physical laws (usually involving some hypothesis on eternal inflation) and an observer selection effect (as the anthropic principle). In doing so, one needs to define a notion of *different universes*

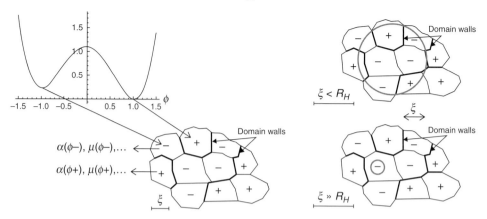

Figure 6.7 (Left) A model accounting for a spatial variation of the fundamental constants (Olive *et al.*, 2011). Given the relative sizes of the correlation length of the field that undergoes the phase transition and the Hubble radius, one can either observe several patches (micro-landscape) or be contained in a single patch (landscape). While the former can be constrained observationally the second is out of reach of experiment. From Uzan (2012).

or *universes similar to ours*. The study of the fundamental constants shows that a typical variation of 10^{-3}–10^{-5} of the masses and couplings leads to universes similar to ours (i.e. in principle able to host life forms similar to ours) but that can be observationally distinguished. This sets some coarse-graining scales in the space of models.

Many models can lead to a spatial distribution of the fundamental constants. Here again, the same microphysics can lead to very different scenarios. For instance, I want to stress that the same model (Olive *et al.*, 2011) can account for a spatial distribution of the constant on sub or super-Hubble scales (see Figure 6.7) while invoking no other universes.

6.5 Conclusion

The interactions between science and philosophy are so rich and numerous that it would be a pity not to stimulate our thoughts reciprocally. As I have argued, being a theoretical physicist and a cosmologist, my first duty is to deliver knowledge that has been validated (i.e. George Ellis's cosmology vs. cosmologia discussion, in this volume) and that may (or may not) set passive constraints on philosophical theories. It also requires an ontology that cannot be found in the theory itself and for which philosophy is of great help. In cosmology, we have to be clear on the level of credence of the models we are considering, and we must make clear the layers of speculative theories they rely on, and confuse the results of our extrapolations to validate results. Once models have been excluded, we also have to make explicit on what piece of observational data.

I have given two examples of theoretical constructions that challenge what we usually assume on the Lorentzian structure of spacetime and on the nature of the fundamental constants. Both can be motivated by theoretical arguments. But, more important, I think they show how local physics can affect the construction of a cosmological model and that the space of possible theories is always larger that what we tend to assume.

References

Ade, P. A. R., *et al.* 2014. Planck intermediate results. XXIV. Constraints on variation of fundamental constants. *arXiv:1406.7482*.

Barnes, L. A. 2012. The Fine-Tuning of the Universe for Intelligent Life. *Publ. Astron. Soc. Austral.*, **29**, 529.

Bars, I. 2001. Survey of two time physics. *Class. Quant. Grav.*, **18**, 3113–30.

Bize, S., Diddams, S. A., Tanaka, U., *et al.* 2003. Testing the Stability of Fundamental Constants with ^{199}Hg$^+$ Single-Ion Optical Clock. *Phys. Rev. Lett.*, **90**, 150802.

Blatt, S., Ludlow, A. D., Campbell, G. K., *et al.* 2008. New Limits on Coupling of Fundamental Constants to Gravity Using ^{87}Sr Optical Lattice Clocks. *Phys. Rev. Lett.*, **100**, 140801.

Caldwell, R. R. and Stebbins, A. 2008. A Test of the Copernican Principle. *Phys. Rev. Lett.*, **100**, 191302.

Carr, B. J. 2007. *Universe or multiverse*. Cambridge: Cambridge University Press.

Chand, H., Petitjean, P., Srianand, R. and Aracil, B. 2004. Probing the cosmological variation of the fine-structure constant: Results based on VLT-UVES sample. *Astron. Astrophys.*, **417**, 853.

Chand, H., Petitjean, P., Srianand, R. and Aracil, B. 2005. Probing the time-variation of the fine-structure constant: Results based on Si IV doublets from a UVES sample. *Astron. Astrophys.*, **430**, 47–58.

Cingöz, A., Lapierre, A., Nguyen, A.-T. *et al.* 2008. Limit on the Temporal Variation of the Fine-Structure Constant Using Atomic Dysprosium. *Phys. Rev. Lett.*, **98**.

Coc, A., Olive, K. A., Uzan, J.-P. and Vangioni, E. 2006. Big bang nucleosynthesis constraints on scalar-tensor theories of gravity. *Phys. Rev. D*, **73**, 083525.

Coc, A., Nunes, N. J., Olive, K. A., Uzan, J.-P., and Vangioni, E. 2007. Coupled variations of the fundamental couplings and primordial nucleosynthesis. *Phys. Rev. D*, **76**, 023511.

Damour, T. and Dyson, F. 1996. The Oklo bound on the time variation of the fine-structure constant revisited. *Nucl. Phys. B*, **480**, 37–54.

Deffayet, C., Gao, X., Steer, D. A. and Zahariade, G. 2011. From k-essence to generalised Galileons. *Phys. Rev. D*, **84**, 064039.

Dent, T. 2007. Composition-dependent long range forces from varying m_p/m_e. *J. Cosmol. Astropart. Phys.*, **2007**(01). 013.

Dent, T. 2008. Eötvös bounds on couplings of fundamental parameters to gravity. *Phys. Rev. Lett.*, **101**, 041102.

Deutsch, D. 1997. *The fabric of reality*. London: Allen Lane.

Dicke, R. H. 1964. Experimental relativity. In C. M. DeWitt and B. S. DeWitt, eds., *Relativity, Groups and Topology. Relativité, Groupes et Topologie*. New York; London: Gordon and Breach, pp. 165–313.

Duff, M. J. 2014. How fundamental are fundamental constants? *arXiv:1412.2040*.

Eddington, A. 1923. *The Mathematical Theory of Relativity*. Cambridge: Cambridge University Press.

Eisenstaedt, J. 1989. In F. W. Meyerstein, ed., *Foundations of Big Bang Cosmology*. Singapore: World Scientific.

Ekström, S., Coc, A., Descouvemont, P., *et al.* 2010. Effects of the variation of fundamental constants on Population III stellar evolution. *Astron. Astrophys. A*, **514**, 62.

Ellis, G. F. R. 2005. Physics, complexity and causality. *Nature*, **435**, 743.

Ellis, G. F. R. 2012. Recognising Top-Down Causation. *arXiv:1212.2275*, .

Ellis, G. F. R. and Uzan, J.-P. 2005. 'c' is the speed of light, isn't it? *Am. J. Phys.*, **73**, 240–47.

Ellis, G. F. R., Maartens, R. and MacCallum, M. A. H. 2012. *Relativistic Cosmology*. Oxford; New York: Cambridge University Press.

Esposito-Farèse, G., Pitrou, C. and Uzan, J.-P. 2010. Vector theories in cosmology. *Phys. Rev. D*, **81**, 063519.

Fabre, O., Prunet, S. and Uzan, J.-P. 2013. Topology beyond the horizon: how far can it be probed? *arXiv:1311.3509*.

Fleury, P., Beltran, J. P. A., Pitrou, C. and Uzan, J.-P. 2014. On the stability and causality of scalar-vector theories. *arXiv:1406.6254*.

Friedman, J. L. 1998. Lorentzian universes from nothing. *Class. Quant. Grav.*, **15**, 2639–44.

Gibbons, G. W. and Hartle, J. B. 1990. Real Tunneling Geometries and the Large Scale Topology of the Universe. *Phys. Rev. D*, **42**, 2458–68.

Gibbons, G. W. and Ishibashi, A. 2004. Topology and signature changes in brane worlds. *Class. Quant. Grav.*, **21**, 2919–36.

Girelli, F., Liberati, S. and Sindoni, L. 2009. Emergence of Lorentzian signature and scalar gravity. *Phys. Rev. D*, **79**, 044019.

Goodman, J. 1995. Geocentrism reexamined. *Phys. Rev. D*, **52**, 1821–7.

Gott, J. Richard, III. and Li, L.-X. 1998. Can the universe create itself? *Phys. Rev. D*, **58**, 023501.

Halsted, G. B. 1982. Four-Fold Space and Two-Fold Time. *Science*, **19**, 319.

Hartle, J. B. and Hawking, S. W. 1983. Wave Function of the Universe. *Phys. Rev. D*, **28**, 2960–75.

Hogan, C. J. 2000. Why the universe is just so. *Rev. Mod. Phys.*, **72**, 1149–61.

Horndeski, G. W. 1974. Second-order scalar-tensor field equations in a four-dimensional space. *Int. J. Theor. Phys.*, **10**, 363–84.

Karshenboim, S. G. 2006. The search for possible variation of the fine-structure constant. *Gen. Relativ. Gravit.*, **38**, 159.

Kehayias, J., Mukohyama, S. and Uzan, J.-P. 2014. Emergent Lorentz Signature, Fermions, and the Standard Model. *Phys. Rev. D*, **89**, 105017.

Kuroda, P. K. 1956. On the nuclear physical stability of uranium mineral. *J. Chem. Phys.*, **25**, 781.

Luo, F., Olive, K. A. and Uzan, J.-P. 2011. Gyromagnetic Factors and Atomic Clock Constraints on the Variation of Fundamental Constants. *Phys. Rev. D*, **84**, 096004.

Mars, M., Senovilla, J. M. M. and Vera, R. 2007. Lorentzian and signature changing branes. *Phys. Rev. D*, **76**, 044029.

Mielczarek, J. 2014. Signature change in loop quantum cosmology. *Springer Proc. Phys.*, **157**, 555–62.

Moore, G. D. and Nelson, A. E. 2001. Lower bound on the propagation speed of gravity from gravitational Cherenkov radiation. *JHEP*, **0109**, 023.

Mukhanov, V. 2005. *Physical foundations of cosmology*. Cambridge; New York: Cambridge University Press.

Mukohyama, S. and Uzan, J.-P. 2013. From configuration to dynamics: Emergence of Lorentz signature in classical field theory. *Phys. Rev. D*, **87**(6), 065020.

Mukohyama, S. and Uzan, J.-P. 2013. Emergence of the Lorentzian structure in classical field theory. *Int. J. Mod. Phys. D*, **22**, 1342018.

Olive, K. A., Pospelov, M., Qian, Y.-Z., *et al.* 2004. Reexamination of the [187]Re bound on the variation of fundamental couplings. *Phys. Rev. D*, **69**, 027701.

Olive, K. A., Peloso, M. and Uzan, J.-P. 2011. The Wall of Fundamental Constants. *Phys. Rev.*, **D83**, 043509.

Peebles, P. J. and Dicke, R. H. 1962. Cosmology and the Radioactive Decay Ages of Terrestrial Rocks and Meteorites. *Phys. Rev.*, **128**, 2006–11.

Peter, P. and Uzan, J.-P. 2009. *Primordial Cosmology*. Oxford; New York: Oxford University Press.

Petitjean, P., Srianand, R., Chand, H., *et al.* 2009. Constraining fundamental constants of physics with quasar absorption line systems. *Space Sci. Rev.*, **148**, 289–300.

Polchinsky, J. 1998. *String theory*. Cambridge: Cambridge University Press.

Uzan, J.-P. 2003. The Fundamental constants and their variation: Observational status and theoretical motivations. *Rev. Mod. Phys.*, **75**, 403.

Uzan, J.-P. 2005. Variation of the constants in the late and early universe. *AIP Conf. Proc.*, **736**, 3–20.

Uzan, J.-P. 2009. Fundamental constants and tests of general relativity: Theoretical and cosmological considerations. *Space Sci. Rev.*, **148**, 249.

Uzan, J.-P. 2010. Dark energy, gravitation and the Copernican principle. In P. Ruiz-Lapuente, ed., *Dark Energy*. Cambridge: Cambridge University Press, pp. 3–47.

Uzan, J.-P. 2011. Varying Constants, Gravitation and Cosmology. *Living Rev. Rel.*, **14**, 2.

Uzan, J.-P. 2012. Variation of fundamental constants on sub- and super-Hubble scales: From the equivalence principle to the multiverse. *AIP Conf. Proc.*, **1514**, 14–20.

Uzan, J.-P. 2013. Models of the cosmos and emergence of complexity. In G. F. R. Ellis, M. Heller and T. Pabjan, eds., *The Causal Universe*. Krakow: Copernicus Center Press, pp. 93–119.

Uzan, J.-P. and Bernardeau, F. 2001. Lensing at cosmological scales: A Test of higher dimensional gravity. *Phys. Rev. D*, **64**, 083004.

Uzan, J.-P. and Lehoucq, L. 2005. *Les constantes fondamentales*. Paris: Belin.

Uzan, J.-P., Clarkson, C. and Ellis, G. F. R. 2008. Time drift of cosmological redshifts as a test of the Copernican principle. *Phys. Rev. Lett.*, **100**, 191303.

Webb, J. K., Flambaum, V. V., Churchill, C. W., Drinkwater, M. J. and Barrow, J. D. 1999. Search for time variation of the fine-structure constant. *Phys. Rev. Lett.*, **82**, 884–7.

Webb, J. K., Murphy, M. T., Flambaum, V. V., *et al.* 2001. Further evidence for cosmological evolution of the fine-structure constant. *Phys. Rev. Lett.*, **87**, 091301.

Weinberg, S. 1983. Overview of theoretical prospects for understanding the values of fundamental constants. *Philos. Trans. R. Soc. London, Ser. A*, **310**, 249.

Will, C. M. 1993. *Theory and Experiment in Gravitational Physics*. 2nd edn. Cambridge; New York: Cambridge University Press.

7

Cosmological Structure Formation

JOEL R. PRIMACK

7.1 Introduction

Cosmology has finally become a mature science during the past decade, with predictions now routinely confirmed by observations. The modern cosmological theory is known as ΛCDM – CDM for cold dark matter, particles that moved sluggishly in the early universe and thereby preserved fluctuations down to small scales (Blumenthal *et al.*, 1984, see Figure 7.1), and Λ for the cosmological constant (e.g. Lahav *et al.*, 1991). A wide variety of large-scale astronomical observations – including the Cosmic Microwave Background (CMB), measurements of baryon acoustic oscillations (BAO), gravitational lensing, the large-scale distribution of galaxies, and the properties of galaxy clusters – agree very well with the predictions of the ΛCDM cosmology.

Like the standard model of particle physics, the ΛCDM standard cosmological model requires the determination of a number of relevant cosmological parameters, and the theory does not attempt to explain why they have the measured values – or to explain the fundamental nature of the dark matter and dark energy. These remain challenges for the future. But the good news is that the key cosmological parameters are now all determined with unprecedented accuracy, and the six-parameter ΛCDM theory provides a very good match to all the observational data including the 2015 Planck temperature and polarization data (Planck Collaboration *et al.*, 2015a). Within uncertainties less than 1 per cent, the Universe has critical cosmic density – i.e., $\Omega_{\text{total}} = 1.00$ and the Universe is Euclidean (or "flat") on large scales. The expansion rate of the Universe is measured by the Hubble parameter $h = 0.6774 \pm 0.0046$, and $\Omega_{\text{matter}} = 0.3089 \pm 0.0062$; this leads to the age of the Universe $t_0 = 13.799 \pm 0.021$ Gyr. The power spectrum normalization parameter is $\sigma_8 = 0.816 \pm 0.009$, and the primordial fluctuations are consistent with a purely adiabatic spectrum of fluctuations with a spectral tilt $n_s = 0.968 \pm 0.006$, as predicted by single-field inflationary models (Planck Collaboration *et al.*, 2015a). The same cosmological parameters that are such a good match to the CMB observations also predict the observed distribution of density fluctuations from small scales probed by the Lyman alpha

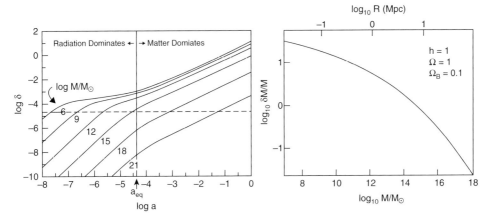

Figure 7.1 The origin of the CDM spectrum of density fluctuations. **Left panel**: Fluctuations corresponding to mass scales $10^6 M_\odot$, $10^9 M_\odot$, etc., grow proportionally to the square of scale factor a when they are outside the horizon, and when they enter the horizon (cross the horizontal dashed line) the growth of the fluctuation amplitude δ is much slower if they enter when the Universe is radiation dominated (i.e. $a < a_{eq}$). Fluctuations on mass scales $> 10^{15} M_\odot$ enter the horizon after it becomes matter dominated, so their growth is proportional to scale factor a; that explains the larger separation between amplitudes for such higher-mass fluctuations. **Right panel**: The resulting CDM fluctuation spectrum ($\kappa^{3/2}|\delta_\kappa| = \Delta M/M$). This calculation assumed that the primordial fluctuations are scale-invariant (Zel'dovich) and that $\Omega_{matter} = 1$ and Hubble parameter $h = 1$. (From a 1983 conference presentation Primack and Blumenthal (1984), reprinted in Primack (1984).)

forest[1] to the entire horizon, as shown in Figure 7.2. The near-power-law galaxy–galaxy correlation function at low redshifts is now known to be a cosmic coincidence (Watson *et al.*, 2011). I was personally particularly impressed that the evolution of the galaxy–galaxy correlations with redshift predicted by ΛCDM (Kravtsov *et al.*, 2004) turned out to be in excellent agreement with the subsequent observations (e.g. Conroy *et al.*, 2006).

For non-astronomers, there should be a more friendly name than ΛCDM for the standard modern cosmology. Since about 95 per cent of the cosmic density is dark energy (either a cosmological constant with $\Omega_\Lambda = 0.69$ or some dynamical field that plays a similar cosmic role) and cold dark matter with $\Omega_{CDM} = 0.31$, I recommend the simple name "Double Dark Theory" for the modern cosmological standard model (Primack and Abrams, 2006; Abrams and Primack, 2011). The contribution of ordinary baryonic matter is only $\Omega_b = 0.05$. Only about 10 per cent of the baryonic matter is in the form of stars or gas clouds that emit electromagnetic radiation, and the contribution of what astronomers call "metals" – chemical elements heavier than helium – to the cosmic density is only $\Omega_{metals} \approx 0.0005$, most of which is in white dwarfs and neutron stars (Fukugita and Peebles, 2004). The contribution of neutrino mass to the cosmic density is $0.002 \leq \Omega_\nu \leq 0.005$, far greater

---

[1] The Lyman forest is the many absorption lines in quasar spectra due to clouds of neutral hydrogen along the line of sight to the quasar.

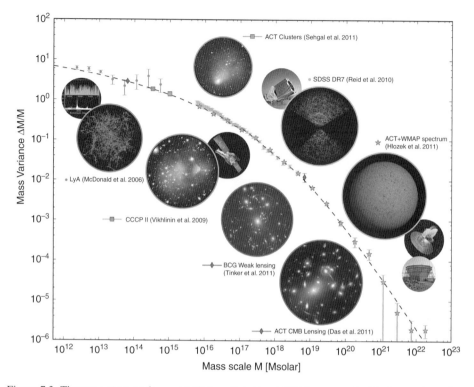

Figure 7.2 The r.m.s. mass variance $\Delta M/M$ predicted by ΛCDM compared with observations, from CMB and the Atacama cosmology telescope (ACT) on large scales, brightest cluster galaxy weak lensing, clusters, the SDSS galaxy distribution, to the Lyman alpha forest on small scales. This figure highlights the consistency of power spectrum measurements by an array of cosmological probes over a large range of scales. (Redrawn from Figure 5 in Hlozek *et al.* (2012), which gives the sources of the data.)

than Ω_{metals}. Thus our bodies and our planet are made of the rarest form of matter in the universe: elements forged in stars and stellar explosions.

Potential challenges to ΛCDM on large scales come from the tails of the predicted distribution functions, such as CMB cold spots and massive clusters at high redshifts. However, the existing observations appear to be consistent thus far with predictions of standard ΛCDM with standard primordial power spectra; non-Gaussian initial conditions are not required (Planck Collaboration *et al.*, 2015b). Larger surveys now underway may provide more stringent tests.

7.2 Large-Scale Structure

Large, high-resolution simulations permit detailed predictions of the distribution and properties of galaxies and clusters. From 2005 to 2010, the benchmark simulations were

Millennium-I (Springel *et al.*, 2005) and Millennium-II (Boylan-Kolchin *et al.*, 2009), which have been the basis for more than 400 papers. However, these simulations used first-year Wilkinson microwave anisotropy probe (WMAP) cosmological parameters, including $\sigma_8 = 0.90$, that are now in serious disagreement with the latest observations. Improved cosmological parameters, simulation codes, and computer power have permitted more accurate simulations (Kuhlen *et al.*, 2012; Skillman *et al.*, 2014) including Bolshoi (Klypin *et al.*, 2011) and BigBolshoi/MultiDark (Prada *et al.*, 2012; Riebe *et al.*, 2013), which have recently been rerun using the Planck cosmological parameters (Klypin *et al.*, 2016).[2]

Dark matter halos can be characterized in a number of ways. A common one is by mass, but the mass attributed to a halo depends on a number of factors including how the outer edge of the halo is defined; popular choices include the spherical radius within which the average density is either 200 times critical density or the virial density (which depend on redshift). Properties of all the halos in many stored time steps of both the Bolshoi and Big-Bolshoi/MultiDark simulations are available on the web in the MultiDark database.[2] For many purposes it is more useful to characterize halos by their maximum circular velocity V_{max}, which is defined as the maximum value of $[GM(< r)/r]^{1/2}$, where G is Newton's constant and $M(< r)$ is the mass enclosed within radius r. The reason this is useful is that V_{max} is reached at a relatively low radius r_{max}, closer to the central region of a halo where stars or gas can be used to trace the velocity of the halo, while most of the halo mass is at larger radii. Moreover, the measured internal velocity of a galaxy (line of sight velocity dispersion for early-type galaxies and rotation velocity for late-type galaxies) is closely related to its luminosity according to the Faber–Jackson and Tully–Fisher relations. In addition, after a subhalo has been accreted by a larger halo, tidal stripping of its outer parts can drastically reduce the halo mass but typically decreases V_{max} much less. (Since the stellar content of a subhalo is thought to be determined before it was accreted, some authors define V_{max} to be the peak value at any redshift for the main progenitor of a halo.) Because of the observational connection between larger halo internal velocity and brighter galaxy luminosity, a common simple method of assigning galaxies to dark matter halos and subhalos is to rank order the galaxies by luminosity and the halos by V_{max}, and then match them such that the number densities are comparable (Kravtsov *et al.*, 2004; Tasitsiomi *et al.*, 2004; Conroy *et al.*, 2006). This is called "halo abundance matching" or (more modestly) "sub-halo abundance matching" (SHAM) (Reddick *et al.*, 2014). Halo abundance matching using the Bolshoi simulation predicts galaxy–galaxy correlations (which are essentially counts of the numbers of pairs of galaxies at different separation distances) that are in good agreement with the Sloan Digital Sky Survey (SDSS) observations (Trujillo-Gomez *et al.*, 2011; Reddick *et al.*, 2013).

Abundance matching with the Bolshoi simulation also predicts galaxy velocity–mass scaling relations consistent with observations (Trujillo-Gomez *et al.*, 2011), and a galaxy

---

[2] The web address for the MultiDark simulation data center is www.cosmosim.org/cms/simulations/ multidark-project/; more detailed analyses of the Bolshoi-Planck and MultiDark-Planck simulations are available at www.hipacc.ucsc.edu/Bolshoi/MergerTrees.html.

velocity function in good agreement with observations for maximum circular velocities $V_{max} \gtrsim 100$ km/s, but higher than the HI Parkes All Sky Survey (HIPASS) and the Arecibo Legacy Fast ALFA (ALFALFA) Survey radio observations (Zwaan *et al.*, 2010; Papastergis *et al.*, 2011) by about a factor of 2 at 80 km/s and a factor of 10 at 50 km/s. This either means that these radio surveys are increasingly incomplete at lower velocities, or else ΛCDM is in trouble because it predicts far more small-V_{max} halos than there are observed low-V galaxies. A deeper optical survey out to 10 Mpc found no disagreement between V_{max} predictions and observations for $V_{max} \geq 60$ km/s, and only a factor of 2 excess of halos compared to galaxies at 40 km/s (Klypin *et al.*, 2015). This may indicate that there is no serious inconsistency with theory, since for $V \lesssim 30$ km/s reionization and feedback can plausibly explain why there are fewer observed galaxies than dark matter halos (Bullock *et al.*, 2000; Somerville, 2002; Benson *et al.*, 2003; Kravtsov, 2010; Wadepuhl and Springel, 2011; Sawala *et al.*, 2012), and also the observed scaling of metallicity with galaxy mass (Dekel and Woo, 2003; Woo *et al.*, 2008; Kirby *et al.*, 2011).

The radial dark matter density distribution in halos can be approximately fit by the simple formula $\rho_{NFW} = 4\rho_s x^{-1}(1 + x)^{-2}$, where $x \equiv r/r_s$ (Navarro *et al.*, 1996), and the "concentration" of a dark matter halo is defined as $C = R_{vir}/R_s$ where R_{vir} is the virial radius of the halo. When we first understood that dark matter halos form with relatively low concentration $C \sim 4$ and evolve to higher concentration, we suggested that "red" galaxies that shine mostly by the light of red giant stars because they have stopped forming stars should be found in high-concentration halos while "blue" galaxies that are still forming stars should be found in younger low-concentration halos (Bullock *et al.*, 2001). This idea was recently rediscovered by Hearin and Watson, who used the Bolshoi simulation to show that this leads to remarkably accurate predictions for the correlation functions of red and blue galaxies (Hearin and Watson, 2013; Hearin *et al.*, 2014).

The Milky Way has two rather bright satellite galaxies, the Large and Small Magellanic Clouds. It is possible using sub-halo abundance matching with the Bolshoi simulation to determine the number of Milky-Way-mass dark matter halos that have subhalos with high enough circular velocity to host such satellites. It turns out that about 55 per cent have no such subhalos, about 28 per cent have one, about 11 per cent have two, and so on (Busha *et al.*, 2011a). Remarkably, these predictions are in excellent agreement with an analysis of observations by the Sloan Digital Sky Survey (SDSS) (Liu *et al.*, 2011). The distribution of the relative velocities of central and bright satellite galaxies from SDSS spectroscopic observations is also in very good agreement with the predictions of the Millennium-II simulation (Tollerud *et al.*, 2011), and the Milky Way's lower-luminosity satellite population is not unusual (Strigari and Wechsler, 2012). Considered in a cosmological context, the Magellanic clouds are likely to have been accreted within about the last Gyr (Besla *et al.*, 2012), and the Milky Way halo mass is $1.2^{+0.7}_{-0.4}$(stat.)± 0.3(sys.)$\times 10^{12} M_\odot$ (Busha *et al.*, 2011b).

7.3 Galaxy Formation

At early times, for example the CMB epoch about 400,000 years after the big bang, or on very large scales at later times, linear calculations starting from the ΛCDM fluctuation

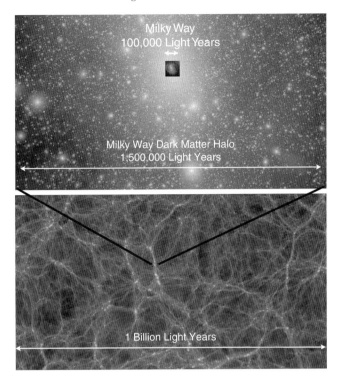

Figure 7.3 The stellar disk of a large spiral galaxy like the Milky Way is about 100,000 light years across, which is tiny compared with the dark matter halo of such a galaxy (from the Aquarius dark matter simulation, Springel *et al.*, 2008), and even much smaller compared with the large-scale cosmic web (from the Bolshoi simulation, Klypin *et al.*, 2011).

spectrum allow accurate predictions. But on scales where structure forms, the fluctuations have grown large enough that they are strongly non-linear, and we must resort to simulations. The basic idea is that regions that start out with slightly higher than average density expand a little more slowly than average because of gravity, and regions that start out with slightly lower density expand a little faster. Non-linear structure forms by the process known by the somewhat misleading name "gravitational collapse" – misleading because what really happens is that when positive fluctuations have grown sufficiently that they are about twice as dense as typical regions their size, they stop expanding while the surrounding universe keeps expanding around them. The result is that regions that collapse earlier are denser than those that collapse later; thus galaxy dark matter halos are denser than cluster halos. The visible galaxies form because the ordinary baryonic matter can radiate away its kinetic energy and fall toward the centers of the dark matter halos; when the ordinary matter becomes dense enough it forms stars. Thus visible galaxies are much smaller than their host dark matter halos, which in turn are much smaller than the large scale structure of the cosmic web, as shown in Figure 7.3.

Astronomical observations represent snapshots of moments long ago when the light we now observe left distant astronomical objects. It is the role of astrophysical theory to produce movies – both metaphorical and actual – that link these snapshots together into a coherent physical picture. To predict cosmological large-scale structures, it has been sufficient to treat all the mass as dark matter in order to calculate the growth of structure and dark matter halo properties. But hydrodynamic simulations – i.e. including baryonic matter – are necessary to treat the formation and evolution of galaxies.

An old criticism of ΛCDM has been that the order of cosmogony is wrong: halos grow from small to large by accretion in a hierarchical formation theory like ΛCDM, but the oldest stellar populations are found in the most massive galaxies – suggesting that these massive galaxies form earliest, a phenomenon known as "downsizing" (Cowie *et al.*, 1996). The key to explaining the downsizing phenomenon is the realization that star formation is most efficient in dark matter halos with masses in the band between about 10^{10} and $10^{12} M_\odot$ (Figure 1 bottom in Behroozi *et al.*, 2013). This goes back at least as far as the original Cold Dark Matter paper (Blumenthal *et al.*, 1984): see Figure 7.4. A dark matter halo that has the total mass of a cluster of galaxies today will have crossed this star-forming mass band at an early epoch, and it will therefore contain galaxies whose stars formed early. These galaxies will be red and dead today. A less massive dark matter halo that is now entering the star-forming band today will just be forming significant numbers of stars, and it will be blue today. The details of the origin of the star-forming band are still being worked out. Back in 1984, we argued that cooling would be inefficient for masses greater than about $10^{12} M_\odot$ because the density would be too low, and inefficient for masses less than about $10^8 M_\odot$ because the gas would not be heated enough by falling into these small potential wells. Now we know that reionization, supernovae (Dekel and Silk, 1986), and other energy input additionally impedes star formation for halo masses below about $10^{10} M_\odot$, and feedback from active galactic nuclei (AGN) additionally impedes star formation for halo masses above about $10^{12} M_\odot$.

Early simulations of disk galaxy formation found that the stellar disks had much lower rotation velocities than observed galaxies (Navarro and Steinmetz, 2000). This problem seemed so serious that it became known as the "angular momentum catastrophe". A major cause of this was excessive cooling of the gas in small halos before they merged to form larger galaxies (Maller and Dekel, 2002). Simulations with better resolution and more physical treatment of feedback from star formation appear to resolve this problem. In particular, the Eris cosmological simulation (Guedes *et al.*, 2011) produced a very realistic spiral galaxy, as have many simulations since then. Somerville and Davé (2015) is an excellent recent review of progress in understanding galaxy formation. In the following I summarize some of the latest developments. There are now two leading approaches to simulating galaxies:

- Low resolution, $\sim$1 kiloparsec. The **advantages** of this approach are that it is possible to simulate many galaxies and study galaxy populations and their interactions with the circumgalactic and intergalactic media, but the **disadvantages** are that we learn

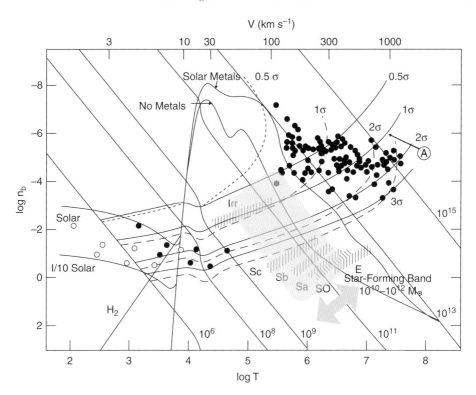

Figure 7.4 The Star-Forming Band on a diagram of baryon density n_b versus the three-dimensional r.m.s. velocity dispersion V and virial temperature T for structures of various sizes in the universe, where $T = \mu V^2/3k$, μ is mean molecular weight (≈ 0.6 for ionized primordial H + He) and k is Boltzmann's constant. Below the No Metals and Solar Metals cooling curves, the cooling timescale is more rapid than the gravitational timescale. Dots are groups and clusters. Diagonal lines show the halo masses in units of $M_{\odot}$. (This is Figure 3 in Blumenthal *et al.*, 1984, with the Star-Forming Band added.)

relatively little about how galaxies themselves form and evolve at high redshifts. The prime examples of this approach now are the *Illustris* (Vogelsberger *et al.*, 2014b) and EAGLE (Schaye *et al.*, 2015) simulations. Like semi-analytic models of galaxy formation (reviewed in Benson, 2010), these projects adjusted the parameters governing star-formation and feedback processes in order to reproduce key properties of galaxies at the present epoch, redshift $z = 0$. The *Illustris* simulation in a volume $(106.5 \text{ Mpc})^3$ forms $\sim$40,000 galaxies at the present epoch with a reasonable mix of elliptical and spiral galaxies that have realistic appearances (Snyder *et al.*, 2015), obey observed scaling relations, and have the observed numbers of galaxies as a function of their luminosity, and were formed with the observed cosmic star formation rate (Vogelsberger *et al.*, 2014a). It forms massive compact galaxies by redshift $z = 2$ via central starbursts in major mergers of gas-rich galaxies or else by assembly at very early times (Wellons *et al.*, 2015). Remarkably, the *Illustris* simulation also predicts a population of damped Lyman α absorbers (DLAs, small-galaxy-size clouds of neutral hydrogen) that agrees with some of the key observational properties of DLAs (Bird *et al.*, 2014, 2015). The EAGLE simulation in volumes up to $(100 \text{ Mpc})^3$ reproduces the observed galaxy mass function from 10^8 to $10^{11} M_\odot$ at a level of agreement close to that attained by semi-analytic models (Schaye *et al.*, 2015), and the observed atomic and molecular hydrogen content of galaxies out to $z \sim 3$ (Rahmati *et al.*, 2015; Lagos *et al.*, 2015).

- High resolution, $\sim$ 10s of parsecs. The **advantages** are that it is possible to compare simulation outputs in detail with high-redshift galaxy images and spectra to discover the drivers of morphological changes as galaxies evolve, but the **disadvantage** is that such simulations are so expensive computationally that it is as yet impossible to run enough cases to create statistical samples. Leading examples of this approach are FIRE simulations led by Phil Hopkins (e.g. Hopkins *et al.*, 2014) and the ART simulation suite led by Avishai Dekel and myself (e.g. Zolotov *et al.*, 2015). We try to compensate for the small number of high-resolution simulations by using simulation outputs to tune semi-analytic models, which in turn use cosmological dark-matter-only simulations like Bolshoi to follow the evolution of $\sim 10^6$ galaxies in their cosmological context (e.g. Porter *et al.*, 2014a, 2014b; Brennan *et al.*, 2015).

The high-resolution FIRE simulations, based on the GIZMO smooth particle hydro-dynamics code (Hopkins, 2015) with supernova and stellar feedback, including radiative feedback (RF) pressure from massive stars, treated with zero adjusted parameters, repro-duce the observed relation between stellar and halo mass up to $M_{\text{halo}} \sim 10^{12} M_\odot$ and the observed star formation rates (Hopkins *et al.*, 2014). FIRE simulations predict covering fractions of neutral hydrogen with column densities from 10^{17}cm^{-2} (Lyman limit sys-tems, LLS) to $> 10^{20.3} \text{cm}^{-2}$ (DLAs) in agreement with observations at redshifts z=2-2.5 (Faucher-Giguère *et al.*, 2015); this success is a consequence of the simulated galactic winds. FIRE simulations also correctly predict the observed evolution of the decrease of metallicity with stellar mass (Ma *et al.*, 2015), and produce dwarf galaxies that appear to agree with observations (Oñorbe *et al.*, 2015) as we will discuss in more detail below.

The high-resolution simulation suite based on the ART adaptive mesh refinement (AMR) approach (Kravtsov *et al.*, 1997; Ceverino and Klypin, 2009) incorporates at the sub-grid level many of the physical processes relevant for galaxy formation. Our initial group of 30 zoom-in simulations of galaxies in dark matter halos of mass $(1 - 30) \times 10^{12} M_\odot$ at redshift $z = 1$ were run at 35–70 pc maximum (physical) resolution (Ceverino *et al.*, 2012, 2015a). The second group of 35 simulations (VELA01 to VELA35) with 17.5 to 35 pc resolution of halos of mass $(2 - 20) \times 10^{11} M_\odot$ at redshift $z = 1$ have now been run three times with varying inclusion of radiative pressure feedback (none, UV, UV+IR), as described in Ceverino *et al.* (2014). RF pressure including the effects of stellar winds (Hopkins *et al.*, 2012, 2014) captures essential features of star formation in our simulations. In particular, RF begins to affect the star-forming region as soon as massive stars form, long before the first supernovae occur, and the amount of energy released in RF greatly exceeds that

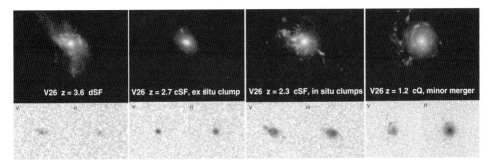

Figure 7.5 Face-on images of Vela26 simulated galaxy with UV radiation pressure feedback, at four redshifts: **(a)** $z = 3.6$ when it is diffuse and star forming (dSF); **(b)** $z = 2.7$ when it has become compact and star forming (cSF) with a red *ex-situ* clump; **(c)** $z = 2.3$ still cSF, now with *in situ* clumps apparent in the V-band image; **(d)** compact and quenched (cQ) during a minor merger, with tidal features visible in the V-band image. Top panels: three-color composite images at high resolution; bottom panels: CANDELized V and H band images. The observed V band images correspond to ultraviolet radiation from massive young stars in the galaxy rest frame, while the observed H band images show optical light from the entire stellar population including old stars. The CANDELS survey took advantage of the infrared capability of the Wide Field Camera 3, installed on the last service visit to HST in 2009.

released by supernovae (Ceverino *et al.*, 2014; Trujillo-Gomez *et al.*, 2015). In addition to radiation pressure, the local UV flux from young star clusters also affects the cooling and heating processes in star-forming regions through photoheating and photoionization. We use our *Sunrise* code (Jonsson, 2006; Jonsson *et al.*, 2006, 2010; Jonsson and Primack, 2010) to make realistic images and spectra of these simulated galaxies in many wavebands and at many times during their evolution, including the effects of stellar evolution and of dust scattering, absorption, and re-emission, to compare with the imaging and photometry from CANDELS[3] and other surveys – see Figure 7.5 for examples including the effect of CANDELization (reducing the resolution and adding noise) to allow direct comparison with holographic spacetime (HST) images.

In comparing our simulations with HST observations, especially those from the CANDELS and 3D-HST surveys, we are finding that the simulations can help us interpret a variety of observed phenomena that we now realize are important in galaxy evolution. One is the **formation of compact galaxies**. Analysis of CANDELS images suggested (Barro *et al.*, 2013, 2014a,b) that diffuse star-forming galaxies become compact galaxies ("blue nuggets") which subsequently quench ("red nuggets"). We see very similar behavior in our VELA simulations with UV radiative feedback (Zolotov *et al.*, 2015, see Figure 2), and we have identified in our simulations several mechanisms that lead to compaction often followed by rapid quenching, including major gas-rich mergers, disk instabilities often

---

[3] CANDELS, the Cosmic Assembly Near-infrared Deep Extragalactic Legacy Survey, was the largest-ever Hubble Space Telescope survey, see http://candels.ucolick.org/.

triggered by minor mergers, and opposing gas flows into the central galaxy (Danovich *et al.*, 2015).

Another aspect of galaxy formation seen in HST observations is **massive star-forming clumps** (Guo *et al.*, 2012; Wuyts *et al.*, 2013, and references therein), which occur in a large fraction of star-forming galaxies at redshifts $z = 1 - 3$ (Guo *et al.*, 2015). In our simulations there are two types of clumps. Some are a stage of minor mergers – we call those *ex situ* clumps. A majority of the clumps originate *in situ* from violent disk instabilities (VDI) in gas-rich galaxies (Ceverino *et al.*, 2012; Moody *et al.*, 2014; Mandelker *et al.*, 2014). Some of these *in situ* clumps are associated with gas instabilities that help to create compact spheroids, and some form after the central spheroid and are associated with the formation of surrounding disks. We find that there is not a clear separation between these processes, since minor mergers often trigger disk instabilities in our simulations (Zolotov *et al.*, 2015).

Star-forming galaxies with stellar masses $M_* \lesssim 3 \times 10^9 M_\odot$ at $z > 1$ have recently been shown to have mostly **elongated (prolate) stellar distributions** (van der Wel *et al.*, 2014) rather than disks or spheroids, based on their observed axis ratio distribution. In our simulations this occurs because most dark matter halos are prolate especially at small radii (Allgood *et al.*, 2006), and the first stars form in these elongated inner halos; at lower redshifts, as the stars begin to dominate the dark matter, the galaxy centers become disky or spheroidal (Ceverino *et al.*, 2015b).

Both the FIRE and ART simulation groups and many others are participating in the Assembling Galaxies of Resolved Anatomy (AGORA) collaboration (Kim *et al.*, 2014) to run high-resolution simulations of the same initial conditions with halos of masses 10^{10}, 10^{11}, 10^{12}, and $10^{13} M_\odot$ at $z = 0$ with as much as possible the same astrophysical assumptions. AGORA cosmological runs using different simulation codes will be systematically compared with each other using a common analysis toolkit and validated against observations to verify that the solutions are robust – i.e. that the astrophysical assumptions are responsible for any success, rather than artifacts of particular implementations. The goals of the AGORA project are, broadly speaking, to raise the realism and predictive power of galaxy simulations and the understanding of the feedback processes that regulate galaxy "metabolism".

It still remains to be seen whether the entire population of galaxies can be explained in the context of ΛCDM. A concern regarding disk galaxies is whether the formation of bulges by both galaxy mergers and secular evolution will prevent the formation of as many pure disk galaxies as we see in the nearby universe (Kormendy and Fisher, 2008). A concern regarding massive galaxies is whether theory can naturally account for the relatively large number of ultra-luminous infrared galaxies. The bright sub-millimeter galaxies were the greatest discrepancy between our semi-analytic model predictions compared with observations out to high redshift (Somerville *et al.*, 2012). This could possibly be explained by a top-heavy stellar initial mass function, or perhaps more plausibly by more realistic simulations including self-consistent treatment of dust (Hayward *et al.*, 2011, 2013).

Clearly, there is much still to be done, both observationally and theoretically. It is possible that all the potential discrepancies between ΛCDM and observations of relatively massive galaxies will be resolved by a better understanding of the complex astrophysics of their formation and evolution. But small galaxies might provide more stringent tests of ΛCDM.

7.4 Smaller Scale Issues: Cusps

Cusps were perhaps the first potential discrepancy pointed out between the dark matter halos predicted by CDM and the observations of small galaxies that appeared to be dominated by dark matter nearly to their centers (Flores and Primack, 1994; Moore, 1994). Pure dark matter simulations predicted that the central density of dark matter halos behaves roughly as $\rho \sim r^{-1}$. As mentioned above, dark matter halos have a density distribution that can be roughly approximated as $\rho_{NFW} = 4\rho_s x^{-1}(1+x)^{-2}$, where $x \equiv r/r_s$ (Navarro *et al.*, 1996). But this predicted r^{-1} central cusp in the dark matter distribution seemed inconsistent with published observations of the rotation velocity of neutral hydrogen as a function of radius.

In small galaxies with significant stellar populations, simulations show that central starbursts can naturally produce relatively flat density profiles (Governato *et al.*, 2010, 2012; Pontzen and Governato, 2012; Teyssier *et al.*, 2013; Brooks, 2014; Brooks and Zolotov, 2014; Madau *et al.*, 2014; Pontzen and Governato, 2014; Oñorbe *et al.*, 2015; Nipoti and Binney, 2015). Gas cools into the galaxy center and becomes gravitationally dominant, adiabatically pulling in some of the dark matter (Blumenthal *et al.*, 1986; Gnedin *et al.*, 2011). But then the gas is driven out very rapidly by supernovae and the entire central region expands, with the density correspondingly dropping. Several such episodes can occur, producing a more or less constant central density consistent with observations, as shown in Figure 7.6. The figure shows that galaxies in the THINGS sample are consistent with ΛCDM hydrodynamic simulations. But simulated galaxies with stellar mass less than about $3 \times 10^6 M_\odot$ may have cusps, although Oñorbe *et al.* (2015) found that stellar effects can soften the cusp in even lower-mass galaxies if the star formation is extended in time. The observational situation is unclear. In Sculptor and Fornax, the brightest dwarf spheroidal satellite galaxies of the Milky Way, stellar motions may imply a flatter central dark matter radial profile than $\rho \sim r^{-1}$ (Walker and Peñarrubia, 2011; Amorisco and Evans, 2012; Jardel and Gebhardt, 2012). However, other papers have questioned this (Jardel and Gebhardt, 2013; Breddels and Helmi, 2013, 2014; Richardson and Fairbairn, 2014).

Will baryonic effects explain the radial density distributions in larger low surface brightness (LSB) galaxies? These are among the most common galaxies. They have a range of masses but many have fairly large rotation velocities indicating fairly deep potential wells, and many of them may not have enough stars for the scenario just described to explain the observed rotation curves (Kuzio de Naray and Spekkens, 2011). Can we understand the observed distribution of the $\Delta_{1/2}$ measure of central density (Alam *et al.*, 2002) and

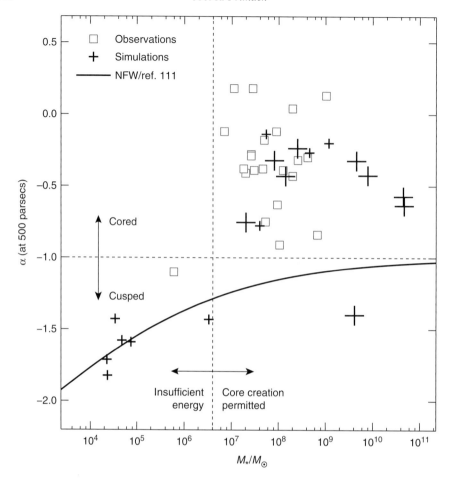

Figure 7.6 Dark matter cores are generated by baryonic effects in galaxies with sufficient stellar mass. The slope α of the dark matter central density profile r^{α} is plotted vs. stellar mass measured at 500 parsecs from simulations described in Pontzen and Governato (2012). The solid NFW curve assumes the halo concentrations given by Macciò *et al.* (2007). Large crosses: halos with $> 5 \times 10^5$ dark matter particles; small crosses: $> 5 \times 10^4$ particles. Squares represent galaxies observed by The HI Nearby Galaxy Survey (THINGS). (Figure 3 in Pontzen and Governato, 2014.)

the observed diversity of rotation curves (Macciò *et al.*, 2012b; Oman *et al.*, 2015)? This is a challenge for galaxy simulators.

Some authors have proposed that warm dark matter (WDM), with initial velocities large enough to prevent formation of small dark matter halos, could solve some of these problems. However, that does not appear to work: the systematics of galactic radial density profiles predicted by WDM do not at all match the observed ones (Kuzio de Naray *et al.*, 2010; Macciò *et al.*, 2012a, 2013). WDM that is warm enough to affect galaxy centers may not permit early enough galaxy formation to reionize the universe (Governato *et al.*, 2015). Yet another constraint on WDM is the evidence for a great deal of dark matter substructure in galaxy halos (Zentner and Bullock, 2003), discussed further below.

7.5 Smaller Scale Issues: Satellite Galaxies

As the top panel of Figure 7.3 shows, ΛCDM predicts that there are many fairly massive subhalos within dark matter halos of galaxies like the Milky Way and the Andromeda galaxy, more than there are observed satellite galaxies (Klypin *et al.*, 1999; Moore *et al.*, 1999). This is not obviously a problem for the theory since reionization, stellar feedback, and other phenomena are likely to suppress gas content and star formation in low-mass satellites. As more faint satellite galaxies have been discovered, especially using multicolor information from SDSS observations, the discrepancy between the predicted and observed satellite population has been alleviated. Many additional satellite galaxies are predicted to be discovered by deeper surveys (e.g. Bullock *et al.*, 2010), including those in the Southern Hemisphere seen by the Dark Energy Survey (The DES Collaboration *et al.*, 2015) and eventually the Large Synoptic Survey Telescope (LSST).

However, a potential discrepancy between theory and observations is the "too big to fail" (TBTF) problem (Boylan-Kolchin *et al.*, 2011, 2012). The Via Lactea-II high-resolution dark-matter-only simulation of a Milky Way size halo (Diemand *et al.*, 2007, 2008) and the six similar Aquarius simulations (Springel *et al.*, 2008) all have several subhalos that are too dense in their centers to host any observed Milky Way satellite galaxy. The brightest observed dwarf spheroidal (dSph) satellites all have $12 \lesssim V_{max} \lesssim 25$ km/s. But the Aquarius simulations predict at least ten subhalos with $V_{max} > 25$ km/s. These halos are also among the most massive at early times, and thus are not expected to have had their star formation greatly suppressed by reionization. They thus appear to be too big to fail to become observable satellites (Boylan-Kolchin *et al.*, 2012).

The TBTF problem is closely related to the cusp-core issue, since TBTF is alleviated by any process that lowers the central density and thus the internal velocity of satellite galaxies. Many of the papers finding that baryonic effects remove central cusps cited in the previous section are thus also arguments against TBTF. A recent simulation of regions like the Local Group found the number, internal velocities, and distribution of the satellite galaxies to be very comparable with observations (Sawala *et al.*, 2014).

Perhaps there is additional physics beyond ΛCDM that comes into play on small scales. One possibility that has been investigated is warm dark matter (WDM). A simulation like Aquarius but with WDM has fewer high-V_{max} halos (Lovell *et al.*, 2012). But it is not clear that such WDM simulations with the lowest WDM particle mass allowed by the Lyman alpha forest and other observations (Viel *et al.*, 2013; Horiuchi *et al.*, 2014) will have enough substructure to account for the observed faint satellite galaxies (e.g. Polisensky and Ricotti, 2011), and as already mentioned WDM does not appear to be consistent with observed systematics of small galaxies.

Another possibility is that the dark matter particles interact with themselves much more strongly than they interact with ordinary matter (Spergel and Steinhardt, 2000). There are strong constraints on such self-interacting dark matter (SIDM) from colliding galaxy clusters (Harvey *et al.*, 2015; Massey *et al.*, 2015), and in hydrodynamic simulations of dwarf galaxies SIDM has similar central cusps to CDM (Bastidas Fry *et al.*, 2015). SIDM can be velocity-dependent, at the cost of adding additional parameters, and if the self-interaction grows with an inverse power of velocity the effects can be strong in dwarf galaxies (Elbert *et al.*, 2015). An Aquarius-type simulation but with velocity-dependent SIDM produced subhalos with inner density structure that may be compatible with the bright dSph satellites of the Milky Way (Vogelsberger *et al.*, 2012). Whether higher-resolution simulations of this type will turn out to be consistent with observations remains to be seen.

7.6 Smaller Scale Issues: Dark Matter Halo Substructure

The first strong indication of galaxy dark matter halo substructure was the flux ratio anomalies seen in quadruply imaged radio quasars ("radio quads") (Metcalf and Madau, 2001; Dalal and Kochanek, 2002; Metcalf and Zhao, 2002). Smooth mass models of lensing galaxies can easily explain the observed positions of the images, but the predictions of such models of the corresponding fluxes are frequently observed to be strongly violated. Optical and X-ray quasars have such small angular sizes that the observed optical and X-ray flux anomalies can be caused by stars ("microlensing"), which has allowed a measurement of the stellar mass along the lines of sight in lensing galaxies (Pooley *et al.*, 2012). But because the quasar radio-emitting region is larger, the observed radio flux anomalies can only be caused by relatively massive objects, with masses of order 10^6 to $10^8 M_\odot$ along the line of sight. After some controversy regarding whether ΛCDM simulations predict enough dark matter substructure to account for the observations, the latest papers concur that the observations are consistent with standard theory, taking into account uncertainty in lens system ellipticity (Metcalf and Amara, 2012) and intervening objects along the line of sight (Xu *et al.*, 2012, 2015). But this analysis is based on a relatively small number of observed systems (Table 2 of Chen *et al.* (2011) lists the ten quads that have been observed in the radio or mid-IR), and further observational and theoretical work would be very helpful.

Another gravitational lensing indication of dark matter halo substructure consistent with ΛCDM simulations comes from detailed analysis of galaxy–galaxy lensing (Vegetti *et al.*, 2010, 2012, 2014), although much more such data will need to be analyzed to get strong constraints. Other gravitational lensing observations including time delays can probe the structure of dark matter halos in new ways (Keeton and Moustakas, 2009). Hezaveh *et al.* (2013, 2014) show that dark matter substructure can be detected using spatially resolved spectroscopy of gravitationally lensed dusty galaxies observed with ALMA. Nierenberg *et al.* (2014) demonstrate that subhalos can be detected using strongly lensed narrow-line quasar emission, as originally proposed by Moustakas and Metcalf (2003).

The great thing about gravitational lensing is that it directly measures mass along the line of sight. This can provide important information that is difficult to obtain in other ways. For example, the absence of anomalous skewness in the distribution of high redshift type Ia supernovae brightnesses compared with low redshift ones implies that massive compact halo objects (MACHOs) in the enormous mass range 10^{-2} to $10^{10} M_\odot$ cannot be the main constituent of dark matter in the universe (Metcalf and Silk, 2007). The low observed rate of gravitational microlensing of stars in the Large and Small Magellanic clouds by foreground compact objects implies that MACHOs in the mass range between 0.6×10^{-7} and $15 M_\odot$ cannot be a significant fraction of the dark matter in the halo of the Milky Way (Tisserand *et al.*, 2007). Gravitational microlensing could even detect free-floating planets down to $10^{-8} M_\odot$, just 1 per cent of the mass of the earth (Strigari *et al.*, 2012).

A completely independent way of determining the amount of dark matter halo substructure is to look carefully at the structure of dynamically cold stellar streams. Such streams come from the tidal disruption of small satellite galaxies or globular clusters. In numerical simulations, the streams suffer many tens of impacts from encounters with dark matter substructures of mass 10^5 to $10^7 M_\odot$ during their lifetimes, which create fluctuations in the stream surface density on scales of a few degrees or less. The observed streams contain just such fluctuations (Yoon *et al.*, 2011; Carlberg, 2012; Carlberg *et al.*, 2012; Carlberg and Grillmair, 2013), so they provide strong evidence that the predicted population of subhalos is present in the halos of galaxies like the Milky Way and M31. Comparing additional observations of dynamically cold stellar streams with fully self-consistent simulations will give more detailed information about the substructure population. The Gaia spacecraft's measurements of the positions and motions of vast numbers of Milky Way stars will be helpful in quantifying the nature of dark matter substructure (Ngan and Carlberg, 2014; Feldmann and Spolyar, 2015).

7.7 Conclusions

ΛCDM appears to be extremely successful in predicting the cosmic microwave background and large-scale structure, including the observed distribution of galaxies both nearby and at high redshift. It has therefore become the standard cosmological framework within which to understand cosmological structure formation, and it continues to teach us about galaxy formation and evolution. For example, I used to think that galaxies are pretty smooth, that they generally grow in size as they evolve, and that they are a combination of disks and spheroids. But as I discussed in Section 7.3, HST observations combined with high-resolution hydrodynamic simulations are showing that most star-forming galaxies are very clumpy; that galaxies often undergo compaction, which reduces their radius and greatly increases their central density; and that most lower-mass galaxies are not spheroids or disks but are instead elongated when their centers are dominated by dark matter.

ΛCDM faces challenges on smaller scales. Although starbursts can rapidly drive gas out of the central regions of galaxies and thereby reduce the central dark matter density, it remains to be seen whether this and/or other baryonic physics can explain the observed rotation curves of the entire population of dwarf and low surface brightness (LSB) galaxies. If not, perhaps more complicated physics such as self-interacting dark matter may be needed. But standard ΛCDM appears to be successful in predicting the dark matter halo substructure that is now observed via gravitational lensing and stellar streams, and any alternative theory must do at least as well.

Acknowledgment

My research is supported by grants from NASA, and I am also grateful for access to NASA Advanced Supercomputing and to NERSC supercomputers.

References

Abrams, N. E. and Primack, J. R. 2011. *The New Universe and the Human Future: How a Shared Cosmology Could Transform the World.* Yale University Press.

Alam, S. M. K., Bullock, J. S. and Weinberg, D. H. 2002. Dark Matter Properties and Halo Central Densities. *ApJ*, **572**(June), 34–40.

Allgood, B., Flores, R. A., Primack, J. R. *et al.* 2006. The shape of dark matter haloes: dependence on mass, redshift, radius and formation. *MNRAS*, **367**(Apr.), 1781–96.

Amorisco, N. C. and Evans, N. W. 2012. Dark matter cores and cusps: the case of multiple stellar populations in dwarf spheroidals. *MNRAS*, **419**(Jan.), 184–96.

Barro, G., Faber, S. M., Pérez-González, P. G., *et al.* 2013. CANDELS: The Progenitors of Compact Quiescent Galaxies at $z \sim 2$. *ApJ*, **765**(Mar.), 104.

Barro, G., Faber, S. M., Pérez-González, P. G., *et al.* 2014a. CANDELS+3D-HST: Compact SFGs at $z \sim 2-3$, the Progenitors of the First Quiescent Galaxies. *ApJ*, **791**(Aug.), 52.

Barro, G., Trump, J. R., Koo, D. C., *et al.* 2014b. Keck-I MOSFIRE Spectroscopy of Compact Star-forming Galaxies at $z \gtrsim 2$: High Velocity Dispersions in Progenitors of Compact Quiescent Galaxies. *ApJ*, **795**(Nov.), 145.

Bastidas Fry, A., Governato, F., Pontzen, A., *et al.* 2015. All about baryons: revisiting SIDM predictions at small halo masses. *MNRAS*, **452**(Sept.), 1468–79.

Behroozi, P. S., Wechsler, R. H. and Conroy, C. 2013. On the Lack of Evolution in Galaxy Star Formation Efficiency. *ApJL*, **762**(Jan.), L31.

Benson, A. J. 2010. Galaxy formation theory. *Physics Reports*, **495**(Oct.), 33–86.

Benson, A. J., Frenk, C. S., Baugh, C. M., Cole, S. and Lacey, C. G. 2003. The effects of photoionization on galaxy formation – III. Environmental dependence in the luminosity function. *MNRAS*, **343**(Aug.), 679–91.

Besla, G., Kallivayalil, N., Hernquist, L., *et al.* 2012. The role of dwarf galaxy interactions in shaping the Magellanic System and implications for Magellanic Irregulars. *MNRAS*, **421**(Apr.), 2109–38.

Bird, S., Vogelsberger, M., Haehnelt, M., *et al.* 2014. Damped Lyman α absorbers as a probe of stellar feedback. *MNRAS*, **445**(Dec.), 2313–24.

Bird, S., Haehnelt, M., Neeleman, M., *et al.* 2015. Reproducing the kinematics of damped Lyman α systems. *MNRAS*, **447**(Feb.), 1834–46.

Blumenthal, G. R., Faber, S. M., Primack, J. R. and Rees, M. J. 1984. Formation of galaxies and large-scale structure with cold dark matter. *Nature*, **311**(Oct.), 517–25.

Blumenthal, G. R., Faber, S. M., Flores, R. and Primack, J. R. 1986. Contraction of dark matter galactic halos due to baryonic infall. *ApJ*, **301**(Feb.), 27–34.

Boylan-Kolchin, M., Springel, V., White, S. D. M., Jenkins, A. and Lemson, G. 2009. Resolving cosmic structure formation with the Millennium-II Simulation. *MNRAS*, **398**(Sept.), 1150–64.

Boylan-Kolchin, M., Bullock, J. S. and Kaplinghat, M. 2011. Too big to fail? The puzzling darkness of massive Milky Way subhaloes. *MNRAS*, **415**(July), L40–4.

Boylan-Kolchin, M., Bullock, J. S. and Kaplinghat, M. 2012. The Milky Way's bright satellites as an apparent failure of ΛCDM. *MNRAS*, **422**(May), 1203–18.

Breddels, M. A. and Helmi, A. 2013. Model comparison of the dark matter profiles of Fornax, Sculptor, Carina and Sextans. *Astronomy & Astrophysics*, **558**(Oct.), A35.

Breddels, M. A. and Helmi, A. 2014. Complexity on Dwarf Galaxy Scales: A Bimodal Distribution Function in Sculptor. *ApJL*, **791**(Aug.), L3.

Brennan, R., Pandya, V., Somerville, R. S., *et al.* 2015. Quenching and morphological transformation in semi-analytic models and CANDELS. *MNRAS*, **451**(Aug.), 2933–56.

Brooks, A. 2014. Re-examining astrophysical constraints on the dark matter model. *Annalen der Physik*, **526**(Aug.), 294–308.

Brooks, A. M. and Zolotov, A. 2014. Why Baryons Matter: The Kinematics of Dwarf Spheroidal Satellites. *ApJ*, **786**(May), 87.

Bullock, J. S., Kravtsov, A. V. and Weinberg, D. H. 2000. Reionization and the Abundance of Galactic Satellites. *ApJ*, **539**(Aug.), 517–21.

Bullock, J. S., Kolatt, T. S., Sigad, Y., *et al.* 2001. Profiles of dark haloes: evolution, scatter and environment. *MNRAS*, **321**(Mar.), 559–75.

Bullock, J. S., Stewart, K. R., Kaplinghat, M., Tollerud, E. J. and Wolf, J. 2010. Stealth Galaxies in the Halo of the Milky Way. *ApJ*, **717**(July), 1043–53.

Busha, M. T., Wechsler, R. H., Behroozi, P. S., *et al.* 2011a. Statistics of Satellite Galaxies around Milky-Way-like Hosts. *ApJ*, **743**(Dec.), 117.

Busha, M. T., Marshall, P. J., Wechsler, R. H., Klypin, A. and Primack, J. 2011b. The Mass Distribution and Assembly of the Milky Way from the Properties of the Magellanic Clouds. *ApJ*, **743**(Dec.), 40.

Carlberg, R. G. 2012. Dark Matter Sub-halo Counts via Star Stream Crossings. *ApJ*, **748**(Mar.), 20.

Carlberg, R. G. and Grillmair, C. J. 2013. Gaps in the GD-1 Star Stream. *ApJ*, **768**(May), 171.

Carlberg, R. G., Grillmair, C. J. and Hetherington, N. 2012. The Pal 5 Star Stream Gaps. *ApJ*, **760**(Nov.), 75.

Ceverino, D. and Klypin, A. 2009. The Role of Stellar Feedback in the Formation of Galaxies. *ApJ*, **695**(Apr.), 292–309.

Ceverino, D., Dekel, A., Mandelker, N., *et al.* 2012. Rotational support of giant clumps in high-z disc galaxies. *MNRAS*, **420**(Mar.), 3490–520.

Ceverino, D., Klypin, A., Klimek, E. S., *et al.* 2014. Radiative feedback and the low efficiency of galaxy formation in low-mass haloes at high redshift. *MNRAS*, **442**(Aug.), 1545–59.

Ceverino, D., Dekel, A., Tweed, D. and Primack, J. 2015a. Early formation of massive, compact, spheroidal galaxies with classical profiles by violent disc instability or mergers. *MNRAS*, **447**(Mar.), 3291–10.

Ceverino, D., Primack, J. and Dekel, A. 2015b. Formation of elongated galaxies with low masses at high redshift. *MNRAS*, **453**(Oct.), 408–13.

Chen, J., Koushiappas, S. M. and Zentner, A. R. 2011. The Effects of Halo-to-halo Variation on Substructure Lensing. *ApJ*, **741**(Nov.), 117.

Conroy, C., Wechsler, R. H. and Kravtsov, A. V. 2006. Modeling Luminosity-dependent Galaxy Clustering through Cosmic Time. *ApJ*, **647**(Aug.), 201–14.

Cowie, L. L., Songaila, A., Hu, E. M. and Cohen, J. G. 1996. New Insight on Galaxy Formation and Evolution From Keck Spectroscopy of the Hawaii Deep Fields. *AJ*, **112**(Sept.), 839.

Dalal, N. and Kochanek, C. S. 2002. Direct Detection of Cold Dark Matter Substructure. *ApJ*, **572**(June), 25–33.

Danovich, M., Dekel, A., Hahn, O., Ceverino, D. and Primack, J. 2015. Four phases of angular-momentum buildup in high-z galaxies: from cosmic-web streams through an extended ring to disc and bulge. *MNRAS*, **449**(May), 2087–111.

Dekel, A. and Silk, J. 1986. The origin of dwarf galaxies, cold dark matter, and biased galaxy formation. *ApJ*, **303**(Apr.), 39–55.

Dekel, A. and Woo, J. 2003. Feedback and the fundamental line of low-luminosity low-surface-brightness/dwarf galaxies. *MNRAS*, **344**(Oct.), 1131–44.

Diemand, J., Kuhlen, M. and Madau, P. 2007. Formation and Evolution of Galaxy Dark Matter Halos and Their Substructure. *ApJ*, **667**(Oct.), 859–77.

Diemand, J., Kuhlen, M., Madau, P., *et al.* 2008. Clumps and streams in the local dark matter distribution. *Nature*, **454**(Aug.), 735–8.

Elbert, O. D., Bullock, J. S., Garrison-Kimmel, S., *et al.* 2015. Core formation in dwarf haloes with self-interacting dark matter: no fine-tuning necessary. *MNRAS*, **453**(Oct.), 29–37.

Faucher-Giguère, C.-A., Hopkins, P. F., Kereš, D., *et al.* 2015. Neutral hydrogen in galaxy haloes at the peak of the cosmic star formation history. *MNRAS*, **449**(May), 987–1003.

Feldmann, R. and Spolyar, D. 2015. Detecting dark matter substructures around the Milky Way with Gaia. *MNRAS*, **446**(Jan.), 1000–12.

Flores, R. A. and Primack, J. R. 1994. Observational and theoretical constraints on singular dark matter halos. *ApJL*, **427**(May), L1–4.

Fukugita, M. and Peebles, P. J. E. 2004. The Cosmic Energy Inventory. *ApJ*, **616**(Dec.), 643–68.

Gnedin, O. Y., Ceverino, D., Gnedin, N. Y., *et al.* 2011. Halo Contraction Effect in Hydrodynamic Simulations of Galaxy Formation. *ArXiv e-prints*, Aug.

Governato, F., Brook, C., Mayer, L., *et al.* 2010. Bulgeless dwarf galaxies and dark matter cores from supernova-driven outflows. *Nature*, **463**(Jan.), 203–6.

Governato, F., Zolotov, A., Pontzen, A., *et al.* 2012. Cuspy no more: how outflows affect the central dark matter and baryon distribution in Λ cold dark matter galaxies. *MNRAS*, **422**(May), 1231–40.

Governato, F., Weisz, D., Pontzen, A., *et al.* 2015. Faint dwarfs as a test of DM models: WDM versus CDM. *MNRAS*, **448**(Mar.), 792–803.

Guedes, J., Callegari, S., Madau, P. and Mayer, L. 2011. Forming Realistic Late-type Spirals in a ΛCDM Universe: The Eris Simulation. *ApJ*, **742**(Dec.), 76.

Guo, Y., Giavalisco, M., Ferguson, H. C., Cassata, P. and Koekemoer, A. M. 2012. Multi-wavelength View of Kiloparsec-scale Clumps in Star-forming Galaxies at z ~ 2. *ApJ*, **757**(Oct.), 120.

Guo, Y., Ferguson, H. C., Bell, E. F., *et al.* 2015. Clumpy Galaxies in CANDELS. I. The Definition of UV Clumps and the Fraction of Clumpy Galaxies at 0.5 < z < 3. *ApJ*, **800**(Feb.), 39.

Harvey, D., Massey, R., Kitching, T., Taylor, A. and Tittley, E. 2015. The nongravitational interactions of dark matter in colliding galaxy clusters. *Science*, **347**(Mar.), 1462–65.

Hayward, C. C., Kereš, D., Jonsson, P., *et al.* 2011. What Does a Submillimeter Galaxy Selection Actually Select? The Dependence of Submillimeter Flux Density on Star Formation Rate and Dust Mass. *ApJ*, **743**(Dec.), 159.

Hayward, C. C., Behroozi, P. S., Somerville, R. S., *et al.* 2013. Spatially unassociated galaxies contribute significantly to the blended submillimetre galaxy population: predictions for follow-up observations of ALMA sources. *MNRAS*, **434**(Sept.), 2572–81.

Hearin, A. P. and Watson, D. F. 2013. The dark side of galaxy colour. *MNRAS*, **435**(Oct.), 1313–24.

Hearin, A. P., Watson, D. F., Becker, M. R., *et al.* 2014. The dark side of galaxy colour: evidence from new SDSS measurements of galaxy clustering and lensing. *MNRAS*, **444**(Oct.), 729–43.

Hezaveh, Y., Dalal, N., Holder, G., *et al.* 2013. Dark Matter Substructure Detection Using Spatially Resolved Spectroscopy of Lensed Dusty Galaxies. *ApJ*, **767**(Apr.), 9.

Hezaveh, Y., Dalal, N., Holder, G., *et al.* 2014. Measuring the power spectrum of dark matter substructure using strong gravitational lensing. *JCAP*, **11**(Nov.), 048.

Hlozek, R., Dunkley, J., Addison, G., *et al.* 2012. The Atacama Cosmology Telescope: A Measurement of the Primordial Power Spectrum. *ApJ*, **749**(Apr.), 90.

Hopkins, P. F. 2015. A new class of accurate, mesh-free hydrodynamic simulation methods. *MNRAS*, **450**(June), 53–110.

Hopkins, P. F., Quataert, E. and Murray, N. 2012. Stellar feedback in galaxies and the origin of galaxy-scale winds. *MNRAS*, **421**(Apr.), 3522–37.

Hopkins, P. F., Kereš, D., Oñorbe, J., *et al.* 2014. Galaxies on FIRE (Feedback In Realistic Environments): stellar feedback explains cosmologically inefficient star formation. *MNRAS*, **445**(Nov.), 581–603.

Horiuchi, S., Humphrey, P. J., Oñorbe, J., *et al.* 2014. Sterile neutrino dark matter bounds from galaxies of the Local Group. *Phys. Rev. D*, **89**(2), 025017.

Jardel, J. R. and Gebhardt, K. 2012. The Dark Matter Density Profile of the Fornax Dwarf. *ApJ*, **746**(Feb.), 89.

Jardel, J. R. and Gebhardt, K. 2013. Variations in a Universal Dark Matter Profile for Dwarf Spheroidals. *ApJL*, **775**(Sept.), L30.

Jonsson, P. 2006. SUNRISE: polychromatic dust radiative transfer in arbitrary geometries. *MNRAS*, **372**(Oct.), 2–20.

Jonsson, P. and Primack, J. R. 2010. Accelerating dust temperature calculations with graphics-processing units. *Nature*, **15**(Aug.), 509–14.

Jonsson, P., Cox, T. J., Primack, J. R. and Somerville, R. S. 2006. Simulations of Dust in Interacting Galaxies. I. Dust Attenuation. *ApJ*, **637**(Jan.), 255–68.

Jonsson, P., Groves, B. A. and Cox, T. J. 2010. High-resolution panchromatic spectral models of galaxies including photoionization and dust. *MNRAS*, **403**(Mar.), 17–44.

Keeton, C. R. and Moustakas, L. A. 2009. A New Channel for Detecting Dark Matter Substructure in Galaxies: Gravitational Lens Time Delays. *ApJ*, **699**(July), 1720–31.

Kim, J.-H., Abel, T., Agertz, O., *et al.* 2014. The AGORA High-resolution Galaxy Simulations Comparison Project. *ApJS*, **210**(Jan.), 14.

Kirby, E. N., Martin, C. L. and Finlator, K. 2011. Metals Removed by Outflows from Milky Way Dwarf Spheroidal Galaxies. *ApJL*, **742**(Dec.), L25.

Klypin, A., Kravtsov, A. V., Valenzuela, O. and Prada, F. 1999. Where Are the Missing Galactic Satellites? *ApJ*, **522**(Sept.), 82–92.

Klypin, A. A., Trujillo-Gomez, S. and Primack, J. 2011. Dark Matter Halos in the Standard Cosmological Model: Results from the Bolshoi Simulation. *ApJ*, **740**(Oct.), 102.

Klypin, A., Yepes, G., Gottlober, S., Prada, F. and Hess, S. 2016. MultiDark simulations: the story of dark matter halo concentrations and density profiles. *MNRAS*, **457**, 4340.

Klypin, A., Karachentsev, I., Makarov, D. and Nasonova, O. 2015. Abundance of field galaxies. *MNRAS*, **454**(Dec.), 1798–810.

Kormendy, J. and Fisher, D. B. 2008 (Oct.). Secular Evolution in Disk Galaxies: Pseudobulge Growth and the Formation of Spheroidal Galaxies. In J. G. Funes and E. M. Corsini, eds., *Formation and Evolution of Galaxy Disks*. Astronomical Society of the Pacific Conference Series, vol. 396, p. 297.

Kravtsov, A. 2010. Dark Matter Substructure and Dwarf Galactic Satellites. *Advances in Astronomy*, **2010**, 1–21.

Kravtsov, A. V., Klypin, A. A. and Khokhlov, A. M. 1997. Adaptive Refinement Tree: A New High-Resolution N-Body Code for Cosmological Simulations. *ApJS*, **111**(July), 73–94.

Kravtsov, A. V., Berlind, A. A., Wechsler, R. H., *et al.* 2004. The Dark Side of the Halo Occupation Distribution. *ApJ*, **609**(July), 35–49.

Kuhlen, M., Vogelsberger, M. and Angulo, R. 2012. Numerical simulations of the dark universe: State of the art and the next decade. *Physics of the Dark Universe*, **1**(Nov.), 50–93.

Kuzio de Naray, R. and Spekkens, K. 2011. Do Baryons Alter the Halos of Low Surface Brightness Galaxies? *ApJL*, **741**(Nov.), L29.

Kuzio de Naray, R., Martinez, G. D., Bullock, J. S. and Kaplinghat, M. 2010. The Case Against Warm or Self-Interacting Dark Matter as Explanations for Cores in Low Surface Brightness Galaxies. *ApJL*, **710**(Feb.), L161–6.

Lagos, C. d. P., Crain, R. A., Schaye, J., *et al.* 2015. Molecular hydrogen abundances of galaxies in the EAGLE simulations. *MNRAS*, **452**(Oct.), 3815–37.

Lahav, O., Lilje, P. B., Primack, J. R. and Rees, M. J. 1991. Dynamical effects of the cosmological constant. *MNRAS*, **251**(July), 128–36.

Liu, L., Gerke, B. F., Wechsler, R. H., Behroozi, P. S. and Busha, M. T. 2011. How Common are the Magellanic Clouds? *ApJ*, **733**(May), 62.

Lovell, M. R., Eke, V., Frenk, C. S., *et al.* 2012. The haloes of bright satellite galaxies in a warm dark matter universe. *MNRAS*, **420**(Mar.), 2318–24.

Ma, X., Hopkins, P. F., Faucher-Giguere, C.-A., *et al.* 2015. The Origin and Evolution of the Galaxy Mass-Metallicity Relation. *MNRAS*, **456**, 2140–56.

Macciò, A. V., Dutton, A. A., van den Bosch, F. C., *et al.* 2007. Concentration, spin and shape of dark matter haloes: scatter and the dependence on mass and environment. *MNRAS*, **378**(June), 55–71.

Macciò, A. V., Paduroiu, S., Anderhalden, D., *et al.* 2012a. Cores in warm dark matter haloes: a Catch 22 problem. *MNRAS*, **424**(Aug.), 1105–12.

Macciò, A. V., Stinson, G., Brook, C. B., Wadsley, J., Couchman, H. M. P., Shen, S., Gibson, B. K., and Quinn, T. 2012b. Halo Expansion in Cosmological Hydro Simulations: Toward a Baryonic Solution of the Cusp/Core Problem in Massive Spirals. *ApJL*, **744**(Jan.), L9.

Macciò, A. V., Paduroiu, S., Anderhalden, D., *et al.* 2013. Erratum: Cores in warm dark matter haloes: a Catch 22 problem. *MNRAS*, **428**(Feb.), 3715–16.

Madau, P., Shen, S. and Governato, F. 2014. Dark Matter Heating and Early Core Formation in Dwarf Galaxies. *ApJL*, **789**(July), L17.

Maller, A. H. and Dekel, A. 2002. Towards a resolution of the galactic spin crisis: mergers, feedback and spin segregation. *MNRAS*, **335**(Sept.), 487–98.

Mandelker, N., Dekel, A., Ceverino, D., *et al.* 2014. The population of giant clumps in simulated high-z galaxies: in situ and ex situ migration and survival. *MNRAS*, **443**(Oct.), 3675–702.

Massey, R., Williams, L., Smit, R., *et al.* 2015. The behaviour of dark matter associated with four bright cluster galaxies in the 10 kpc core of Abell 3827. *MNRAS*, **449**(June), 3393–406.

Metcalf, R. B. and Amara, A. 2012. Small-scale structures of dark matter and flux anomalies in quasar gravitational lenses. *MNRAS*, **419**(Feb.), 3414–25.

Metcalf, R. B. and Madau, P. 2001. Compound Gravitational Lensing as a Probe of Dark Matter Substructure within Galaxy Halos. *ApJ*, **563**(Dec.), 9–20.

Metcalf, R. B. and Silk, J. 2007. New Constraints on Macroscopic Compact Objects as Dark Matter Candidates from Gravitational Lensing of Type Ia Supernovae. *Physical Review Letters*, **98**(7), 071302.

Metcalf, R. B. and Zhao, H. 2002. Flux Ratios as a Probe of Dark Substructures in Quadruple-Image Gravitational Lenses. *ApJL*, **567**(Mar.), L5–8.

Moody, C. E., Guo, Y., Mandelker, N., *et al.* 2014. Star formation and clumps in cosmological galaxy simulations with radiation pressure feedback. *MNRAS*, **444**(Oct.), 1389–99.

Moore, B. 1994. Evidence against dissipation-less dark matter from observations of galaxy haloes. *Nature*, **370**(Aug.), 629–31.

Moore, B., Ghigna, S., Governato, F., *et al.* 1999. Dark Matter Substructure within Galactic Halos. *ApJL*, **524**(Oct.), L19–22.

Moustakas, L. A. and Metcalf, R. B. 2003. Detecting dark matter substructure spectroscopically in strong gravitational lenses. *MNRAS*, **339**(Mar.), 607–15.

Navarro, J. F. and Steinmetz, M. 2000. The Core Density of Dark Matter Halos: A Critical Challenge to the ΛCDM Paradigm? *ApJ*, **528**(Jan.), 607–11.

Navarro, J. F., Frenk, C. S. and White, S. D. M. 1996. The Structure of Cold Dark Matter Halos. *ApJ*, **462**(May), 563.

Ngan, W. H. W. and Carlberg, R. G. 2014. Using Gaps in N-body Tidal Streams to Probe Missing Satellites. *ApJ*, **788**(June), 181.

Nierenberg, A. M., Treu, T., Wright, S. A., Fassnacht, C. D. and Auger, M. W. 2014. Detection of substructure with adaptive optics integral field spectroscopy of the gravitational lens B1422+231. *MNRAS*, **442**(Aug.), 2434–45.

Nipoti, C. and Binney, J. 2015. Early flattening of dark matter cusps in dwarf spheroidal galaxies. *MNRAS*, **446**(Jan.), 1820–8.

Oman, K. A., Navarro, J. F., Fattahi, A., *et al.* 2015. The unexpected diversity of dwarf galaxy rotation curves. *MNRAS*, **452**(Oct.), 3650–65.

Oñorbe, J., Boylan-Kolchin, M., Bullock, J. S., *et al.* 2015. Forged in FIRE: cusps, cores, and baryons in low-mass dwarf galaxies. *MNRAS*, **454**, 2092–106.

Papastergis, E., Martin, A. M., Giovanelli, R. and Haynes, M. P. 2011. The Velocity Width Function of Galaxies from the 40% ALFALFA Survey: Shedding Light on the Cold Dark Matter Overabundance Problem. *ApJ*, **739**(Sept.), 38.

Planck Collaboration, Ade, P. A. R., Aghanim, N., *et al.* 2015a. Planck 2015 results. XIII. Cosmological parameters. *A&A*, **594A**, 14.

Planck Collaboration, Ade, P. A. R., Aghanim, N., Arnaud, M., *et al.* 2015b. Planck 2015 results. XVII. Constraints on primordial non-Gaussianity. *A&A*, **594A**, 17.

Polisensky, E. and Ricotti, M. 2011. Constraints on the dark matter particle mass from the number of Milky Way satellites. *Phys. Rev. D*, **83**(4), 043506.

Pontzen, A. and Governato, F. 2012. How supernova feedback turns dark matter cusps into cores. *MNRAS*, Mar., 2641.

Pontzen, A. and Governato, F. 2014. Cold dark matter heats up. *Nature*, **506**(Feb.), 171–8.

Pooley, D., Rappaport, S., Blackburne, J. A., Schechter, P. L. and Wambsganss, J. 2012. X-Ray and Optical Flux Ratio Anomalies in Quadruply Lensed Quasars. II. Mapping the Dark Matter Content in Elliptical Galaxies. *ApJ*, **744**(Jan.), 111.

Porter, L. A., Somerville, R. S., Primack, J. R. *et al.* 2014a. Modelling the ages and metallicities of early-type galaxies in Fundamental Plane space. *MNRAS*, **445**(Dec.), 3092–104.

Porter, L. A., Somerville, R. S., Primack, J. R., and Johansson, P. H. 2014b. Understanding the structural scaling relations of early-type galaxies. *MNRAS*, **444**(Oct.), 942–960.

Prada, F., Klypin, A. A., Cuesta, A. J., Betancort-Rijo, J. E. and Primack, J. 2012. Halo concentrations in the standard Λ cold dark matter cosmology. *MNRAS*, **423**(July), 3018–30.

Primack, J. R. 1984. Dark Matter, Galaxies and Large Scale Structure in the Universe. *Proc. Int. Sch. Phys. Fermi*, **92**, 140.

Primack, J. R. and Abrams, N. E. 2006. *The View from the Center of the Universe: Discovering Our Extraordinary Place in the Cosmos*. New York: Riverhead.

Primack, J. R. and Blumenthal, G. R. 1984. Growth of Perturbations between Horizon Crossing and Matter Dominance: Implications for Galaxy Formation. *Astrophys. Space Sci. Libr.*, **111**, 435–40.

Rahmati, A., Schaye, J., Bower, R. G., *et al.* 2015. The distribution of neutral hydrogen around high-redshift galaxies and quasars in the EAGLE simulation. *MNRAS*, **452**, 2034–56.

Reddick, R. M., Wechsler, R. H., Tinker, J. L. and Behroozi, P. S. 2013. The Connection between Galaxies and Dark Matter Structures in the Local Universe. *ApJ*, **771**(July), 30.

Reddick, R. M., Tinker, J. L., Wechsler, R. H. and Lu, Y. 2014. Cosmological Constraints from Galaxy Clustering and the Mass-to-number Ratio of Galaxy Clusters: Marginalizing over the Physics of Galaxy Formation. *ApJ*, **783**(Mar.), 118.

Richardson, T. and Fairbairn, M. 2014. On the dark matter profile in Sculptor: breaking the β degeneracy with Virial shape parameters. *MNRAS*, **441**(June), 1584–600.

Riebe, K., Partl, A. M., Enke, H., *et al.* 2013. The MultiDark Database: Release of the Bolshoi and MultiDark cosmological simulations. *Astronomische Nachrichten*, **334**(Aug.), 691–708.

Sawala, T., Scannapieco, C. and White, S. 2012. Local Group dwarf galaxies: nature and nurture. *MNRAS*, **420**(Feb.), 1714–30.

Sawala, T., Frenk, C. S., Fattahi, A., *et al.* 2014. Local Group galaxies emerge from the dark. *ArXiv e-prints*, Dec.

Schaye, J., Crain, R. A., Bower, R. G., *et al.* 2015. The EAGLE project: simulating the evolution and assembly of galaxies and their environments. *MNRAS*, **446**(Jan.), 521–54.

Skillman, S. W., Warren, M. S., Turk, M. J., *et al.* 2014. Dark Sky Simulations: Early Data Release. *ArXiv e-prints*, July.

Snyder, G. F., Torrey, P., Lotz, J. M., *et al.* 2015. Galaxy Morphology and Star Formation in the Illustris Simulation at z=0. *MNRAS*, **454**, 1886–908.

Somerville, R. S. 2002. Can Photoionization Squelching Resolve the Substructure Crisis? *ApJL*, **572**(June), L23–6.

Somerville, R. S. and Davé, R. 2015. Physical Models of Galaxy Formation in a Cosmological Framework. *ARAA*, **53**, 51–113.

Somerville, R. S., Gilmore, R. C., Primack, J. R. and Domínguez, A. 2012. Galaxy properties from the ultraviolet to the far-infrared: Λ cold dark matter models confront observations. *MNRAS*, **423**(July), 1992–2015.

Spergel, D. N. and Steinhardt, P. J. 2000. Observational Evidence for Self-Interacting Cold Dark Matter. *Physical Review Letters*, **84**(Apr.), 3760–63.

Springel, V., White, S. D. M., Jenkins, A., *et al.* 2005. Simulations of the formation, evolution and clustering of galaxies and quasars. *Nature*, **435**(June), 629–36.

Springel, V., Wang, J., Vogelsberger, M., *et al.* 2008. The Aquarius Project: the subhaloes of galactic haloes. *MNRAS*, **391**(Dec.), 1685–711.

Strigari, L. E. and Wechsler, R. H. 2012. The Cosmic Abundance of Classical Milky Way Satellites. *ApJ*, **749**(Apr.), 75.

Strigari, L. E., Barnabè, M., Marshall, P. J. and Blandford, R. D. 2012. Nomads of the Galaxy. *MNRAS*, **423**(June), 1856–65.

Tasitsiomi, A., Kravtsov, A. V., Wechsler, R. H. and Primack, J. R. 2004. Modeling Galaxy-Mass Correlations in Dissipationless Simulations. *ApJ*, **614**(Oct.), 533–46.

Teyssier, R., Pontzen, A., Dubois, Y. and Read, J. I. 2013. Cusp-core transformations in dwarf galaxies: observational predictions. *MNRAS*, **429**(Mar.), 3068–78.

The DES Collaboration, Bechtol, K., Drlica-Wagner, A., *et al.* 2015. Eight New Milky Way Companions Discovered in First-Year Dark Energy Survey Data. *ApJ*, **807**(July), 50–66.

Tisserand, P., Le Guillou, L., Afonso, C., *et al.* 2007. Limits on the Macho content of the Galactic Halo from the EROS-2 Survey of the Magellanic Clouds. *Astronomy & Astrophysics*, **469**(July), 387–404.

Tollerud, E. J., Boylan-Kolchin, M., Barton, E. J., Bullock, J. S. and Trinh, C. Q. 2011. Small-scale Structure in the Sloan Digital Sky Survey and ΛCDM: Isolated $\tilde{L}_*$ Galaxies with Bright Satellites. *ApJ*, **738**(Sept.), 102.

Trujillo-Gomez, S., Klypin, A., Primack, J. and Romanowsky, A. J. 2011. Galaxies in ΛCDM with Halo Abundance Matching: Luminosity-Velocity Relation, Baryonic Mass-Velocity Relation, Velocity Function, and Clustering. *ApJ*, **742**(Nov.), 16.

Trujillo-Gomez, S., Klypin, A., Colín, P., *et al.* 2015. Low-mass galaxy assembly in simulations: regulation of early star formation by radiation from massive stars. *MNRAS*, **446**(Jan.), 1140–62.

van der Wel, A., Chang, Y.-Y., Bell, E. F., *et al.* 2014. Geometry of Star-forming Galaxies from SDSS, 3D-HST, and CANDELS. *ApJL*, **792**(Sept.), L6.

Vegetti, S., Koopmans, L. V. E., Bolton, A., Treu, T. and Gavazzi, R. 2010. Detection of a dark substructure through gravitational imaging. *MNRAS*, **408**(Nov.), 1969–81.

Vegetti, S., Lagattuta, D. J., McKean, J. P., *et al.* 2012. Gravitational detection of a low-mass dark satellite galaxy at cosmological distance. *Nature*, **481**(Jan.), 341–3.

Vegetti, S., Koopmans, L. V. E., Auger, M. W., *et al.* 2014. Inference of the cold dark matter substructure mass function at z = 0.2 using strong gravitational lenses. *MNRAS*, **442**(Aug.), 2017–35.

Viel, M., Becker, G. D., Bolton, J. S. and Haehnelt, M. G. 2013. Warm dark matter as a solution to the small scale crisis: New constraints from high redshift Lyman-α forest data. *Phys. Rev. D*, **88**(4), 043502.

Vogelsberger, M., Zavala, J. and Loeb, A. 2012. Subhaloes in self-interacting galactic dark matter haloes. *MNRAS*, **423**(July), 3740–52.

Vogelsberger, M., Genel, S., Springel, V., *et al.* 2014a. Introducing the Illustris Project: simulating the coevolution of dark and visible matter in the Universe. *MNRAS*, **444**(Oct.), 1518–1547.

Vogelsberger, M., Genel, S., Springel, V., *et al.* 2014b. Properties of galaxies reproduced by a hydrodynamic simulation. *Nature*, **509**(May), 177–82.

Wadepuhl, M. and Springel, V. 2011. Satellite galaxies in hydrodynamical simulations of Milky Way-sized galaxies. *MNRAS*, **410**(Jan.), 1975–92.

Walker, M. G. and Peñarrubia, J. 2011. A Method for Measuring (Slopes of) the Mass Profiles of Dwarf Spheroidal Galaxies. *ApJ*, **742**(Nov.), 20.

Watson, D. F., Berlind, A. A. and Zentner, A. R. 2011. A Cosmic Coincidence: The Power-law Galaxy Correlation Function. *ApJ*, **738**(Sept.), 22.

Wellons, S., Torrey, P., Ma, C.-P., *et al.* 2015. The formation of massive, compact galaxies at z = 2 in the Illustris simulation. *MNRAS*, **449**(May), 361–72.

Woo, J., Courteau, S. and Dekel, A. 2008. Scaling relations and the fundamental line of the local group dwarf galaxies. *MNRAS*, **390**(Nov.), 1453–1469.

Wuyts, S., Förster Schreiber, N. M., Nelson, E. J., *et al.* 2013. A CANDELS-3D-HST synergy: Resolved Star Formation Patterns at $0.7 < z < 1.5$. *ApJ*, **779**(Dec.), 135.

Xu, D., Mao, S., Cooper, A. P., *et al.* 2012. On the effects of line-of-sight structures on lensing flux-ratio anomalies in a ΛCDM universe. *MNRAS*, Feb., 2426.

Xu, D., Sluse, D., Gao, L., *et al.* 2015. How well can cold dark matter substructures account for the observed radio flux-ratio anomalies? *MNRAS*, **447**(Mar.), 3189–206.

Yoon, J. H., Johnston, K. V. and Hogg, D. W. 2011. Clumpy Streams from Clumpy Halos: Detecting Missing Satellites with Cold Stellar Structures. *ApJ*, **731**(Apr.), 58.

Zentner, A. R. and Bullock, J. S. 2003. Halo Substructure and the Power Spectrum. *ApJ*, **598**(Nov.), 49–72.

Zolotov, A., Dekel, A., Mandelker, N., *et al.* 2015. Compaction and Quenching of High-z Galaxies in Cosmological Simulations: Blue and Red Nuggets. *MNRAS*, **450**, 2327–53.

Zwaan, M. A., Meyer, M. J. and Staveley-Smith, L. 2010. The velocity function of gas-rich galaxies. *MNRAS*, **403**(Apr.), 1969–77.

8

Formation of Galaxies

JOSEPH SILK

8.1 Introduction

Galaxies are key elements of the universe. They probe cosmology, they control our existence. The broad lines of their formation and evolution are clear. Beginning as infinitesimal density fluctuations, in the early universe, leaving the observed relic pattern of temperature fluctuations on the last scattering surface of the CMB, galaxy halos grew via gravitational instability of cold weakly interacting dark matter, within which baryons dissipated and cooled into the observed galaxies. We are piecing together the missing steps, that involve assembly of massive halos from a hierarchy of merging subhalos. Memory remains in massive halos of the substructure forged by gravity: this has been one of the major revelations to come from computer simulations of structure formation in the expanding universe.

A major advance in demonstrating that galaxies formed via gravitational instability in the early universe came with the discovery of fine-scale angular fluctuations in the CMB. These were predicted as essential relics if galaxies had indeed formed by the conjectured instability. Prior to their discovery, one had no idea of the initial conditions for seeding structure formation. Thermal fluctuations in standard cosmology were known to be too small.

Observations provided the seeds. The breakthrough came with the COBE satellite in 1990. This provided the proof that temperature fluctuations are present and monitor the existence of large-scale density fluctuations. These had little to do, however, with the search for the precursors of galaxies, other than giving confidence that the latter are present under the assumption of a scale-invariant density fluctuation spectrum as advocated by inflationary cosmology.

The key theoretical insight indeed preceded the data. Inflationary cosmology was developed in the 1980s, and provided for the first time a coherent understanding of the size of the universe, its near-Euclidean geometry, and of the origin of the seed density fluctuations from quantum fluctuations at the Planck epoch. Even inflationary cosmology remains incomplete, since there is still no theory of quantum gravity required in order to connect Planck-scale physics rigorously with the Einstein–Friedmann–Lemaître cosmology that successfully describes the evolution of the universe. Such a connection is essential in order

to understand the nature of the acceleration that couples inflation to the current acceleration of the universe, with an associated decrease in vacuum energy of some 120 orders of
magnitude.

It was to take another decade before the search for the elusive fluctuations that seeded
the large-scale galaxy distribution were found. Hints emerged with a series of sub-orbital
experiments, culminating in the detection of fluctuations that probed the horizon angular
scale at matter-radiation decoupling, demonstrating conclusively that acausal fluctuations
were present of sufficient strength to seed galaxy formation. Another important conclusion
was that the geometry of the universe is Euclidean, as effectively measured by the horizon
angular scale. The immediate consequence was that most of the matter in the universe has
to be dark and weakly interacting.

Nor did the mysteries end there. It rapidly became apparent that there was not enough
dark matter to maintain the flatness of space without leading to other anomalies such
as excessive peculiar velocities of galaxies. Finally, the acceleration of the universe was
discovered from the use of type Ia supernovae as standard candles, to provide us with
another challenge, the dominance of dark energy, capable of inducing recent cosmological
acceleration via its constant energy density with equation of state $p = -\rho c^2$.

Dark energy and dark matter represent the infrastructure of modern cosmology. The
nature and low value of the dark energy, that is of the cosmological constant, is a strong
theoretical challenge. Conventional cosmology offers no explanation. The dark matter also
remains an outstanding mystery. Here the challenge is more observational than theoretical,
however. Many experiments world-wide are seeking to find evidence for the elusive weakly
interacting particle that is the favoured dark matter candidate. Should these attempts fail,
and there is already a full programme for the next decade and beyond, one may need to
reconsider the dark matter paradigm.

The building blocks of cosmology are the galaxies. These too contain significant
unknowns. The detailed theory of galaxy formation remains a mystery. Semi-analytical
and fully numerical simulations fail to reproduce both local and early universe observations of galaxies (Silk *et al.*, 2013; Somerville and Davé, 2014). Star formation is regulated
by feedback; if not, star formation occurs too rapidly, too efficiently and too early. However, as more and more detailed physics is put into ever larger simulations, the final result
is inevitably that either star formation occurs too early or too late, because feedback is
either too weak or too strong. Simulations that reproduce the Milky Way galaxy fail disas-
trously at high redshift. Those tuned to high redshift do poorly at reproducing low redshift
star-forming galaxies.

Of course many phenomena can be 'explained': these are considered to be the successes
of the theory of galaxy formation. Phenomena that can be reasonably well understood
include individual morphologies of galaxies and clusters as well as statistical correlations
for large samples. The former includes spheroidal and disk galaxy morphologies and angular momentum, star formation histories, dwarf galaxy profiles, intracluster gas abundances
and pressure, and the ubiquity of intergalactic hydrogen clouds. The latter includes environmental correlations, chemical evolution, cosmic star formation history, the main sequence

of star formation, the correlation between star formation rate and gas surface density, the massive black-hole spheroid velocity dispersion correlation.

These are major accomplishments. Moreover, all are verifiable via formulating predictions to be made with ever larger telescopes at ever higher redshifts, at epochs when galaxies were being assembled. So why the negative attitude? The point is that no single model explains everything. The dynamical range needed is beyond the reach of any foreseeable computers at least in the next decade. Approximations to the input physics are inevitable. All of the interesting astrophysical output lies in the inevitable deviations from any necessarily imperfect canonical model.

8.2 Understanding Opacity-Limited Fragmentation

How stars form is a difficult question to answer, since an incredibly broad dynamic range is needed. Local observations are useful, but offer no guarantee of compliance under the wildly different conditions in the early universe.

The holy grail of star formation reduces to answering the following questions:

- Can we predict the masses of stars?
- Can we predict the efficiency of star formation?
- Can we predict the rate of star formation?

Sadly, in any fundamental sense, the answers are negative on all three counts. But phenomenology is remarkably useful in that it allows us to study these questions in detail in the nearby universe. Armed with this experience we can try to tackle the more extreme conditions in the distant universe when galaxies were young.

Let us begin from the beginning, at least as is appropriate for the formation of stars. The story begins with Newton who wrote

if the matter was evenly disposed throughout an infinite space, it could never convene into one mass; but some of it would convene into one mass and some into another, so as to make an infinite number of great masses, scattered at great distances from one to another throughout all that infinite space. And thus might the sun and fixed stars be formed.

(Isaac Newton, letter to Richard Bentley, 10 December, 1692).

This is essentially the notion of gravitational instability. Newton abandoned his argument however, when he realised that

If the sun at rest were an opaque body like the planets or the planets lucid bodies like the sun, how he alone should be changed into a shining body whilst all they continue opaque, or all they be changed into opaque ones whilst he remains unchanged, I do not think explicable by mere natural causes, but am forced to ascribe it to the counsel and contrivance of a voluntary Agent.

It was James Jeans, some two centuries later, who first quantified gravitational instability in 1902:

We have found that as Newton first conjectured. All celestial bodies originate by a process of fragmentation of nebulae out of chaos, of stars out of nebulae, of planets out of stars and satellites out of planets. From the intrinsic evidence of his creation, the Great Architect of the Universe now begins to appear as a pure mathematician.

The Jeans criterion for gravitational stability is based on the concept that pressure gradients oppose collapse: sound waves must cross regions to communicate pressure changes to inhibit collapse.

Consider a molecular cloud that is unstable to gravitational fragmentation. Under what conditions will it naturally fragment into stellar mass clumps? Let me first attempt to make an *a priori* estimate. The goal is to write the typical mass of a star in terms of fundamental units. Can one predict the mass of a star from fundamental physics arguments?

I start with a cold interstellar gas cloud, the typical nursery of star formation. I assume the cloud is spherical and uniform. This may seem to adopt what has been denigratingly described as the spherical cow approximation, but it suffices for estimates of the relevant orders of magnitude of physical quantities.

One can first derive the collapse time of a collapsing cold gas sphere of mass M where the gas density satisfies $\rho(r) = \sigma^2/(2\pi G r^2)$, and the velocity dispersion σ is equivalent to the kinetic temperature T and the square of the sound speed v_s^2 by writing $\sigma^2 = kT/m_p = v_s^2$, as in a self-gravitating isothermal sphere. Some useful scaling relations are radius $R = v_s t_{dyn} = M^{1/3}\rho^{-1/3} = GMv_s^{-2}$, and density $\rho = M/R^3 = v_s^6 M^{-2} G^{-3}$. Here t_{dyn} is the free-fall time.

As the sphere collapses, let me suppose that it cools freely. The cloud stays approximately isothermal. A cold cloud is subject to Jeans gravitational instability, hence it fragments. As the density increases, the Jeans mass decreases. This means that the fragments subfragment into ever smaller masses. The process stops only when a fragment is dense enough to become opaque. It can no longer radiate freely. It is said to be optically thick to radiation. The temperature rises and the Jeans scale no longer decreases. Fragmentation stops.

The luminosity of an opaque fragment satisfies

$$L_{rad} = \sigma 4\pi T^4 R^2 = \sigma 4\pi T^4 G^2 M^2 v_s^{-4}$$

The fragments now accrete surrounding diffuse matter and grow in mass. The increasing gravity field of a fragment means that it slowly contracts in order to support its growing mass. We call the central region the protostellar core, since it is destined to become hot enough to be the core of a star.

One can now derive the mass limit from opacity-limited fragmentation. The accretion rate of the surrounding envelope onto a protostellar core controls the rate of gravitational energy release. The gravitational energy release from collapse is

$$L_g = GM^2 R^{-1} t_{dyn}^{-1} = GM^2 R^{-2} v_s = v_s^5 G^{-1}$$

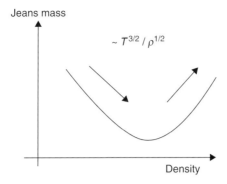

Figure 8.1 Sketch of Jeans mass versus density., for a series of isothermal clouds of increasing density. The minimal Jeans mass sets the opacity limit of fragmentation. This is generically found to be about $0.01 M_\odot$.

Hence following the derivation by Rees (1976), one may equate the luminosity L_{rad} with the rate of release of gravitational energy L_g, so that $\sigma 4\pi T^4 G^2 M^2 v_s^{-4} = v_s^5 G^{-1}$. This leads to an expression for the typical mass of a fragment,

$$M = v_s^{9/2} T^{-2} G^{-3/2} (\sigma 4\pi)^{-1/2} = const. T^{1/4}$$

Our goal is to express this result in terms of fundamental physics parameters. To convert to fundamental units as first done in Low and Lynden-Bell (1976), I use $\sigma = (2\pi^5/15)k^4 h^{-3} c^{-2}$ and write T dimensionlessly as $kT/m_p c^2$ to obtain $M/m_p = \alpha_g^{-3/2}\alpha^{-1}(kT/m_p c^2)^{1/4}$. Also, I write T in units of Rydbergs, 1 Ry $= \alpha^2 m_e c^2/2$, and obtain $M = \alpha_g^{-3/2}\alpha^{-1}(m_e/m_p)^{1/4}(kT/1\mathrm{Ry})^{1/4}$. Note that I have introduced the gravitational fine-structure constant $\alpha_g = Gm_p^2/e^2 = 7.5 \times 10^{-37}$. This constant represents the ratio between gravitational and electromagnetic forces between a pair of protons. It is fundamental to understanding what forces determine the mass of a star. Any fundamental stellar mass, including the Chandrasekhar mass or the Eddington mass, can be expressed in terms of $\alpha_g^{-3/2}$ in combination with the fine-structure constant.

Attainable temperatures are 10K in nearby molecular cloud cores and 1000K in primordial clouds where there are only trace amounts of H_2 as coolants. The sound speed and accretion rate vary considerably over this temperature range. However one always finds that the minimum mass limited by opacity is $M \sim 0.01 M_\odot$. This is true for primordial star formation with cooling via H_2 (Palla *et al.*, 1983) as well as via dust (Omukai *et al.*, 2005). In other words, the minimum mass of a fragment is limited to be around $0.01 M_\odot$, or well below the mass of even the smallest star.

In the primordial case, the parent clouds are typically in the mass range $10^6 - 10^7 M_\odot$ and have trace amounts of molecular hydrogen. Radiative trapping raises the minimum fragment mass in the case of more massive primordial clouds that are purely atomic and cool by Lyman alpha emission (Latif *et al.*, 2011). But it still stays small and remarkably close to the opacity-limited fragmentation value.

Fundamental theory applied to a diffuse interstellar cloud that is collapsing under self-gravity yields a minimum and small fragment mass. This is a robust but wrong result in terms of star formation! Such a result at first sight seems irrelevant to stars. In fact, it provides the building blocks of stars. The derived scale is central to star formation theory, and vindicated by numerical simulations. The point is that additional physics is needed, as we see below, to build on these fragments. The missing physics is that of fragment growth by accretion.

8.3 The First Stars

To build up the stellar mass function, from fragments to stars, the crucial ingredient is supplementing fragmentation by continuing accretion of cold gas. To avoid making only massive stars, however, one has to limit accretion. This is achieved by feedback that taps stellar energy via magnetic turbulence and outflows for low mass stars and ionisation fronts (HII regions) and winds for massive stars. The accretion rate $c_s^3/G \propto T^{3/2}/G$ for a temperature of around 10K today for nearby star-forming cold clouds is $\sim 10^{-6} M_\odot$/year. However long ago, in the absence of heavy elements, the sound speed was high, up to ~ 10km/s at $T \sim 10^4$K and the accretion rate exceeded $\sim 10^{-3} M_\odot$/year. This means that the first stars were relatively massive. In fact there is feedback from the forming massive stars, due to ionisation fronts and outflows. The net result is that there is a wide range of stellar masses with a relatively flat mass function, for the first generation of stars (Hirano *et al.*, 2015).

If the central accretion rate is sufficiently high, there is little time for the infalling gas to fragment into stars before a central star forms that rapidly undergoes direct collapse into a black hole (Inayoshi and Haiman, 2014). There is no time in this case for radiative feedback from the central object to halt the collapse.

8.4 The Typical Mass of a Star

However Jeans did not address the question posed by Newton as to what distinguishes planets from stars. The physics that differentiates stars from planets was first laid down by Arthur Eddington. He was the first to understand Newton's dilemma, as to how gravitational instability and fragmentation could distinguish opaque objects that we now know shine by reflected sunlight (planets) from intrinsically luminous objects (stars). Eddington wrote

We can imagine a physicist on a cloud-bound planet who has never heard tell of the stars. (She) calculates the ratio of radiation pressure to gas pressure for a series of globes of gas of various sizes, starting, say, with a globe of mass 10 gm, then 10^2 gm, 10^3gm, and so on, so that (her) nth globe contains 10^ngm. Regarded as a tussle between gas pressure and radiation pressure, the contest is overwhelmingly one sided except between Nos 33 to 35, where we may expect something interesting to happen. What happens is the stars. We draw aside the veil of clouds beneath which our physicist has been working and let (her) look up at the sky. There she will find a thousand million globes of gas nearly all between 1.2 and 59 times the sun's mass.

(Sir Arthur Eddington, *The Internal Constitution of the Stars*, 1926).

The step from planets to stars involves the force of gravity requiring enough heat under the enormous central pressure that the object shines. Indeed the critical step is that it maintains a source of central heat by thermonuclear reactions, otherwise stars would be too short-lived. Remarkably, when Eddington deduced the main sequence range of stars he did not know about the burning of hydrogen to helium as the central energy supply for sun-like stars. This only came decades later with the work of nuclear physicist Hans Bethe.

To understand Eddington's conclusions better, one can cast the minimum mass of a star in terms of fundamental units. I review and reassess the arguments of Krumholz (2011), in which he re-evaluates the gravitational fragmentation mass for a protostar by using accretion to set the protostellar luminosity, and the onset of deuterium burning to set the central temperature. He argues that approximately half of the Bonnor–Ebert mass forms the protostar, namely

$$M_* \approx 0.6 \sqrt{\left(\frac{k_b T_e}{\mu_{H_2} M_H G}\right)^3 \frac{1}{\rho_e}}$$

Here T_e is the gas temperature at the edge of the gas accreting to form the star, and ρ_e is the local gas density, averaged over the collapsing region. He fixes T_e from the protostellar luminosity assuming dust opacity dominates, and obtains the luminosity from the energy released by accretion onto the central protostar. The scalings are $L \propto T_c \dot{M}_* \propto M T_c \rho_e^{1/2}$ and the central temperature T_c is fixed by deuterium burning to be constant (in terms of fundamental constants, it is proportional to the Gamow energy, E_G, for the D burning reaction, itself $\propto \alpha^2 m_H c^2$). The temperature T_e is determined by the central luminosity and, for the case of fiducial Milky Way-type dust cooling in a $n = 3/2$ polytrope, scales as $(\Sigma L/M)^{1/4}$. The interstellar pressure enters via the assumption that surface density (more physically, dust opacity) is constant, and consequently $p_{ism} \propto \Sigma^2$.

The resultant scaling for stellar mass is $M_* \propto \alpha_g^{-3/2} \alpha^{-4/3} p^{-1/18} \Theta_c^{-4/3}$, where $\Theta_c = (E_G/4kT_c)^{1/3} \sim 10$ and $T_c \approx 10^6$K at the onset of the deuterium burning that determines the onset of the formation of the star. Krumholz in fact normalises the interstellar pressure to the Planck pressure $c^7/\hbar G^2$ and thereby introduces what is a misleading dependence of the characteristic stellar mass on the gravitational fine structure parameter, Gm_H^2/e^2, obtaining

$$M_* = A_K m_H \left(\frac{\alpha^{16}}{\alpha_G^{25}}\right)^{1/18} \left(\frac{p}{p_{Planck}}\right)^{-1/18},$$

where A_k is a dimensionless constant.

In fact, it is more appropriate to normalise the temperature to natural atomic units, namely to a Rydberg $kT_{ryd} = \alpha^2 m_e c^2/2$. After all, quantum gravity has little relevance for star formation. Only atomic or molecular processes, specifically Lyman alpha emission or molecular H_2 rotational excitations, are effective for primordial star formation, with $T \sim 1000$ K, and of course in the present epoch interstellar medium, star formation occurs at $T \sim 10$ K. The pressure at the boundary of the protostar can be written to within dimensionless factors of order unity as $p = \rho kT/m_p = v_s^8/G^3M^2$. For purposes of normalisation, I use a fiducial value $v_s = \alpha c$, equivalent to the Rydberg energy kT_{Ryd} of an electron. Now rewriting the expression for the characteristic mass, I find that

$$M_* = A_S m_H \alpha_g^{-3/2} \alpha^{-4/3} \left(\frac{kT_c}{m_p c^2}\right)^{4/9} \left(\frac{c^8 m_p^4}{e^6 p}\right),$$

where A_S is a dimensionless factor of order unity. After inserting the previous expression for the pressure in terms of v_s, this reduces to

$$M_* = A_S m_H \alpha_g^{-3/2} \alpha^{-2} \left(\frac{m_p}{m_e}\right)^{1/4} \left(\frac{kT_c}{m_p c^2}\right)^{1/2} \left(\frac{T_{Ryd}}{T}\right)^{1/4}$$

Krumholz's conclusion, as corrected here, is essentially unchanged: the only dependence of M_* on astrophysical parameters is via the interstellar pressure, and the dependence is exceedingly weak. There is still an explicit dependence on the deuterium burning temperature. However, there is no longer any scaling with ambient pressure, rather with ambient temperature. Note that the temperature scaling is the inverse of that found for opacity-limited fragmentation. The scaling in fundamental units gives a mass scale of around a solar mass. This is the typical mass of a star in today's universe.

In the early universe, conditions were very different. There were no heavy elements, since the stars that produce and disperse them had not yet formed. This meant that in the absence of the cooling transitions from low-lying fine-structure states of atoms and ions such as carbon and silicon, the only coolant was hydrogen. Hence the temperature was relatively high, thousands of degrees Kelvin rather than tens of degrees Kelvin.

Since the sound speed was high, accretion rates were high. Massive stars predominated. Trace amounts of molecular hydrogen allowed temperatures of order 1000K. This was optimal for massive stars to form. However if the molecular hydrogen, a relatively fragile component of the gas, was destroyed, cooling by atomic hydrogen only occurred at a temperature of order 10,000K. Indeed if the temperature was this high, as expected from atomic hydrogen cooling, the enhanced accretion rate $\sim v_s^3/G$, led to the optimal situation for black hole formation. In such a situation, ultraviolet radiation, produced by the first massive stars, maintains the suppression of any trace amounts of molecular hydrogen and cooling to lower temperatures cannot occur. One needs some cooling of course, in order for the gas to collapse to high density under the action of self-gravity, and this is supplied by Lyman alpha emission. The resulting huge accretion rate quenches any fragmentation, and the outcome is likely to be direct collapse to a black hole, and possibly a massive black hole.

While such objects formed long ago, there are various ways one can probe their existence. For example the short-lived massive stars of the now extinct Population III leave nucleosynthetic tracers in the atmospheres of the oldest low mass members of Population II. These stars can be seen in the halo by virtue of their low metallicities, rare stars being found with an iron content lower than that of the Sun by 10^6 or more. Another characteristic is their nebular emission: intense lines of HeII are predicted with no accompanying emission by heavier ions such as [OII]. Tentative evidence for such objects has recently been reported (Heap *et al.*, 2015; Sobral *et al.*, 2015). The first black holes could provide the required seeds for the growth of supermassive black holes seen as quasars in the early universe. The need for such seeds is inferred from the high black hole masses, in excess of $\sim 10^9 M_\odot$, inferred at very high redshift when there is inadequate time for Eddington-limited growth. An alternative solution may appeal to a duty cycle with many intense but short-lived periods of super-Eddington growth.

Population III stars and black hole seeds may also contribute to the ionising photon budget responsible for reionisation of the universe. While dwarf galaxies are often considered to be the dominant contributor (Robertson *et al.*, 2015), considerable doubt exists as to whether the escape fraction of ionising photons is high enough, according to simulations that include gas infall via cosmological accretion (Ma *et al.*, 2015). Nor is it likely from diffuse X-ray background constraints that black hole seeds, acting as mini-quasars, can account for more than ~ 10 per cent of the ionising photon budget (Haardt and Salvaterra, 2015).

Perhaps the ultimate solution is the combination of mini-quasars, essentially generated by accretion onto black hole seeds of mass $10^3 - 10^5 M_\odot$, that drive cavities in the accreting gas, combined with bursts of star formation from the associated dwarf galaxies.

Indeed there is increasing evidence that dwarfs contain central massive black holes. The AGN fraction may be as large as ~ 10 per cent for the lowest mass nearby infrared-detected dwarfs (Satyapal *et al.*, 2014) and ~ 1 per cent if optically selected (Moran *et al.*, 2014) with active AGN found for black holes in the mass range $10^4 - 10^6 M_\odot$, infrared excesses selecting the lowest mass MBH. X-ray selected dwarfs are rarer, with central massive BH candidates amounting to $\sim 0.1 - 1$ per cent (Lemons *et al.*, 2015; Pardo *et al.*, 2016).

8.5 Modelling the Galaxy Luminosity Function

The initial condition problem plagued all early studies of galaxy formation: pioneers included Lemaître, Gamow, Harrison, Peebles, Zel'dovich, Ozernoy and others. This was first solved schematically with COBE in 1992, in the sense that ultra-large large-scale fluctuations were detected that had no direct match in observed features in the galaxy distribution, but were nevertheless a compelling indicator of the presence of the nearly scale-invariant fluctuations needed for large-scale structure formation. The definitive all-sky detection of the elusive fossil seeds awaited the WMAP/PLANCK satellites some two decades later, when the precursors of galaxy clusters were directly mapped as infinitesimal cosmic microwave background radiation fluctuations.

The linear theory of density fluctuation growth in the expanding universe is well justified by virtue of the measured temperature fluctuations in the CMB. These provide the elusive initial conditions for structure formation, with only a modest extrapolation required from cluster down to galaxy scales. Scale invariance and Gaussianity are usually invoked to make this last step unique. These are effectively measured in the CMB fluctuations over a broad range of scales. The fully non-linear endpoint is the mass function of galaxies, calculated as the assembly of newly non-linear objects in the rare peaks of the Gaussian tail that move above a certain threshold as the fluctuations grow in amplitude by gravitational instability.

It is notoriously difficult to transform the mass function of galaxies, understood from theory, into the observed luminosity or stellar mass function. This is because of the complexity of baryonic physics. The subdominant baryons, initially tracing the cold dark matter, dissipate energy via atomic cooling and cool to form dense self-gravitating clouds that fragment into stars. However, neither the efficiency of star formation nor the initial mass function of the stars nor the rate of star formation can be predicted in any fundamental way. So we cannot easily predict the stellar mass function of galaxies if we fail at the first hurdle, namely that of forming the stars.

What we do understand is the mass function of dark matter halos. Moreover while the ratio of dark matter to gravitationally bound baryons is known, the mass fraction in stars is unknown, *a priori*. So one simple starting point is to renormalise the dark matter mass function to fit the observed galaxy luminosity function. We can then calculate the ratio of dark to luminous matter required for the two functions to overlap. In fact they overlap at only one point. Fortunately this point is well defined, as the mean galaxy luminosity and the mean dark mass, both mass and luminosity functions being fully convergent. What we infer is M_*/L_*, where L_* is the Schechter luminosity of about $3 \times 10^{10} M_\odot$, and M_* is the required dark mass corresponding to the halo of a galaxy of luminosity L_* in order for the mass and luminosity functions to overlap at their mean values. We can measure this ratio observationally: it is approximately 30 in solar units $M_\odot$ and $L_\odot$.

Remarkably, we can calculate M_* from first principles. A necessary condition for star formation in a protogalaxy is gas cooling. This does not guarantee star formation unless self-gravity also plays a role. It is instructive to compare the two key timescales, the free-fall time with the cooling time for an isothermal sphere. In this way one can estimate the maximum mass for a galaxy, or more realistically a characteristic mass (Binney, 1977; Rees and Ostriker, 1977; Silk, 1977).

In order to describe how the precursor cloud destined to form a galaxy can contract under gravity I need to introduce the atomic cooling function for a primordial plasma of hydrogen and helium, written as $\Lambda_H = \alpha \sigma_T c^2 v_c^{-1} E_\gamma \left(v_c/v_{ref}\right)^{2\beta}$, where v_c is the circular velocity of a test particle in the galaxy at the half-mass radius, $v_{ref} = \alpha c \left(m_e/m_p\right)^{1/2}$ and $E_\gamma = \alpha^2 c^2 m_e$. Here β is a function of temperature, where $\beta = 0.5$ describes bremsstrahlung (e.g. braking radiation) and $\beta = -0.5$ approximates bound-free cooling of hydrogen. This is a very simplified formulation and more accurate cooling rates are given by Sutherland

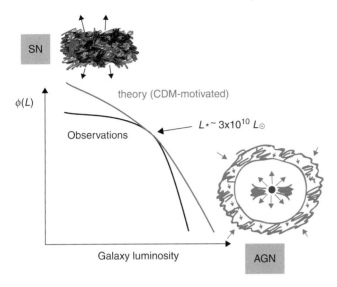

Figure 8.2 Comparison of observed galaxy luminosity function and predicted galaxy mass function in a cold dark matter-dominated universe with initial conditions set by generic inflationary cosmology. The mass function has been renormalised to match the luminosity function at the mean mass (and luminosity) of a galaxy. The inferred ratio of cooled baryonic (stellar) mass to luminosity is approximately 30, as expected for an old stellar population.

and Dopita (1993), with an update including on-line tables by Gnat and Sternberg (2007). In fact, for a hydrogen–helium plasma at $10^5 - 10^7$ K, the relevant temperature for typical galaxy halos, we can bypass these details and use $\beta \approx -0.5$ as a reasonable approximation. I set $t_c = 3kT/(2\Lambda_H n)$ and $t_d = GM_g/(2v_c^3)$. Here M_g is the dynamical mass. With a slightly more general parametrisation of the cooling function, the characteristic galaxy mass is now

$$M_g = \frac{\alpha}{\alpha_g^2} \left(\frac{m_p}{m_e}\right)^{3/2} \left(\frac{v_s}{c}\right)^2 \frac{t_c}{t_d}$$

Finally let us require the cooling time to be less than the dynamical time as a necessary (if not sufficient) condition for galaxy formation. The magic mass, the typical mass of a galaxy, is $\sim 3 \times 10^{11} M_\odot$, and is obtained from simple fundamental physics reasoning, namely that a necessary condition to form a galaxy is that a protogalactic cloud cool within a gravitational collapse time. Of course this may not be a sufficient condition, but exploring this caveat would take us into the complex and messy area of baryonic feedback. For an old stellar population, this stellar mass gives the mean galaxy luminosity, or the Schechter luminosity, in visible light of $\sim 3 \times 10^{10} L_\odot$. This is true for H plus He cooling, but remains more or less the case even when metal cooling is included.

In fact, gas accretion continues over a Hubble time unless feedback and/or environmental effects intervene. To explore this regime, I replace the galaxy dynamical time in the previous derivation by the age of the universe, i.e. set gas cooling time equal to Hubble time. This provides the maximum mass of a galaxy. The additional factor is $t_H/t_d = (12\pi G\rho/H_0^2)^{1/2} = (\frac{9}{2}\rho_g/\bar{\rho}(z))^{1/2} \approx 25$. This would result in an excess of the most massive galaxies. Feedback must intervene, and this is generally accepted to be via quasar (supermassive black hole outflow)-driven heating of the intergalactic medium.

8.6 Massive Black Holes and Quasars

Quasars are powered by accretion onto supermassive black holes and are detected about as far back in time as we can observe galaxies, that is to a redshift of 7 or 8. In fact, the first galaxies are in place somewhat earlier, but growth both for galaxies and quasars peaks much later, at $z \sim 2$. We measure the growth of the stellar component of galaxies by the star formation rate, and the accretion rate onto black holes by the X-ray luminosity of quasars.

We observe quasars powered by black holes of mass up to $\sim 10^{10} M_\odot$ at high redshift, but we do not understand how such massive black holes grow so rapidly in the early universe. The presence of even a few billion solar mass MBHs at high redshift challenges conventional Eddington-limited growth models for MBHs.

Assuming Eddington-limited accretion, a black hole mass grows as

$$M = M_0 \exp\{[(1-\epsilon)/\epsilon](t/t_s)\},$$

where the Salpeter timescale $t_s = \epsilon\sigma_T c/(4\pi G m_p) = 0.45\,\epsilon$ Gyr, and ϵ is the radiative efficiency, normally related to black hole spin in radiatively efficient accretion disks, ranging from $\epsilon = 0.057$ to $\epsilon = 0.32$ for spin parameters ranging from 0 to 0.998. Given the age of the Universe at $z = 6 - 7$ and the estimated MBH masses, $> 10^9 M_\odot$, even with quasi-continuous Eddington-limited accretion, we would require seeds of $10^3 < M_0 < 10^5 M_\odot$. Such seeds are not easily found.

The causal connection between massive black hole growth and associated quasar activity and galaxy growth by star formation is unknown, although co-evolution of massive black holes and galaxies certainly occurs as evidenced by the similar cosmic history of star formation rate and black hole accretion rate in Madau and Dickinson (2014). The evidence for co-evolution, and most likely self-regulation, comes from the interpretation of the empirical correlation between black hole mass and spheroid mass, more specifically the velocity dispersion of spheroid stars which is effectively a proxy for the mass.

To examine co-evolution, and in particular the competition between accretion-driven black hole growth and black hole-powered outflows, I begin with the Eddington limit

$$L_{Edd} = 4\pi cGM_{BH}\frac{m_p}{\sigma_T},$$

and set the corresponding mechanical energy flux $\frac{1}{2}\dot{M}_{Edd}v_w^2 = \eta L_{Edd}/2$, where η is the radiative efficiency of the black hole, $L/(\dot{M}_{BH}c^2)$. Note that $v_w = \eta c$, if $L = L_{Edd}$, as observed in quasar broad emission line regions for $\eta \approx 0.1$, as expected for canonical values of black hole spin. Now balance the outward energy with that of the gravitational energy flux of bound and accreting gas, $\sim 16 f_g \sigma^5/G$, in a marginally escaping shell, where σ is the gas velocity dispersion and the escape velocity for an isothermal distribution is taken to be 2σ.

If the black hole is massive enough that the Eddington energy release exceeds that of the infalling gas, the gas reservoir is ejected, star formation is quenched, and the black hole stops growing. In the case of energy feedback, as would apply for a quasar wind, I find that (Silk and Rees, 1998)

$$M_{BH} = f_g \frac{\alpha \sigma^5}{\pi \eta c G^2} \frac{\sigma_T}{m_p},$$

where $\alpha \approx 8$. This would correspond to very strong feedback, that however neglects energy loss by cooling.

Let us consider momentum feedback, as would apply for a radiatively driven outflow. The radiative momentum flux is L/c. This is mostly trapped if photons undergo single scatterings. I compare this with the required momentum flux to eject the gas that has not yet formed stars. The momentum flux of the gas-rich protospheroid, if in a shell at radius r ejected at velocity σ, is $4 f_g \sigma^4/G$. Identify L with the Eddington luminosity and one has

$$M_{BH} = f_g \frac{\sigma^4}{\pi G^2} \frac{\sigma_T}{m_p}$$

Momentum feedback fails by a factor ~ 10 to give the required normalisation of the $M_{BH} - \sigma$ relation, whereas energy feedback works for a momentum boost factor inferred from energy conservation ($E_{outflow} \propto 1/v_{shock}$) provided that the thermalisation of the kinetic energy of the central ultrafast outflows is effective over scales from that of the subparsec accretion disk surrounding the central SMBH, where the outflow originates, to galactic scales. Observational evidence supports this connection (Feruglio *et al.*, 2015; Tombesi *et al.*, 2015).

8.7 Feedback Can Be Positive

The fundamental recipe for star formation, involving a density threshold, has had a certain amount of success. Sub-grid formulations of star formation are essential, given the broad dynamic range needed to tackle cosmological volumes. The Kennicutt–Schmidt law is treated as the gold standard for testing sub-grid models of star formation in cosmological simulations, as in Hopkins *et al.* (2014). The observations can be matched for a wide range of galaxy masses and surface densities. Control of the gas supply plays a crucial role, since the gas reservoir, accreted or *in situ*, controls star formation.

It is generally argued that quenching of star formation is needed for massive galaxies. Many of these are red and dead, in the sense that there is little ongoing star formation. Yet the ongoing accretion of cold dark matter and accompanying gas that seems inevitable in typical environments should result in star formation.

There are three basic ingredients to understanding the rate of galaxy formation: gas accretion, gas outflow and star formation. These are likely to be interconnected and even self-regulated. One resolution is to argue that jets from radio galaxies heat up the circumgalactic medium sufficiently so that gas infall and cooling are inhibited. On smaller scales and at earlier times, vigorous winds from quasars eject large amounts of gas beyond the escape velocity from the galaxy. Ejection is possible for sufficiently massive black holes. Hence black hole growth plays an important role. Star formation is saturated once the gas supply is inhibited. This generates the correlation between black hole and spheroid mass and also, if outflow rate balances accretion rate approximately, accounts for the roughly 50 per cent shortfall of baryons in massive galaxies.

Yet all is not well. For example, state-of-the-art cosmological simulations fail to reproduce enough starbursts (Sparre *et al.*, 2015). It is argued that elevated density thresholds and turbulence suppress star formation, as observed in some giant molecular clouds (Rathborne *et al.*, 2014). Unfortunately, in more extreme situations of density and turbulence, star formation is actually enhanced, as for GMCs undergoing nuclear starbursts (Leroy *et al.*, 2015).

One solution may be to argue that feedback is not necessarily always negative. Especially if the massive black holes form first, they can initially trigger star formation. Here is how this might work.

A vigorous outflow is normally disruptive with respect to star formation. The gas reservoir for star formation is disrupted. Cold gas is entrained in the outflow and ejected. Molecular clouds are disrupted by Kelvin–Helmholtz instabilities and evaporation. These effects are seen in simulations. On the observational side, outflows of cold gas that are comparable to or exceed the star formation rate are frequently detected, and attributed to the impact of the AGN wind and/or jet. All of this amounts to negative feedback.

However, if molecular clouds are sufficiently massive, they will be compressed rather than disrupted as the outflow sweeps around them. Compression leads to collapse and star formation. Moreover since the outflows are observed at thousands of kilometres per second, one has the potential of stimulating more coherent and intense star formation than would be achievable by instability of a self-gravitating gas-rich galactic disk. The enhancement in star formation rate should be of the order $v_{outflow}/v_{circ}$, where $v_{outflow}$ is the outflow velocity and v_{circ} is the rotation velocity. This effect could provide the needed boost in star formation.

The usual argument appeals to galaxy mergers, in order to explain the ultra-luminous starbursts. However, mergers only can augment the gas reservoir available for star formation. Another ingredient is needed to account for enhanced conversion of gas into stars. What remains to be determined is whether this additional physics input is no more than the merger-driven enhanced gas turbulence or might involve the output from the inevitably present AGN, itself undoubtedly provoked by enhanced gas feeding as a consequence of the merger.

8.8 Outstanding Questions

We do not yet have a complete theory of galaxy formation. There are more questions than answers. Observations are ahead of theory, which is hard pressed to confront the latest observational discoveries. I conclude with a list of outstanding questions in galaxy formation theory.

- **Are stars inevitable?**
  This is essential from the fundamental physics perspective. We would like to know, for example, given the six parameters that fit the initial conditions of cosmology and the large-scale structure of the universe, as demonstrated by the Planck satellite, whether stars inevitably form. This is the essence of anthropic arguments that seek to explain the cosmological parameters.
- **Does nearby star formation provide a robust template for first or even second generation star formation?**
  Conditions were very different in the early universe. The systems were gas-rich, there was more turbulence, and the assembled halo and baryonic masses were lower. We have the advantage of exquisite angular resolution in developing a theory of local star formation. Even here we are still far from the ultimate theory. Theorists commonly apply the local rules to modelling the distant, more violent universe.
- **Are galaxies inevitable?**
  Galaxies are where stars form. There is a characteristic mass for a galaxy; both the dwarfs and the giants contribute relatively little, or at least are subdominant, in terms of the stellar mass budget. But the smallest galaxies provide clues on how galaxies formed hierarchically. Their number is important, for it controls the ionisation of the universe.
- **Does the physics that controls the global evolution of galaxies, including mass assembly, star formation rates and gas accretion in nearby galaxies, apply in the early universe?**
  We observe exceptionally large star formation rates in the early universe, and rapid timescales for star formation. Is it more of the same at early epochs? i.e. just like the Milky Way but more gas. Massive outflows are observed, driven by active galactic nuclei. Is this all there is to quenching of star formation? Or is something radically new needed that involves positive as well as negative feedback from supermassive black holes?
- **Do the monsters in the middle (supermassive black holes) form by gas accretion or by swallowing stars and smaller massive black holes?**
  There is barely time available to form supermassive black holes in the early universe. One needs monster seeds, themselves massive black holes of $10^3 - 10^4 M_\odot$. How these seeds form is vigorously debated, with direct collapse being a favoured but by no means unique option.
- **Active galactic nuclei: aftermath, precursor or co-eval to star formation?**
  We cannot be sure of the nature of the first objects in the universe. Most would bet on galaxies, but black holes might be an option. The most distant quasar is at a redshift of

about 7, whereas galaxies are detected to a redshift as large as 10. However, these first galaxies most likely contain supermassive black holes.

- **Will computer simulations eventually resolve the problem of galaxy formation?**
State-of-the-art numerical simulations of galaxy formation provide impressive images that to the observer's eye are indistinguishable from the genuine article. The difficulty is that the dynamic range required to span star formation and massive black hole accretion in a cosmological volume is well beyond current capabilities. The theorist has to add *ad hoc* subgrid physics as a black box to galactic or supergalactic scale simulations. And even the resolved physics is subject to assumptions about the initial conditions or even the theory of gravity that can profoundly modify the outcome. The resulting images are beautiful but incomplete, and certainly cannot be used for robust predictions of what to look for next. Nevertheless they are useful indicators, and simulations have proved to be essential for constructing mock catalogues as a tool for implementing and analysing large-scale structure surveys. There is no doubt that simulations will eventually rise to the task of forming galaxies, but it promises to be a long wait.

References

Binney, J. 1977. The physics of dissipational galaxy formation. *Astrophys. J.*, **215**(July), 483–91.

Feruglio, C., Fiore, F., Carniani, S., *et al.* 2015. AGN-driven winds on all scales in Markarian 231: from hot nuclear ultra-fast up to kpc-extended molecular outflow. *Astron. Astrophys.*, **583**(Nov), 99–105.

Gnat, O. and Sternberg, A. 2007. Time-dependent Ionization in Radiatively Cooling Gas. *Astrophys. J. Supp.*, **168**(Feb.), 213–30.

Haardt, F. and Salvaterra, R. 2015. High-redshift active galactic nuclei and H I reionisation: limits from the unresolved X-ray background. *Astron. Astrophys.*, **575**(Mar.), L16.

Heap, S., Bouret, J.-C. and Hubeny, I. 2015. Population III Stars in I Zw 18. *ArXiv e-prints*, Apr.

Hirano, S., Hosokawa, T., Yoshida, N., Omukai, K. and Yorke, H. W. 2015. Primordial star formation under the influence of far ultraviolet radiation: 1540 cosmological haloes and the stellar mass distribution. *Mon. Not. R. Astron. Soc.*, **448**(Mar.), 568–87.

Hopkins, P. F., Kereš, D., Oñorbe, J., *et al.* 2014. Galaxies on FIRE (Feedback In Realistic Environments): stellar feedback explains cosmologically inefficient star formation. *Mon. Not. R. Astron. Soc.*, **445**(Nov.), 581–603.

Inayoshi, K. and Haiman, Z. 2014. Does disc fragmentation prevent the formation of supermassive stars in protogalaxies? *Mon. Not. R. Astron. Soc.*, **445**(Dec.), 1549–57.

Krumholz, M. R. 2011. On the Origin of Stellar Masses. *Astrophys. J.*, **743**(Dec.), 110.

Latif, M. A., Zaroubi, S. and Spaans, M. 2011. The impact of Lyman α trapping on the formation of primordial objects. *Mon. Not. R. Astron. Soc.*, **411**(Mar.), 1659–70.

Lemons, S., Reines, A., Plotkin, R., Gallo, E. and Greene, J. 2015. An X-ray Selected Sample of Candidate Black Holes in Dwarf Galaxies. *Astrophys. J.*, **805**(May), 12–22.

Leroy, A. K., Bolatto, A. D., Ostriker, E. C., *et al.* 2015. ALMA Reveals the Molecular Medium Fueling the Nearest Nuclear Starburst. *Astrophys. J.*, **801**(Mar.), 25.

Low, C. and Lynden-Bell, D. 1976. The minimum Jeans mass or when fragmentation must stop. *MNRAS*, **176**(Aug.), 367–90.

Ma, X., Kasen, D., Hopkins, P. F., *et al.* 2015. The Difficulty Getting High Escape Fractions of Ionizing Photons from High-redshift Galaxies: a View from the FIRE Cosmological Simulations. *Mon. Not. R. Astr. Soc.*, **453**(Oct), 960–75.

Madau, P. and Dickinson, M. 2014. Cosmic Star-Formation History. *Ann. Rev. Astron. Astrophys.*, **52**(Aug.), 415–86.

Moran, E. C., Shahinyan, K., Sugarman, H. R., *et al.*, 2014. Black Holes at the Centers of Nearby Dwarf Galaxies. *Astron. J.*, **148**(Nov.), 136–58.

Omukai, K., Tsuribe, T., Schneider, R. and Ferrara, A. 2005. Thermal and Fragmentation Properties of Star-forming Clouds in Low-Metallicity Environments. *Astrophys. J.*, **626**(June), 627–43.

Pardo, K., Goulding, A. D., Greene, J. E., *et al.*, 2016. X-ray Detected Active Galactic Nuclei in Dwarf Galaxies at $0<z<1$. *Astrophys. J.*, **831**(Nov.), 203–18.

Palla, F., Salpeter, E. E. and Stahler, S. W. 1983. Primordial star formation – The role of molecular hydrogen. *Astrophys. J.*, **271**(Aug.), 632–41.

Rathborne, J. M., Longmore, S. N., Jackson, J. M., *et al.* 2014. Turbulence Sets the Initial Conditions for Star Formation in High-pressure Environments. *Astrophys. J. Lett.*, **795**(Nov.), L25.

Rees, M. J. 1976. Opacity-limited hierarchical fragmentation and the masses of protostars. *Mon. Not. R. Astron. Soc.*, **176**(Sept.), 483–86.

Rees, M. J. and Ostriker, J. P. 1977. Cooling, dynamics and fragmentation of massive gas clouds – Clues to the masses and radii of galaxies and clusters. *Mon. Not. R. Astron. Soc.*, **179**(June), 541–59.

Robertson, B. E., Ellis, R. S., Furlanetto, S. R. and Dunlop, J. S. 2015. Cosmic Reionization and Early Star-forming Galaxies: A Joint Analysis of New Constraints from Planck and the Hubble Space Telescope. *Astrophys. J. Lett.*, **802**(Apr.), L19.

Satyapal, S., Secrest, N. J., McAlpine, W., *et al.* 2014. Discovery of a Population of Bulgeless Galaxies with Extremely Red Mid-IR Colors: Obscured AGN Activity in the Low-mass Regime? *Astrophys. J.*, **784**(Apr.), 113.

Silk, J. 1977. On the fragmentation of cosmic gas clouds. II – Opacity-limited star formation. *Astrophys. J.*, **214**(May), 152–60.

Silk, J. and Rees, M. J. 1998. Quasars and galaxy formation. *Astron. Astrophys.*, **331**(Mar.), L1–4.

Silk, J., Di Cintio, A. and Dvorkin, I. 2013. Galaxy formation. *ArXiv e-prints*, Nov.

Sobral, D., Matthee, J., Darvish, B., *et al.* 2015. Evidence for PopIII-like stellar populations in the most luminous Lyman-α emitters at the epoch of re-ionisation: spectroscopic confirmation. *Astrophys. J.,* **808**(Aug), 139–53.

Somerville, R. S. and Davé, R. 2014. Physical Models of Galaxy Formation in a Cosmological Framework. *Ann. Revs. Astron. Astrophys.*, **53**(Aug), 51–113.

Sparre, M., Hayward, C. C., Springel, V., *et al.* 2015. The star formation main sequence and stellar mass assembly of galaxies in the Illustris simulation. *Mon. Not. R. Astron. Soc.*, **447**(Mar.), 3548–63.

Sutherland, R. S. and Dopita, M. A. 1993. Cooling functions for low-density astrophysical plasmas. *Astrophys. J. Supp.*, **88**(Sept.), 253–327.

Tombesi, F., Meléndez, M., Veilleux, S., *et al.* 2015. Wind from the black-hole accretion disk driving a molecular outflow in an active galaxy. *Nature (London)*, **519**(Mar.), 436–8.

Part III

Foundations of Cosmology: Gravity and the Quantum

9

The Observer Strikes Back

JAMES HARTLE AND THOMAS HERTOG

9.1 Introduction

The context for this chapter is our universe – the whole closed system at all times containing all the galaxies, stars, planets, biota, human societies, you, us, etc. There is nothing outside. Two kinds of description of the universe can be distinguished:

Third person descriptions: Descriptions of what the universe contains and how that evolves – histories of what occurs.

First person descriptions: Descriptions of what we as the collection of human scientists observe of the universe and use to test cosmological models.

The connection between these two kinds of description is the subject of this chapter.

Quantum mechanical theories, and also classical ones, provide probabilities for the different descriptions. Correspondingly we can distinguish two different kinds of probabilities for any observable $\mathcal{O}$. Third person probabilities[1]

$$p(\mathcal{O}) \tag{9.1a}$$

for what values of $\mathcal{O}$ occur, and first person probabilities

$$p^{(1p)}(\mathcal{O}) \tag{9.1b}$$

for what values of $\mathcal{O}$ we observe. The connection between these two kinds of probabilities is the focus of this chapter.

We will consider theories whose direct outputs are third person probabilities for histories of what occurs in the universe. These include probabilities for the existence, evolution, and functioning of any observing subsystems such as ourselves. Such subsystems play no special role in formulating the theory – they are just one kind of subsystem among many. We refer to such theories as "third person" theories. Our most successful theories of cosmology are of this kind. Classical physics in general is an example of a third person theory. In quantum mechanics the extensions of the ideas of Everett [2, 32] are third person theories, including those used in this chapter.

---

[1] Following Hawking and Hertog [1], in most of our previous work we have used "bottom-up probabilities" and "top-down probabilities" for what are called here "third-person probabilities" and "first person probabilities".

Observers do play a preferred role in calculating first person probabilities for observations from third person ones. First person probabilities are third person ones conditioned on a description of the observational situation. We shall use a variety of specific models to infer the following general conclusions:

- What is most probable to occur is not necessarily what is most probable to be observed.
- Anthropic selection is an automatic consequence of first person probabilities.
- In universes large enough that we may be duplicated as physical subsystems elsewhere, the description of the observational situation needed to compute first person probabilities must also specify whether our particular situation is typical of all the others. It is the combination of the model of the observer, including this typicality assumption, and the third person theory which is tested by observation.

Observers and their observations are of central importance in the formulation of Copenhagen quantum mechanics. In the quantum mechanics of closed systems observers might seem to have been demoted to the status of one subsystem among many. Indeed, they have little effect on third person probabilities. But, as a consequence of the conclusions above, the observer returns to importance in the calculation of first person probabilities for observations by which the theory is used and tested. The observer strikes back.

Section 9.2 sketches the framework of the third person theory we will employ. Section 9.3 discusses issues involved with first person probabilities. Section 9.4 uses a model universe to make more concrete the notions of first and third person descriptions of the universe and their associated probabilities. Here we also describe the connection between third and first person probabilities in a set of simple models. Section 9.5 describes how anthropic selection emerges automatically as a feature of a certain class of first person probabilities. Section 9.6 shows how first person probabilities can sometimes be calculated directly from the theory with an appropriate coarse graining. Section 9.7 discusses the first person predictions of a number of cosmological observables, such as the cosmological constant, in cosmological models based on the no-boundary quantum state of the universe and a dynamical landscape theory in which these observables can take a range of values. In Section 9.9 we try to set our results in a more general view of physical theories.

9.2 Third Person Quantum Cosmology

This section briefly describes the elements of the theory that predicts third person probabilities for histories of the universe which we use to reach the general conclusions mentioned in the introduction.[2]

---

[2] For more details the reader can consult the authors' papers on which this chapter is implicitly based and through them find further references. For the quantum mechanics of closed systems see e.g. Hartle [3, 4]. For quantum cosmology see e.g. Hawking and Hertog [1]; Hartle *et al.* [5, 9]. There is a little more detail about the no-boundary quantum state of the universe [10] in Appendix 9.2.

We view the universe as a closed quantum mechanical system. It contains everything. Galaxies, stars, planets, their biota, observers (including us!) etc. are physical subsystems of the universe subject to its quantum mechanical laws consistent with an Everettian point of view [2]. The basic variables describing the universe and its contents are four-dimensional cosmological spacetime geometries and four-dimensional configurations of matter fields. The basic ingredients of the theory are an action I describing the dynamics of geometry coupled to matter fields, and a quantum state of the universe Ψ. We denote the theory as (I, Ψ).

The theory (I, Ψ) predicts third person probabilities for the individual members of sets of alternative four-dimensional histories of the universe, including those histories that describe its classical evolution. In this way (I, Ψ) can supply third person probabilities for such large scale features of the universe as the amount of inflation, the approximate homogeneity and isotropy, the pattern of cosmic microwave background (CMB) variations, and the formation and evolution of the distribution of galaxies. In principle (I, Ψ) also supplies third person probabilities for the accidents of biological evolution, the existence of observers like ourselves, etc. that are well beyond our power to compute or even estimate. For the examples in this chapter we will mostly use histories in which we are at a single moment of time: the time approximately 14 Gyr after the big bang when our observations of the universe are made. The theory (I, Ψ) is an example of what we will call a third person theory.

We test and utilize a theory not by its third person probabilities for what occurs, but by its first person probabilities for what we observe. To compute first person probabilities we first need to model the observational situation, which we do next.

9.3 First Person Quantum Cosmology

We begin by recalling the definition of a Hubble volume. We cannot see further in the universe than the distance that light travels to us from the big bang, roughly 14 Gyr ago. A present volume of the same size as this distance is called a "Hubble volume". In order of magnitude the distance is c/H_0 where H_0 is the present Hubble constant. This size is approximately 4000 Megaparsec or 10^{23} km. This is the largest scale over which we can currently observe the universe.

As observers we are physical systems within the universe with only a probability to have evolved in any one Hubble volume and a probability to be replicated in other Hubble volumes if the universe has a very large number of them. An observer is a very special kind of fluctuation D_0 in the universe. It is a fluctuation that is not singled out by quantum theory from, say, density fluctuations that produced the CMB. But the probability for observers is very difficult to compute. We will therefore employ a highly simplified model of observers: All observers are alike (copies of us) and either exist in any Hubble volume with a third person probability $p_E(D_0)$ or do not exist with a probability $1 - p_E$. Realistically the probability p_E incorporates the probability of the accidents of several billion years of biological evolution. Therefore, whatever its value is, it is very, very, very small. This is a very crude model of complex observers, but better than many treatments where the probability that observing systems evolved as part of the universe's evolution is not considered at all.

Since both we and what we observe are part of the universe, first person probabilities can be computed from the third person ones. If we are unique as physical systems within the universe the first person probabilities are simply third person ones for what is observed conditioned on a description of the observational situation like the one given above, in terms of probabilities p_E for data D_0. But if we are not unique then a more careful specification of the observer is required which includes an assumption about which instance of D_0 is us, or more generally a probability distribution on the set of copies called a xerographic distribution [11]. In this chapter we will make the minimal assumption that we are equally likely to be any of the incidences of D_0 that the third person theory (I, Ψ) predicts. The way to make other assumptions is discussed in Appendix 9.1.

It is a common intuition that the presence or absence of observers is unimportant for the behavior of the universe on cosmological scales because observers are generally small subsystems. That intuition is correct for third person probabilities, but it is not correct for first person ones. As physical systems we have only a tiny probability to exist in any one Hubble volume. That is why third person probabilities are little affected by the presence or absence of us. But this also means that we have a greater probability to live in a larger universe than a smaller one, because in the larger there are more Hubble volumes in which to be. Therefore, even if the third person probabilities favor smaller universes the first person ones may favor larger ones. As we will show by example, what is most probable to occur (third person) is not necessarily what is most probable to be observed (first person). It is in this way that the observer returns to importance in cosmology.

9.4 Toy Model Universes

This section derives the connection between third and first person probabilities for a very simple class of models that provide elementary examples showing what is most probable to be observed is not necessarily what is most probable to occur. In the trade these are called "box models" [12].

9.4.1 Third Person Description and Probabilities

Histories of the universe at one time are modeled as a collection of N boxes representing Hubble volumes at that time. Each box has a color – say white or gray – modeling different CMB maps. Each box may or may not contain an observer. A third person description of a history specifies the number of boxes, their color, and whether each box contains an observer or does not. A third person quantum theory specifies probabilities for these histories – a probability for the number of boxes, a probability for the color of each box, and the probability p_E for whether an observer exists in each.

We illustrate this with the very simple set of single time histories [12] shown in Figure 9.1. At one time, only two possible histories of N and color are possible – N_1 boxes

Figure 9.1 Two histories of the simple box model universe described in the text. The boxes model Hubble volumes. Their color models an observable like the CMB. An "E" means that an observer is in the box observing its color. A blank means there is no observer in the box. The third person probabilities for these histories to occur are at the right. (BLU = Blue color; see arXiv:1503.07205 for color figures).

all red that occurs with a third person probability p_1, and N_2 boxes all blue that occurs with third person probability $p_2 = 1 - p_1$. The probability that there is an observer in any box is p_E – the same for all boxes. The probability that there is no observer in a box is $1 - p_E$. The third person probability of a history with a specific set of n_E of N_k boxes occupied by observers is

$$p_k p_E^{n_E} (1 - p_E)^{N_k - n_E} \tag{9.2}$$

where $k = 1$ (all red) or $k = 2$ (all blue). The complete set of histories consists of all red and all blue histories, with different numbers of boxes in each, and the various possible ways the boxes can be occupied by observers.

9.4.2 First Person Probabilities for Observation

We are one of the observers in one of the histories. We now ask for the theory's prediction for the first person probability that we observe red (WOR). To calculate that, assume that in either history we are equally likely to be any one of the occurrences of E. Then the probability that we observe white (WOR) is evidently the probability that we are in the history with all red boxes ($k = 1$).

The probability that WOR is *not* the probability p_1 that the history with all white boxes occurs because that could happen with no boxes occupied by observers. Rather the probability for WOR is the probability that history $k = 1$ occurs with *at least one observer*. Denoting "at least one E" by $E^{\geq}$ we have

$$p(E^{\geq} \text{ in } 1) = 1 - p(\text{no E in } 1)$$
$$= 1 - (1 - p_E)^{N_1}. \tag{9.3}$$

The normalized probability that we observe red (WOR) is then

$$p^{(1p)}(WOR) = \frac{p_1[1 - (1 - p_E)^{N_1}]}{\sum_k p_k[1 - (1 - p_E)^{N_k}]}. \tag{9.4a}$$

Similarly, the probability that we observe blue is

$$p^{(1p)}(WOB) = \frac{p_2[1 - (1 - p_E)^{N_2}]}{\sum_k p_k[1 - (1 - p_E)^{N_k}]} . \tag{9.4b}$$

We now discuss important limiting cases.

Rare in all histories: When both $p_E N_1 \ll 1$ and $p_E N_2 \ll 1$ physical systems like us occur only rarely in each of the two possible histories. Therefore we can assume that as a physical subsystem we are unique in the universe. The probabilities for color observation (9.4) then become

$$p^{(1p)}(WOR) \approx \frac{N_1 p_1}{N_1 p_1 + N_2 p_2}, \tag{9.5a}$$

$$p^{(1p)}(WOB) \approx \frac{N_2 p_2}{N_1 p_1 + N_2 p_2} . \tag{9.5b}$$

The first person probabilities for our observation of red or blue are the third person probabilities of the all red and all blue histories *weighted* by the number of Hubble volumes in each. This is called "volume weighting". It favors larger universes where there are more places for us to occur as has been extensively discussed in cosmology (e.g. Hartle *et al.* [5]; Page [13]; Hawking [14]).

It is important to emphasize that volume weighting is not an extra assumption in addition to the theory of the histories. Rather it is a straightforward consequence of that theory in models where we are rare in all histories.

In a third person description of this model blue is more likely to occur if $p_2 > p_1$. But red is more likely to be observed if $N_1 p_1 > N_2 p_2$. Thus we have an elementary example of what is most probable to be observed is not necessarily what is most likely to occur. That is the return of the observer.

Other limiting cases are also interesting:

Common in both histories: When both $p_E N_1 \gg 1$ and $p_E N_2 \gg 1$, copies of us as physical systems are common in both histories. Then the probabilities (9.4) are

$$p^{(1p)}(WOR) \approx p_1, \tag{9.6a}$$

$$p^{(1p)}(WOB) \approx p_2. \tag{9.6b}$$

Thus, when all histories in the ensemble are very large, universes what is predicted to be observed is also what is predicted to occur.

Rare in one history, common in the other: When, say, $p_E N_1 \gg 1$ but $p_E N_2 \ll 1$, copies of us as physical systems are common in the all-red history and rare in the blue one. We have $[1 - (1 - p_E)^{N_1}] \approx 1$ and $[1 - (1 - p_E)^{N_2}] \approx N_2 p_E \approx 0$ since p_E is very, very small.

Then,

$$p^{(1p)}(WOR) \approx 1, \tag{9.7a}$$

$$p^{(1p)}(WOB) \approx 0. \tag{9.7b}$$

This is the case when the return of the observer has its most striking effect. The first person probabilities select large histories over small ones even when the latter have larger third person probabilities. We will see a concrete example of this in a more realistic model in Section 9.8.

9.4.3 A Crisis of Computability?

Box models divide the theory for predicting first person probabilities for what we observe into two parts. First, there is the specification of the third person probabilities p_k for the large scale features of the models – the number of Hubble volumes and the color of each. Second, there are the third person probabilities for the occurrence of observers inside each Hubble volume, summarized by the one number p_E, which are used to describe and to condition on the observational situation.

As we will see in Section 9.7, it is possible to make computationally tractable calculations of p_k in simple models. But the probability p_E would naturally include the probability that human observers evolved in a Hubble volume. To calculate this, or even estimate it, would involve considering several billion years of the chance accidents of biological evolution. This is well beyond our power to compute even assuming that we have a theory that is well enough formulated to define the task.

It is therefore fortunate that in all of the interesting limiting cases discussed above the probability p_E cancels out. Thus, it is in the regime of universes so small that we are unique, or universes so large that we are common, that observations are easily and objectively calculated.

9.5 Anthropic Selection is Automatic in Quantum Cosmology

Consider again the simple box model in Section 9.4.2 but suppose that the probability for an observer to occupy a red box p_E^R is different from the probability p_E^B to occupy a blue box. Suppose further that for some reason red is necessary for observers so that the probability p_E^B to occupy a blue box is exceedingly small. It is then obvious, and easily worked out, that the first person probability that we observe red is very near unity. This is a very simple example of anthropic selection: The all-red history is selected because observers do not exist in the alternative.

The key point here is that anthropic selection is automatic. No additional assumptions or principles were needed beyond the probabilities p_1, p_2, p_E^R and p_E^B which are all third person probabilities following from the underlying theory. Anthropic selection emerges as an intrinsic feature of the first person probabilities for the observer's observations [15].

This is the case also in more general and more realistic cosmological models. As an example consider an ensemble of single time histories which we take to be at the present age of the universe $t_0 \approx 14$ Gyr. Assume that the theory (I, Ψ) predicts third person probabilities for what we may call a set of background histories each with the same number of Hubble volumes but differing in the value of the cosmological constant Λ, which is assumed positive and the same in all Hubble volumes of one history. The theory thus allows the cosmological constant to vary. The histories can be labeled by the value of Λ and their third person probabilities written $p(\Lambda)$. As a fluctuation on these backgrounds the theory (I, Ψ) also predicts a probability that in any Hubble volume there occur data D_0 that describe our observational situation (but not including any record we might have of the value of Λ). These probabilities depend on Λ. The complete set of histories is thus labeled by the backgrounds and which Hubble volumes in them are occupied by D_0.

Assume that we are typical of the incidences of D_0 in any one history. The first person probability that we observe a value of Λ (WOΛ) is the third person probability for the history Λ conditioned on the existence of at least one instance of D_0 (cf. Eq. (9.3) *et seq.*). Using the Bayes identity we have

$$p^{(1p)}(WO\Lambda) \equiv p(\Lambda | D_0^{\geq 1}) = \frac{p(D_0^{\geq 1} | \Lambda) p(\Lambda)}{\sum_\Lambda p(D_0^{\geq 1} | \Lambda) p(\Lambda)}. \tag{9.8}$$

For values of Λ for which $p(D_0^{\geq 1} | \Lambda)$ is negligibly small the probability that we observe that value will also be negligibly small. Hence anthropic selection of values of Λ that are consistent with observers is automatic. This kind of argument assumes that the third person theory allows Λ to vary over an anthropically allowed range – a range consistent with the evolution of D_0. We give an example of such a model in Section 9.8. If however the theory determines a unique value of Λ, then either that must be consistent with $D_0^{\geq 1}$ or the theory is incorrect (e.g. Hartle [21]).

Barrow and Tipler [16] and Weinberg [17] have argued that the observed value of the cosmological constant could not be much larger than $\Lambda \sim 10^{-122}$ in Planck units, not far from its observed value. Were Λ larger the universe would expand too rapidly for galaxies to have formed by the present age $t_0 \sim 14$ Gyr and human observers would not be here. In our scheme the probability $p(D_0^{\geq 1} | \Lambda)$ would be near zero.

This kind of argument is an example of what one could call traditional anthropic reasoning. In this, the anthropically allowed range of values of an observable like Λ is determined from classical arguments like those involving galaxy formation mentioned above. It is then assumed that there is some unknown mechanism for Λ to vary over this range. In the absence of a specific model of this mechanism a uniform distribution over the range is often assumed and the most probable Λ predicted on the basis of purely anthropic arguments. Impressively detailed calculations were carried out in this way with mixed results, e. g. Tegmark *et al.* [19]. But this program suffers from several uncertainties. For example, different results are obtained if different combinations of constants were assumed to vary

[20]. In addition, traditional anthropic reasoning is not part of any theoretical framework. Rather anthropic selection arises from an additional assumption or "principle".

The chief difference with anthropic selection in quantum cosmology is that (1) there the theory (I, Ψ) provides a mechanism for what constants vary and how they vary, (2) the observer is a physical subsystem with a certain probability predicted by the theory to evolve in any Hubble volume, and (3) anthropic selection emerges automatically as a property of the first person probabilities by which we use and test the theory. To summarize, anthropic selection in quantum cosmology:

- Is not a choice.
- Does not require invoking some "anthropic principle".
- Does not change the objective nature of the underlying third person theory.
- Does require a typicality assumption if there is a chance we are replicated.

The observer is important for first person probabilities for observations through anthropic selection. In Section 9.8.3 we will describe a more precise calculation of $p^{(1p)}(WO\Lambda)$ in which the probabilities $p(\Lambda)$ in Eq. (9.8) are obtained from a concrete model of the universe's quantum state Ψ in combination with a dynamical theory I in which Λ can vary.

9.6 A Remark on Coarse Graining

Our observations of the universe extend at most over a Hubble volume. But this may only be a tiny region in a vastly larger inflationary universe of the kind contemplated in contemporary cosmology. Indeed, some calculations [24] suggest that the universe typically becomes spatially infinite as a consequence of eternal inflation. Much of the third person information about what occurs in the universe on very large scales is irrelevant for the first person predictions for our observations in our Hubble volume, and perhaps not even well defined. In this section we illustrate how, in certain models, first person probabilities for our observations can be calculated directly using coarse grainings that ignore most of the structure outside our Hubble volume. Coarse graining is not an *ad hoc* assumption. It is central and inevitable in quantum mechanics, statistical physics, complexity, and many other areas of science (e.g. Gell-Mann and Hartle [29]).

We illustrate how coarse graining works with a simple box model of the kind used in Section 9.4. Consider a universe with an infinite set of boxes as illustrated in Figure 9.2. Each box has a color. This is either yellow with probability p_Y or green with probability $p_G = 1 - p_Y$. There is a probability p_E for an observer to be in any box observing its color. Thus all the boxes are statistically identical. In this sense this universe has a discrete translation symmetry.

Now we ask for the first person probability that we observe yellow. The answer is obvious. We are in one box. All the boxes are statistically the same. The first person probability $p^{(1p)}(Y)$ for observing yellow is the same as the probability that any of the boxes is

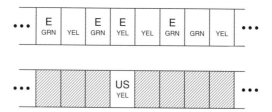

Figure 9.2 Fine and coarse-grained histories of a box model discussed in the text. The boxes model Hubble volumes. Their color models an observable like the CMB in this case either yellow or green. An "E" means that an observer is in the box observing its color. The top history is fine grained with a color and E or not E in every box. The bottom history is coarse grained. The possibilities have been summed over for every box except one – our box. That enables a straightforward calculation of the first person probability that we see one color or the other. The details outside our box are irrelevant for this. (GRN = green and YEL = yellow colors; see arXiv:1503.07205 for color figures).

yellow, viz.

$$p^{(1p)}(Y) = p_Y. \tag{9.9}$$

Although obvious, it is instructive to see how this result follows from our general framework through coarse graining starting from third person probabilities. A fine-grained history would specify the color (Y, G) of each box, and whether there exists an observer in it or not $(E, \bar{E})$ (top in Figure 9.2). The third person probability for one particular history having n_Y white boxes, n_G gray boxes, n_E boxes with observers, and $n_{\bar{E}}$ without, is

$$(p_Y)^{n_Y}(1 - p_Y)^{n_G}(p_E)^{n_E}(1 - p_E)^{n_{\bar{E}}}. \tag{9.10}$$

But these probabilities tend to zero in a universe with an infinite number of boxes. The probabilities for these fine-grained histories are not well defined.

Finite probabilities can be obtained by coarse graining. That is, they can be obtained by summing Eq. (9.10) over what's irrelevant for our observations. To get first person probabilities for our observations we can coarse grain over the alternatives in every box but ours. That means sum the probabilities Eq. (9.10) over the alternatives (Y, G) and $(E, \bar{E})$ giving a factor of unity for every box but ours (bottom in Figure 9.2). That gives the first person probabilities.

This result generalizes straightforwardly to any finite number of kinds (colors) of Hubble volumes, to models where the probabilities p_E depend on the color, and to more than one configuration (history) of boxes with third person probabilities like the $p(k)$ of Section 9.4.1. In Section 9.8 we review an example of coarse graining used in realistic cosmology (see also Hartle *et al.* [9]).

To summarize, in infinite or just very large universes focusing on our observations in our Hubble volume motivates coarse grainings that directly lead to well defined first person probabilities for observations. It is an intriguing open question whether such a local framework for prediction can be achieved more generally in quantum cosmology. Such a framework would truly return the observer to importance.

9.7 Inflation in Quantum Cosmology

We now turn from illustrative but artificial toy models to more realistic cosmological models. In this and the next section we consider two examples in which the return of the observer is important – where the probabilities for what we observe are significantly different from the probabilities for what occurs.

The two examples share a common theoretical framework (I, Ψ). For the dynamics we assume a spatially closed cosmological spacetime metric g coupled to a number of scalar fields $\vec{\phi}$. The dynamics is specified by an (Euclidean) action $I[g, \vec{\phi}]$ consisting of the action for general relativity plus the action for the scalar fields $\vec{\phi}$ coupled to the metric g. For the state Ψ we assume the no-boundary wave function of the universe (NBWF) [10]. This is the natural analog of the notion of "ground state" for closed cosmologies.

Many of our large scale observations are of properties of our universe's classical history. The rate of the universe's expansion and the distribution of galaxies in our Hubble volume are examples. A history of the universe behaves classically when the quantum probability is high that it exhibits correlations in time governed by the Lorentzian Einstein equation and the classical field equations. The NBWF predicts an ensemble of alternative classical histories along with third person probabilities for which history in the ensemble occurs. For a little more on what the NBWF is, and how it predicts probabilities see Appendix 9.2. For much more see Hartle *et al.* [6].

The first example is concerned with the classical cosmological histories predicted by the NBWF when the matter consists of a single scalar field moving in a quadratic potential

$$V(\phi) = \frac{1}{2}m^2\phi^2 \qquad (9.11)$$

with $m^2 \ll 1$ and zero cosmological constant. As in the rest of this chapter we are using Planck units where $\hbar = 8\pi G = c = 1$. Geometry and field are restricted to be homogeneous and isotropic thus defining a minisuperspace model. Lorentz-signatured homogeneous and isotropic spacetime geometries can be described by metrics of the form

$$ds^2 = -dt^2 + a^2(t)d\Omega_3^2. \qquad (9.12)$$

Here, $d\Omega_3^2$ is the metric on a unit round three-sphere. The time-dependence of the scale factor $a(t)$ describes how this closed universe expands and contracts. Standard closed FLRW cosmological models have metrics of this form that satisfy the Einstein equation. The homogeneous field is a function only of time, viz $\phi = \phi(t)$. A quantum history of this model universe is therefore specified by $(a(t), \phi(t))$. Classical histories are described by a pair $(a(t), \phi(t))$ that obey the Einstein equation and the classical equations of motion for the field.

As sketched in Appendix 9.2 the NBWF in this minisuperspace model predicts a one-parameter ensemble of possible classical histories. These classical histories can be labeled by a parameter ϕ_0 that can be roughly thought of as the value of the scalar field from which it starts to roll down to the bottom of the potential Eq. (9.11). By examining any one classical $a(t)$ we can find the number of efolds N_e of slow roll inflation it has. Remarkably all histories in the NBWF classical ensemble turn out to have some inflation at early times [6]. Inflation and the emergence of a classical universe in the NBWF are therefore profoundly connected [6, 28].

Does our theory (I, Ψ) predict a significant probability for an extended period of inflation in the early universe? As stated, this question is ambiguous. It could mean "Are the third-person probabilities from (I, Ψ) high for classical histories with an early period of inflation?" But it could also mean "Are the first person probabilities high that the classical history we observe has a significant period of early inflation?" We will display the answers to both questions in our model and find that they are significantly different.

To evaluate the first person probabilities for the amount of inflation we can assume that we are rare physical systems in any of the classical histories predicted by the NBWF. This is because the number of Hubble volumes in all histories of the classical ensemble with matter densities below the Planck density is much smaller than any realistic value of p_E^{-1} [7]. Hence volume weighting Eq. (9.5) connects the first person probabilities for our observations with the third person probabilities for what occurs.

The third person probabilities $p = p(\phi_0)$ for the histories predicted by the NBWF are given approximately by Eq. (9.29)

$$p(\phi_0) \propto \exp[3\pi/V(\phi_0)]. \qquad (9.13)$$

It turns out in this model that roughly $N_e \approx (3/2)\phi_0^2$ starting at $N_e \approx 1$. The third person NBWF probabilities for classical histories Eq. (9.13) thus imply probabilities $p(N_e)$ for the number of efolds that occur, starting roughly at $N_e \approx 1$,

$$p(N_e) \propto \exp\left(\frac{9\pi}{m^2 N_e}\right). \qquad (9.14)$$

Third person probabilities are therefore much larger for a small number of efolds than for the minimal number ~ 60 required for agreement with observation, as illustrated in Figure 9.3.

The situation is much different for first person probabilities for the amount of inflation in *our* history – the one we observe. Since it is assumed that we are rare, the first person probabilities are the volume weighted third person probabilities as in Eq. (9.5). During a period of inflation the volume of the universe expands by a factor $\exp(3N_e)$. The first person probabilities are then approximately

$$p^{(1p)}(N_e) \propto \exp\left(3N_e + \frac{9\pi}{m^2 N_e}\right). \qquad (9.15)$$

Thus the NBWF predicts a significant probability for us to find ourselves in a universe with a large number of efolds (Figure 9.3). This implies there is a significant probability for us

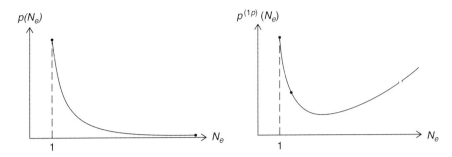

Figure 9.3 Third (left) and first person (right) probabilities for the number of efolds N_e of matter-driven inflation in the early universe in the no-boundary quantum state Ψ. The first person probabilities favor universes with a large amount of inflation because in the larger universes that result from an extended period of early inflation there are more Hubble volumes for us to be.

to observe the consequences of inflation – approximate spatial flatness, a scale invariant spectrum of density fluctuations, etc.

There is a higher probability for us to live in a larger universe because we are a physical subsystem within the universe with a very small probability p_E to have evolved in any Hubble volume. In the larger universes that result from an extended period of early inflation there are more Hubble volumes for us to be. Thus the observer returns to importance in the probabilities for observing the consequences of inflation.

9.8 Eternal Inflation and Anthropic Selection

We now extend the above model in three crucial ways to arrive, at last, at a realistic model for the early universe:

(1) *Fluctuations:* We include fluctuations away from homogeneity and isotropy. These fluctuations provide the necessary degrees of freedom to describe e.g. the pattern of temperature variations in the CMB. A consequence is that a range of classical histories exhibit eternal inflation becoming spatially very large and highly inhomogeneous on the largest scales.

(2) *Landscape potential:* We no longer assume one scalar field in a potential with a single minimum like Eq. (9.11). Rather we assume many scalar fields in a multi-field potential that has many different minima with different directions of approach, as a toy model for the string landscape [27]. As a consequence the dynamical theory provides a mechanism for the observable parameters of the histories to vary. Automatic anthropic selection can then be explicitly illustrated.

(3) *Not rare but common:* The above assumption that we are rare in all histories is no longer tenable with eternal inflation. The probability p_E for us to exist in any Hubble volume is very small. Nevertheless, we will be common in a range of histories where the universe becomes sufficiently large. In that range the connection between third and first

person probabilities will no longer involve volume weighting but instead be given by Eq. (9.6).

We now consider the meaning and implications of these extensions in more detail.

9.8.1 Eternal Inflation: Histories where Observers are Common

Cosmological perturbation theory extends the homogeneous and isotropic models of Section 9.7 straightforwardly to include linear fluctuations away from these symmetries for each classical history. The early period of slow roll inflation in the classical NBWF histories stretches and amplifies quantum vacuum fluctuations and generates a pattern of classical perturbations on scales larger than the horizon, which much later produce the small temperature fluctuations we observe in the microwave sky. However, it turns out that very long-wavelength fluctuations, those which leave the horizon and become classical at values of ϕ where

$$V^3 \geq |dV/d\phi|^2 \,, \tag{9.16}$$

have a large expected amplitude. There is non-perturbative evidence that histories in which ϕ is initially in this regime reach very large (or even infinite) spatial volumes, because these large very long-wavelength fluctuations tend to make them highly inhomogeneous on scales much larger than our Hubble volume [22–24]. This dynamical process is known as eternal inflation [25, 26], and Eq. (9.16) defines the regime of field values where eternal inflation occurs.

For quadratic potentials like Eq. (9.11), the condition for eternal inflation Eq. (9.16) is met for sufficiently large $\phi_0 > \phi_{\rm ei}$ where

$$\phi_{\rm ei} \sim 1/\sqrt{m}, \tag{9.17}$$

well below the Planck scale $\phi_{\rm pl} \sim 1/m$ for realistic values $m \sim 10^{-5}$.

Therefore, when we include fluctuations the set of classical NBWF histories divides into two parts: Those with $\phi_0 \gtrsim \phi_{\rm ei}$ which are inhomogeneous on the largest scales and have a great many Hubble volumes N due to eternal inflation, and those with $\phi_0 \lesssim \phi_{\rm ei}$ which are much smaller and approximately homogeneous.

This division has significant implications for the first person probabilities for observation. We are now effectively in the box model case Eq. (9.7) where (assuming typicality)

$$[1 - (1 - p_E)^N] \approx 1, \quad \phi_0 > \phi_{\rm ei}, \tag{9.18a}$$

$$[1 - (1 - p_E)^N] \approx p_E N \ll 1, \quad \phi_0 < \phi_{\rm ei}. \tag{9.18b}$$

Eternally inflating histories are thus strongly selected, whereas histories with slow roll inflation only are strongly suppressed by the very small value of p_E. In selecting for eternally inflating universes as the ones we observe, the observer has returned in force.

9.8.2 Landscapes: A Mechanism for Cosmological Parameters to Vary

As discussed in Section 9.5, the anthropic selection of observed cosmological parameters requires a third person theory (I, Ψ) that allows the parameters to vary. Theories with landscape potentials are a very simple example.

To illustrate what landscape potentials are, consider a third person theory with two scalar fields ϕ_1 and ϕ_2 moving in a potential $V(\phi_1, \phi_2)$. A three-dimensional plot of this potential could be made using ϕ_1 and ϕ_2 as the x and y-axes and plotting V along the z-axis. The plot might resemble a mountainous landscape on Earth whence the name "landscape potential".

Suppose the potential has a number of different minima each surrounded by a number of different valleys leading to it.[3] In our past history the fields "rolled down" a particular valley ("our" valley) to a particular minimum ("our" minimum). The value of the potential at our minimum is the value of the cosmological constant we would observe. The shape of our valley near our minimum determines the spectrum of density fluctuations in the CMB we would see. A third person theory that predicts probabilities for which of the possible histories occurs is thus a starting point for calculating the first person probabilities for the values of these parameters we observe.

The notion of a landscape potential is easily extended to many scalar fields $\vec{\phi}$ so that $V = V(\vec{\phi})$. Assume that each minimum in $V(\vec{\phi})$ is surrounded by effectively one-dimensional valleys. Suppose that these valleys are separated by large barriers so that transitions between valleys are negligible. Then we have, in effect, an ensemble of one-dimensional potentials $V_K(\phi_K)$ $K = 1, 2, \cdots$ – one for each valley.

The NBWF predicts the probabilities for classical inflationary histories for each one-dimensional potential in this landscape. The total classical ensemble is the union of all these. An individual history can therefore be labeled by (K, ϕ_{0K}) where, as explained before, ϕ_{0K} is roughly the value of the field ϕ_K at the start of its roll down the potential V_K. Thus the NBWF predicts third person probabilities $p(K, \phi_{0K})$ that our universe rolled down the potential from ϕ_{0K} in the valley K to its minimum.

Generically a landscape of the kind under discussion contains some valleys K where the potential $V_K(\phi_K)$ has a regime of eternal inflation. Eq. (9.18) implies that histories in these valleys will dominate the first person probabilities. In the presence of eternal inflation the first person probabilities are

$$p^{(1p)}(K, \phi_{K0}) \approx p(K, \phi_{K0}) \approx \exp[3\pi/V_K(\phi_{K0})], \quad \phi_{K0} \gtrsim \phi_{K\mathrm{ei}}$$
$$\approx 0, \quad \phi_{K0} \lesssim \phi_{K\mathrm{ei}} \tag{9.19}$$

When the potentials are increasing with field, the most probable NBWF history in a given valley will be that where the field starts around the exit of eternal inflation, i.e. $\phi_{K0} = \phi_{K\mathrm{ei}}$. Thus, to good approximation, the probability that we rolled down in valley K is

$$p^{(1p)}(K) \approx p(K) \approx \exp[3\pi/V_K(\phi_{K\mathrm{ei}})]. \tag{9.20}$$

---

[3] Valleys were called "channels" in Hartle *et al.* [9] and other places.

We now apply this result in a concrete model landscape.

9.8.3 First Person Predictions of the No-Boundary State in a Landscape Model

We now discuss an example of a first person prediction in a landscape model with valleys where the eternal inflation condition Eq. (9.16) is satisfied. This example relates directly to the historical effort described briefly in Section 9.5 to determine the anthropically allowed ranges of cosmological parameters that are consistent with our existence as observers (see, e.g. Barrow and Tipler [16]; Weinberg [17]; Tegmark *et al.* [19]; Livio and Rees [20]).

Specifically, we calculate the NBWF predictions for the first person probability of a correlation between three observed numbers: First is the observed value of the cosmological constant $\Lambda \sim 10^{-123}$ (in Planck units). Second is the observed value of the amplitude of primordial density fluctuations $Q \sim 10^{-5}$ in the CMB. And third is the part of our data D_{loc} on the scales of our galaxy and nearby ones, which, together with Λ, determine the present age of the universe $t_0(\Lambda, D_{\text{loc}}) \sim 14 \text{Gyr}$. Thus we are interested in $p^{(1p)}(\Lambda, Q, D_{\text{loc}})$ – the first person probability for this correlation to be observed. Many other such correlations could be investigated to test a given theory.[4] But this example will nicely illustrate several implications of the previous discussion.

We consider a landscape potential in which the parameters Λ and Q vary so the theory (I, Ψ) will predict probabilities for their values. Specifically, we assume a landscape of different one-dimensional valleys with potentials of the form[5]

$$V_K(\phi) = \Lambda_K + \frac{1}{2}m_K^2\phi_K^2, \quad K = 1, 2, \cdots. \tag{9.21}$$

A valley in this landscape is therefore specified by the values (Λ, m). (From now on we will drop the subscripts K to simplify the notation.)

The NBWF predicts an ensemble of inflationary universes in this landscape. The observed classical history of our universe rolled down one of the valleys – "our valley". The values of Λ and m that characterize our valley can be determined by observation. Measurement of the expansion history of the universe determines Λ and CMB measurements determine m, because the amplitude of the primordial temperature fluctuations is given by Hartle *et al.* [8] and Hartle and Hertog [15]

$$Q \approx N_*^2 m^2. \tag{9.22}$$

Here $N_* \sim \mathcal{O}(60)$ is the number of efolds before the end of inflation that the COBE scale left the horizon during inflation. Observations indicate that $Q \sim 10^{-5}$.

The first person probability that we observe specific values of Λ and Q is the first person probability that our past history rolled down the valley which has those values. Since each

---

[4] For a recent discussion of observables associated with features of the CMB fluctuations see e.g. Hartle and Hertog [15].

[5] For examples of more general landscapes see Hertog [28].

of the valleys Eq. (9.21) has a regime of eternal inflation for $\phi > \phi_{ei} \sim 1/\sqrt{m}$ the first person probabilities Eq. (9.20) that we observe values of Λ and Q can be written as

$$p^{(1p)}(\Lambda, Q) \approx p(\Lambda, Q) \propto \exp[3\pi/(\Lambda + cQ/N_*)]. \qquad (9.23)$$

where $c \sim \mathcal{O}(1)$. For fixed Λ this distribution favors small Q. As mentioned above, we are interested in the joint probability $p^{(1p)}(\Lambda, Q, D_{loc}) \approx p(\Lambda, Q, D_{loc})$ (first equals third in a regime of eternal inflation). The latter can be rewritten[6]

$$p(\Lambda, Q, D_{loc}) = p(D_{loc}|\Lambda, Q)p(\Lambda, Q). \qquad (9.24)$$

The second factor in the expression above is given by Eq. (9.23). The first will be proportional to the number of habitable galaxies that have formed in our Hubble volume by the present age $t_0(\Lambda, D_{loc})$. This is plausibly proportional to the fraction of baryons in the form of galaxies by the present age t_0 assuming that is bigger than the collapse time to form a galaxy. Denoting this by $f(\Lambda, Q, t_0)$ we have from Eq. (9.23)

$$p^{(1p)}(\Lambda, Q, D_{loc}) \approx p(\Lambda, Q, D_{loc}) \propto f(\Lambda, Q, t_0) \exp[3\pi/(\Lambda + cQ/N_*)]. \qquad (9.25)$$

All the quantities in Eq. (9.25) are determinable in one Hubble volume – ours. The rest of the calculation can therefore be carried out in that volume. We do not need to ask in which of the vast number of Hubble volumes in the third person eternally inflating universe we dwell. In this model they are all statistically the same as far as the quantities in Eq. (9.25) go. We are effectively in the situation of the yellow–green box model in Section 9.6. We can coarse grain over all other Hubble volumes but ours. In this way we are able to make contact with – and use – calculations in traditional anthropic reasoning.

Figure 9.4 is a contour plot of f at $t_0 \sim 11$ Gyr — earlier than the present age 14 Gyr because galaxies have been around for a while. This was adapted from the detailed astrophysical calculations of Tegmark *et al.* [19]. For values of (Λ, Q) to the right of the diagonal dotted line the universe accelerates too quickly for pre-galactic halos to collapse. Fluctuations have to be large enough to collapse into galaxies a bit before t_0 and produce D_{loc}. That determines the bottom boundary. If the fluctuations are too large (top boundary) the bound systems are mostly large black holes inconsistent with D_{loc}. The central region where $f \geq 0.6$ is the anthropically allowed region.

The cosmological constant Λ is negligible compared to Q in the anthropically allowed (white) region of Figure 9.4. The exponential dependence $\exp[3\pi N_*/cQ]$ implied by the NBWF means that the probabilities Eq. (9.25) are sharply confined to the smallest allowed values consistent with galaxies by t_0, i.e. $Q \sim 10^{-5}$. The resulting marginal distribution for Λ is shown in Figure 9.5 and is peaked about $\Lambda \sim 10^{-123}$ close to the observed value.

---

[6] In traditional anthropic reasoning the first factor in Eq. (9.24) is called the "selection probability" which can be calculated; the second is the "prior" which is assumed. See, e.g. Eq. (1) in Tegmark *et al.* [19]. The "prior" is typically assumed to be uniform in the anthropically allowed range reflecting ignorance of part of the theory. Here both factors follow from the same theory (I, Ψ).

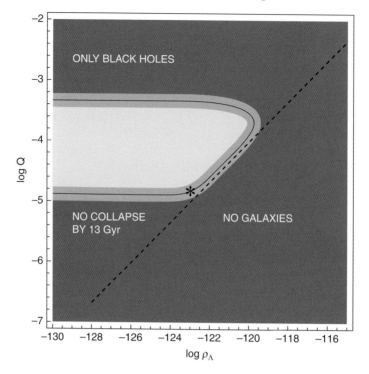

Figure 9.4 A contour plot of the fraction $f(\Lambda, Q, D_{\mathrm{loc}})$ of baryons in the form of galaxies by the time $t_0 \sim 11$ Gyr, adapted from the calculation of Tegmark *et al.* [19]. As discussed in the text, the fraction is negligible in the dark (gray) region either because gravitationally bound systems do not collapse or because they collapse to black holes. The $*$ locates the observed values of Q and Λ.

However, the agreement with observation is not the most important conclusion. The model is still too simplified for that. What is important is how the example illustrates the previous discussion. Specifically, how anthropic reasoning emerges automatically in quantum cosmology, how it can be sharpened by a theory of the universes's quantum state, how first person probabilities select for large eternally inflating universes, and how what is most probable to be observed is not necessarily what is most probable to occur – the return of the observer.

9.9 Conclusion

Fundamental physical theories have generally been organized into a third person theory of what occurs and a first person theory of what we observe. In quantum mechanics the output of the two parts are third person probabilities and first person probabilities. The character of these two parts, their relative importance, and the relation between them has changed over time as new data on new scales of observation required new theory and as

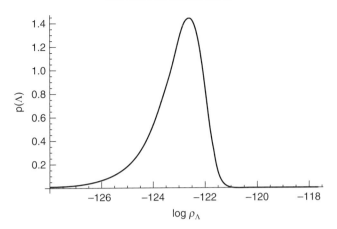

Figure 9.5 The marginal distribution for the cosmological constant Λ obtained by integrating Eq. (9.25) over Q.

the understanding of our observational situation in the universe evolved. The history of the transitions from classical physics to Copenhagen quantum mechanics, and then to the quantum mechanics of closed systems that was briefly sketched in the Introduction is part of this evolution.

As we discussed in this chapter, modern cosmology implies new requirements on both the first person theory of what occurs in the universe and the third person theory of what we observe of the universe.

A third person theory of the whole universe requires a quantum mechanics of closed systems including quantum spacetime as in Hartle [4] for example. It requires a quantum mechanics that predicts probabilities for what happens in the early universe when there were no observers around and no measurements being made. It requires a quantum mechanics that predicts probabilities for the emergence of classical spacetime in a quantum theory of gravity. It requires a quantum mechanics that can explain the origin of the rest of classical predictability in distant realms of the universe that we will never visit.

Quantum cosmology also has implications for first person theory. As observers, individually and collectively, we are physical systems within the universe with only a probability to have evolved in any one Hubble volume and a probability to be replicated in many if the universe is very large. What we have seen in this chapter is that this implies that the observer returns to importance in at least the following ways:

- By generally showing that what is most probable to be observed is not necessarily the most probable to occur.
- By favoring larger universes over small ones if we are rare or if we are common and thus favoring observations of a significant amount of inflation.
- By making anthropic selection automatic. We will not observe what is where we cannot exist.

- By requiring the addition of an assumption of typicality to the theory of what occurs that is made concrete by the xerographic distribution – an addition that can be tested and used to improve prediction.
- By leading to an understanding of how to compute probabilities for our observations of fundamental constants in a landscape that allows them to vary.

This list is a brief summary of the results of this chapter, but these lead naturally to a number of questions that we discuss here.

A natural question is why the observer was not important in classical cosmology when it is in quantum cosmology. An answer can be traced to differences in starting points and objectives. The starting point for classical cosmology was the assumption that the universe had a single spacetime geometry. The goal was to infer the geometry of that spacetime from large scale observation. Is it approximately homogeneous and isotropic on large scales, open or closed ...? What are the values of the cosmological parameters that characterize it – the cosmological constant, the Hubble constant, the amount of radiation, the amount of baryons, the amplitude of the density fluctuations, etc.? Is there evidence that the spacetime had an early period of inflation? Observers were presumed to exist but had a negligible influence on the answers to these questions.

Quantum cosmology does not start by assuming classical spacetime. Rather it starts from a theory of the universe's quantum state and dynamics. From that it seeks to explain when spacetime is classical and predict probabilities for what different possible classical spacetimes there are – questions that cannot even be asked in the context of classical cosmology. It therefore answers the classical questions with probabilities about large scale geometry.

A second natural question is, *why* do fundamental physical theories encompass both a third person theory of what occurs and a first person theory of what is observed? The historical success of theories organized in this way is indisputable. That success tells us something about the world. It supports the idea of some form of realism, perhaps along the lines of what Putnam called 'realism with a human face' [31]. To paraphrase J. A. Wheeler "In a quantum world, the universe is a grand synthesis, putting itself together all the time as a whole. Its history is ... a totality which includes us and in which what happens now gives reality to what happened then" [30].

A third natural question concerns the scale on which we have to know something of the universe to make first person predictions. An IGUS[7] like us is confined to a very local region inside of a vast Hubble volume which itself is typically but a small part of a much larger universe. Yet the present formulation of the first person theory requires information beyond our Hubble volume to determine whether we are rare, or common, or other. It remains to be seen whether a more local computation of probabilities for observation in quantum cosmology can be found. That would be a further way the observer and universe are unified.

---

[7] Information gathering and utilizing system.

Everett's insight was that, as observers of the universe, we are physical systems within it, not outside it. We are subject to the laws of quantum mechanics but play no special role in its formulation. We are negligible perturbations on a third person description of the universe. But, as shown in this chapter in several different ways , we return to importance for calculating first person probabilities for our observations precisely because we are physical systems in the universe. We may have only begun to appreciate the ramifications of that insight.

Acknowledgments

The authors have had the benefit of collaboration and discussions with a great many scientists in this work. The cited papers have those acknowledgments. However, our collaborators on the papers that underlie this work should be thanked. They are Murray Gell-Mann, Stephen Hawking, and Mark Srednicki. Gregory Benford suggested the title. Comments of Don Page and Simon Saunders were useful. The work of J.H. was supported in part by the US NSF grant PHY12-05500. The work of T.H. was supported in part by the National Science Foundation of Belgium (FWO) grant G.001.12 Odysseus and by the European Research Council grant no. ERC-2013-CoG 616732 HoloQosmos.

Appendix 9.1: Typicality – The Xerographic Distribution

Consider for a moment a third person theory – classical or quantum – that predicts a large number of Hubble volumes, some with one kind of observable property, some with another. A box model like those of Section 9.4 with one history and different colored volumes would be a very simple example.

Suppose that the data D_0 that describe our observational situation (including us) occur in many different Hubble volumes. One of them is ours, but in other volumes the result of the observation specified by D_0 could be different. A third person theory (I, Ψ) would predict what is observed in all of the instances of D_0. But it does not predict which one is ours. Indeed, it has no notion of "we" or "us".

We do not know which of these copies of D_0 are us. It could be any one of them. To make first person predictions for our observations, the third person theory must therefore be augmented by an assumption about which instance of D_0 is us, or more generally with a probability distribution on the set of copies. If there is no such assumption there are no predictions. Put differently, predictions for our observations require a statement on what exactly we mean by "us" – a specification of how we think our particular situation relates to other instances of D_0 in the universe.

In the body of this chapter we have consistently assumed that we are equally likely to be any of the incidences of our data D_0 that the third person theory (I, Ψ) predicts. This is the simplest and least informative assumption but it is not the only possible one. Other, more informative assumptions may lead to better agreement between theory and observation, be more justified by fact, and be more predictive.[8]

We call a distribution that gives the probability that we are any of the incidences of D_0 or a subset of it a *xerographic distribution* [11]. It is usually written ξ_i where the index ranges over all incidences of D_0. A xerographic distribution is effectively a formal expression of the assumption about how typical we are in the universe in the set of all other incidences of D_0.

It is the theoretical structure consisting of (I, Ψ) and ξ that yields first person predictions for observation. Each of the elements in this combination is therefore testable by experiment and observation. Just as we can compete different (I, Ψ) by their predictions for observation we can also compete different ξ.

Appendix 9.2: The No-Boundary Wave Function in Homogeneous and Isotropic Minisuperspace

The example in Section 9.7 assumed the no-boundary wave function (NBWF) for the quantum state Ψ in the third person theory (I, Ψ). The NBWF is the analog of the ground state for closed cosmologies and therefore a natural candidate for the wave function of our universe. Its predictions for observations are in good agreement with the results of actual observations. For example, it predicts that fluctuations away from homogeneity and isotropy start out near the big bang in their ground state [33]. Combined with its prediction that our universe underwent an early period of inflation [5] that leads to good agreement with the observed fluctuations in the CMB. This Appendix provides the reader a little more explanation of what that wave function is and how its consequences used in the examples are derived. For more detail see Hartle and Hawking [10].

A quantum state of the universe like the NBWF is represented by a wave function on a configuration space of three geometries and matter field configurations on a spacelike surface Σ. On the minisuperspace of homogeneous and isotropic geometries, Eq. (9.12), and a single homogeneous scalar field we write

$$\Psi = \Psi(b, \chi). \tag{9.26}$$

Here b is the scale factor of the homogeneous, isotropic metric Eq. (9.12) on Σ and χ is the value of the homogeneous scalar field. The no boundary wave function is a particular wave function of this form.

The NBWF is formally defined by a sum over homogeneous and isotropic Euclidean geometries that are regular on a topological four-disk and match b and χ on its boundary [10], weighted by $\exp(-I)$ where I is the Euclidean action of the configurations. We will not need to go into this construction. We will only need its leading order in $\hbar$ semiclassical

---

[8] As for the Boltzmann brain problem [12]. Boltzmann brains are not a problem if we assume that our observations are not typical of deluded observers who only imagine that they have the data D_0.

approximation. That is given by the saddle points (extrema) of the action $I[a(\tau), \phi(\tau)]$ on this disk for metric coupled to scalar field. There is one dominant saddle point for each (b, χ). Generally, these saddle points will have complex values of both metric and field. The semiclassical NBWF is a sum over saddle points terms of the form (in units where $\hbar = 1$)

$$\Psi(b, \chi) \propto \exp[-I(b, \chi)/\hbar] \propto \exp\left[-I_R(b, \chi) + iS(b, \chi)\right] \qquad (9.27)$$

where $I(b, \chi)$ is the action evaluated at the saddle point and I_R and $-S$ are its real and imaginary parts.

The wave function Eq. (9.27) has a standard WKB semiclassical form. As in non-relativistic quantum mechanics, an ensemble of classical histories is predicted in regions of configuration space where S varies rapidly in comparison with I_R. The histories are the integral curves of S defined by solving the Hamilton–Jacobi relations relating the momenta p_b and p_χ conjugate to a and ϕ to the gradients of S

$$p_b = \nabla_b S, \qquad p_\chi = \nabla_\chi S. \qquad (9.28)$$

There is a one-parameter family of classical histories – one for each saddle point. It is convenient to label them by the magnitude of the scalar field ϕ_0 at the center of the saddle point. One can think crudely of ϕ_0 as the value of the scalar field at which it starts to roll down to the bottom of the potential in a classical history.

The third person probabilities for these histories are proportional to the absolute square of the wave function Eq. (9.27)

$$p(\phi_0) \propto \exp[-2I_R(\phi_0)] \approx \exp[-3\pi/V(\phi_0)]. \qquad (9.29)$$

The last term is a crude approximation to the action which is useful in rough estimates [18].

This is a very quick summary of a lot of work. For more details see Hartle *et al.* [6]. In short, in the leading semiclassical approximation the NBWF predicts an ensemble of possible classical spacetimes obeying Eq. (9.28) with third person probabilities (9.13).

References

[1] S.W. Hawking and T. Hertog, *Populating the Landscape: A Top Down Approach,* Phys. Rev. D **73**, 123527 (2006), arXiv:hep-th/0602091.

[2] S. Saunders, J. Barrett, A Kent, and D. Wallace, eds, *Many Worlds*, (Oxford: OUP, 2010).

[3] J.B. Hartle, The Quantum Mechanics of Closed Systems. In B.-L. Hu, M. P. Ryan and C. V. Vishveshwara, eds., *Directions in General Relativity, Volume 1*, (Cambridge: Cambridge University Press, 1993), arXiv: gr-qc/9210006.

[4] J.B. Hartle, Spacetime Quantum Mechanics and the Quantum Mechanics of Spacetime. In B. Julia and J. Zinn-Justin, eds., *Gravitation and Quantizations*, Proceedings of the 1992 Les Houches Summer School, (Amsterdam: North Holland, 1995), gr-qc/9304006.

[5] J. B. Hartle, S. W. Hawking and T. Hertog, *The no-boundary measure of the universe,* Phys. Rev. Lett. **100**, 202301 (2008), arXiv:0711:4630.

[6] J. B. Hartle, S. W. Hawking and T. Hertog, *Classical universes of the no-boundary quantum state,* Phys. Rev. D **77**, 123537 (2008), arXiv:0803.1663.

[7] J.B. Hartle and T. Hertog, *Replication regulates volume weighting in quantum cosmology,* Phys. Rev. D **80**, 063531 (2009), arXiv:0906.0042.

[8] J. B. Hartle, S. W. Hawking and T. Hertog, *The No-Boundary Measure in the Regime of Eternal Inflation,* Phys. Rev. D **82**, 063510 (2010), arXiv:1001.0262.

[9] J. B. Hartle, S. W. Hawking and T. Hertog, *Local Observation in Eternal Inflation,* Phys. Rev. Lett. **106**, 141302 (2011), arXiv:1009.252.

[10] J. B. Hartle and S. W. Hawking, *The Wave Function of the Universe,* Phys. Rev. D **28**, 2960–75 (1983).

[11] M. Srednicki and J.B. Hartle, *Science in a Very Large Universe*, Phys. Rev. D **81** 123524 (2010), arXiv:0906.0042.

[12] J.B. Hartle and M. Srednicki, *Are We Typical?,* Phys. Rev. D **75**, 123523 (2007), arXiv:0704.2630.

[13] D. Page, *Space for both no-boundary and tunneling quantum states of the universe,* Phys. Rev. D **56**, 2065 (1997).

[14] S.W. Hawking, *Volume weighting in the no boundary proposal,* (2007), arXiv:0710.2029.

[15] J.B. Hartle and T. Hertog, *Anthropic Bounds on Lambda from the No-Boundary Quantum State*, Phys. Rev. D **88**, 123516 (2013).

[16] J. Barrow, F. Tipler, *The Anthropic Cosmological Principle,* (Oxford: Oxford University Press, 1986).

[17] S. Weinberg, *Anthropic Bound on the Cosmological Constant,* Phys. Rev. Lett., **62**, 485 (1989).

[18] G.W. Lyons, *Complex solutions for the scalar field model of the universe*, Phys. Rev. D, **46**, 1546–50 (1992).

[19] M. Tegmark, A. Aguirre, M.J. Rees and F. Wilczek, *Dimensionless constants, cosmology, and other dark matters,* Phys. Rev. D **73**, 023505 (2006).

[20] M. Livio and M.J. Rees, *Anthropic Reasoning,* Science, **309**, 1022 (2005).

[21] J. B. Hartle, *Anthropic Reasoning and Quantum Cosmology,* AIP Conf. Proc. **743**, 298 (2005), arXiv:gr-qc/0406104.

[22] A. Starobinsky, *Stochastic de Sitter (Inflationary) state in the early universe.* In H. de Vega and N. Sanchez, eds. *Field theory, quantum gravity and strings*, (Berlin: Springer-Verlag, 1986).

[23] A. D. Linde, D. A. Linde and A. Mezhlumian, *From the big bang theory to the theory of a stationary universe*, Phys. Rev. D **49**, 1783 (1986).

[24] P. Creminelli, S. Dubovsky, A. Nicolis, L. Senatore and M. Zaldarriaga, *The phase transition to slow-roll eternal inflation,* JHEP **0809**, 036 (2008), arXiv:0802.1067.

[25] A. Vilenkin, *Birth of inflationary universes*, Phys. Rev. D **27**, 2848 (1983).

[26] A.D. Linde, *Eternally existing self-reproducing chaotic inflationary universe*, Phys. Lett. B **175**, 395 (1986).

[27] L. Susskind, *The Anthropic landscape of string theory*, In Carr, B. ed. *Universe or multiverse?* (Cambridge: Cambridge University Press, 2007), pp. 247–66.

[28] T. Hertog, *Predicting a Prior for Planck,* JCAP **02** (2014) 043; arXiv:1305.6135.

[29] M. Gell-Mann and J. Hartle, *Quasiclassical Coarse Graining and Thermodynamic Entropy*, Phys. Rev. A, **76**, 022104 (2007), arXiv:quant-ph/0609190.

[30] J.A. Wheeler and K. Ford, *Geons, Black Holes and Quantum Foam*, New York: Norton & Company, (1998).

[31] H. Putnam, *Realism with a Human Face*, (Cambridge, MA: Harvard University Press, 1990).

[32] H. Everett III, *Relative State Formulation of Quantum Mechanics*, Rev. Mod. Phys., **29**, 454, (1957).

[33] J. J. Halliwell and S. W. Hawking, *Origin of Structure in the Universe,* Phys. Rev. D **31**, 1777 (1985).

10

Testing Inflation

CHRIS SMEENK

10.1 Introduction

Over the last 30 years, inflationary cosmology has been the dominant theoretical framework for the study of the very early universe. The central idea of inflation is that the
universe passed through an impressive growth spurt, a transient phase of quasi-exponential
("inflationary") expansion which sets the stage for subsequent evolution described by the
standard big bang model of cosmology. This inflationary phase leaves an imprint on various observable features of the universe. Observations can then constrain the fundamental
physics driving inflation, typically described in terms of an "inflaton" field. Traces of an
inflationary stage left in the form of temperature variations and polarization of the cosmic
microwave background radiation (CMB) are particularly revealing regarding the inflationary phase. There are currently many models of inflation compatible with the available data,
including the precise data sets generated by assiduous observations of the CMB. Yet there
are ongoing debates regarding how strongly these data support inflation. Critics of inflation
argue, among other things, that its compatibility with the data reflects little more than the
enormous flexibility of inflationary model-building. These concerns have become particularly pressing in light of the widespread acceptance of eternal inflation, which seems to
imply that all possible observations are realized somewhere in a vast multiverse.

Whether inflation can be empirically justified – whether it is "falsifiable" – is a leitmotif
of these debates. There has been little agreement among cosmologists about how to define
falsifiability, and whether it demarcates science from the rest as Popper intended.[1] The
question at issue is how to characterize a theory's empirical success, and to what extent success, so characterized, justifies accepting a theory's claims, including those that extend far
beyond its evidential basis. Success defined as merely making correct predictions, merely
"saving the phenomena", does not provide sufficient justification, for familiar reasons.
False theories can make correct predictions, and predictive success alone is not sufficient to
distinguish among rival theories that happen to agree in domains we have access to. Facile
arguments along these lines do not identify legitimate limits on the scope of scientific

---

[1] The question of inflation's falsifiability is discussed by several of the contributors to Turok (1997); Barrow and
Liddle (1997) argue that inflation can be falsified. For recent discussions from opposing viewpoints see, for
example, Ellis and Silk (2014) and Carroll (2014). These debates use Sir Karl's terminology but do not engage
in detail with his views about scientific method.

knowledge; instead, they indicate the need for a more careful analysis of how evidence supports theory. Philosophers have long acknowledged this need, and physicists have historically demanded much more of their theories than mere predictive success. Below I will focus on two historical cases exemplifying strong evidential support. The strategies illustrated in these cases generalize, and inspire an account of how theory and data should be related for a theory to meet a higher standard of empirical success. A theory that is successful by this standard arguably makes a stable contribution to our understanding of nature, in the sense that it will be recovered as a valid approximation within a restricted domain according to any subsequent theory.

Both strategies focus on mitigating the risk associated with accepting a theory. In the initial stages of inquiry, a theory is often accepted based on its promise for extending our epistemic reach. Theories allow us to use relatively accessible data to answer questions about some other domain; they provide an epistemic handle on entities or phenomena that are otherwise beyond our grasp. Inflationary cosmology allows us to gain access to the very early universe, and high energy physics, in just this sense: if inflation occurred, then observable features of the CMB reflect the dynamical evolution of the inflaton in the very early universe. Using theory to gain access to unobservable phenomena poses an obvious risk. The theory provides the connections between data and the target phenomena, and the data provide relevant evidence when interpreted in light of the theory. How does one avoid accepting a just-so story, in the form of an incorrect theory that fits the data? Demanding strong evidence at the outset of inquiry would be counter-productive, because the best evidence is typically developed through a period of theory-guided exploration. The detailed quantitative assessment of a theory is a long-term achievement. The discussion of historical cases in Section 10.2 illustrates how a theory can be tightly constrained by independent measurements, and subjected to ongoing tests as a research program develops.

These considerations suggest reformulating debates regarding the falsifiability of inflation with an assessment of two questions (Section 10.4). To what extent do observations of the early universe provide multiple, independent constraints on the physics underlying inflation? And has inflation made it possible to identify new physical features of the early universe that can be checked independently? Focusing on these questions allows for a clearer assessment of the challenges faced by cosmologists in developing evidence of comparable strength to that in other areas of physics, going beyond compatibility of inflationary models with available observations. I will argue that the main challenge to the program of reconstructing the inflaton field is a lack of independent lines of evidence. But if inflation is generically eternal, I will briefly argue that the challenges are insurmountable: eternality undermines the evidence taken to support inflation, and blocks any possibility of making a stronger empirical case.

10.2 The Determination of Theory by Evidence

Assessment of the degree of evidential support for theories, drawing distinctions among theories that all "save the phenomena", has long been a focus of epistemological discussions in physics. On one extreme, some theories are merely compatible with the phenomena, in that their success may reflect ingenuity and flexibility rather than accuracy. Although models constructed by fitting the data can be useful for a variety of purposes, they are not regarded as revealing regularities that can be reliably projected to other cases. On the other extreme, the new laws and fundamental quantities introduced by a theory are as tightly tied down by the evidence as Gulliver by the Lilliputian's ropes. Even though such theories make claims about the structure of the natural world that go far beyond the data used to support them, physicists take them to accurately capture the relevant quantities and law-like relations among them, which can then be projected to other cases and used as the starting point for further work. In numerous historical cases physicists regard a given body of evidence as strong enough to determine the correct theory.[2]

Judgments of the strength of evidence are as difficult to analyze as they are central to the practice of physics. To borrow an analogy from Howard Stein, the situation is akin to that in nineteenth century mathematics: the notion of an adequate mathematical proof, despite its significance to mathematical practice, had not yet been given a systematic treatment. Successful mathematical reasoning did not await the development of classical logic, however, just as the evaluation of physical theories proceeds without a canonical inductive logic. Below I will highlight two styles of argument that have been employed effectively in the history of physics to establish that theories have strong evidential support. Although I will not attempt to analyze these in detail, I hold that any proposed systematic account of inductive reasoning should be judged in part by its treatment of cases like these.

One style of argument exploits a theory's unification of diverse phenomena, exemplified by Perrin's famous case for atomism. Perrin argued for the existence of atoms based on agreement among 13 different ways of determining Avogadro's number N, drawing on phenomena ranging from Brownian motion to the sky's color. This case is particularly striking due to the diversity of phenomena used to constrain the value of N, and also to the ease of comparison of different results, all characterized in terms of the numerical value of a single parameter. This argument was only possible due to refinements of the atomic hypothesis, and extensions of statistical mechanics, that allowed precise formulations of relationships between the physical properties of atoms or molecules and measurable quantities. Perrin focused on N (the number of atoms or molecules in a mole of a given substance) in particular as a useful invariant quantity, and measured N in a series of experiments on Brownian motion, drawing on theoretical advances due to Einstein and others. (See Nye (1972) for a historical study of Perrin's work.) By roughly 1912, Perrin's arguments had succeeded in convincing the scientific community of the reality of atoms, decisively settling what had previously been regarded as an inherently intractable, "metaphysical" question.

This kind of overdetermination argument has been used repeatedly in the history of physics (see, in particular, Norton, 2000). One common skeptical line of thought holds that theories are inherently precarious because they introduce new entities, such as atoms,

---

[2] My approach to these issues is indebted to the work of several philosophers, in particular Harper (2012), Norton (1993) (from whom I've borrowed the title of this section), Norton (2000), Smith (2014), and Stein (1994).

in order to unify phenomena. Success fitting a body of data, the skeptic contends, merely reflects the flexibility of these novel theoretical constructs. The consistent determination of theoretical parameters from diverse phenomena counters the worry that the theory only succeeds due to a judicious tuning of free parameters. The overdetermination argument shows that, rather than the piecemeal success the skeptic expects, the theory succeeds with a single choice of parameters. The strength of this reply to the skeptic depends on the extent to which the phenomena probe the underlying theoretical assumptions in distinct ways. Furthermore, the diversity of phenomena minimizes the impact of systematic errors in the measurement of the parameters. The sources of systematic error relevant to Perrin's study of Brownian motion have little to do with those related to measurements of N based on radioactivity, for example. As the number of independent methods increases, the probability that the striking agreement can be attributed to systematic errors decreases.

The conclusion to be drawn from the overdetermination argument depends upon how unlikely the agreement is antecedently taken to be. The truth of the atomic hypothesis and kinetic theory implies an equation relating N to a number of quantities measurable by experimental study of Brownian motion, a second equation relating N to radioactivity, and so on. If the atomic hypothesis were false, there is no reason to expect these combinations of measurable quantities from different domains to all yield the same numerical value, within experimental error. This claim reflects an assessment of competing theories: what is the probability of a numerical agreement of this sort, granting the truth of a competing theory regarding the constitution of matter? The overdetermination argument has little impact if there is a competing theory which predicts the same numerical agreements. In Perrin's case, by contrast, the probability assigned to the agreeing measurements of N, were the atomic hypothesis to be false, is arguably very low. In arguing for a low antecedent probability of agreement, Perrin emphasized the independence and diversity of the phenomena used to determine the value of N. (Obviously this brief account highlights only one aspect of Perrin's argument; see Chalmers (2011); Psillos (2011) for more thorough treatments.)

There is a second respect in which the conclusions to be drawn from an overdetermination argument must be qualified. These arguments typically bear on only part of a theory, namely whatever is needed to derive the connections between theoretical parameters and measurable quantities. Perrin's case is unusual in that the evidence bears directly on the central question in the dispute regarding atomism, unlike other historical cases in which this style of argument was not as decisive. The strength of an overdetermination argument depends on whether there is sufficient evidence to constrain all of a theory's novel components, or at least the ones at issue in a particular debate. The argument only directly supports parts of the theory needed to establish connections between measurable quantities and theoretical parameters. Identifying the distinct components of a theory and clarifying their contribution to its empirical success is often quite challenging, as the acceptance of the aether based on success of electromagnetic theory in the nineteenth century illustrates.

The second style of argument focuses on evidence that accumulates over time as a theory supports ongoing inquiry. A physical theory introduces a set of fundamental quantities and laws holding among them that provide the means for explicating some domain of phenomena. Accepting a theory implies a commitment to account for phenomena within this domain on the theory's terms, under the pressure of new discoveries and improving standards of experimental and observational precision. Often this involves treating complex phenomena via a series of successive approximations, with further refinements driven by discrepancies between current theoretical descriptions and observations. Resolving discrepancies by adding further details, without abandoning basic commitments, provides evidence that the theory accurately captures the fundamental physical relationships. The evidence is particularly strong when this process uncovers new features of the system that can be independently confirmed.

Newtonian gravitational theory supported centuries of research in celestial mechanics in just this sense. With the benefit of gravitational theory, one could approach enormously complicated orbital motions, such as that of the Moon, via a series of idealizations that incorporate physical details thought to be relevant. Throughout the history of celestial mechanics, there have nearly always been systematic discrepancies between observations and trajectories calcuated based on all the relevant details known at a given time. Subsequent efforts then focused on identifying details left out of the calculation that might resolve the discrepancy. Leverrier's inference that an undiscovered planet was the source of discrepancies in Uranus's orbit is perhaps the most famous example of this type of reasoning. But in most cases the physical source that was eventually identified was not as concrete as an additional planet; the secular acceleration of the Moon, for example, results from the slowing rotation of the Earth due to tidal friction. The new details are then incorporated in a more elaborate model, and the search for discrepancies continues. By the early twentieth century, calculations of orbital motion included an enormous number of details. The theory was sufficiently precise to reveal very subtle discrepancies, such as systematic errors in determining sidereal time due to a periodic fluctuation in the Earth's rotational speed.

Smith (2014) convincingly argues that the success of this line of inquiry provides much stronger support for Newtonian gravity than is apparent if the theory is simply treated as making a series of successful predictions. Theoretical models of celestial motions had to be in place to even identify the small discrepancies that were the target of analysis, and in that sense the theory itself underwrites its detailed comparison with observation. The core commitments of the theory place stringent constraints on the kind of new physical details that can be introduced to account for discrepancies. Furthermore, these additions could usually be checked using methods that did not depend on gravity – as with the discovery of Neptune in the location predicted by Leverrier, or measurements of the periodic fluctuation in the Earth's rotation, initially detected by astronomers, using atomic clocks. These independent checks on the details incorporated in ever more elaborate models support the theory's claim to have accurately identified the appropriate quantities and laws. It would be an enormous coincidence for a fundamentally incorrect theory to be so useful in discovering new features of the solar system.

These two historical cases illustrate strategies, arguably used throughout physics, to provide an effective response to skepticism regarding theoretical knowledge. This skepticism is inspired by the apparent circularity of relying so heavily on the very theory in question to support detailed comparison with observations. When interpreted with the aid of the theory, the phenomena yield constraints on the parameters of the theory, and discrepancies that can be the target of further work; yet neither the constraints nor the discrepancies are readily available without the guidance provided by theory. How do we guard against accepting a theory that is self-certifying, sufficiently flexible to avoid refutation despite its flaws? In both examples above, the response to this worry relies upon using multiple, independent lines of evidence, while acknowledging the theory-dependence of the reasoning. The prosaic point underlying this response is that using multiple sources of information, dependent on theory in different ways and with different sources of systematic error, minimizes epistemic risk. This response shifts the burden of proof onto the skeptic: if the underlying theory were false, it would be an enormous coincidence for all of the multiple ways of measuring a parameter to coincide, or for new features added to resolve discrepancies to be independently confirmed.

A second skeptical objection regards the nature of the claims supported by these arguments: can they be regarded as stable contributions to our knowledge of the natural world? Perrin made the case for atomism prior to the advent of quantum theory, and the reasoning in celestial mechanics described above precedes Einstein. Are these arguments undermined by quantum mechanics and general relativity, respectively? As a brief reply to this Kuhnian worry, consider the nature of the claims that are supported by the arguments above. These are claims that specific law-like relations hold between different physical quantities within some domain. Perrin's case depends upon the relations between atomic scale properties and macroscopic, measurable properties. The development of celestial mechanics supports a variety of claims about what features of the solar system are relevant to planetary motions. In these two cases, the claims in question are arguably preserved through theory change, in the sense that there are lawlike relations in the successor theory which are approximated, within a restricted domain, by corresponding relations in the preceding theory.[3] This is true in spite of the dramatic conceptual differences between classical mechanics and quantum mechanics, and between Newtonian gravity and general relativity. As a consequence, the reasoning employed in arguing for the atomic hypothesis and in the

---

[3] Determining how to recover the preceding theory in an appropriate limit is often surprisingly subtle, and I do not have the space to explore the issue fully here. In the case of Perrin's argument, for example, the central assumptions of kinetic theory Perrin used in his study of Brownian motion are good approximations to a quantum statistical mechanical treatment, except in cases of low temperatures or high densities; a full discussion would consider the approximations involved in the other methods for measuring N as well. For celestial mechanics, the claims in question regard the impact of, for example, Neptune on Uranus's orbit. The current practice of modeling the solar system using Newtonian physics with general relativistic corrections presupposes that the Newtonian description is a valid approximation. Finally, this claim regarding continuity does not require that the earlier theory provides a full account of the phenomena. Perrin did not have a complete theory of the nature of molecules, for example; he was well aware that the problem of specific heats identified by Maxwell had not been solved.

development of celestial mechanics is validated rather than undermined by the successor theories.

There may be cases in which a new theory recovers only the predictions of an earlier theory, without establishing the validity of its evidential reasoning in this stronger sense. I do not intend to rule out that possibility by fiat; the evidence may simply unravel. But the burden of proof rests with the new theory to explain away the apparent successes of an old theory when the latter is not recovered as an approximation. In the cases described above, no one has imagined a credible alternative theory that matches the successes of the atomic hypothesis or Newtonian gravity without recovering aspects of these theories as limiting cases. This is the qualified sense in which theories supported by arguments like those described above constitute a stable contribution to our understanding of nature.

10.3 The Standard Model and Inflation

The standard model of cosmology (SMC) is based on bold extrapolations of theories that have been well tested by Earth-bound experiments and astronomical observations. The interpretation of cosmological data depends, to varying degrees, on a background cosmological model, and hence assumes the validity of extrapolating general relativity to length scales roughly 14 orders of magnitude greater than those where the theory is subject to high precision tests. The SMC describes the contents of the universe and their evolution based on the Standard Model of particle physics, supplemented with two distinctive types of matter – dark matter and dark energy – that have so far only been detected due to their large-scale gravitational effects. Cosmological observations performed over the last few decades substantiate the enormous extrapolations and novel assumptions of the SMC.

The development of a precise cosmological model compatible with the rich set of cosmological data currently available is an impressive achievement. Yet cosmology clearly relies very heavily on theory; the cosmological parameters that have been the target of observational campaigns are only defined within a background cosmological model. The SMC includes several free parameters, such as the density parameters characterizing the abundance of different types of matter, each of which can be measured by a variety of different types of observations.[4] CMB observations, in particular, place powerful constraints on many cosmological parameters. (Inferences to parameter values from observations of the CMB typically require prior assumptions regarding the nature of the primordial power spectrum, and there are several parameter degeneracies that cannot be resolved based solely on CMB observations.) There is a variety of independent ways of measuring the cosmological parameters that depend on different aspects of theory and have different sources of

---

[4] See Beringer *et al.* (2012) for a review of evidence bearing on the cosmological parameters. The total number of parameters used to specify a cosmological model varies in different studies, but typically five to ten fundamental parameters are used to determine the best fit to a given data set. (Specific models often require a variety of further "nuisance parameters" to account for astrophysical processes.)

observational error. For example, the abundance of deuterium produced during big bang nucleosynthesis depends sensitively on the baryon density. Nucleosynthesis is described using well-tested nuclear physics, and the light element abundances are frozen in within the "first three minutes". The amplitudes of the acoustic peaks in the CMB angular power spectrum can be used to determine the baryon density at a later time (recombination, at $t \approx 400,000$ years), based on quite different theoretical assumptions and observational techniques. Current measurements fix the baryon density to an accuracy of 1 per cent, and the values determined by these two methods agree within observational error. This agreement (augmented by other consistent measurements) is an important consistency check for the SMC.

The strongest case for accepting the SMC rests on the evidence in favor of the underlying physics in concert with the overdetermination of cosmological parameters. (See, e.g. Peebles *et al.* (2009), §5.4 for a brief discussion of tests of the SMC emphasizing the importance of independent measurements of the parameters.) The overdetermination argument has a similar structure to Perrin's argument for atomism described above. The case for the SMC does not yet have the diversity of independent lines of measurement that made Perrin's case so powerful; there are unexplained discrepancies among some measurements; individual measurements are not as precise as those available in many other areas of physics; and there are theoretical loopholes related to each measurement. But the essential epistemic point is the same: due to the diversity of measurements, it is unlikely that evaluation of the SMC has been entirely misguided due to incorrect theoretical assumptions or systematic observational errors. Several lines of observational and theoretical work currently being pursued promise to substantially strengthen the evidential support for the SMC.

Several of the cosmological parameters characterize the universe's "initial" state.[5] The SMC describes the large-scale structure of the universe as a perturbed Friedmann–Lemaître–Robertson–Walker (FLRW) model. The FLRW models are homogeneous and isotropic solutions of Einstein's field equations (EFE). The models are topologically $\Sigma \times \Re$, visualizable as a "stack" of three-dimensional spatial surfaces $\Sigma(t)$ labeled by cosmic time t. The worldlines of "fundamental observers", at rest with respect to matter, are orthogonal to these surfaces, and the cosmic time corresponds to their proper time. EFE simplify to two equations governing $R(t)$, the spatial distance between fundamental observers.

One cosmological parameter – the spatial curvature, Ω_k – characterizes which of the FLRW models best fits observations. It is constrained by observations to be very close to zero, corresponding to the "flat" FLRW model whose spatial hypersurfaces $\Sigma(t)$ have zero curvature. For the flat model, the total energy density takes exactly the value ($\Omega = 1$)

---

[5] This is often taken to be the state as specified at the "boundary of the domain of applicability of classical GR" — e.g. at the Planck time, $t \approx 10^{-43} s$. Appropriate initial data for EFE specify the full solution (for globally hyperbolic spacetimes), so the choice of a specific cosmic time at which to characterize the initial state is a matter of convention. But this conventional choice is significant if a given solution is treated as a perturbed FLRW solution, since dynamical evolution modifies the power spectrum of perturbations.

needed to counteract the initial velocity of expansion, $\dot{R} \to 0$ as $t \to \infty$. $\Omega =: \frac{\rho}{\rho_c}$, where ρ is the mass-energy density and the critical density is defined as $\rho_c = \frac{3}{8\pi} \left(H^2 - \frac{\Lambda}{3} \right)$. H is the Hubble "constant" (which in fact varies with cosmic time), defined as $H = \frac{\dot{R}}{R}$, and Λ is the cosmological constant. Other parameters characterize perturbations to the underlying FLRW model, which are fluctuations in mass density needed to provide the seeds for structure formation via gravitational clumping. If these fluctuations obey Gaussian statistics, they can be fully characterized in terms of a dimensionless power spectrum $\mathcal{P}(k)$. The power spectrum of the primeval mass distribution in the SMC takes the simple form of a power law, $\mathcal{P}(k) \propto k^{n_s}$. This power spectrum is parametrized in the SMC by two numbers. The first, the spectral index n_s, is equal to unity if there is no preferred scale in the power spectrum; observations currently favor $n_s = 0.96$, indicating a slight "blue tilt" in the power spectrum, with less power on smaller length scales. A second number is needed to specify the amplitude of the perturbations. (There are a few different ways of doing so. For example, σ_8 is the mass variance of the primordial distribution within a given radius (defined in terms of another parameter, the distance scale h: $8h^{-1} \approx 11$ *Mpc*, given current estimates of h), projected forward to the current time using linear perturbation theory).

The initial state required by the SMC has three particularly puzzling features. First, it is surprising that the simple, uniform FLRW models can be used at all in describing the early universe. These models have a finite horizon distance, much smaller than the scales at which we observe the CMB.[6] The observed isotropy of the early universe – revealed most strikingly by the temperature of the CMB – supports the use of the FLRW models; yet these observations cover thousands of causally disjoint regions. Why did the universe start off with such a glorious pre-established harmony? Second, an FLRW model close to the "flat" model, with nearly critical density at some specified early time is driven rapidly away from critical density under FLRW dynamics; the flat model is an unstable fixed point under dynamical evolution. In order for observations at late times to be compatible with a flat model, the initial state has to be *very* close to the flat model (or, equivalently, *very* close to critical density, $\Omega = 1$). (It follows from the FLRW dynamics that $\frac{|\Omega - 1|}{\Omega} \propto R^{3\gamma - 2}(t)$. $\gamma > 2/3$ if the strong energy condition holds, and in that case an initial value of Ω not equal to 1 is driven rapidly away from 1. Observational constraints on $\Omega(t_0)$ can be extrapolated back to a constraint on the total energy density of the Planck time, namely $|\Omega(t_p) - 1| \leq 10^{-59}$.)

Finally, the perturbations to the flat FLRW model postulated in the SMC are challenging to explain physically. It is not clear what physical processes could account for the amplitude of the perturbations. Suppose, for example, that one takes the "initial" perturbation spectrum to be imprinted at $t_i \approx 10^{-35}$s. Observations imply that at this time the initial perturbations would be far, far smaller than thermal fluctuations. (Blau and Guth (1987)

---

[6] A horizon is the surface in a time slice t_0 separating particles moving along geodesics that could have been observed from a worldline γ by t_0 from those which could not. For a radiation-dominated FLRW model, the expression for horizon distance d_h is finite; the horizon distance at decoupling corresponds to an angular separation of $\approx 1°$ on the surface of last scattering.

calculate that observations imply a density contrast $\frac{\delta\rho}{\rho} \approx 10^{-49}$ at t_i, nine orders of magnitude *smaller* than thermal fluctuations.) In addition, the perturbations of the appropriate scale to eventually form galaxies would, in the early universe, be coherent at scales that seem to conflict with the causal structure of the FLRW models. A simple scaling argument shows that the wavelength λ of a given perturbation "crosses the horizon" with expansion, at the time when $\lambda \approx H^{-1}$ (where H^{-1} is the Hubble radius). Assuming that the perturbation spectrum is scale invariant, and for a simple model with $R(t) \propto t^n$ ($n < 1$) the wavelength of a given mode simply scales with the expansion $\lambda \propto t^n$. H^{-1} scales as $H^{-1} \propto t$; as a result, the Hubble radius "crosses over" perturbation modes with expansion. Prior to horizon-crossing, the perturbation would have been coherent on length scales greater than the Hubble radius. The Hubble radius is typically regarded as marking the limit of causal interactions, and as a result it is puzzling how normal physics operating in the early universe could produce coherent perturbations at such scales.[7]

Since the late 1970s, cosmologists have sought a physical understanding of how such an unusual "initial state" came about. On a more phenomenological approach, the gravitational degrees of freedom of the initial state could be chosen to fit with later observations. Inflation in effect replaces such a specification with a hypothesis regarding the initial conditions and dynamical evolution of a proposed "inflaton" field (or fields). In the simplest inflationary models, a single field ϕ, trapped in a false vacuum state, triggers a phase of exponential expansion. If the inflaton field ϕ is homogeneous, then the false vacuum state contributes an effective cosmological constant to EFE, leading to quasi-de Sitter expansion.[8]

The resulting spurt of inflationary expansion can provide a simple physical account of the SMC's starting point, as emphasized with sufficient clarity to launch a field by Guth (1981). Inflationary expansion stretches the horizon length; for N "e-foldings" of expansion the horizon distance d_h is multiplied by e^N. For $N > 65$ the horizon distance, while still finite, encompasses the observed universe. The observed universe could then have evolved from a single pre-inflationary patch, rather than encompassing an enormous number of causally disjoint regions. (This pre-inflationary patch is larger than the Hubble radius (Vachaspati and Trodden, 2000), however, so inflation does not dispense with pre-established harmony.) During an inflationary phase the density parameter Ω is driven *towards* one. (This is apparent from the equation above, given that $\gamma = 0$ during inflation.) An inflationary stage long enough to solve the horizon problem drives a large range of

---

[7] The Hubble radius is defined in terms of the instantaneous expansion rate $R(t)$, by contrast with the horizon distance, which depends upon the expansion history over some interval (the particle horizon, e.g. depends on the full expansion history). For radiation or matter-dominated solutions, the two quantities have the same order of magnitude.

[8] The stress-energy tensor is given by $T_{ab} = \nabla_a\phi\nabla_b\phi - \frac{1}{2}g_{ab}\left(g^{cd}\nabla_c\nabla_d\phi - V(\phi)\right)$, where $V(\phi)$ is the effective potential; for a homogeneous state, such that $V(\phi) >> g^{cd}\nabla_c\nabla_d\phi$, $T_{ab} \approx -V(\phi)g_{ab}$, leading to $R(t) \propto e^{\xi t}$ with $\xi^2 = \frac{8\pi V(\phi)}{3}$.

pre-inflationary values of $\Omega(t_i)$ close enough to 1 by the end of inflation to be compatible with observations.

The most remarkable feature of inflation, widely recognized shortly after Guth's paper, was its ability to generate a nearly scale-invariant spectrum of density perturbations with correlations on length scales larger than the Hubble radius.[9] Inflation produces density perturbations by amplifying vacuum fluctuations in a scalar field ϕ, with characteristic features due to the scaling behavior of the field through an inflationary phase. Start with a massless, minimally coupled scalar field ϕ evolving in a background FLRW model. The Fourier modes ϕ_k of linear perturbations to a background solution are uncoupled, with evolution like that of a damped harmonic oscillator. For modes such that $\frac{k}{R} \ll H$, the damping term is negligible, whereas those with $\frac{k}{R} \gg H$ evolve like an over-damped oscillator and "freeze in" with a fixed amplitude. The inflationary account runs very roughly as follows. (This behavior follows from the Klein–Gordon equation in an FLRW spacetime, considering linearized perturbations around a background solution; see Mukhanov *et al.* (1992) for a comprehensive review of the evolution of perturbations through inflation, or Liddle and Lyth (2000) for a textbook treatment). Prior to inflation one assumes a vacuum state, i.e. the modes ϕ_k are initially in their ground state. For $\frac{k}{R} \ll H$ the modes evolve adiabatically, remaining in their ground states. This account is not sensitive to exactly when a given mode is assumed to be born in its ground state. During inflation the modes scale with the exponential expansion whereas H is approximately constant. Due to this scaling behavior, modes will reach the horizon scale $\frac{k}{R} \approx H$ — "horizon exit". The damping term is then no longer negligible and the modes "freeze in" as they cross the horizon. Modes then "re-enter" the horizon after inflation has ended, because in standard FLRW expansion the scaling behavior is reversed. Finally, these modes are treated as classical density perturbations upon re-entering the horizon. (This is a quantum to classical transition; whether it can be justified by appeals to squeezing of the quantum state and decoherence is contentious.) This evolution leads to a nearly scale invariant spectrum, with the amplitude of the pertubations fixed by the energy scale of inflation at horizon exit (as discussed below). (The spectrum is not *exactly* scale invariant because the Hubble constant is not truly constant throughout inflation.)

To provide an account of the SMC's initial state, the inflationary phase has to be followed by a stage called "re-heating". Any matter or radiation present prior to inflation is rapidly diluted away during inflationary expansion, leaving a universe that is essentially empty except for the inflaton field. Reheating is required to fill the universe with matter and radiation, with temperature and densities appropriate for subsequent evolution within the standard big bang model.

---

[9] This is sometimes regarded as a successful prediction of inflation, since this feature of inflation was initially not known to many researchers (despite early results, including Mukhanov and Chibisov, 1981; Starobinsky, 1979, 1980). Yet the initial prediction based on specific models under discussion (with inflation driven by a Higgs field in a grand unified theory) was incorrect, as the amplitude of pertubations was far too large. Insuring the correct amplitude leads to one of the "fine-tuning" problems of inflation, since the coupling of the scalar field driving inflation has to be very small.

Inflation provides a physical account of otherwise puzzling features of the starting point for the SMC. This is often described as solving "fine-tuning" problems of the SMC. Imagine choosing a cosmological model at random from among the space of solutions of EFE. Even without a well-defined measure on this space of solutions, it seems obvious that an FLRW model (or a perturbed FLRW model) must be an incredibly "improbable" choice. According to the SMC alone, what we observe is incredibly improbable; according to the SMC *plus inflation*, on the other hand, what we observe is to be expected, because "generic" pre-inflationary states lead to an appropriate starting point for the SMC. There are several objections to this line of argument, some of which go back to an incisive early criticism by Penrose.[10] Perhaps the most fundamental objection regards the starting point for the argument: why should we treat the initial state as "generic", a "random choice" from among all possible states? (Possible according to which theory?) It is also not clear that inflation succeeds by its own lights: Penrose, in particular, argued that a pre-inflationary patch with an appropriate state to trigger the onset of inflation should be *less likely* than an initial state for the SMC (without inflation). I will not explore these issues further here, in part because many proponents of inflation apparently regard the emphasis on fine-tuning as part of the initial motivation for inflation that can now be replaced with a more powerful empirical argument (see, e.g. Liddle and Lyth 2000, p. 5), to which I now turn.

10.4 Assessing Inflation

Inflation provides a promising account of the origins of the initial state for the SMC, and at the same time opens up the prospect of using observations of the CMB and large scale structure to constrain physics at an energy scale of $\approx 10^{15} - 10^{16}$ GeV. Unlike other competing theories, it has not been ruled out as the observational picture of the early universe has come into sharper focus over the last 30 years. Observations have led to a remarkably simple picture of the state of the early universe, which is well described by a flat FLRW model, with Gaussian, adiabatic, linear, nearly scale invariant density pertubations. Inflation generates primordial fluctuations in the very early universe, at length scales larger than the Hubble radius. As they cross the Hubble radius, they set up coherent oscillations leading to acoustic peaks in the CMB power spectrum. Observations of acoustic peaks support the primordial nature of the fluctuations, contrasting with predictions from competing models of structure formation based on active sources for fluctuations (such as topological defects). (See, e.g. Durrer et al. (2002) for a review of structure formation via topological defects, and its contrasting predictions for CMB anisotropies.) That inflation is compatible with this observational picture of the early universe is an important success.[11] Does this amount to mere compatibility with the data, or does inflation fulfill its promise of providing

[10] See Penrose 2004, Chapter 28 for a recent exposition of the arguments he first made in the early 1980s; see also Albrecht (2004); Earman and Mosterin (1999); Carroll and Tam (2010); Gibbons and Turok (2008); Hollands and Wald (2002b) for related discussions.

[11] Here I will not address other conceptual and theoretical problems related to inflation, discussed in, e.g. Brandenberger (2014); Earman and Mosterin (1999); Hollands and Wald (2002b); Turok (2002).

a physical understanding of the early state? Here I will briefly assess challenges to providing stronger evidence in favor of inflation, based on following the strategies described above.

The inflaton is typically treated as a new field to be added to the Standard Model of particle physics. Michael Turner called inflation a "paradigm without a theory" to emphasize the resulting flexibility of inflation. A bewildering variety of different inflationary models has been proposed, so many that theorists complain of the difficulty in finding an unused name for a new model. Many of the models can be characterized in terms of the Lagrangian proposed for the inflaton field:

$$\mathcal{L} = -\frac{1}{2}g^{ab}\partial_a\phi\partial_b\phi - V(\phi) + \mathcal{L}_I(\phi, A_a, \psi, ...),\tag{10.1}$$

where $V(\phi)$ is the effective potential, and $\mathcal{L}_I$ is an interaction term, specifying interactions with other fields in the Standard Model. Assuming that inflation is driven by a single field with a Lagrangian with this form already reflects some simplifications. Inflationary models with multiple scalar fields have been developed, motivated by proposals in high-energy physics that include many light scalar fields expected to be dynamically relevant in the early universe. But Planck observations support restricting attention to simple single-field models. Planck 2015 data provides strong evidence that the perturbations are adiabatic, which is compatible with simple single-field models; the failure to detect non-Gaussianities further supports the use of single-field models, and the choice of a standard kinetic term (the first term) in the Lagrangian (see, e.g. the discussions in Ade *et al.* (2015), §10, and Martin (2015)). A model from this class is characterized by a choice of the effective potential $V(\phi)$ and interaction term $\mathcal{L}_I$, along with assumptions regarding initial conditions for the field.

Observations of the CMB and large scale structure constrain the Lagrangian in two main ways. The primordial fluctuations place constraints on the effective potential well before the end of the inflationary phase. Inflation generates scalar and tensor perturbations whose physical properties depend on the features of the effective potential $V(\phi)$ at horizon exit, with $\frac{k}{R} \approx H$. Perturbations relevant to CMB observations typically crossed the horizon at ≈ 60 e-foldings before the end of inflation, whereas those that are re-entering the horizon now were produced a few e-foldings later. (The calculation is model-dependent and depends on assumptions regarding the reheating temperature; this estimate holds for a variety of slow-roll models with plausible further assumptions.) The features of scalar and tensor perturbations amplified through the inflationary phase can be described, in some cases, with equations relating the perturbation spectra to the value of $V(\phi)$ and its derivatives (at the scale when the perturbations crossed the horizon). Equations have been derived for models satisfying the slow-roll approximation, although it is not possible in general to calculate the perturbation spectra for an arbitrarily chosen $V(\phi)$. "Slow-roll" models feature a flat effective potential, such that (roughly) $V', V'' << V$, leading to a long inflationary phase, sufficient to solve the horizon and flatness problems. (Here $'$ is the derivative $\frac{d}{d\phi}$. The slow-roll conditions are constraints on V', V'' which insure that the damping term $(3H\dot{\phi})$ dominates over $\ddot{\phi}$ in the equation of motion for the inflation field: $\ddot{\phi} + 3H\dot{\phi} + V'(\phi) = 0$.) For these models there are simple expressions for the amplitude, spectral index, and "running" of the spectral index for both scalar and tensor perturbations in terms of V, V', V''. The scalar and tensor perturbation spectra are not independent, and a consistency relation, relating the spectral index of the tensor perturbations to the ratio of amplitudes of scalar and tensor perturbations, r, can be obtained by solving to eliminate V. (More generally, in a perturbative treatment there is a hierarchy of consistency relations. There are a few different parametrizations used in relating the effective potential to observable features of the perturbations, including slow-roll parameters and Hubble flow parameters.)

A successful account of reheating depends on a different part of $V(\phi)$, along with the interaction term $\mathcal{L}_I$. Early accounts of inflation treated reheating as occurring when the inflaton field oscillated near the true minima of $V(\phi)$, assumed to be much steeper than the flat plateau needed for slow-roll, transferring energy to other particle species. Subsequent work has focused on energy transfer from the inflaton field to other particle species via coherent oscillations with parametric resonance. Observational constraints on the details of reheating are weaker than those related to the generation of primordial fluctuations (see, e.g. Martin *et al.*, 2015).

There are several different approaches to reconstructing the inflaton potential from observations, and evaluating competing inflationary models (see, e.g. Lidsey *et al.*, 1997; Martin *et al.*, 2013). Given the wealth of observational data already available, upcoming observations, and the large variety of inflationary models, these techniques naturally focus on determining which models best fit the data. Martin *et al.* (2013), for example, adopt a Bayesian approach to analyze 193 single-field slow-roll inflationary models, concluding that nine models with "plateau"-shaped potentials are preferred. The method relies on statistical tools optimized for determining the best model given the inherent noise and uncertainty of observational data. The ranking weighs the closeness of fit to the data a given model achieves against the complexity of the model, to avoid the pitfall of overfitting the data, but it is not designed to assess physical plausibility of a given model. Although this issue deserves further scrutiny, the epistemic point addressed in the model selection literature is distinct from the question addressed by the historical strategies discussed above. Finding the best fit model, granting the general framework used for interpreting the data, is not the same as evaluating the framework itself, although obviously the existence of a successful model – or lack of one – is relevant to this second task. The historical strategies aim to assess the validity of the underlying framework, to guard against being systematically misled by accepting an incorrect framework that nevertheless accommodates the data.

Turning to the first strategy discussed above, observations do provide independent constraints on the underlying inflationary mechanism for amplifying perturbations. A scale-invariant spectrum of scalar perturbations was proposed well before inflation on general grounds (Harrison, 1970; Peebles and Yu, 1970; Zel'dovich, 1972). But there is not a similar argument in favor of a scale-invariant tensor perturbation spectrum, or any theory-independent reason to expect the two spectra to be linked as reflected in the consistency relation. Furthermore, measurements of the tensor perturbations directly constrain $V(\phi)$ at the point where a given length scale crossed the horizon. Measuring the tensor perturbation spectrum at different length scales, if it were feasible observationally, would give a direct reconstruction of $V(\phi)$.[12] Detection of CMB B-mode polarization, leading to a measurement of r, along with a measurement of the spectral index for tensor perturbations, n_t, directly tests the consistency relation. Measuring r is the target of a number of post-Planck missions, but the follow-up measurement of n_t is particularly challenging for small values of r. The possibility of nailing down the inflationary mechanism for amplifying perturbations in this fashion is certainly one of inflation's most appealing features.

There are, however, several contrasts with overdetermination arguments such as Perrin's. The first contrast regards the target of the argument, the Lagrangian for the inflaton field – and in particular the function $V(\phi)$ and the various couplings included in the interaction term $\mathcal{L}_I$. It is obviously much more challenging to provide a compelling overdetermination argument for the Lagrangian as opposed to a single number N. Furthermore, the existing observational constraints apply to two distinct dynamical regimes of the inflaton's evolution: the amplification of quantum fluctuations at horizon crossing, ≈ 60 e-folds before the end of inflation, compared with the decay of the inflaton and reheating at the very end of the inflationary phase. Inflationary models are a package deal rather than a single ticket: without theoretical constraints on the properties of the inflaton field, one can choose the shape of the potential relevant to amplification of perturbations, and then separately choose the shape of the potential near the true minimum and couplings in the interaction term. As long as this remains a relatively free choice, with weak constraints imposed in either direction, success in these two distinct dynamical regimes does not provide overlapping constraints.

The evidential situation changes substantially for an inflaton Lagrangian that is identified within a specific particle physics model. In such a case, the parameters appearing in the Lagrangian are constrained by cosmological data related to the details of inflation, as well as whatever experimental data are relevant to the particle physics model. This would provide a compelling set of independent constraints. Furthermore, since the inflaton model would be a single ticket item in this case, different cosmological measurements provide overlapping constraints on the Lagrangian. Yet the promise of directly identifying a canonical candidate for the inflaton field has not been fulfilled; instead, there has been a proliferation of toy models of inflation. Constructing physically plausible models for the inflaton has been difficult because $V(\phi)$ has to be very flat. The prospects for re-establishing a tighter link through direct experimental study are extremely bleak: the

---

[12] This may even include constraints based on solar-system scale gravitational wave observatories. Boyle *et al.* (2014) argue that if there is a tilt in the tensor perturbation spectrum, as suggested by the BICEP2 initial results, the proposed Big Bang Observatory would provide a second set of measurements at a scale 10^{18} smaller than those relevant for the CMB, providing an enormous lever arm for more precise tests of inflation. See also Alvarez *et al.* (2014) for an overview of tests of inflation based on large-scale structure.

properties required for an inflaton field in a slow-roll model insure that it can only feasibly be studied observationally through its impact on the early universe.

Even without resolving the identity of the inflaton, the case for inflation can be strengthened by imposing other constraints on the class of allowed models. In practice this is reflected in assessments of the plausibility of different inflationary models, given assumptions about physics at the appropriate energy scale. There have also been proposals to characterize how inflationary predictions depend upon the amount of fine-tuning of the potential, which lead to constraints on the parameter ranges compatible with less finely-tuned potentials. (Boyle *et al.* (2006), for example, characterize fine-tuning in terms of the number of zeroes appearing in the slow-roll parameter η and its first derivative (with respect to the number of e-folds), which is intended as a measure of the number of "features" added to the effective potential; inflaton models with little fine-tuning in this sense favor specific parameter ranges for n_s, r.) On either of these approaches, considerable weight is put on the further constraints imposed in the name of plausibility or simplicity. Past debates regarding the viability of different types of models make the challenges to achieving consensus on these questions clear.[13]

A final contrast regards the assessment of alternatives. Perrin argued that the agreeing measurements of N would be an enormous coincidence if the atomic hypothesis were false. How likely is the simple early state required by the SMC, if inflation did not occur? Turok (2002), for example, remarks that "*The success of the simplest inflationary models is perhaps more of a success for simplicity than it is for inflation*" (p. 3458, emphasis original). Any early universe theory that generates a nearly scale invariant spectrum of primordial fluctuations will match many of inflation's successes. Theorists have discovered several different ways of generating such fluctuations, ranging from alternative ways of modifying causal structure (varying speed of light theories) to "bounce" models, which replace the Big Bang singularity with a Big Bounce. They treat the primordial fluctuations as generated prior to the bounce, although details of implementation, and the physics used to construct the model, differ substantially among the different models. (See Brandenberger (2014) for an overview of the matter bounce and string gas models, and Lehners (2008) for a review of ekpyrotic and cyclic models.) There is no reason to expect a consistency relation between tensor and scalar perturbations in a "simple" initial state, and this relation also discriminates between inflation and other models for the generation of primordial fluctuations. Further observational work, in particular detection of primordial gravitational waves and tests of the consistency relation, would lead to a much stronger case that the observed properties of the early universe would be an enormous coincidence if inflation were false.

---

[13] For example, inflation is commonly taken to predict a flat universe with $\Omega_0 = 1$. There were heated debates in the mid to late 1990s regarding so-called "open inflation" models that yield a value of $\Omega_0 \approx 0.2 - 0.3$, which was at that time favored by observations. Insofar as these were regarded as plausible models, inflation no longer predicts flatness, and the value of Ω_0 instead provides a constraint on the parameter space for models (see, e.g. the discussions in Turok, 1997).

Regarding the second strategy, I am unaware of any case in which inflation has been used to uncover a new feature of the early universe that can be independently checked. There is a variety of ways in which inflationary model-building has become more sophisticated, with a much clearer understanding of what needs to be in place in a full account of inflation. There is no shortage of theoretical innovation in building inflationary models. Inflation has also guided observational work by identifying features that can be used to constrain inflationary models and contrast inflation with competing theories. But the question is whether inflation has allowed cosmologists to identify robust physical features of the early universe that can be tested in ways that do not assume inflationary theory itself. There are no analogs, as far as I am aware, of adding a new physical feature as part of the model that can, like the existence of Neptune, be easily checked by other means. This is in part due to the observational inaccessibility of the early universe, but also to the lack of a canonical choice of the inflaton field. Given a fixed choice for the inflaton field, discrepancies with observations would force theorists to elaborate the model, possibly identifying new features of the early universe in the process. At present the choice of inflationary models is too flexible to support this kind of approach.

Above I emphasized the need for multiple, independent lines of evidence, in order to mitigate the theory-dependence of evidential reasoning. The challenges to pursuing the two historical strategies in cosmology both reflect our lack of accessibility to the early universe and to the energy scales of inflation. The observed state of the universe is compatible with inflationary models, but we have not yet developed a more detailed account of how inflation occurred. In the historical cases described above, it was ultimately the development of detailed accounts of the nature of atoms, and of the motions in the solar system and their causes, that provided confidence in the theories employed along the way. The alternative to regarding these theories as stable contributions to our knowledge of the natural world is to accept that, for example, measurements of the Earth's slowing rotation by atomic clocks simply happened to agree with measurements of the Moon's motion, by an astronomical coincidence. A successful, detailed account of inflation, going beyond the initial step of using CMB data to constrain the inflaton Lagrangian, could support an argument of this sort. The challenge to taking the next steps is our lack of access to energy scales associated with the inflaton, and to specific quantities that discriminate among models.

The challenges to the observational program of further constraining the inflaton field have little to do directly with the distinctive features of cosmology, such as the uniqueness of the universe. Neither of the strategies described above require that the system under study can be experimentally manipulated. It is also not essential to consider a repeatable phenomenon, with multiple instances subject to study. The inability to conduct relevant experiments, and lack of multiple instances, are often taken to distinguish cosmology from other areas of inquiry, leading to limits on what can be established (e.g. Munitz, 1962). To make the contrast between limitations that are inherent to cosmology and problems of accessibility more vivid, imagine that an alien civilization provided us with an accelerator able to probe physics at $10^{16} GeV$. Access to the physics at this energy scale, to determine the properties of the inflaton (if it exists), would enable thorough development and testing of a detailed account of the universe's early history.

There is a more interesting challenge in cosmology regarding how to deal with initial conditions, and potential trade-offs between assumptions regarding initial conditions and dynamics (cf. Smolin, 2015). Early discussions of inflation often emphasized its ability to "wash away" dependence on the pre-inflationary state of the universe, doing away with the need for assumptions about the initial state. (Collins and Stewart (1971) noted in response to a precursor to inflation that dynamics cannot completely "wash away" the initial state, however. Given fairly weak assumptions about the dynamics, it follows from standard existence and uniqueness theorems for differential equations that one can always find a pre-inflationary state that will lead to any given post-inflationary value of Ω_0 (for example).) But it is clear that inflation requires assumptions regarding the initial state of the inflaton field (homogeneous, with an appropriate value of $V(\phi)$, in a spacetime region larger than the Hubble radius), along with an appropriate form of the potential $V(\phi)$. These are sometimes called inflation's fine-tuning problems. Assumptions about what is a plausible initial state are also relevant to assessing the account of structure formation. Hollands and Wald (2002b) construct a simple model that produces a similar spectrum of density perturbations without an inflationary phase based on a different *Ansatz* for the initial conditions. Their model describes quantized sound waves in a perfect fluid, with the same "overdamping" of modes with $\lambda >> H^{-1}$ as in inflation. By contrast with inflation, there is no horizon crossing, so it is significant precisely when the modes are taken to be in a vacuum state.[14] The fine-tuning problems of inflation are often thought to be resolved within the context of eternal inflation, to which I now turn.

Many cosmologists hold that inflation is "generically eternal", in the sense that inflation produces a multiverse consisting of "pocket universes", where inflation has ended, even as inflationary expansion continues elsewhere. (See, e.g., Aguirre (2007) for an introduction to eternal inflation.) The mechanism leading to this multiverse structure is also assumed to lead to some variation in the physical parameters among the different pocket universes. The solution to the fine-tuning problems of the original inflationary models is based on invoking an anthropic selection effect: pocket universes featuring observers will be ones in which various necessary preconditions for the existence of life like us hold. These preconditions plausibly include the existence of structures like galaxies; the formation of galaxies depends upon the presence of small fluctuations in an expanding FLRW model; and the small fluctuations themselves are ultimately traced back to an initial state for ϕ and form of the effective potential $V(\phi)$ appropriate to trigger an inflationary phase.

---

[14] Hollands and Wald (2002a) propose to take the modes to be "born" in a ground state when their proper wavelength is equal to the Planck scale, motivated by considerations of the domain of applicability of semi-classical quantum gravity. The modes will be "born" at different times, continually "emerging out of the spacetime foam", with the modes relevant to large-scale structure born at times much earlier than the Planck time. By way of contrast, in the usual approach the modes at all length scales are specified to be in a ground state at a particular time, such as the Planck time. But the precise time at which one stipulates the field modes to be in a vacuum state does not matter given that the sub-horizon modes evolve adiabatically.

Accepting eternal inflation undermines the observational program of attempting to con-strain and fix the features of the inflaton field, in two senses.[15] First, the appeal to anthropic selection undercuts the motivation for introducing a specific dynamical mechanism for generating a multiverse. The anthropic argument is intended to counter the objection that ϕ and $V(\phi)$ probably do not have appropriate properties to initiate inflation. While that may be true in the multiverse as a whole, it is, the argument goes, not the case in the habitable pocket universes, which are expected to have undergone inflation in order to produce galaxies (for example). However, the argument works just as well with other pro-posed ensembles, as long as the observed universe is compatible with the underlying laws. Rather than the inflationary multiverse, why not simply consider a relativistic cosmolog-ical model with infinite spatial sections, and some variation among different regions? By parity of reasoning, even if a region with properties like our observed Hubble volume is incredibly improbable in general, it may be highly probable within the anthropic subset. I do not see a plausible way to refine the argument to draw a distinction between these two cases, so that one can preserve the original motivations for inflation while accepting eternal inflation.

The second challenge raised by eternal inflation regards the prospects for using evidence to constrain theory along the lines outlined above. Briefly put, the two strategies above both rely on the exactness of theory in order to develop strong evidence – either in the form of connections among different types of phenomena, or in the form of rigidity as a theory is extended to give a more detailed account. Eternal inflation is anything but exact. Deriving "predictions" from eternal inflation requires specifying the ensemble of pocket universes under consideration; a measure over this ensemble that is well motivated; and a specification of the subset of the ensemble within which observers can be located. Each of these raises a number of technical and conceptual problems. But even if these are resolved, there are then several substantive auxiliary assumptions standing between the predictions of eternal inflation and the comparison with observations. Rather than using observations to directly constrain and probe the physics behind the formation of structure, we would instead be delimiting the anthropic subset of the multiverse.

10.5 Conclusion

Science often proceeds by making substantial theoretical assumptions that allow us to extend our reach into a new domain. My approach above has been to focus on asking how evidence can accumulate in favor of these assumptions as the research based on them advances. In many historical cases, subsequent research has established that a theory has to be accepted, at least as a good approximation, with the only alternative being to accept an enormously implausible set of coincidences. Based on the methodological insights gleaned from these historical cases, I have argued that the main problems with establishing inflation

---

[15] See Ijjas et al. (2013b, 2014) for an assessment of related problems, along with the response by Guth *et al.* (2014). I discuss these issues at greater length in Smeenk (2014).

with the same degree of confidence stem from our lack of independent lines of access. In addition, eternal inflation undercuts the observational program devoted to constraining the inflaton field.

References

Ade, P. A. R., *et al.* (2015). Planck 2015. XX. Constraints on Inflation. *arXiv preprint arXiv:1502.02114* .

Aguirre, A. (2007). Eternal inflation, past and future. *arXiv preprint arXiv:0712.0571* .

Albrecht, A. (2004). Cosmic inflation and the arrow of time. In J. D. Barrow, P. C. W. Davies and Charles Harper, eds. *Science and Ultimate Reality*, Cambridge: Cambridge University Press, pp. 363–401.

Alvarez, M., *et al.* (2014). Testing inflation with large scale structure: Connecting hopes with reality. *arXiv preprint arXiv:1412.4671* .

Barrow, J. D. and Liddle, A. R. (1997). Can inflation be falsified? *General Relativity and Gravitation* **29**, 1503–10.

Beringer, J., *et al.* (2012). Review of particle physics. *Physical Review D* **86**.

Blau, S. K. and Guth, A. (1987). Inflationary cosmology. In S. W. Hawking and Werner Israel, eds. *300 Years of gravitation*, Cambridge: Cambridge University Press, pp. 524–603.

Boyle, L., Smith, K. M., Dvorkin, C. and Turok, N. (2014). On testing and extending the inflationary consistency relation for tensor modes. *arXiv preprint arXiv:1408.3129* .

Boyle, L. A., Steinhardt, P. J. and Turok, N. (2006). Inflationary predictions for scalar and tensor fluctuations reconsidered. *Physical review letters* **96**, 111301.

Brandenberger, R. (2014). Do we have a theory of early universe cosmology? *Studies in History and Philosophy of Science Part B: Studies in History and Philosophy of Modern Physics* **46**, 109–21.

Carroll, S. (2014). What scientific idea is ready for retirement? Falsifiability. URL http://edge.org/response-detail/25322.

Carroll, S. M. and Tam, H. (2010). Unitary Evolution and Cosmological Fine-Tuning. *ArXiv e-prints* 1007.1417.

Chalmers, A. (2011). Drawing philosophical lessons from Perrin's experiments on Brownian motion. *The British Journal for the Philosophy of Science* **62**, 711–32.

Collins, C. B. and Stewart, J. M. (1971). Qualitative cosmology. *Monthly Notices of the Royal Astronomical Society* **153**, 419–34.

Durrer, R., Kunz, M.N. and Melchiorri, A. (2002). Cosmic structure formation with topological defects. *Physics Reports* **364**, 1–81.

Earman, J. and Mosterin, J. (1999). A critical analysis of inflationary cosmology. *Philosophy of Science* **66**, 1–49.

Ellis, G. F. R. and Silk, J. (2014). Scientific method: Defend the integrity of physics. *Nature* **516**, 321.

Gibbons, G. W. and Turok, N. (2008). Measure problem in cosmology. *Physical Review D* **77**, 063516.

Guth, A. (1981). Inflationary universe: A possible solution for the horizon and flatness problems. *Physical Review D* **23**, 347–56.

Guth, A., Kaiser, D. and Nomura, Y. (2014). Inflationary paradigm after Planck 2013. *Physics Letters B* **733**, 112–9.

Harper, W. L. (2012). *Isaac Newton's Scientific Method: Turning Data Into Evidence about Gravity and Cosmology*. Oxford: Oxford University Press.

Harrison, E. R. (1970). Fluctuations at the threshold of classical cosmology. *Physics Review* D **1**, 2726–30.

Hollands, S. and Wald, R. (2002a). Comment on inflation and alternative cosmology. Hep-th/0210001.

Hollands, S. and Wald, R. (2002b). Essay: An alternative to inflation. *General Relativity and Gravitation* **34**, 2043–55.

Ijjas, A., Steinhardt, P. J. and Loeb, A. (2013b). Inflationary paradigm in trouble after Planck 2013. *Physics Letters B* **723**, 261–6.

Ijjas, A., Steinhardt, P. J. and Loeb, A. (2014). Inflationary schism. *Physics Letters B* **736**, 142–6.

Lehners, J. (2008). Ekpyrotic and cyclic cosmology. *Physics Reports* **465**, 223–263.

Liddle, A. and Lyth, D. (2000). *Cosmological Inflation and Large-Scale Structure*. Cambridge: Cambridge University Press.

Lidsey, J. E. *et al.,* (1997). Reconstructing the inflaton potentialÂŮan overview. *Reviews of Modern Physics* **69**, 373.

Martin, J. (2015). The observational status of cosmic inflation after Planck. *arXiv preprint arXiv:1502.05733* .

Martin, J., Ringeval, C. and Vennin, V. (2013). Encyclopaedia inflationaris. *arXiv preprint arXiv:1303.3787.*

Martin, J., Ringeval, C. and Vennin, V. (2015). Observing inflationary reheating. *Physical Review Letters* **114**, 081303.

Mukhanov, V. F. and Chibisov, G. V. (1981). Quantum fluctuations and a nonsingular universe. *JETP Letters* **33**, 532–5.

Mukhanov, V. F., Feldman, H. A. and Brandenberger, R. H. (1992). Theory of cosmological perturbations. part 1. Classical perturbations. part 2. Quantum theory of perturbations. part 3. extensions. *Physics Reports* **215**, 203–333.

Munitz, M. K. (1962). The logic of cosmology. *British Journal for the Philosophy of Science* **13**, 34–50.

Norton, J. (1993). Determination of theory by evidence: How Einstein discovered general relativity. *Synthase*, **13**, 1–31.

Norton, J. (2000). How we know about electrons. In R. Nola and H. Sankey, eds. *After Popper, Kuhn and Feyerabend: Recent Issues in Theories of Scientific Method*, Dordrecht: Kluwer Academic Publishers, pp. 67–97.

Nye, M. J. (1972). *Molecular reality: A perspective on the scientific work of Jean Perrin*. New York: American Elsevier.

Peebles, P. J. E. and Yu, J. T. (1970). Primeval adiabatic perturbation in an expanding universe. *Astrophysics Journal* **162**, 815–36.

Peebles, P. J. E., Page Jr, L. A. and Partridge, R. B. (2009). *Finding the big bang*. Cambridge: Cambridge University Press.

Penrose, R. (2004). *The road to reality*. London: Jonathan Cape.

Psillos, S. (2011). Moving molecules above the scientific horizon: On Perrin's case for realism. *Journal for General Philosophy of Science* **42**, 339–63.

Smeenk, C. (2014). Predictability crisis in early universe cosmology. *Studies in History and Philosophy of Science Part B: Studies in History and Philosophy of Modern Physics* **46**, 122–33.

Smith, G. E. (2014). Closing the loop. In Z. Biener and E. Schliesser, eds. *Newton and Empiricism*, Oxford: Oxford University Press, pp. 262–351.

Smolin, L. (2015). Temporal naturalism. *Studies in History and Philosophy of Science Part B: Studies in History and Philosophy of Modern Physics*, **52A**, 86–102.

Starobinsky, A. (1979). Spectrum of relict gravitational radiation and the early state of the universe. *JETP Letters* **30**, 682–5.

Starobinsky, A. (1980). A new type of isotropic cosmological models without singularity. *Physics Letters B* **91**, 99–102.

Stein, H. (1994). Some reflections on the structure of our knowledge in physics. In *Logic, Metholodogy and Philosophy of Science, Proceedings of the Ninth International Congress of Logic, Methodology and Philosophy of Science*, pp. 633–55.

Turok, N., ed. (1997). *Critical dialogues in cosmology*. Singapore: World Scientific.

Turok, N. (2002). A critical review of inflation. *Classical and Quantum Gravity* **19**, 3449.

Vachaspati, T. and Trodden, M. (2000). Causality and cosmic inflation. *Physical Review D* **61**, 3502–6. Gr-qc/9811037.

Zel'dovich, Ya. B. (1972). A hypothesis, unifying the structure and the entropy of the universe. *Monthly Notices of the Royal Astronomical Society* **160**, 1–3.

11

Why Boltzmann Brains do not Fluctuate into Existence from the de Sitter Vacuum

KIMBERLY K. BODDY, SEAN M. CARROLL AND JASON POLLACK

11.1 Introduction

The Boltzmann Brain problem [2, 6, 7] is a novel constraint on cosmological models. It arises when there are thought to be very large spacetime volumes in a de Sitter vacuum state – empty space with a positive cosmological constant Λ. This could apply to theories of eternal inflation and the cosmological multiverse, but it also arises in the future of our current universe, according to the popular ΛCDM cosmology.

Observers in de Sitter are surrounded by a cosmological horizon at a distance $R = H^{-1}$, where $H = \sqrt{\Lambda/3}$ is the (fixed) Hubble parameter. Such horizons are associated with a finite entropy $S = 3\pi/G\Lambda$ and temperature $T = H/2\pi$ [11]. With a finite temperature and spatial volume, and an infinite amount of time, it has been suggested that we should expect quantum/thermal fluctuations into all allowed configurations. In this context, any particular kind of anthropically interesting situation (such as an individual conscious "brain", or the current macrostate of the room you are now in, or the Earth and its biosphere) will fluctuate into existence many times. With very high probability, when we conditionalize on the appearance of some local situation, the rest of the state of the universe will be generic – close to thermal equilibrium – and both the past and future will be higher-entropy states.[1] These features are wildly different from the universe we think we actually live in, featuring a low-entropy Big Bang state approximately 13.8 billion years ago. Therefore, the story goes, our universe must not be one with sufficiently large de Sitter regions to allow such fluctuations to dominate.

In this chapter we summarize and amplify a previous paper in which we argued that the Boltzmann Brain problem is less generic (and therefore more easily avoided) than is often supposed [5]. Our argument involves a more precise understanding of the informal notion of "quantum fluctuations". This term is used in ambiguous ways when we are talking about conventional laboratory physics: it might refer to Boltzmann (thermal)

---

[1] The real problem with an eternally fluctuating universe is not that it would look very different from ours. It is that it would contain observers who see exactly what we see, but have no reason to take any of their observations as reliable indicators of external reality, since the mental impressions of those observations are likely to have randomly fluctuated into their brains.

228

fluctuations, where the microstate of the system is truly time-dependent; or measurement-induced fluctuations, where repeated observation of a quantum system returns stochastic results; or time-independent "fluctuations" of particles or fields in the vacuum, which are really just a poetic way of distinguishing between quantum and classical behavior. In the de Sitter vacuum, which is a stationary state, there are time-independent vacuum fluctuations, but there are no dynamical processes that could actually bring Boltzmann Brains (or related configurations) into existence. Working in the Everett (Many-Worlds) formulation of quantum mechanics, we argue that the kinds of events where something may be said to "fluctuate into existence" are dynamical processes in which branches of the wave function decohere. Having a non-zero *amplitude* for a certain quantum event should not be directly associated with the probability that such an event will *happen*; things only happen when the wave function branches into worlds in which those things occur.

Given this understanding, the Boltzmann Brain problem is avoided when horizon-sized patches of the universe evolve toward the de Sitter vacuum state. This is generically to be expected in the context of quantum field theory in curved spacetime, according to the cosmic no-hair theorems [16, 19, 28]. It would not be expected in the context of horizon complementarity in a theory with a true de Sitter minimum; there, the whole theory is described by a finite-dimensional Hilbert space, and we should expect Poincaré recurrences and Boltzmann fluctuations [2–4, 6, 7, 21, 26, 27]. Such theories do have a Boltzmann Brain problem. However, if we consider a $\Lambda > 0$ false-vacuum state in a theory where there is also a $\Lambda = 0$ state, the theory as a whole has an infinite-dimensional Hilbert space. Then we would expect the false-vacuum state, considered by itself, to dissipate toward a quiescent state, free of dynamical fluctuations. Hence, the Boltzmann Brain problem is easier to avoid than conventionally imagined.

Our argument raises an interesting issue concerning what "really happens" in the Everettian wave function. We briefly discuss this issue in Section 11.4.

11.2 Quantum Fluctuations in Everettian Quantum Mechanics

The existence of Boltzmann Brain fluctuations is a rare example of a question whose answer depends sensitively on one's preferred formulation of quantum theory. Here we investigate the issue in the context of Everettian quantum mechanics (EQM) [8, 24, 29]. The underlying ontology of EQM is extremely simple, coming down to two postulates:

1. The world is fully represented by quantum states $|\psi\rangle$ that are elements of a Hilbert space $\mathcal{H}$.
2. States evolve with time according to the Schrödinger equation,

$$\hat{H}|\psi(t)\rangle = i\partial_t|\psi(t)\rangle, \tag{11.1}$$

for some self-adjoint Hamiltonian operator $\hat{H}$.

The challenge, of course, is matching this austere framework onto empirical reality. In EQM, our task is to derive, rather than posit, features such as the apparent collapse of the wave function (even though the true dynamics are completely unitary) and the Born Rule for probabilities (even though the full theory is completely deterministic). We will not delve into these issues here, but only emphasize that in EQM the quantum state and its unitary evolution are assumed to give a complete description of reality. No other physical variables or measurement postulates are required.

Within this framework, consider a toy system such as a one-dimensional simple harmonic oscillator with potential $V(x) = \frac{1}{2}\omega^2 x^2$. Its ground state is a Gaussian wave function whose only time-dependence is an overall phase factor,

$$\psi(x, t) \propto e^{-iE_0 t} e^{-E_0 x^2}. \tag{11.2}$$

The phase is of course physically irrelevant; one way of seeing that is to note that equivalent information is encoded in the pure-state density operator,

$$\rho(x, t) = |\psi(x, t)\rangle\langle\psi(x, t)| = |\psi(x, 0)\rangle\langle\psi(x, 0)|, \tag{11.3}$$

which is manifestly independent of time. We will refer to such states, which of course would include any energy eigenstate of any system with a time-independent Hamiltonian, as "stationary".

In a stationary state, there is nothing about the isolated quantum system that is true at one time but not true at another time. There is no sense, therefore, in which anything is dynamically fluctuating into existence. Nevertheless, we often informally talk about "quantum fluctuations" in such contexts, whether we are considering a simple harmonic oscillator, an electron in its lowest-energy atomic orbital, or vacuum fluctuations in quantum field theory. Clearly it is important to separate this casual notion of fluctuations from true time-dependent processes.

To that end, it is useful to distinguish between different concepts that are related to the informal notion of "quantum fluctuations". We can identify three such ideas:

- **Vacuum fluctuations** are the differences in properties of a quantum and its classical analogue, and exist even in stationary states.
- **Boltzmann fluctuations** are dynamical processes that arise when the microstate of a system is time-dependent even if its coarse-grained macrostate may not be.
- **Measurement-induced fluctuations** are the stochastic observational outcomes resulting from the interaction of a system with a measurement device, followed by decoherence and branching.

Let us amplify these definitions a bit. By "vacuum fluctuations" we mean the simple fact that quantum states, even while stationary, generally have non-zero variance for observable properties. Given some observable $\hat{\mathcal{O}}$, we expect expectation values in a state $|\psi\rangle$ to satisfy $\langle\hat{\mathcal{O}}^2\rangle_\psi > \langle\hat{\mathcal{O}}\rangle^2_\psi$. The fact that the position of the harmonic oscillator is not localized to the origin in its ground state is a consequence of this kind of fluctuation. Other manifestations include the Casimir effect, the Lamb shift, and radiative corrections due to virtual particles in quantum field theory. Nothing in our analysis denies the existence of these kinds of fluctuation; we are merely pointing out that they are non-dynamical, and therefore not associated with anything literally fluctuating into existence.

This is in contrast with "Boltzmann fluctuations", which are true dynamical processes. In classical statistical mechanics, we might have a system in equilibrium described by a canonical ensemble, where macroscopic quantities such as temperature and density are time-independent. Nevertheless, any particular realization of such a system is represented by a microstate with true time-dependence; the molecules in a box of gas are moving around, even in equilibrium. From a Boltzmannian perspective, we coarse-grain phase space into macrostates, and associate to each microstate and entropy $S = k_B \log \Omega$, where Ω is the volume of the macrostate in which the microstate lives. We then expect rare fluctuations into lower-entropy states, with a probability that scales as $P(\Delta S) \sim e^{-\Delta S}$, where ΔS is the decrease in entropy. Such Boltzmann fluctuations are *not* expected to occur in stationary quantum states, where there is no microscopic property that is actually fluctuating beneath the surface (at least in EQM).

This can even be true in "thermal" states in quantum mechanics. Consider a composite system AB, with weak coupling between the two factors, and A much smaller than B. When the composite system is in a stationary pure state $|\psi\rangle$, we expect the reduced density matrix of the subsystem to look thermal,

$$\rho_A = \text{Tr}_B \, |\psi\rangle\langle\psi| \sim \exp(-\beta \hat{H}_A) = \sum_n e^{-\beta E_n} \, |E_n\rangle\langle E_n| \,, \qquad (11.4)$$

where $\hat{H}_A$ is the Hamiltonian for A, β is the inverse temperature, and the states $|E_n\rangle$ are energy eigenstates of $\hat{H}_A$. Despite the thermal nature of this density operator, it is strictly time-independent, and there are no dynamical fluctuations.

Finally, we have measurement-induced fluctuations: processes in which we repeatedly measure a quantum system and obtain "fluctuating" results. In EQM, the measurement process consists of unitary dynamics creating entanglement between the observed system and a macroscopic apparatus, followed by decoherence and branching. We can decompose Hilbert space into factors representing the system, the apparatus (a macroscopic configuration that may or may not include observing agents), and the environment (a large set of degrees of freedom that we do not keep track of):

$$\mathcal{H} = \mathcal{H}_S \otimes \mathcal{H}_A \otimes \mathcal{H}_E \,. \qquad (11.5)$$

We assume that the apparatus begins in a specific "ready" state, and both the apparatus and environment are initially unentangled with the system to be observed.[2] For simplicity, imagine that the system is a single qubit in a superposition of up and down. Unitary

---

[2] The justification for these assumptions can ultimately be traced to the low-entropy state of the early universe.

evolution then proceeds as follows:

$$|\Psi\rangle = (|+\rangle_S + |-\rangle_S)|a_0\rangle_A|e_0\rangle_E \tag{11.6}$$

$$\rightarrow (|+\rangle_S|a_+\rangle_A + |-\rangle_S|a_-\rangle_A)|e_0\rangle_E \tag{11.7}$$

$$\rightarrow |+\rangle_S|a_+\rangle_A|e_+\rangle_E + |-\rangle_S|a_-\rangle_A|e_-\rangle_E. \tag{11.8}$$

The first line represents the system in some superposition of $|+\rangle$ and $|-\rangle$, while the apparatus and environment are unentangled with it. In the second line (pre-measurement), the apparatus has interacted with the system; its readout value "+" is entangled with the + state of the qubit, and likewise for "−". In the final line, the apparatus has become entangled with the environment. This is the decoherence step; generically, the environment states will quickly become very close to orthogonal, $\langle e_+|e_-\rangle \approx 0$, after which the two branches of the wave function will evolve essentially independently. If we imagine setting up a system in some stationary state, performing a measurement, re-setting it, and repeating the process, the resulting record of measurement readouts will form a stochastic series of quantities obeying the statistics of the Born Rule. This is the kind of "fluctuation" that would arise from the measurement process.

There are several things to note about this description of the measurement process in EQM. First, the reduced density matrix $\rho_{SA} = \mathrm{Tr}_E |\Psi\rangle\langle\Psi|$ obtained by tracing over the environment is diagonal in a very specific basis, the "pointer basis" for the apparatus [23, 30–33]. The pointer states making up this basis are those that are robust with respect to continual monitoring by the environment; in realistic situations, this amounts to states that have a definite spatial profile (such as the pointer on a measuring device indicating some specific result). Second, branching is necessarily an out-of-equilibrium process. The initial state is highly non-generic; one way of seeing this is that the reduced density matrix has an initially low entropy $S_{SA} = \mathrm{Tr}\, \rho_{SA} \log \rho_{SA}$. Third, this entropy increases during the measurement process, in accordance with the thermodynamic arrow of time. Given sufficient time to evolve, the system will approach equilibrium and the entropy will be maximal. At this point the density matrix will no longer be diagonal in the pointer basis (it will be thermal, and hence diagonal in the energy eigenbasis). This process is portrayed in Figure 11.1. Needless to say, none of these features – a special, out-of-equilibrium initial state, in which entropy increases as the system becomes increasingly entangled with the environment over time – apply to isolated stationary systems.

The relationship of fluctuations and observations is worth emphasizing. Consider again the one-dimensional harmonic oscillator. We can imagine constructing a projection operator onto the positive values of the coordinate,

$$\hat{P}_+ = \int_{x>0} dx\, |x\rangle\langle x|. \tag{11.9}$$

Now in some state $|\psi\rangle$, we can consider the quantity

$$p_+ = \langle\psi|\hat{P}_+|\psi\rangle. \tag{11.10}$$

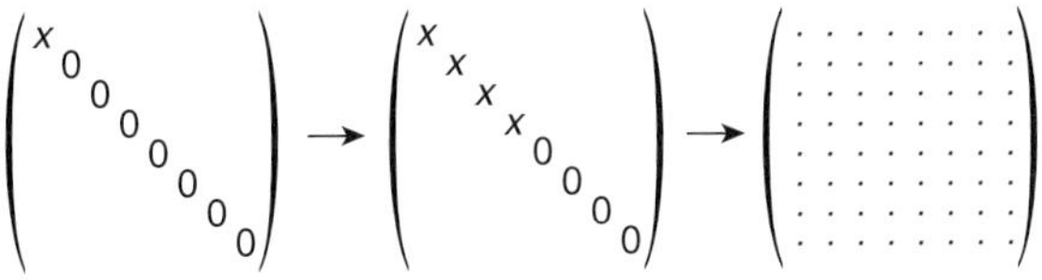

$$\begin{pmatrix} x & 0 & & & \\ & 0 & 0 & & \\ & & 0 & 0 & \\ & & & 0 & 0 \\ & & & & 0 \end{pmatrix} \rightarrow \begin{pmatrix} x & 0 & & & \\ & x & x & & \\ & & x & 0 & \\ & & & 0 & 0 \\ & & & & 0 \end{pmatrix} \rightarrow \begin{pmatrix} \cdot & \cdot & \cdot & \cdot & \cdot & \cdot \\ \cdot & \cdot & \cdot & \cdot & \cdot & \cdot \\ \cdot & \cdot & \cdot & \cdot & \cdot & \cdot \\ \cdot & \cdot & \cdot & \cdot & \cdot & \cdot \\ \cdot & \cdot & \cdot & \cdot & \cdot & \cdot \\ \cdot & \cdot & \cdot & \cdot & \cdot & \cdot \end{pmatrix}$$

Figure 11.1   Schematic evolution of a reduced density matrix in the pointer basis. The density matrix on the left represents a low-entropy situation, where only a few states are represented in the wave function. There are no off-diagonal terms, since the pointer states rapidly decohere. The second matrix represents the situation after the wave function has branched a few times. In the third matrix, the system has reached equilibrium; the density matrix would be diagonal in an energy eigenbasis, but in the pointer basis, decoherence has disappeared and the off-diagonal terms are non-zero.

In conventional laboratory settings, it makes sense to think of this as "the probability that the particle is at $x > 0$". But that is not strictly correct in EQM. There is no such thing as "where the particle is"; rather, the state of the particle is described by its entire wave function. The quantity p_+ is the probability that we would *observe* the particle at $x > 0$ were we to measure its position. Quantum variables are not equivalent to classical stochastic variables. They may behave similarly when measured repeatedly over time, in which case it is sensible to identify the non-zero variance of a quantum-mechanical observable with the physical fluctuations of a classical variable, but the state in EQM is simply the wave function, not the collection of possible measurement outcomes.

11.3 Boltzmann Brains and de Sitter Space

With this setup in mind, the application to de Sitter space is straightforward. As mentioned in the Introduction, observers in de Sitter are surrounded by a horizon with a finite entropy. In the vacuum, the quantum state in any horizon patch is given by a time-independent thermal density matrix,

$$\rho_{\text{patch}} \propto e^{-\beta \hat{H}}, \tag{11.11}$$

where the Hamiltonian is the static Hamiltonian for the fields in that patch and $\beta \propto 1/\sqrt{\Lambda}$.

According to the analysis in the previous section, this kind of thermal state does *not* exhibit dynamical fluctuations of any sort, including into Boltzmann Brains. It is a stationary state, so there is no time-dependence in any process. In particular, the entropy is maximal for the thermal density matrix, so there are no processes corresponding to decoherence and branching.[3] There may be non-zero overlap between some state |brain⟩ and the de Sitter vacuum, but there is no dynamics that brings that state into existence on a decoherent branch of the wave function. Indeed, one way of establishing the thermal nature of the

---

[3] The idea that quantum fluctuations during inflation are responsible for the density perturbations in our current universe is unaffected by this reasoning. During inflation the state is nearly stationary, with non-dynamical vacuum fluctuations as defined above; branching and decoherence occur when the entropy ultimately increases, for example during reheating.

state is to notice that a particle detector placed in de Sitter space will come to equilibrium and then stop evolving [25]. Therefore, Boltzmann Brains do not fluctuate into existence in such a state, and should not be counted among observers in the cosmological multiverse.

It is useful to contrast this situation with that of a black hole in a Minkowski background. There, as in de Sitter space, we have a horizon with a non-zero temperature and finite entropy. However, real-world black holes are not stationary states. They evaporate by emitting radiation, and the entropy increases along the way. A particle detector placed in orbit around a black hole will not simply come to equilibrium and stop evolving; it will detect particles being emitted from the direction of the hole, with a gradually increasing temperature as the hole shrinks. This is a very different situation than the equilibrium de Sitter vacuum.

It remains to determine whether the stationary vacuum state is actually attained in the course of cosmological evolution. There are classical and quantum versions of the cosmic no-hair theorem [16, 19, 28]. Classically, the spacetime geometry of each horizon-sized patch of a universe with $\Lambda > 0$ asymptotically approaches that of de Sitter space, as long as it does not contain a perturbation so large that it collapses to a singularity. In the context of quantum field theory in curved spacetime, analogous results show that each patch approaches the de Sitter vacuum state. Intuitively, this behavior can be thought of as excitations leaving the horizon and not coming back, as portrayed in the first part of Figure 11.2. The timescale for this process is parametrically set by the Hubble time, and will generally be enormously faster than the rate of Boltzmann fluctuations in states that have not quite reached the vacuum. Hence, if we think of conventional ΛCDM cosmology in terms of semiclassical quantum gravity, it seems reasonable to suppose that the model does not suffer from a Boltzmann Brain problem.

The situation is somewhat different if we take quantum gravity into account. In this case we are lacking a fully well-defined theory, and any statements we make must have a conjectural aspect. A clue, however, is provided by the idea of horizon complementarity [3, 4, 21, 26, 27]. According to this idea, we should only attribute a local spacetime description to regions on one side of any horizon at a time, rather than globally. For example, we could describe the spacetime outside of a black hole, or as seen by an observer who has fallen into the black hole, but should not use both descriptions simultaneously; the rest of the quantum state can be thought of as living on a timelike "stretched horizon" just outside the null horizon itself. Applying this philosophy to de Sitter space leads to the idea that the whole theory should be thought of as that of a single horizon patch, with everything normally associated with the rest of the universe actually encoded on the stretched cosmological horizon. Since the patch has a finite entropy (approximately 10^{123} for the measured value of Λ), the corresponding quantum theory is (plausibly) finite-dimensional, with dim $\mathcal{H} = e^S$.

Hence, applying horizon complementarity to a universe with a single true de Sitter vacuum, the intuitive picture behind the cosmic no-hair theorem no longer applies. There is no outside world for perturbations to escape to; rather, they are absorbed by the stretched horizon, and will eventually be emitted back into the bulk spacetime, as shown in the middle part of Figure 11.2. This is consistent with our expectations for a quantum theory on

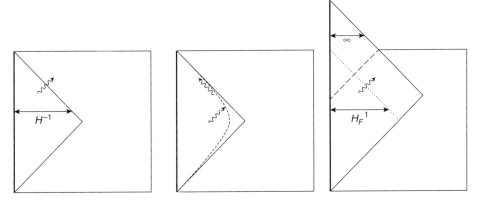

Figure 11.2 Conformal diagrams for de Sitter space in different scenarios. We consider an observer at the north pole, represented by the line on the left boundary and their causal diamond (solid triangle). The wavy line represents excitations of the vacuum approaching the horizon. In QFT in curved spacetime, portrayed on the left, the excitation exits and the state inside the diamond approaches the vacuum, in accordance with the cosmic no-hair theorems. In contrast, horizon complementarity implies that excitations are effectively absorbed at the stretched horizon (dashed curve just inside the true horizon) and eventually return to the bulk, as shown in the middle diagram. The third diagram portrays the situation when the de Sitter minimum is a false vacuum, and the full theory contains a state with $\Lambda = 0$; the upper triangle represents nucleation of a bubble of this Minkowski vacuum. In that case, excitations can leave the apparent horizon of the false vacuum while remaining inside the true horizon; we then expect there to be no dynamical Boltzmann fluctuations.

a finite-dimensional Hilbert space, which should exhibit fluctuations and Poincaré recurrences. This was the case originally considered by Dyson, Kleban and Susskind [2, 7] in their examination of what is now known as the Boltzmann Brain problem. Nothing in our analysis changes this expectation; if Hilbert space is finite-dimensional and time evolves eternally, it is natural to expect that the Boltzmann Brain problem is real. (Although see Albrecht [1] for one attempt at escaping this conclusion.)

The situation changes if the de Sitter vacuum state is only metastable, and is embedded in a larger theory with a $\Lambda = 0$ minimum. In that case the underlying quantum theory will be infinite-dimensional, since the entropy of Minkowski space is infinite. In a semiclassical solution based on the de Sitter vacuum, the dynamics will not be completely unitary, since there will be interactions (such as bubble nucleations) connecting different sectors of the theory. Although a full understanding is lacking, intuitively we expect the dynamics in such states to be dissipative, much as higher-energy excitations of metastable minima decay away faster in ordinary quantum mechanics. The Poincaré recurrence time is infinite, so there is no necessity for Boltzmann fluctuations or recurrences. The spacetime viewpoint relevant to this case is portrayed in the third panel of Figure 11.2. The existence of Coleman–De Luccia transitions to the $\Lambda = 0$ vacua permits the true horizon size to be larger than the de Sitter radius, so perturbations can appear to leave the horizon and never return, even under complementarity.

The complete picture we suggest is therefore straightforward. If we are dealing with de Sitter vacua in a theory with an infinite-dimensional Hilbert space, we expect horizon patches to evolve to a stationary quantum vacuum state, and there to be no dynamical fluctuations, and the Boltzmann Brain problem is avoided. This applies to QFT in curved spacetime, or to complementarity in the presence of $\Lambda = 0$ vacua. If, on the other hand, the Hilbert space is finite, fluctuations are very natural, and the Boltzmann Brain problem is potentially very real.

11.4 What Happens in the Wave Function?

We have been careful to distinguish between vacuum fluctuations in a quantum state, which can be present even if the state is stationary, and true dynamical processes, such as Boltzmann and measurement-induced (branching) fluctuations. One may ask, however, whether our interpretation of stationary states in EQM is the right one. More specifically: is it potentially legitimate to think of a stationary quantum state as a superposition of many time-dependent states? This is a particular aspect of a seemingly broader question: what "happens" inside the quantum wave function?

One way to address this question is by using the decoherent (or consistent) histories formalism [10, 12–15, 20]. This approach allows us to ask when two possible histories of a quantum system decohere from each other and can be assigned probabilities. We might want to say that an event (such as a fluctuation into a Boltzmann Brain) "happens" in the wave function if that event occurs as part of a history that decoheres from other histories in some consistent set. (Although we will argue that, in fact, this criterion is too forgiving.)

Consider a closed system described by a density operator $\rho(t_0)$ at an initial time t_0. We want to consider possible coarse-grained histories of the system, described by sequences of projection operators $\{\hat{P}_\alpha\}$. These operators partition the state of the system at some time into mutually exclusive alternatives and obey

$$\sum_\alpha \hat{P}_\alpha = \mathbb{1} \,, \qquad \hat{P}_\alpha \hat{P}_\beta = \delta_{\alpha\beta}\hat{P}_\alpha \,. \qquad (11.12)$$

A history is described by a sequence of such alternatives, given by a sequence of projectors at specified times, $\{\hat{P}_{\vec{\alpha}_1}^{(1)}(t_1), \ldots \hat{P}_{\vec{\alpha}_n}^{(n)}(t_n)\}$. At each time t_i, we have a distinct set of projectors $\hat{P}_\alpha^{(i)}$, and the particular history is described by a vector of specific projectors labeled by $\vec{\alpha}$. The decoherence functional of two histories $\vec{\alpha}$ and $\vec{\alpha}'$ is

$$D(\vec{\alpha}, \vec{\alpha}') = \text{Tr}[\hat{P}_{\vec{\alpha}_n}^{(n)}(t_n) \cdots \hat{P}_{\vec{\alpha}_1}^{(1)}(t_1)\rho(t_0)\hat{P}_{\vec{\alpha}_1'}^{(1)}(t_1) \cdots \hat{P}_{\vec{\alpha}_n'}^{(n)}(t_n)] \,, \qquad (11.13)$$

where the trace is taken over the complete Hilbert space. If the decoherence functional vanishes for two histories, we say that those histories are consistent or decoherent, and they can be treated according to the rules of classical probability theory.

Following a suggestion by Lloyd [18], we can apply the decoherent histories formalism to a simple harmonic oscillator in its ground state. One choice of projectors are those given by the energy eigenstates $|E_n\rangle$ themselves,

$$\hat{P}_n = |E_n\rangle\langle E_n|. \tag{11.14}$$

It is easy to check that the corresponding histories trivially decohere. This simply reflects the fact that the system begins in an energy eigenstate and stays there.

But we are free to consider other sets of projectors as well. Let us restrict our attention to an N-dimensional subspace of the infinite-dimensional oscillator Hilbert space, consisting of the span of the energy eigenstates $|E_n\rangle$ with n between 0 and $N-1$. Then Lloyd [18] points out that we can define phase states

$$|\phi_m\rangle = \frac{1}{\sqrt{N}} \sum_{n=0}^{N-1} e^{2\pi imn/N} |E_n\rangle. \tag{11.15}$$

These have the property that they evolve into each other after timesteps $\Delta t = 2\pi/N\omega$,

$$e^{-iH\Delta t}|\phi_m\rangle = |\phi_{m+1}\rangle. \tag{11.16}$$

Now we can consider histories defined by the phase projectors

$$\tilde{P}_m = |\phi_m\rangle\langle\phi_m|, \tag{11.17}$$

evaluated at times separated by Δt. These histories, it is again simple to check, *also* mutually decohere with each other (although, of course, not with the original energy-eigenstate histories). Each such history describes a time-dependent system, whose phase rotates around, analogous to a classical oscillator rocking back and forth in its potential.[4]

We therefore have two (and actually, many more) sets of histories, which decohere within the sets, but are mutually inconsistent with each other. In some sets there is no time-dependence, while in others there is. In the stationary thermal state of a de Sitter horizon patch, there is no obstacle in principle to defining a set of decoherent histories with the properties that some of them describe Boltzmann Brains fluctuating into existence. On the other hand, we are not *forced* to consider such histories; there are also consistent sets in which the states remain perfectly stationary.

This situation raises a fundamental puzzle. When we are doing multiverse cosmology, we often want to ask what is seen by observers satisfying certain criteria (which may be as general as "all intelligent observers" or as specific as "observers in precisely defined local conditions"). To answer that question, we want to know whether an amplitude corresponding to such an observer is actually physically realized in the quantum state of the universe. The decoherent histories formalism seems to give an ambiguous answer to the question:

---

[4] It was not necessary to carefully choose the phase states. In the decoherent histories formalism, we have the freedom to choose projection operators separately at each time. Given some initial projectors, we can always define projectors at later times by simply evolving them forward by an appropriate amount; the resulting histories will decohere. We thank Mark Srednicki for pointing this out.

the number of observers who physically appear in the universe depends on the projection operators we choose to define our coarse-grained histories. This seems to introduce an unacceptably subjective element into a purportedly objective calculation. (A closely related problem has been emphasized by Kent [17].)

Our own conclusion from this analysis is simple: the existence of decoherent histories describing certain dynamical processes is *not sufficient* to conclude that those processes "really happen". Note that something somewhat stronger is going on in the standard description of branching and decoherence in EQM. There, the explicit factorization of Hilbert space into system+apparatus and environment directly implies a certain appropriate coarse-graining for the macroscopic variables (by tracing over the environment). Of course, there is arguably a subjective element in how we define the environment in the first place. That choice, however, relies on physical properties of the theory, in particular the specific Hamiltonian and its low-energy excitations around some particular background state. There have been suggestions that the decoherence properties of realistic systems can be defined objectively, by demanding that records of the macroscopic configuration be stored redundantly in the environment [22].

We conjecture, at least tentatively, that the right way to think about observers fluctuating into existence in quantum cosmology is to define an objective division of the variables into "macroscopic system" and "environment", based on the physical properties of the system under consideration, and to look for true branching events where the reduced density matrix of the system decoheres in the pointer basis.[5] Work clearly remains to be done in order to turn this idea into a well-defined program, as well as to justify why such a procedure is the appropriate one. In this context, it is useful to keep in mind that Boltzmann Brains are a difficulty, not a desirable feature, of a given cosmological model. We suggest that the analysis presented here should at the very least shift the burden of proof onto those who believe that Boltzmann Brains are a generic problem.

Acknowledgments

S.M.C. would like to thank the organizers of the Philosophy of Cosmology conference in Tenerife – Joe Silk, Simon Saunders, Khalil Chamcham, John Barrow, Barry Loewer, and David Albert – for putting together such a unique and stimulating meeting. We are grateful to Jim Hartle, Seth Lloyd, and Mark Srednicki for conversations on the issues discussed in Section 11.4. This research is funded in part by the Walter Burke Institute for Theoretical Physics at Caltech, by DOE grant DE-SC0011632, and by the Gordon and Betty Moore Foundation through grant 776 to the Caltech Moore Center for Theoretical Cosmology and Physics.

---

[5] A possible alternative strategy is to look for histories that obey classical equations of motion, as in Gell-Mann and Hartle [9, 10]. Such an approach seems unable to resolve the ambiguity presented by the simple harmonic oscillator.

References

[1] Albrecht, A. 2015. Tuning, Ergodicity, Equilibrium and Cosmology. *Phys. Rev.*, **D91**, 103510

[2] Albrecht, A. and Sorbo, L. 2004. Can the universe afford inflation? *Phys. Rev.*, **D70**, 063528.

[3] Banks, T. 2001. Cosmological breaking of supersymmetry? or Little lambda goes back to the future 2. *Int. J. Mod. Phys.*, **A16**, 910.

[4] Banks, T. and Fischler, W. 2001. M theory observables for cosmological space-times. eprint arXiv: hep-th/0102077.

[5] Boddy, K. K., Carroll, S. M. and Pollack, J. 2014. De Sitter Space Without Dynamical Quantum Fluctuations. eprint arXiv:1405.0298.

[6] Bousso, R. and Freivogel, B. 2007. A Paradox in the global description of the multiverse. *JHEP*, **06**, 018.

[7] Dyson, L., Kleban, M. and Susskind, L. 2002. Disturbing implications of a cosmological constant. *JHEP*, **10**, 011.

[8] Everett, H. 1957. Relative state formulation of quantum mechanics. *Rev. Mod. Phys.*, **29**, 454.

[9] Gell-Mann, M. and Hartle, J. 2007. Quasiclassical Coarse Graining and Thermodynamic Entropy. *Phys. Rev.*, **A76**, 022104.

[10] Gell-Mann, M. and Hartle, J. B. 1993. Classical equations for quantum systems. *Phys. Rev.*, **D47**, 3345.

[11] Gibbons, G. W. and Hawking, S. W. 1977. Cosmological Event Horizons, Thermodynamics, and Particle Creation. *Phys. Rev.*, **D15**, 2738.

[12] Griffiths, R. B. 1984. Consistent histories and the interpretation of quantum mechanics. *J. Statist. Phys.*, **36**, 219.

[13] Halliwell, J. J. 1995. A Review of the decoherent histories approach to quantum mechanics. *Ann. N.Y. Acad. Sci.*, **755**, 726.

[14] Hartle, J. B. 1993. The quantum mechanics of closed systems. In Hu, B. L., Ryan, Jr., M. P. and Vishveshwara, C. V., eds. *Directions in General Relativity*. Cambridge: Cambridge University Press, p. 104.

[15] Hartle, J. B., Laflamme, R. and Marolf, D. 1995. Conservation laws in the quantum mechanics of closed systems. *Phys. Rev.*, **D51**, 7007.

[16] Hollands, S. 2013. Correlators, Feynman diagrams, and quantum no-hair in deSitter spacetime. *Commun. Math. Phys.*, **319**, 1.

[17] Kent, A. 1997. Consistent sets contradict. *Phys. Rev. Lett.*, **78**, 2874–2877.

[18] Lloyd, S. 2014. Decoherent histories approach to the cosmological measure problem. eprint arXiv:1608.05672.

[19] Marolf, D. and Morrison, I. A. 2011. The IR stability of de Sitter QFT: results at all orders. *Phys. Rev.*, **D84**, 044040.

[20] Omnès, R. 1992. Consistent interpretations of quantum mechanics. *Rev. Mod. Phys.*, **64**, 339.

[21] Parikh, M. K., Savonije, I. and Verlinde, E. P. 2003. Elliptic de Sitter space: dS/Z(2). *Phys. Rev.*, **D67**, 064005.

[22] Riedel, C. J., Zurek, W. H. and Zwolak, M. 2016. The Objective Past of a Quantum Universe: Redundant Records of Consistent Histories. *Phys. Rev. A*, **93**, 032126.

[23] Schlosshauer, M. 2004. Decoherence, the measurement problem, and interpretations of quantum mechanics. *Rev. Mod. Phys.*, **76**, 1267.

[24] Sebens, C. T. and Carroll, S. M. 2014. Self-Locating Uncertainty and the Origin of Probability in Everettian Quantum Mechanics. eprint arXiv:1405.7577.

[25] Spradlin, M., Strominger, A. and Volovich, A. 2001. Les Houches lectures on de Sitter space. 423–53. eprint arXiv:hep-th/0110007.

[26] Stephens, C. R., 't Hooft, G. and Whiting, B. F. 1994. Black hole evaporation without information loss. *Class. Quant. Grav.*, **11**, 621.

[27] Susskind, L., Thorlacius, L. and Uglum, J. 1993. The stretched horizon and black hole complementarity. *Phys. Rev.*, **D48**, 3743.

[28] Wald, R. M. 1983. Asymptotic behavior of homogeneous cosmological models in the presence of a positive cosmological constant. *Phys. Rev.*, **D28**, 2118.

[29] Wallace, D. 2012. *The Emergent Multiverse*. Oxford: Oxford University Press.

[30] Zurek, W. H. 2003. Decoherence, einselection, and the quantum origins of the classical. *Rev. Mod. Phys.*, **75**, 715.

[31] Zurek, W. H. 1981. Pointer Basis of Quantum Apparatus: Into What Mixture Does the Wave Packet Collapse? *Phys. Rev.*, **D24**, 1516.

[32] Zurek, W. H. 1993. Preferred States, Predictability, Classicality and the Environment-Induced Decoherence. *Prog. Theor. Phys.*, **89**, 281.

[33] Zurek, W. H. 1998. Decoherence, Einselection, and the existential interpretation: The Rough guide. *Phil. Trans. Roy. Soc. Lond.*, **A356**, 1793.

12

Holographic Inflation Revised

TOM BANKS

12.1 Introduction

The formalism of holographic space-time (HST) is an attempt to write down a theory of quantum gravity which can treat space-times more general than those accessible to traditional string theory. String theory, roughly speaking treats space-times which are asymptotically flat or anti-de Sitter (AdS). As classical space-times, these contain infinite area causal diamonds on which strict boundary conditions are imposed. The corresponding quantum theory has a unique ground state, and the existing formalism describes the evolution of small fluctuations around that ground state in terms of evolution operators involving the infinite set of possible small fluctuations at the boundary. Local physics is obscure in the fundamental formulation of the theory. It emerges only by matching the fundamental amplitudes to those of an effective quantum field theory, in a restricted kinematic regime. In the AdS case, one must also work in a regime where the AdS radius is much larger than the length scale defined by the string tension. That string length scale is bounded below by the Planck length. In regimes where the two scales are close, there are no elementary stringy excitations.

The HST formalism works directly with local quantities. Its important properties are summarized as follows:

- The fundamental geometrical object, a time-like trajectory in space-time, is described by a quantum system with a time dependent Hamiltonian. Four times the logarithm of the dimension (= entropy) of the Hilbert space of the system is viewed as the quantum avatar of the area of the holographic screen of the maximal causal diamond along the trajectory. The causal diamond associated with a segment of a time-like trajectory is the intersection of the interior of the backward light cone of the future endpoint of the segment, with that of the future light cone of the past point. The holographic screen is the maximal area surface on the boundary of the diamond.
  When the entropy of Hilbert spaces is large, space-time geometry is emergent. The case of infinite dimension must be treated by taking a careful limit. The Hilbert space comes with a built-in nested tensor factorization: $\mathcal{H} = \mathcal{H}_{in}(t) \otimes \mathcal{H}_{out}(t)$. t is a discrete parameter, which labels the length of a proper time interval along the trajectory. $\mathcal{H}_{in}(t)$ is the Hilbert space describing the causal diamond of that interval. $\mathcal{H}_{out}(t)$ describes all

241

operators, which commute with those in the causal diamond. We adopt the prescription from quantum field theory (QFT) that space-like separation is encoded in commutivity of the operator algebra of a causal diamond, with that associated to the region of space-time that is space-like separated from that diamond. The holographic bound on the entropy of a finite area diamond allows us to state this in terms of simple tensor factorization. The Hamiltonian of the system *must* be time dependent, in order to couple together only degrees of freedom (DOF) in $\mathcal{H}_{in}(t)$ for time intervals shorter than t.

- Time is fundamental but relative in HST, while space-time is emergent. By relativity of time, we mean that each time-like line has its own quantum description of the world. Space-time is knit together from the causal diamonds of all intervals along a sufficiently rich sampling of trajectories. For each pair of diamonds, Lorentzian geometry gives a maximal area diamond in the intersection. The quantum version of this notion is the identification of a common tensor factor in the Hilbert spaces of these two quantum systems. The initial conditions and Hamiltonians of the two systems must be such that the density matrices prescribed by the two systems for that common tensor factor are unitarity equivalent. This is an infinite number of constraints on the dynamics, and the requirement is quite restrictive. For example, in asymptotically flat space-time, the requirement of Lorentz invariance of the scattering matrix is a consequence of these consistency conditions. Note that in HST geometry is an emergent property of quantum systems, but the metric is not a fluctuating quantum variable. The causal structure and conformal factor (which determine the metric) are determined by the area law and the overlap rules, which are not operators in the Hilbert space.
- The fact that geometry is not a quantum variable fits very nicely with Jacobson's derivation of Einstein's equations [2] as the hydrodynamics of a quantum system whose equation of state ties entropy to geometry via the area law for causal diamonds. Hydrodynamic equations are classical equations valid in high entropy quantum states of systems whose fundamental variables have nothing to do with the hydrodynamic variables (the latter are ensemble expectation values of quantum operators). There is only one situation in which quantized hydrodynamics makes sense: small, low energy fluctuations around the ground state (of a system that has a ground state). This accounts for the success of quantum field theory in reproducing certain limiting boundary correlation functions in asymptotically flat and AdS space-times. A notable feature of Jacobson's derivation of Einstein's equations from hydrodynamics is that it does *not* get the cosmological constant (CC) term. Fischler and I have long argued that the CC is an asymptotic boundary condition, relating the asymptotics of proper time and area in a causal diamond. In the quantum theory it is a regulator for the number of states. If the CC is positive, the number of states is finite. If the CC is negative, it determines the asymptotic growth of the density of states at high energy. High energy states are all black holes of large Schwarzschild radius. In HST, the value of the CC is one of the characteristics that determines different *models* of quantum gravity, with very different Hamiltonians and fundamental DOF.[1]

---

[1] See the discussion of meta-cosmology below for a model that incorporates many different values of the CC into one quantum system.

- In four non-compact dimensions, the variables of quantum gravity are spinor functions $\psi_i^A(p)$ (p is a discrete finite label, which enumerates a cutoff set of eigenfunctions of the Dirac operator on compact extra dimensions of space). These label the states, described by local flows of asymptotically conserved quantum numbers, through the conformal boundary of Minkowski spacetime. The Holographic/Covariant Entropy Principle is implemented on a causal diamond with finite area holographic screen by cutting off the Dirac eigenvalue/angular momentum on the two sphere. The fundamental relation is

$$\pi (RM_P)^2 = LN(N+1),$$

where R is the radius of the screen, L the number of values of p and N the angular momentum cutoff. i and A range from 1 to N and $N+1$, respectively.

- We have constructed [11, 19, 20] a class of models describing scattering theory in Minkowski space. The basic idea of those models is that localized objects are described by constrained states, on which of order EN, with $E \ll N \to \infty$, matrix elements of the square matrices $M_i^j(p,q) \equiv \psi_A^{\dagger\,j}(p)\psi_i^A(q)$ vanish, as the size of the diamond goes to infinity. The Hamiltonian has the form

$$H_{in}(N) = E + \frac{1}{N^2}\mathrm{Tr}\,P(M), \tag{12.1}$$

where P is a polynomial of N independent order ≥ 7. The quantity E defining the constraint is an asymptotically conserved quantum number. The constraints imply that the matrices can be block diagonalized and, when combined with the large N scaling of the Hamiltonian, the blocks are free objects: The Hamiltonian is a sum of commuting terms, describing individual blocks. It can be shown that in the limit in which the individual blocks have large size, these objects are supersymmetric particles. For generic choices of $P(M)$ the long range interactions have Newtonian scaling with energy and impact parameter. All of these models have meta-stable excitations with the properties of black holes, which are produced in particle scattering and decay into particles. Scattering amplitudes in which black holes are not produced can be described by Feynman-like diagrams, with vertices localized on the Planck scale. We have not yet succeeded in imposing the HST consistency conditions for trajectories in relative motion, which we believe will put strong constraints on P and on the spectrum of allowed particles.[2]

- It is easy to convert the models above into models of de Sitter space, by simply keeping N finite. This automatically explains both the de Sitter entropy and temperature, the latter because the definition of a particle state of energy E involves constraining EN of the $o(N^2)$ DOF on the horizon. The probability of having such a state, within the random ensemble, we call the dS vacuum state[3] is e^{-EN}. This says dS space has a temperature

---

[2] The spectrum is encoded in the commutation relations of the variables $\psi_i^A(p)$.

[3] It has been known since the seminal work of Gibbons and Hawking [3], that the dS "vacuum state" is actually a high entropy density matrix.

proportional to the inverse of its Hubble radius. The fact that localized energy corresponds to a constraint on the degrees of freedom in dS space is already evident from the form of the Schwarzschild–de Sitter black hole solution. It has two horizons, given by the roots of $(1 - 2GM/r - r^2/R^2)$. It is easy to verify that the sum of the areas of the two horizons is minimized at $M = 0$, and when M is small, the entropy deficit is that expected from a Boltzmann factor $e^{-2\pi RM}$.

- We have constructed [4–7] a fully consistent quantum model of cosmology, in which the universe is described by a flat Friedmann–Robertson–Walker (FRW) metric with a stress tensor that is the sum of a term with equation of state $p = \rho$ and a term with $p = -\rho$. The geometry, as anticipated by Jacobson, is a coarse grained thermodynamic description, valid in the large entropy limit, of the quantum mechanics of the system. Homogeneity, isotropy and flatness are realized for arbitrary initial states. Homogeneity and isotropy are the only obvious ways to satisfy the consistency conditions between the descriptions of physics along different trajectories, when each trajectory is experiencing randomizing dynamics. Flatness follows from an assumption of asymptotic scale invariance for causal diamonds much larger than the Planck scale but much smaller than the Hubble scale of the CC. The CC itself is an input, basically a declaration that we stop the growth of the Hilbert space, but allow time evolution to proceed forever, with a fixed Hamiltonian, which has entered the scaling regime describing the asymptotics of the $\Lambda = 0$ model. Note by the way that the initial singularity does not appear. The geometric description is valid only in the limit of large entropy/large causal diamonds and the singularity is an extrapolation of this limiting behavior back to a time when the causal diamond is Planck size. We have called this model Everlasting Holographic Inflation (EHI). It has an infinite number of copies of a space-time which is asymptotically a single horizon volume of dS. Unlike field theoretic models of eternal inflation, the different horizon volumes are constrained to have identical initial conditions and may be viewed as gauge copies of each other. In HST stable dS space is a quantum system with a finite dimensional Hilbert space [14, 15]. This model has no local excitations, except those which arise as thermal fluctuations in the infinite dS era. It is important to note that the EHI model and the more realistic model described below are not the same as the HST model of stable dS space. The latter model has infinite, rather than semi-infinite proper time intervals, and a Hamiltonian, for each trajectory, which satisfies $H(T) = H(-T)$ for each proper time. In the limit of infinite dS radius, it approaches the HST model for Minkowski space. Both the EHI model and our semi-realistic Holographic Inflation model [1] use the same time dependent Hamiltonian $H(T)$, where T runs over a semi-infinite interval. There is no time reflection symmetry in the system. EHI and HI differ only in their boundary conditions, with the latter having fine-tuned boundary conditions, which guarantee an era where localized excitations are approximately decoupled from horizon DOF. In EHI, *despite its intrinsic time asymmetry*, the universe is always in a generic state of its Hilbert space at all times, and local excitations, which are of low entropy, because they are defined by constrained states on which large numbers of $\psi_i^A(p)$ vanish, arise only as sporadic thermal fluctuations.

• *Perhaps the most important difference between HST and QFT lies in the counting of entropy. In HST the generic state of the variables in a causal diamond of holoscreen area $\sim R^2$, has no localized excitations: all of the action takes place on the horizon, with a Hamiltonian that is not local on the holographic screen. Bulk localized states are constrained states in which of order ER of the $o(R^2)$ variables are set equal to zero. R is the holoscreen radius, and $E \ll R$ (all these equations are in Planck units) is the approximately conserved energy (it becomes conserved in the limit R goes to infinity). It is only for these constrained states that QFT gives a good description of some transition amplitudes.*

Our model of the universe we live in, the topic of this chapter, proceeds from a starting point identical to that of the model discussed in the penultimate bullet point. However, we consider a much larger Hilbert space, for a single trajectory, which includes many copies of a single inflationary horizon volume. This corresponds to the growth of the apparent horizon, after inflation, in conventional inflationary models. This model contains elements of the conventional narrative about cosmological inflation, so we call it the Holographic Inflation model. The purpose of the inflationary era in this model is quite different from that which inflation serves in field theory models. Homogeneity, isotropy and flatness are natural in HST. We need fine tuning of initial conditions in order to get local excitations, and this is the purpose that the inflationary era serves. I will comment below on the degree of fine tuning required compared to field theoretic inflation, but the most important point is that our fine tuning is the minimal amount required to get localized excitations, so the very crudest kind of anthropic reasoning, a *topikèsthrophic* restriction (from the Greek word for locality), says that, within the HST formalism, a universe with this amount of fine tuning of initial conditions is the only kind that can ever be observed. The only assumption that goes into this is that any kind of observer will require the approximate validity of local bulk physics. It does not require the existence of human beings, or anything like conventional biology.

The derivation of a macroscopic world from quantum mechanics, a world in which the ordinary rules of logic and the notion of decoherent histories are valid, relies on the existence of macroscopic objects with macroscopic moving parts. In the HST model of quantum cosmology, a typical macroscopic object is the entire apparent horizon; slightly less typical ones are localized black holes. None of these have complex webs of semi-classical collective coordinates. We know how to derive the existence of complex macro-objects in quantum field theory, even with an ultra-violet cutoff, but in HST quantum states approximately describable by QFT are highly atypical. If we want a universe in which such states appear as anything but ephemeral thermal fluctuations, we *must* impose constraints on the initial conditions. Thus, in HST, the reason the universe began in a low entropy state, is that this is the only way in which the model produces a complex, approximately classical world.

12.2 The Holographic Inflation Model

Our cosmology begins on a space-like hypersurface, called the Big Bang, which has the topology of three-dimensional flat space. A sampling of time-like trajectories in the emergent space-time are labelled by a regular lattice on the space-time, whose geometry is irrelevant. Probably a general three-dimensional simplicial complex, with the simplicial homology of flat space, is sufficient. The quantum avatar of each trajectory is an independent quantum system, whose Hilbert space is finite dimensional if the ultimate value of the CC is positive.

We incorporate causality into our quantum system, by insisting that the Hamiltonian is proper time *dependent* along the trajectory and has the form

$$H(t_n) = H_{in}(t_n) + H_{out}(t_n). \tag{12.2}$$

At proper time t_n the horizon has area

$$\pi (R_n M_p)^2 = Ln(n+1). \tag{12.3}$$

$H_{in}(t_n)$ is constructed from the matrices M built out of the subalgebra of ψ_i^A with indices $i \leq n$ and $A \leq n+1$. $H_{out}(t_n)$ depends only on the rest of the variables. In our simplest models, we will not have to specify H_{out}, but it will be crucial to our Holographic Inflation Model.

All of our models choose exactly the same sequence of Hamiltonians, and the same initial state, for each trajectory in our lattice. The initial state is, however, completely arbitrary, so there is no fine tuning of initial conditions. Along each trajectory, the Hamiltonian $H_{in}(t_n)$ is the trace of a polynomial, P, in the matrices $M(t_n)$. The coefficients in P are chosen randomly at each time t_n. Systems like this almost certainly have fast scrambling behavior [16]. The maximal order of the polynomial is fixed and $\ll N$, but we do not yet know what it is beyond a bound $\deg P \geq 7$. This bound is required to get the proper scaling of Newton's law in regimes that are approximately flat space [17]. Finally, we require that, when n is large, but $\ll N$, the Hamiltonian approach that of a $1 + 1$ dimensional C (onformal) F(ield) T(heory) on a half line, with central charge of order n^2. The system is obviously finite dimensional, so this statement can only make sense in the presence of UV and IR cutoffs on the CFT . We choose them to be $\Lambda_{UV} \sim 1/n$, $V_{IR} \sim Ln$, and also insist that $L \gg 1$ so that the CFT behavior will be manifest in the presence of the cutoffs.

The fast scrambling nature of the dynamics implies that we can make thermodynamic estimates of the expectation value of H_{in} and the entropy of the time-averaged density matrix

$$E \sim Ln,$$

$$S \sim Ln^2.$$

The volume of the bulk region inside the horizon at this time scales like n^3, so the energy and entropy densities scale like

$$\rho \sim n^{-2},$$

$$\sigma \sim n^{-1} = \sqrt{\rho}.$$

If we recall that in a flat FRW metric the horizon size n, scales like the cosmological time t we recognize the first of these equations as the Friedmann equation and the second as the relation between entropy and energy densities for an equation of state $p = \rho$. This geometric description was to be expected, from Jacobson's argument [2] showing that Einstein's equations are the hydrodynamics of systems obeying the area/entropy connection. Note that the geometric/hydrodynamic description should not be extrapolated into the low entropy regime of small n, so that the cosmological singularity of the FRW cosmology is irrelevant. Note also that the quantum mechanics of this system is in no sense that of the quantized Einstein equations. Quantized hydrodynamics is a valid approximation for describing small excitations of a system around its ground state. In the HST models, the early universe is very far from its ground state, and does not even have a time-independent Hamiltonian.[4]

In the EHI models, the quantum systems along different trajectories are knit together into a space-time, by specifying at each time t_n that the maximal causal diamond in the intersection between causal diamonds of two trajectories that are D steps apart on the lattice is the tensor factor in each Hilbert space, on which $H_{in}(t_n - D)$ acted. The choice of identical Hamiltonians and initial state for each trajectory insures that the density matrix on this tensor factor at time t_n is independent of which trajectory we choose to view it from. This infinite set of consistency conditions is the fundamental dynamical principle of HST. The locus of all points D steps away on the lattice is identified by this choice as a set of points on the surface of a sphere in the emergent space-time, because of the causal relations. The fact that the dynamics is independent of the point is rotation invariance on that sphere, and this is consistent with the fact that our fundamental variables transform as a representation of $SU(2)$ and the Hamiltonian is rotation invariant. We see that homogeneity, isotropy and flatness are properties of the space-time of this model, which are independent of the choice of initial state. In our more realistic models of the universe, in which local objects emerge from a choice of fine tuned initial conditions, we do not yet have a solution to the consistency conditions for trajectories with Planck scale spacing, but homogeneity and isotropy play a crucial role in satisfying a coarse grained verse of the consistency conditions for trajectories whose spatial separation is of order the inflationary horizon size.

As $n \to N$, we need to change the rules only slightly. Proper time is now decoupled from the growth of the horizon. We model this by allowing the system to propagate forever with the Hamiltonian $H_{in}(N)$. In addition, we do an (approximate) conformal transformation on the Hamiltonian, rescaling the UV and IR cutoffs so that the total Hamiltonian is bounded by $1/n \to 1/N$. This is analogous to the transformation from FRW to static observer coordinates [18] in an asymptotically dS universe, and is appropriate because we are postulating

---

[4] In passing we note that the only known quantum gravitational systems with a ground state are asymptotically flat and anti-de Sitter space-times. The proper description of these is String Theory-AdS/CFT, and certain amplitudes in the quantum theory *are* well approximated by QFT. IMHO, string theorists and conventional inflation theorists make a mistake in trying to extrapolate that approximation to the early universe.

a time independent Hamiltonian. Finally we have to modify the overlap rules, to be consistent with the fact that individual trajectories have a finite dimensional Hilbert space. The new overlaps are the same as the old ones, except that points that are more than N steps apart on the lattice never have any overlap at any time. The asymptotic causal structure is then that of dS space, with Hubble radius satisfying $\pi(R_H M_P)^2 = LN(N+1)$.

The simplest metric interpolating between the $p = \rho$ and $p = -\rho$ equations of state is $a(t) = \sinh^{1/3}(3t/R_H)$, which is an exact solution of Einstein's equations with a fluid that has two components with these equations of state. All of the geometric information in our model is consistent with this ansatz for the metric. The space-time of our model contains no localized excitations. At all times, all DOF are localized on the apparent horizon, in a completely thermalized state, obeying none of the constraints which characterize bulk localized systems in HST [11, 19, 20]. This EHI model is not a good model of the universe we inhabit, although it is, according to the rules of HST, a perfectly good model of quantum gravity. Localized excitations will occur only as ephemeral thermal fluctuations in the eternal dS phase of this cosmology.

12.2.1 A More Realistic Description of the Universe

The description of localized objects in HST was worked out in a series of papers [11, 19, 20] devoted to scattering theory in Minkowski space-time. Here we summarize the results and explain how they are used to construct the HST version of inflationary cosmology. The variables $\psi_i^A(p)$ are sections of the Dirac-cutoff spinor bundle on the two sphere, so the matrices $M_i^j(p,q)$ are sections of the bundle of differential forms on the sphere. Two forms can be integrated over the sphere and the fuzzy analog of integration is the trace, which we have used in constructing our model Hamiltonians. Expressions involving a trace of a polynomial in M are invariant under unitary conjugation and this invariance converges to the group of measure preserving transformations on the sphere (we do not require them to be smooth). Saying that a matrix is block diagonal in some basis can be interpreted as saying that the corresponding forms vanish outside of some localized region on the sphere. In our quantum mechanics, the matrix elements of M are operators, so a statement that some of them vanish is a *constraint on the states*. If this constraint is *approximately* preserved under the action of the Hamiltonian taking us from one causal diamond to the next, then this defines the track of a localized object in space-time.

This connection between localization, and constraints that put the system in a low entropy state, is supported by a piece of evidence from classical GR. An object in dS space, localized in a region much smaller than the dS radius, will have a dS Schwarzschild field. The local entropy will be maximized if the object is a black hole, filling the localization region, but independently of that choice, the entropy of the horizon shrinks, so that the total entropy of the system is less than that of empty dS space. Recall that empty dS space is a thermal system, with DOF that must be considered to live on the cosmological horizon. The idea that a localized object of energy E is a constraint on a system with $o(N^2)$

DOF, which freezes $o(EN)$ of them, explains the scaling of the temperature with dS radius. In Banks and colleagues [11, 19, 20] we showed that the same idea explains the conservation of energy in Minkowski space, as well as the critical impact parameter at which particle scattering leads to black hole production, the temperature/mass/entropy of black holes, and even the scaling of large impact parameter scattering with energy and impact parameter (Newton's Law).

All of these results are quite explicit calculations in our quantum mechanical matrix models, and they mirror scaling laws usually derived from classical GR. It is crucial to all of them that the block diagonal constraint on matrices gives us a finite number of blocks of size K_i, and one large block of size $N - K \gg 1$, where $K = \sum K_i$, and turns out to be the conserved energy. The individual K_i represent the amount of energy going out through the horizon in different angular directions. Energy is only conserved in the limit $N \to \infty$, since in that limit the Hamiltonian cannot remove $O(KN)$ constraints. A final result from the matrix model shows that large impact parameter scattering amplitudes scale with energy and impact parameter as one would expect from Newton's law. Again, the existence of a vast set of very low $(o(1/N))$ energy DOF, which do not have a particle description, is crucial to this result. In passing, I remark that these DOF resolve the firewall paradox of AMPS [9–12].

The correlation between localization and low entropy is the key to answering Penrose's question[5] of why the initial state of the universe had low entropy. We have already seen that an unconstrained initial state of the universe leads to the EHI cosmology, whose only localized excitations are low entropy thermal fluctuations late in the dS phase of the universe.

The maximal entropy state with localized excitations, is one in which those excitations are black holes. As the horizon expands one may encounter more black holes or the process may stop at some fixed horizon size. The maximal set of black holes for a fixed horizon size R in Planck units is constrained by the inequality

$$\sum K_i \leq R. \tag{12.4}$$

Note that this constraint can be derived *either* from the geometric requirement that the Schwarzschild radius of the total black hole mass be smaller than the horizon size, or from the matrix model constraint that the black hole blocks of the matrix fit inside the full matrix available at that value of the horizon size.

At this point, we must recognize the distinction between time in the matrix model, which represents the time along a particular trajectory, and FRW time. The time slices in the matrix model always correspond to "hyperboloids" lying between the boundaries of two successive causal diamonds, while FRW time corresponds to horizontal lines. If we say that at some fixed causal diamond size, smaller than the dS radius of the ultimate CC, the process of new black holes coming into the horizon stops, then trajectories far enough

---

[5] Which of course goes back to Boltzmann. Penrose was the first to raise the issue in the context of the General Relativistic theory of gravity.

from the one we have been discussing will see the region around our preferred trajectory as a collection of black holes localized in a given region. The system will not be, even approximately, homogeneous on the would be FRW slices. Some of the black holes may decay into radiation, but many will be gravitationally bound, and coalesce into large black holes, which have lower probability of decay. The universe will not look anything like our own, and it is unlikely to ever produce complex structures, which could play the role of "observers". It is often said that once a single galaxy forms, the question of the evolution of observers is independent of the rest of the cosmos, but that statement depends on defining a galaxy as a gravitationally bound structure whose constituents are primarily composed of baryonic matter. In these HST models of the universe, the primordial constituents of the universe are black holes. Matter, baryonic or otherwise, must be produced in black hole decay.

Although we have not explored all such inhomogeneous scenarios, it seems clear that a model with a fairly homogeneous black hole gas is much more likely to produce an observer-ready cosmology than an inhomogeneous one. If the gas is sufficiently dilute, and the black holes sufficiently small compared to the size of the ultimate cosmological horizon, they will all decay before they can coalesce into larger black holes. That decay is the hot Big Bang in the HST models.

It is also, as we will see, much easier to solve the HST constraints relating the descriptions along different time-like trajectories when the universe is homogeneous and isotropic in a coarse grained way. Exact homogeneity and isotropy are incompatible with quantum mechanics. Our black holes are really quantum systems with finite dimensional Hilbert spaces, with a quantum state that is varying by order one on a time scale nL_P, the Schwarzschild radius. Occasionally the Hamiltonian will put us in a state where the value of n is effectively smaller, because some of the matrix elements vanish. Thus, the black hole radius is to be thought of as an expectation value of an operator which is the trace of some polynomial in the matrices M. It will have statistical fluctuations in the time averaged density matrix of the system. By the usual rules of statistical mechanics these will be small and approximately Gaussian, with size

$$\frac{\delta n}{n} \sim \frac{1}{n}.$$

Similarly, although the expectation value of the black hole angular momentum is obviously zero, it will have Gaussian fluctuations of the same order of magnitude. Since these are fluctuations in collective coordinates of a large chaotic quantum system, and time averaged fluctuations at that, it is obvious that quantum interference effects in the statistics of these variables are negligible, of order e^{-n^2}.

Fluctuations of the mass and angular momentum of a black hole are fluctuations of the spin zero and spin two parts of the Weyl Curvature tensor, and thus have the properties of scalar and tensor fluctuations in cosmological perturbation theory. We will postpone an analysis of the phenomenological implications of this remark until we have completed the sketch of our model.

Now let us return to the matrix model description of our model. For some period of time, which we call the era of inflation, the horizon size remains fixed at n and the system evolves with a time independent Hamiltonian

$$H_{in}(n) = \frac{1}{n^2} P(M_n), \qquad (12.5)$$

which is the asymptotic Hamiltonian of our HEI model. Then, the horizon begins to expand, eventually reaching the cosmological horizon N. At time t_k, when the horizon size is k, some number of black holes will have come into the horizon. If the average black hole size is n, the $k \times k$ matrices M_k, which appear in the Hamiltonian $H_{in}(t_k)$ must be in a block diagonal state, with some number of blocks $p_k \leq k/n$ of size n. The black hole energy density is thus

$$\rho_{BH} = \frac{p_k n}{k^3} \leq \frac{1}{k^2}. \qquad (12.6)$$

If we choose $t_k \propto k$, as we expect for any flat FRW model, then the RHS is just the Friedmann equation for the energy density if we choose $p_k = n_{BH} \frac{k}{n}$, with n_{BH} interpreted as an initial black hole number density.

I do not have space here to sketch the full matrix model treatment of this system, but will instead rely on the reader's familiarity with semi-classical black hole dynamics, together with the evidence from Banks and colleagues [11, 19, 20] that thermalized block diagonal matrices have qualitative behavior very similar to that of black holes. The initial number density of black holes on FRW slices must be $< 1/n^3$, so that the black holes are further apart than their Schwarzschild radii. What happens next is a competition between two processes: The growth of fluctuations in a universe dominated by an almost homogeneous gas of black holes, and the decay of the black holes. As we will recall below, the size of the primordial scalar (mass) fluctuations in the black hole energy density is $Q = \frac{\delta\rho}{\rho} \sim \frac{1}{n\epsilon}$, where ϵ is a "slow roll parameter", currently bounded above by ~ 0.1 by observation. In the black hole dominated universe, these will grow to $o(1)$ when the scale factor has grown by $n\epsilon$, which occurs in FRW time $t_{FRW} \sim (n\epsilon)^{3/2}$, which is much less than the black hole evaporation time $t_{evap} \sim n^3$. Black holes will thus begin to combine on this time scale, potentially shutting off the evaporation process and leaving us with a universe dominated by large black holes, forever. However, since black holes are separated by distances *much* larger than their Schwarzschild radii by this time, the fluid approximation does not capture the full coalescence process. We must also estimate the infall time for two widely separated black holes in a local over-density to actually collide. It turns out that this time is longer than the decay time as long as $n_{BH} < n^{-3}$. So our model produces a radiation dominated universe, with a reheat temperature given roughly by the redshifted black hole energy density at the decay time

$$gT^4 \sim n n_{BH} \frac{g^2}{n^6} < \frac{g^2}{n^8}. \qquad (12.7)$$

Here g is the effective number of massless species into which the black holes can decay. The standard model gives $g \sim 10^2$. Supersymmetry, particularly with a low energy SUSY breaking sector added on boosts this to $g \sim 10^3$. The relation between n and primordial fluctuations gives

$$T_{RH} < g^{1/4}\epsilon^2 Q^2 = 10^{-9.25}\epsilon^2 < 10^{-11}. \tag{12.8}$$

This estimate, which takes into account the observations of Q and the observational bound on ϵ is in Planck units and corresponds to $T_{RH} < 10^7 - 10^8$ GeV.

What does all of this have to do with inflation? In order to answer that, we have to return to the time t_k, and ask what the Hamiltonian $H_{out}(t_k)$ must be doing in order to be consistent with the fact that a new system with $o(n^2)$ DOF is about to be added to $H_{in}(t_k)$, and also with the overlap conditions of HST. The first of these constraints says that, at the time it comes into "our" trajectory's horizon, this system must have been completely decoupled from the rest of the DOF in the universe. This is consistent with dynamics along the trajectory that has contained those DOF in its causal diamond since the horizon size grew to n, *if that trajectory is still experiencing inflation.* Here again we must recall the differences between the time slicings of individual trajectories and the FRW slicing. The future tip of a given trajectory's causal diamond lies on a particular FRW slice, but its intersection with a spatially remote trajectory is at a correspondingly remote FRW time in the past. In an asymptotically dS universe, where the scale factor $a(\eta)$ has a pole at some conformal time η_0, the last bit of information that comes into the horizon of a given trajectory is on an FRW slice with conformal time $\eta_0/2$. Thus, for an approximately homogeneous and isotropic collection of black holes, the universe had to undergo inflation up to this conformal time. Along each trajectory, until the time $t*$ when the future tip of its diamond hits the FRW surface $\eta_0/2$, the Hamiltonian $H_{in}(t < t*)$ is the Hamiltonian of the EHI universe.

The shortest wavelength fluctuations that occur in this model will have wavelength of order na_{NOW}/a_I, where a_I is the scale factor at the end of inflation. These are fluctuations that came into the horizon just after inflation ends. The largest wavelength fluctuations are those which cover the entire sky and have wavelength of order N. In a conventional inflation theory we would write $e^{N_e} \geq \frac{Na_I}{na_{NOW}}$, whereas in the HST model this is a strict equality. There are only as many e-folds as we can see. A crude estimate, given the cosmological history we have sketched, gives $N_e \sim 80$. This is larger than the conventional lower bound $N_e > 60$, because our cosmology has a long period in which the universe is dominated by a dilute black hole gas. Reheating does not occur immediately after inflation.

12.2.2 SO(1, 4) Invariance

Work of Maldacena and others [21–28] has shown that current data on the CMB can be explained in a very simple framework, with no assumptions about particular models. Indeed, it was shown in Banks *et al.* [29] that even the assumption that fluctuations originate from quantized fields is unnecessary. All one needs is the approximate $SO(1,4)$ invariance we will demonstrate below.

If we consider general perturbations of an FRW metric, then as long as the vorticity of the perturbed fluid vanishes (which is an automatic consequence of the symmetry assumptions below), we can go to co-moving gauge, where the perturbations are in the scalar and spin two components of the Weyl tensor. The conventional gauge invariant scalar perturbation ζ is equal, in this gauge, to the local proper time difference between co-moving time slices, in units of the background Hubble scale

$$\zeta = H\delta\tau = \frac{H^2}{\dot{H}}\frac{\delta H}{H} \equiv \frac{\delta H}{\epsilon H} \, . \tag{12.9}$$

This equation alone, if ϵ is small, explains why we have so far seen scalar, but not tensor curvature fluctuations. The gauge invariant measure of scalar fluctuations is suppressed if the background FRW goes through a period in which the variation of the Hubble parameter is smaller than the Hubble scale, which is the essential definition of an inflationary period. The rest of the data on scalar fluctuations essentially correspond to fitting the background $H(t)$. The HST and field theory based models have a different formula [29, 30] relating $H(t)$ to the form of the scalar two point function, but if both models have an approximate $SO(1,4)$ symmetry, they can both fit the data.

Much is made in the inflation literature of the fact that inflationary fluctuations are approximately Gaussian, and this is supposed to be evidence that free quantum field theory is a good approximation to the underlying model. However, Gaussian fluctuations are a good approximation to almost any large quantum system in a regime where the entropy is large. Higher order correlation functions are suppressed by powers of $S^{-1/2}$. The HST model has approximately Gaussian fluctuations for these general entropic reasons, rather than the special form of its ground state. The insistence that a huge quantum system be in its ground state is a monumental fine tuning of initial conditions, and should count as a strike against conventional inflation models.

There is a further suppression of non-Gaussian correlations involving at least one scalar curvature fluctuation, which follows from Maldacena's squeezed limit theorem [31] and approximate $SO(1,4)$ invariance. Maldacena's theorem says that in the limit of zero momentum for the scalar curvature fluctuation, the three point correlator reduces to something proportional to the violation of scale invariance in the two point correlator. The symmetry allows us to argue that this suppression is present for all momenta. Since all three point functions are suppressed relative to two point functions by a nominal factor of 10^{-5}, and this theorem gives us an extra power of 10^{-2}, we should not expect to see these non-Gaussian fluctuations if the world is described by any one of a large class of models, including both HST and many slow roll inflation models.

The tensor two point function is the most likely quantity to be measured in the near future. HST and slow roll models make different predictions for the tilt of the tensor spectrum. HST predicts exact scale invariance while slow roll inflation models have a tilt of $r/8$, where r is the tensor to scalar ratio. Since we now know that $r < 0.1$ with 95 per cent confidence, it will be difficult though not necessarily impossible to observe this difference. Of course, if r is much smaller than its observational upper bound, this observation will not be feasible. We should however point out that HST has a second source of gravitational waves, the decay of black holes. This will have a spatial distribution that mirrors the scalar fluctuations, and so should have the same tilt as the scalars, again disagreeing with the predictions of standard slow roll models. It is suppressed by a factor $1/g$, the number of effective massless species into which the black holes can decay, but with no suppression for small ϵ. Finally, I would like to point out that the PIXIE mission [32], will test for short wavelength primordial gravitational waves and can probably distinguish even a small tilt of the spectrum.

The tensor three point function, in all models having approximate $SO(1, 4)$ invariance, might be the largest of the non-Gaussian fluctuations, unless $r \sim 0.04$ or less. It does not have the $n_S - 1$ factor from Maldacena's squeezed limit theorem. It is by far the most interesting correlation function that humans might someday observe, since symmetry allows three distinct forms for the three point function. Quantum field theory models predict that one of the three dominates the other two by a factor of $n > 10^6$ while the third vanishes to all orders in inverse powers of n. HST models predict two of the three form factors are of comparable magnitude, while the third appears to vanish only if a certain space-time reflection symmetry is imposed as an assumption. Unfortunately, we will probably not be able to measure this three point function in the lifetime of any of my auditors at this conference.

We turn briefly to the derivation of approximate $SO(1, 4)$ invariance of the fluctuations in the HST model. As discussed above, the Hamiltonian acting on the Hilbert space of entropy Ln^2 in the HEI model is the Cartan generator of an approximate $SL(2)$ algebra. This is the statement that the early universe is described approximately by a $1 + 1$ dimensional CFT. As DOF come into the horizon, the initial state must be constrained so that the matrices $M(kn; p, q)$, which appear in $H_{in}(kn)$, are all block diagonal, with blocks of approximate size n. In order for this to be consistent $H_{out}(kn)$ must act on all of the blocks that have not yet come into the horizon, but will in the future, as a sum of independent copies of the $SL(2)$ Cartan generator, $L_0[a]$. This insures that these DOF will have dynamics that mirrors a horizon of fixed size. Once the horizon size has expanded to $N = Kn$, corresponding to the observed value of the CC, all of these DOF are embedded in an $SU(2)$ covariant system, with entropy LN^2. We can organize the DOF, so that $SU(2)$ invariance is preserved at all times, by choosing to define the action of $SU(2)$ so that DOF, which have come into the horizon when its size is kn, transform in the $[kn] \otimes [kn + 1]$ representation of $SU(2)$.

Once the apparent horizon coincides with the cosmological horizon, we can divide the entire set of variables up into variables localized at various angles. To visualize this, take the sphere of radius $\sim Kn = N$ and draw a spherical grid: an icosahedron with triangular faces, each of which is tiled by equilateral triangles of area n^2. We call Ω_i the solid angular coordinate of the *ith* small triangle. Consider the n^2 most localized linearly independent functions we can construct from $o(N^2)$ spherical harmonics, and make a basis which consists of these localized functions in one particular tile, and all rotations of them to different tiles.

The total number of black holes that come into the horizon is certainly no greater than $K = N/n$, and the number which have come in when the horizon size is kn is $< k$. As we have said, the initial state wave function must be such that the matrices $M(kn)$ are block diagonal, with a number of small blocks $< k$. We choose the wave function such that the black hole (if any) which comes in when the horizon grows from kn to $(k+1)n$ is a linear superposition of wave functions in which the black hole DOF are taken from each of the independent tiles on the cosmological horizon. This makes a state which is rotation invariant. We also insist that the rate at which black holes are added is uniform in time. This rate is a parameter of the model which, in FRW slicing, is determined by the initial black hole number density, n_{BH}, at the end of inflation. The fact that the objects being superposed are large quantum systems with a fast scrambler Hamiltonian and a time scale n, means that quantum interference terms in this superposition are negligible, of order e^{-n^2}, so the prediction of the model is that there is a classical probability distribution for finding black holes at various positions in the emergent FRW space-time.

The homogeneous and isotropic nature of the black hole distribution, from the point of view of one trajectory, makes it easy to satisfy the HST consistency conditions between trajectories. We simply choose the same sequence of both in and out Hamiltonians, and the same initial state for each trajectory, and let the geometry of FRW tell us what the overlap Hilbert spaces are. This is only a coarse grained solution of the consistency conditions because both our time steps and the spatial separations of the various trajectories are of order n, rather than the Planck scale. Note that apart from the fine tuning necessary to guarantee a certain number of localized excitations the conditions of homogeneity and isotropy arise naturally, and do not require any extra tuning of the initial state. We can certainly find other solutions of the consistency conditions with inhomogeneous distributions of black holes. However, we have argued that these will typically lead to cosmologies in which the entire content of the universe is a few large black holes, which slowly decay back into the horizon of empty dS space.

At any rate, we can combine the local $SL(2)(a)$ groups with the generators of rotations to construct an algebra that approximates $SO(1,4)$. In flat coordinates, the $SO(1,4)$ algebra splits into seven generators whose action is geometrically obvious, and three which act in a non-intuitive way. The geometric generators are the rotations and translations of the flat coordinates, and the rescaling of the flat space coordinates combined with the translation of FRW time (rescaling of conformal time). In terms of familiar rotations and boosts in five-dimensional Minkowski space, the time translation generator is J_{04}, the rotations are J_{ij} and the translations are J_{+i}, where the $+$ refers to light front time, $X^0 + X^4$ in the embedding coordinates of dS space in five-dimensional Minkowski space. The action of the remaining J_{-i} generators is non-linear in the flat coordinates.

In the matrix model we define $J_{ij} = \epsilon_{ijk} J_k$ to be the obvious rotation generator. The rest of the $SO(1,4)$ generators are defined in terms of the local $SL(2)[a]$ generators.

$$J_{04} = \sum L_0(\Omega[a]),$$

and[6]

$$J_{\pm i} = \sum L_{\pm}(\Omega[a])\Omega[\mathbf{a}]_i.$$

These operators operate only on the tensor factor $\mathcal{H}_{out}(t_k)$ of the Hilbert space, describing DOF which have not yet come into the horizon. The Hamiltonian of these DOF is a sum of non-interacting terms, as one would expect for disjoint horizon volumes in dS space.

The operators described above satisfy the commutation relations of $SO(1,4)$ with accuracy $1/n$, when N/n is large. The density matrix of the system outside the horizon is the tensor product of density matrices $\rho(\Omega_a)$ for the individual, angularly localized, systems of n^2 variables. This is approximately invariant under the individual $SL(2)$ generators at fixed angle, because of the fact that we have gone through a large number of e-folds of evolution with the Hamiltonian $L_0(a)$. It is also invariant under permutation of the individual blocks of n^2 variables. We have argued that these systems enter the horizon as an $SU(2)$ invariant distribution of black holes, if we want the model universe to have a radiation dominated era. It follows that the distribution of black hole fluctuations on the sky of each trajectory is approximately $SO(1,4)$ invariant. This is what we need, to fit the data on the CMB, Large Scale Structure, and galaxy formation.

12.3 Meta-Cosmology and Anthropic Arguments

It is already obvious that our resolution of many of the problems of cosmology relies to a certain extent on what are commonly called anthropic arguments. The resolution of the Boltzmann/Penrose conundrum of why the universe began in a low entropy state is that typical initial states evolve under the influence of the HST Hamiltonian into states which consist entirely of apparent horizon filling black holes, and that the system asymptotes to the density matrix of empty dS space[7] without ever producing localized excitations. It is, I would claim, obvious, that any such model cannot have an era with any kind of organized behavior that we could call an observer.

Before proceeding further, I should elaborate on what my "philosophical" stance is towards anthropic arguments. As the inventor of one of the first models, which was designed to explain the value of the CC on anthropic grounds [33–37], I am certainly not someone who rejects the scientific validity of such arguments outright. However, I do believe in Albrecht's razor: "The physicist who has the smallest number of anthropic arguments in her/his model of the world, wins" [38]. More importantly, I believe that it is clear that many of the values of parameters in the standard model *cannot* be explained anthropically [39], especially if one allows for the existence of scales between the Planck scale and the scale of electroweak interactions. In particular, anthropic arguments cannot explain the existence of multiple generations of quarks and leptons and the bizarre pattern of couplings that determine their properties.

---

[6] $\lfloor a \rfloor$ is a label for a tile in the spherical grid described above. The sums are over all tiles.
[7] We will discuss more elaborate models, in which the CC itself is selected anthropically, below.

Some authors attempt to get around these phenomenological problems by combining anthropic arguments with traditional symmetry arguments, but it is not at all clear that this makes sense. In particular, in the most popular model for a distribution of parameters in the standard model, The String Landscape, the rules seem to imply that models with extra symmetries beyond the standard model gauge group are exponentially improbable [8]. My own feeling is that, within the context of models where we insist on the standard model gauge group, the only way we could have a sensible anthropic explanation of what we see would be if there were only one generation of quarks and leptons, and we relied on anthropic arguments to determine the weak scale[8] *and* insisted that a QCD axion with a GUT scale decay constant were the dark matter. Apart from the problem of generations, this framework has depressing implications for high energy physics. We will not find anything in accelerators we can imagine building.

Even this framework is not immune to criticism. All known anthropic arguments, which rely on detailed properties of the particle physics we know, are about the properties of nuclear physics. This is physics at the MeV scale, and we should really be formulating our arguments in terms of an effective field theory at that scale. One might imagine arguing that nuclear physics would be irreversibly damaged if the underlying gauge theory did not consist of $SU(3) \times U(1)$ with the up and down quark masses and the QCD scale having values close to those in the real world. However, one *cannot* imagine that the weak inter-actions affect nuclear and stellar physics in a way that cannot be mimicked by a host of four fermion interactions which are different than those in the standard model. Thus, an honest anthropic argument, even one that makes the *a priori* assumption of our standard strong and electromagnetic gauge theory, cannot determine the form of the standard model Lagrangian.

Once one gives up the assumption of life based on the physics and chemistry we know, almost all bets are off. We know too little about how physics determines biology, the pos-sibility of radically different forms of organization and intelligence, or a host of other questions, to even pretend to make anthropic arguments in this wider context, except those which rely only on general properties of thermodynamics and gravitation.

In HST, the questions like the nature of the low energy gauge group and the number of generations are determined by the fundamental commutation relations of the variables, and are not subject to anthropic selection among possible states in a given model. Parameters like n, ϵ and n_{BH}, which characterize our cosmology, may well be anthropically selected, subject to inequalities like $1 \ll n \ll N$ and $n_{BH} < n^{-3}$, but in the models studied so far the cosmological horizon size is an input parameter. In Banks and Fischler [13] a more general model was proposed, in which N is a variable. That model is based on the existence of classical solutions of Einstein's equations in which a single horizon volume of de Sitter space is joined onto a stable black hole in the $p = \rho$ FRW model. The black hole and cosmological horizons coincide. There is a completely explicit quantum mechanical HST model, whose coarse grained properties match those of this solution. One can also construct

---

[8] And this leads to unsuccessful predictions of the mass of the Higgs boson.

more general solutions, in which such dS black holes, with varying horizon size, relative positions and velocities, move in the $p = \rho$ background. This allows for anthropic selection of all of the parameters N, n, ϵ, n_{BH}. The nature of the anthropic arguments is quite different from conventional ones, because the HST formalism suggests a relation between N and the scale of supersymmetry breaking. I do not have space to discuss this in detail here, but the arguments give a plausible explanation of the data.

Another "problem" that this generalized model solves is the existence of an infinite number of late time observers, "Boltzmann Brains", whose experience is very different from our own. One can arrange the initial positions and velocities of the black holes with different interior CC, such that collisions always occur on time scales much longer than the current age of the universe, but much shorter than the time scale for production of Boltzmann Brains. I do not consider this a major triumph, for the same reason that I do not think the BB problem is a serious one. The difference between those two time scales is so huge that one can invent an infinite number of changes to the theory, which will eliminate the BBs without changing anything that we will, in principle, be able to measure. BBs are a problem only to a theory which posits that the explanation for the low entropy initial conditions of the universe was a spontaneous fluctuation in a finite system. In HST there is an entirely different resolution of the Boltzmann–Penrose question, so BBs are a silly distraction, which can be disposed of in a way that will never be testable. Indeed, much of the structure of the HST model that allows for anthropic selection will remain forever beyond our reach, since it depends on initial conditions whose consequences are not causally accessible to us until our dS black hole collides with another. Our universe then undergoes a catastrophe, on a time-scale of order $10^{61}L_P \sim 10^{10}$ years, somewhat analogous to what happens when Coleman–deLuccia bubbles collide. After that time, the low energy effective field theory has changed and we will not be around to see a subsequent collision, if one occurs.

As far as I can tell, any model which implements the anthropic principle will suffer from similar problems. It must predict many things, which no local observer can ever observe. This is the reason that I subscribe to Albrecht's razor: a non-anthropic explanation for a fact about the universe can be tested more thoroughly than an anthropic model can. This does not mean that one can ignore the possibility that some of what we observe depends on accidental properties of particular meta-stable states of a system larger than anything we can observe. I have spent years trying to find a more satisfactory explanation of the value of the CC, and I conclude that this will not be possible. The HST model also suggests that n, n_{BH} and the precise form of the early FRW metric during the transition from inflation to the dilute black hole gas phase are also free parameters, which characterize the initial state. They are subject to a combination of entropic and anthropic pressures, and I believe they can be pretty well pinned down by these arguments.

On the other hand, in HST the low energy particle content of the model is determined by the super-algebra of the variables and is fixed once and for all. Different spectra correspond to different candidate models. My hope is that very few of these candidate models actually lead to mathematical consistency. We are familiar from string theory and low energy effective field theory that most candidate models of quantum gravity are inconsistent. Even string models with four-dimensional $N = 1$ supersymmetry, many of which are consistent to all orders in perturbation theory, are expected to be actually consistent for at most discrete values of the various continuous parameters that label these models in perturbation theory. Indeed, generically, there is no reason to believe that the perturbation series is accurate at any of the consistent points. The string perturbation series determines quantities, the scattering amplitudes in Minkowski space, which simply do not exist unless the point in parameter space at which the model exists preserves both supersymmetry and a discrete phase rotation (R) symmetry, which acts on the supersymmetry generators. Such points are very sparse in the space of all parameters.

The theory of SUSY breaking in HST [14, 40] implies that in the limit that the CC is taken to zero, the model becomes super-Poincare invariant, has a discrete R symmetry,[9] and no continuous moduli. There can be no examples of such a model in perturbative string theory, since the string coupling itself appears to be a continuous parameter. General arguments in effective super-gravity imply that such models are rare, corresponding to solving $p + 1$ equations for p unknowns. Furthermore, in perturbative string theory, one can look at the analogous problem of fixing all parameters besides the string coupling and finding a model with a discrete R symmetry. Again, such models are rare. It is thus plausible to guess that consistent HST models with vanishing CC are rare, and the gauge groups and matter content of these models, as well as their parameters, highly constrained. It is not out of the question that we will be able to find arguments that the standard model is the unique low energy gauge theory, which can arise at such a point.

12.4 Conclusions

On an intuitive level, HST models of cosmology are quite simple. The basic principle behind them is that bulk localized excitations in a finite area causal diamond are constrained states of variables living on the boundaries of the diamond. The low entropy of the state of the early universe is explained by the necessity of having such localized excitations in an observable universe: the *topikèsthropic principle*. The initial state with maximal *localized* entropy is a collection of black holes. A competition then ensues between black hole coalescence and black hole decay, which must end in most of the black holes decaying if the universe is ever to develop complex structures. This requires a fairly uniform dilute gas of black holes. Absolute uniformity is impossible, because the black holes are finite quantum systems undergoing fast scrambling dynamics. This leads to fluctuations in the mass and angular momenta of the black holes, of order

$$\frac{\delta M}{M} \sim \frac{\delta L}{L} \sim S^{-1/2}, \qquad (12.10)$$

---

[9] A discrete R symmetry is a discrete symmetry group which acts on the fermionic generators of the SUSY algebra.

where S is the black hole entropy. These are the fluctuations we see in the sky.

What is remarkable is that this model actually has all the features traditionally associated with inflation. The two crucial ingredients in this conflation of apparently different models are the different time slicings associated with the HST model and FRW space-time, and the necessity, in the HST model, to describe the evolution of the black holes, before they enter the horizon, as decoupled quantum systems of fixed entropy. Each decoupled system is the same as the HST model of dS space (with the inflationary Hubble radius, nL_P), so, if we want a homogeneous distribution of black holes, then FRW time slices up to conformal time $\eta_0/2$ must be described as a collection of decoupled quantum systems of fixed entropy. η_0 is the point in conformal time to which our universe converges as the proper time along trajectories goes to infinity. This system thus corresponds to the conventional picture of an inflationary universe as many independent horizon volumes of dS space. The number of e-folds is fixed, with the precise value of N_e dependent on n_{BH}, the primordial density of black holes at the end of inflation. That number density also determines the reheat temperature of the universe after black hole decay. It is bounded from above by $n_{BH} < n^{-3}$. The size of primordial scalar fluctuations is $Q \sim (n\epsilon)^{-1}$, where $\epsilon = \frac{\dot{H}}{H^2}$ is a slow roll parameter. For $\epsilon \sim 0.1$ the CMB data tell us that $n \sim 10^6$ and this implies that the reheat temperature is less than $10^7 - 10^8$ GeV.

I also sketched the argument that the HST curvature fluctuations should be approximately $SO(1,4)$ invariant, which is enough to account for the data, with detailed features of the scalar spectrum fit to $H(t)$ the background FRW metric at the end of inflation. This is just as in conventional inflation models, but those models make much more specific assumptions about the state of the quantum system under discussion, assumptions which amount to a massive fine tuning of initial conditions. They also have to go through an elaborate discussion, rarely touched on in the mainstream inflation literature, to justify why quantum fluctuations in a quantum ground state decohere into a probability distribution for the classical curvature fluctuations.

In contrast, the HST model has fluctuations arising from localized quantum systems with a huge number of states, with the curvature interpreted as in Jacobson [2] as a hydrodynamic average property. In more familiar terms, the fluctuations are fluctuations in mass and angular momentum of mesoscopic black holes.[10] This model also has fine tuning of initial conditions, but it is the minimal tuning necessary to obtain a universe with localized excitations which are not black holes.

Unfortunately, current cosmological data do not allow us to distinguish between these two very different models or many other more exotic field theory models with multiple fields, strange forms of kinetic energy, etc. The most likely observational distinctions to be measured in the near future will be short wavelength gravitational waves. In the distant future, measurement of the tensor three point function might definitively rule out quantum field theory as the source of CMB fluctuations.

---

[10] Black holes for which quantum fluctuations are not entirely negligible, although already decoherent. $1/n$ is not negligible, but e^{-n^2} is.

References

[1] T. Banks and W. Fischler, Holographic Inflation Revised, arXiv: 1501.01686 [hep-th], SO(1,4) Invariance, and the Low Entropy of the Early Universe.

[2] T. Jacobson, Thermodynamics of space-time: The Einstein equation of state. *Phys. Rev. Lett.* **75**, 1260 (1995) [gr-qc/9504004].

[3] G. W. Gibbons and S. W. Hawking, Cosmological Event Horizons, Thermodynamics, and Particle Creation. *Phys. Rev. D* **15**, 2738 (1977).

[4] T. Banks, W. Fischler and L. Mannelli, Microscopic quantum mechanics of the p = rho universe. *Phys. Rev. D* **71**, 123514 (2005) [hep-th/0408076].

[5] T. Banks and W. Fischler, Holographic Theories of Inflation and Fluctuations. (2011) arXiv:1111.4948 [hep-th].

[6] T. Banks, W. Fischler, T. J. Torres and C. L. Wainwright, Holographic Fluctuations from Unitary de Sitter Invariant Field Theory. (2013) arXiv:1306.3999 [hep-th].

[7] T. Banks, W. Fischler, Holographic Inflation Revised. (2015). arXiv:1501.01686 [hep-th].

[8] M. Dine and Z. Sun, R symmetries in the landscape. *JHEP* **0601**, 129 (2006) [hep-th/0506246].

[9] S. L. Braunstein, S. Pirandola and K. Zyczkowski, Better Late than Never: Information Retrieval from Black Holes. *Phys. Rev. Lett.* **110**, no. 10, 101301 (2013) [arXiv:0907.1190 [quant-ph]].

[10] N. Itzhaki, Is the black hole complementarity principle really necessary?. hep-th/9607028 (1996).

[11] A. Almheiri, D. Marolf, J. Polchinski and J. Sully, Black Holes: Complementarity or Firewalls?. *JHEP* **1302**, 062 (2013) [arXiv:1207.3123 [hep-th]].

[12] T. Banks, W. Fischler, S. Kundu and J. F. Pedraza, Holographic Space-time and Black Holes: Mirages As Alternate Reality. (2014) arXiv:1401.3341 [hep-th].

[13] T. Banks and W. Fischler, Holographic Theories of Inflation and Fluctuations. (2011) arXiv:1111.4948 [hep-th].

[14] T. Banks, Cosmological breaking of supersymmetry? or Little lambda goes back to the future 2. (2000) hep-th/0007146.

[15] W. Fischler, Taking de Sitter seriously. Talk given at Role of Scaling Laws in Physics and Biology (Celebrating the 60th Birthday of Geoffrey West). Santa Fe, Dec. 2000.

[16] N. Lashkari, D. Stanford, M. Hastings, T. Osborne and P. Hayden, Towards the Fast Scrambling Conjecture. *JHEP* **1304**, 022 (2013) [arXiv:1111.6580 [hep-th]].

[17] T. Banks and W. Fischler, Holographic Space-time and Newton's Law. (2013) arXiv:1310.6052 [hep-th].

[18] N. Goheer, M. Kleban and L. Susskind, The Trouble with de Sitter space. *JHEP* **0307**, 056 (2003) [hep-th/0212209].

[19] T. Banks and W. Fischler, Holographic Theory of Accelerated Observers, the S-matrix, and the Emergence of Effective Field Theory. (2013) arXiv:1301.5924 [hep-th].

[20] T. Banks, W. Fischler, S. Kundu and J. F. Pedraza, Holographic Space-time and Black Holes: Mirages As Alternate Reality. (2014) arXiv:1401.3341 [hep-th].

[21] J. M. Maldacena, Non-Gaussian features of primordial fluctuations in single field inflationary models. *JHEP* **0305**, 013 (2003) [astro-ph/0210603].

[22] J. M. Maldacena and G. L. Pimentel, On graviton non-Gaussianities during inflation. *JHEP* **1109**, 045 (2011) [arXiv:1104.2846 [hep-th]].

[23] J. Soda, H. Kodama and M. Nozawa, Parity Violation in Graviton Non-gaussianity. *JHEP* **1108**, 067 (2011) [arXiv:1106.3228 [hep-th]].

[24] C. Cheung, P. Creminelli, A. L. Fitzpatrick, J. Kaplan and L. Senatore, The Effective Field Theory of Inflation. *JHEP* **0803**, 014 (2008) [arXiv:0709.0293 [hep-th]].

[25] C. Cheung, A. L. Fitzpatrick, J. Kaplan and L. Senatore, On the consistency relation of the 3-point function in single field inflation. *JCAP* **0802**, 021 (2008) [arXiv:0709.0295 [hep-th]].

[26] P. McFadden and K. Skenderis, Holography for Cosmology. *Phys. Rev. D* **81**, 021301 (2010) [arXiv:0907.5542 [hep-th]].

[27] I. Antoniadis, P. O. Mazur and E. Mottola, Conformal Invariance, Dark Energy, and CMB Non-Gaussianity. *JCAP* **1209**, 024 (2012) [arXiv:1103.4164 [gr-qc]].

[28] N. Kundu, A. Shukla and S. P. Trivedi, Constraints from Conformal Symmetry on the Three Point Scalar Correlator in Inflation. (2015) arXiv:1410.2606 [hep-th].

[29] T. Banks, W. Fischler, T. J. Torres and C. L. Wainwright, Holographic Fluctuations from Unitary de Sitter Invariant Field Theory. (2013) arXiv:1306.3999 [hep-th].

[30] T. Banks, W. Fischler nad T. J. Torres, The Holographic Theory of Inflation and Cosmological Data (forthcoming).

[31] J. M. Maldacena, Non-Gaussian features of primordial fluctuations in single field inflationary models. *JHEP* **0305**, 013 (2003) [astro-ph/0210603].

[32] A. Kogut, D. T. Chuss, *et al.* The Primordial Inflation Explorer (PIXIE) . Proc. SPIE 9143, Space Telescopes and Instrumentation 2014: Optical, Infrared, and Millimeter Wave, 91431E (August 2, 2014); doi:10.1117/12.2056840.

[33] P. C. W. Davies and S. D. Unwin, Why is the Cosmological Constant So Small. *Proc. Roy. Soc.* London, A377, 1769, 147–9, (1981).

[34] A. D. Linde, The inflationary universe, *Rep. Prog. Phys.*, **47**, 925–86, 1984.

[35] A. D. Sakharov, Cosmological transitions with changes in the signature of the metric, *Zh. Eksp. Teor. Fiz.* **87**, 375–83 (August 1984).

[36] T. Banks, Relaxation of the Cosmological Constant. *Phys. Rev. Lett.* **52**, 1461 (1984).

[37] T. Banks, T. C. P., Quantum Gravity, the Cosmological Constant and All That... *Nucl. Phys. B* **249**, 332 (1985).

[38] A. Albrecht, Cosmic inflation and the arrow of time. In Barrow, J. D., et al., eds. *Science and ultimate reality*. Cambridge: Cambridge University Press, pp. 363–401 [astro-ph/0210527].

[39] T. Banks, M. Dine and E. Gorbatov, Is there a string theory landscape?. *JHEP* **0408**, 058 (2004) [hep-th/0309170].

[40] T. Banks, Cosmological breaking of supersymmetry?. *Int. J. Mod. Phys. A* **16**, 910 (2001). SUSY and the holographic screens. (2003) hep-th/0305163.

13

Progress and Gravity:
Overcoming Divisions Between General Relativity and
Particle Physics and Between Physics and HPS

J. BRIAN PITTS

13.1 Introduction: Science and the Philosophy of Science

The ancient "problem of the criterion" is a chicken-or-the-egg problem regarding knowledge and criteria for knowledge, a problem that arises more specifically in relation to science and the philosophy of science. How does one identify reliable knowledge without a reliable method in hand? But how does one identify a reliable method without reliable examples of knowledge in hand? Three possible responses to this problem were entertained by Roderick Chisholm: one can be a skeptic, or identify a reliable method(s) ("methodism"), or identify reliable particular cases of knowledge ("particularism") (Chisholm, 1973). But why should the best resources be all of the same type? Might not some methods and some particular cases be far more secure than all other methods and all other particular cases? Must anything be completely certain anyway? Why not mix and match, letting putative examples and methods tug at each other until one reaches (a personal?) reflective equilibrium?

This problem arises for knowledge and epistemology, more specifically for science and the philosophy of science, and somewhere in between, for inductive inference. Reflective equilibrium is Nelson Goodman's method for induction (as expressed in John Rawls's terminology). One need not agree with Goodman about deduction or take his treatment of induction to be both necessary and *sufficient* to benefit from it. He writes:

A rule is amended if it yields an inference we are unwilling to accept; an inference is rejected if it violates a rule we are unwilling to amend. The process of justification is the delicate one of making mutual adjustments between rules and accepted inferences; and in the agreement achieved lies the only justification needed for either.

All this applies equally well to induction. An inductive inference, too, is justified by conformity to general rules, and a general rule by conformity to accepted inductive inferences. Predictions are justified if they conform to valid canons of induction; and the canons are valid if they accurately codify accepted inductive practice. (Goodman, 1983, p. 64, emphasis in the original)

Most scientists and (more surprisingly) even many philosophers do not take Hume's problem of induction very seriously, although philosophers talk about it a lot. As Colin Howson notes, philosophers often declare it to be insoluble and then proceed as though it were solved (Howson, 2000). I agree with Howson and Hans Reichenbach (Reichenbach, 1938, pp. 346, 347) that one should not let oneself off the hook so easily. That seems especially true in cosmology (Norton, 2011). Whether harmonizing one's rules and examples is sufficient is less clear to me than it was to Goodman, but such reflective equilibrium surely is *necessary* – although difficult and perhaps rare.

My present purpose, however, is partly to apply Goodman-esque reasoning only to a *special case* of the problem of the criterion, as well as to counsel unification within physical inquiry. What is the relationship between philosophy of science (not epistemology in general) on the one hand, and scientific cosmology and its associated fundamental physics, especially gravitation and space-time theory (not knowledge in general) on the other? Neither dictation from philosopher-kings to scientists (the analogue of methodism) nor complete deference to scientists by philosophers (the analogue of particularism) is Goodman's method. It is not popular for philosophy to give orders to science, but it once was. The reverse is more fashionable, a form of scientism or at least a variety of naturalism. I hope to show by examples how sometimes each side should learn from the other.

While Goodman's philosophy has a free-wheeling relativist feel that might make many scientists and philosophers of science nervous, one finds similar views expressed by a law-and-order philosopher of scientific progress, Imre Lakatos. According to him, we should seek

a pluralistic system of authority, partly because the wisdom of the scientific jury and its case law has not been, and cannot be, fully articulated by the philosopher's statute law, and partly because the philosopher's statute law may occasionally be right when the scientists' judgment fails. (Lakatos, 1971, p. 121)

Thus there seems to be no irresistible pull toward relativism in seeking reflective equilibrium rather than picking one side always to win automatically.

13.2 Healing the GR *vs.* Particle Physics Split

A second division that should be overcome to facilitate the progress of knowledge about gravitation and space-time is the general relativist *vs.* particle physicist split. Carlo Rovelli discusses

... the different understanding of the world that the particle physics community on the one hand and the relativity community on the other hand, have. The two communities have made repeated and sincere efforts to talk to each other and understand each other. But the divide remains, and, with the divide, the feeling, on both sides, that the other side is incapable of appreciating something basic and essential.... (Rovelli, 2002)

This split has a fairly long history going back to Einstein's withdrawing from mainstream fundamental physics from the 1920s – that largely being quantum mechanics, relativistic quantum mechanics and quantum field theory. A further issue pertains to the gulf between how Einstein actually found his field equations (as uncovered by recent historical work (Renn, 2005; Renn and Sauer, 1999, 2007)) and the much better known story that Einstein told retrospectively. Work by Jürgen Renn *et al.* has recovered the importance of Einstein's "physical strategy" involving a Newtonian limit, an analogy to electromagnetism, and a quest for energy-momentum conservation; this strategy ran alongside the better advertised mathematical strategy emphasizing his principles (generalized relativity, general covariance, equivalence, etc.). Einstein's reconstruction of his own past is at least in part a persuasive device in defense of his somewhat lonely quest for unified field theories (van Dongen, 2010). Readers with an eye for particle physics will not miss the similarity to the later successful derivations of Einstein's equations as the field equations of a massless spin-2 field assumed initially to live in flat Minkowski space-time (Feynman *et al.*, 1995), in which the resulting dynamics merges the gravitational potentials with the flat space-time geometry such that only an effective curved geometry appears in the Euler–Lagrange equations. One rogue general relativist has recently opined:

HOW MUCH OF AN ADVANTAGE did Einstein gain over his colleagues by his mistakes? Typically, about ten or twenty years. For instance, if Einstein had not introduced the mistaken Principle of Equivalence and approached the theory of general relativity (GR) via this twisted path, other physicists would have discovered the theory of general relativity some twenty years later, via a path originating in relativistic quantum mechanics. (Ohanian, 2008, p. 334, capitalization in the original).

It is much clearer that these derivations work to give Einstein's equation than it is what they mean. Do they imply that one need not and perhaps should not ever have given up flat space-time? Do they, on the contrary, show that theories of gravity in flat space-time could not succeed, because their best effort turns out to give curved space-time after all (Ehlers, 1973)? Such an argument is clearly incomplete without contemplation of massive spin-2 gravity (Freund *et al.*, 1969; Ogievetsky and Polubarinov, 1965). But it might be persuasive if massive spin-2 gravity failed – as it seemed to do roughly when Ehlers wrote (not that he seems to have been watching). But since 2010 massive spin-2 gravity seems potentially viable again (de Rham *et al.*, 2011; Hassan and Rosen, 2011; Maheshwari, 1972) (though some new issues exist). Do the spin-2 derivations of Einstein's equations suggest a conventionalist view that there is no fact of the matter about the true geometry (Feynman *et al.*, 1995, pp. 112, 113)? Much of one's assessment of conventionalism will depend on what one takes the modal scope of the discussion to be: Should one consider only one's best theory (hence the question is largely a matter of exegeting General Relativity, which will favor curved space-time), or should one consider a variety of theories? According to John Norton, the philosophy of geometry is not an enterprise rightly devoted to giving a spurious air of necessity to whatever theory is presently our best (Norton, 1993, pp. 848, 849). Such a view suggests the value of a broader modal scope for the discussion than just our best current theory. On the other hand, the claim has been made that the transition from Special Relativity to General Relativity is as unlikely to be reversed as the transition from classical to quantum mechanics (Ehlers, 1973, pp. 84, 85). If one aspires to proportion belief to evidence, that is a startling claim. The transition from classical to quantum mechanics was motivated by grave empirical problems; there now exist theorems (no local hidden variables) showing how far any empirically adequate physics *must* diverge from classical.

But a constructive derivation of Einstein's equations from a massless spin-2 shows that one can naturally recover the phenomena of GR without giving up a special relativistic frame-work in a sense. The cases differ as twilight and day. Ehlers's remarks are useful, however, in alerting one for Hegelian undercurrents or other doctrines of inevitable progress in the general relativity literature. A classic study of doctrines of progress is Bury (1920).

13.3 Bayesianism, Simplicity, and Scalar *vs.* Tensor Gravity

While Bayesianism has made considerable inroads in the sciences lately, it is helpful to provide a brief sketch before casting further discussion in such terms. I will sketch a rather simple version – one that might well be inadequate for science, in which one sometimes wants uniform probabilities over infinite intervals and hence might want infinitesimals, for example. Abner Shimony's tempered personalism discusses useful features for a scientifi-cally usable form of Bayesianism, including open-mindedness (avoiding prior probabilities so close to 0 or 1 that evidence cannot realistically make much difference (Shimony, 1970)) and assigning non-negligible prior probabilities to seriously proposed hypotheses.

With such qualifications in mind, one can proceed to the sketch of Bayesianism. One is not equally sure of everything that one believes, so why not have degrees of belief, and make them be real numbers between 0 and 1? Thus one can hope to mathematize logic in shades of gray *via* the probability calculus. Bayes's theorem can be applied to a theory T and evidence E:

$$P(T|E) = P(T)\frac{P(E|T)}{P(E)}. \tag{13.1}$$

One wakes up with degrees of belief in all theories (!), "prior probabilities". One opens one's eyes, beholds evidence E, and goes to bed again. While asleep one revises degrees of belief from priors $P(T)$ to posterior probabilities $P(T|E)$. Today's $P(T|E)$ becomes tomor-row's prior $P(T)'$. Then one does the same thing tomorrow, getting some new evidence E', etc. Now the priors $P(T)$ might be partly subjective. If there are no empirically equivalent theories and everyone is open-minded, then eventually evidence should bring convergence of opinion over time (though maybe not soon).

A further wrinkle in the relation between evidence and theory comes from looking at the denominator of Bayes's theorem, $P(E) = P(E|T)P(T) + P(E|T_1)P(T_1) + P(E|T_2) P(T_2) + \ldots$. While one might have hoped to evaluate evidence theory T simply in light of evidence E, this expansion of $P(E)$ shows that such an evaluation is typically undefined, because one must spread degree of belief $1 - P(T)$ among the competitors T_1, T_2, etc. Hence the predictive likelihoods $P(E|T_1)$, etc., subjectively weighted, appear unbidden in the test of T by E. Theory testing generically is *comparative*, making essential reference to rival theories. This fact is sometimes recognized in scientific practice, but Bayesianism can alert one to attend to the question more systematically.

Scientists and philosophers tend to like simplicity. Simplicity might not be objective, but there is significant agreement regarding scientific examples. That is a good thing, because there are lots of theories, especially lots of complicated ones, way too many to handle.

If degrees of belief are real numbers (not infinitesimals), then normalization $\Sigma_i P_i = 1$ requires lots of 0s and or getting ever closer to 0 on some ordering (Earman, 1992, pp. 209, 210). There is no clear reason for prior plausibility to peak away from the simple end. Plausibly, other things equal, simpler theories are more plausible *a priori*, getting a higher prior $P(T)$ in a Bayesian context. Such considerations are vague, but the alternatives are even less principled.

One can now apply Bayesian considerations to gravitational theory choice in the 1910s. One recalls that Einstein had some arguments against a scalar theory of gravity, which motivated his generalization to a tensor theory. Unfortunately they do not work. As Domenico Giulini has said,

On his way to General Relativity, Einstein gave several arguments as to why a special-relativistic theory of gravity based on a massless scalar field could be ruled out merely on grounds of theoretical considerations. We re-investigate his two main arguments, which relate to energy conservation and some form of the principle of the universality of free fall. We find such a theory-based *a priori* abandonment not to be justified. Rather, the theory seems formally perfectly viable, though in clear contradiction with (later) experiments. (Giulini, 2008, emphasis in original)

Scalar (spin-0) gravity is simpler than rank-2 tensor (spin-2). Having one potential is simpler than having ten, especially if they are self-interacting. With Einstein's help, Gunnar Nordström eventually proposed a scalar theory that avoided the theoretical problems mentioned by Giulini. Given simplicity considerations, Nordström's theory was more probable than Einstein's *a priori*: $P(T_N) > P(T_{GR})$. Einstein's further criticisms are generally matters of taste. So prior to evidence for General Relativity, it was more reasonable to favor Nordström's theory. As it actually happened, Einstein's "final" theory and the evidence from Mercury both appeared in November 1915, leaving little time for this logical moment in actual history. Einstein's earlier *Entwurf* theory (Einstein and Grossmann, 1996) could be faulted for having negative-energy degrees of freedom and hence likely being unstable (a problem with roots in Lagrange and Dirichlet (Morrison, 1998)), although apparently no one did so.

Where was the progress of scientific knowledge–truth held for good reasons? Mercury's perihelion gave non-coercive evidence confirming GR and disconfirming Nordström's theory. It was possible to save Nordström's theory using something like dark matter, matter (even if not dark – Seeliger's zodiacal light) of which the mass had been neglected (Roseveare, 1982). Hence there was scope for rational disagreement because Nordström's theory was antecedently more plausible

$$P(T_N) > P(T_{GR})$$

but evidence favored Einstein's non-coercively

$$0 < P(E_{Merc}|T_N) < P(E_{Merc}|T_{GR}).$$

The scene changed in 1919 with the bending of light, which falsified Nordström's theory: $P(E_L|T_N) = 0$. There were not then other plausible theories that predicted light bending, so $P(E_L|T_{GR}) \approx 1 >> P(E_L)$. It is possible to exaggerate the significance of this result, as happened popularly but perhaps less so academically (Brush, 1989), where a search for plausible rival theories that also predicted light bending was made. (Bertrand Russell may have considered Whitehead's to be an example (Russell, 1927, pp. 75–80).) Unfortunately many authors wrongly take Einstein's arguments against scalar gravity seriously (Giulini, 2008). In the long run one does not make *reliable rational* progress by siding with genius as soon as possible: Einstein made many mistakes (often correcting them himself), some of them lucky (Ohanian, 2008) (such as early rejection of scalar theories), followed by barren decades. Given this Bayesian sketch, it was rational to prefer GR over Nordström's scalar theory only when evidence from Mercury was taken into account, and not necessarily even then. The bending of light excluded scalar theories but did not exclude possible rival tensor theories.

13.4 General Relativity Makes Sense About Energy

Resolving conceptual problems is a key part of scientific progress (Laudan, 1977). In the 1910s and again in the 1950s controversy arose over the status of energy-momentum conservation laws of General Relativity. Given Einstein's frequent invocation of energy-momentum conservation in his process of discovery leading to General Relativity (Brading, 2005; Einstein and Grossmann, 1996; Renn and Sauer, 2007, 1999), as well as his retrospective satisfaction (Einstein, 1916), this is ironic. Partly in response to Felix Klein's dissatisfaction, Emmy Noether's theorems appeared (Noether, 1918). Her first theorem says that a rigid symmetry yields a continuity equation. Her second says that a wiggly symmetry yields an identity among Euler–Lagrange equations, making them not all independent. For General Relativity there are four wiggly symmetries, yielding the contracted Bianchi identities $\nabla_\mu \mathcal{G}^\mu_\nu \equiv 0$. In the wake of the conservation law controversies there emerged the widespread view that gravitational energy exists, but it "is not localized". This phrase appears to mean that gravitational energy is not anywhere in particular, although descriptions of it often do have locations. That puzzling conclusion is motivated by mathematical results suggesting that where gravitational energy is depends on an arbitrary conventional choice (a coordinate system), and other results that the total energy/mass does not.

While the energy non-localization lore is harmless enough as long as one knows the mathematical results on which it is based, it has self-toxifying quality. Having accepted that gravitational energy is not localized, one is likely to look askance at the Noether-theoretic calculations that yield it: pseudotensors. The next generation of textbooks might then dispense with the calculations while retaining the lore verbally. Because the purely verbal lore is mystifying, at that point one formally gives license to a variety of doubtful conclusions. Among these are that because General Relativity lacks conservation laws, it is false – a claim at the origins of the just-deceased Soviet/Russian academician A. A. Logunov's high-profile dissent (Logunov and Folomeshkin, 1977). One also hears (for references see Pitts (2010)) that the expansion of the universe, by virtue of violating conservation laws, is false (a special case of Logunov's claim). One hears that the expansion of the universe is a resource for creation science by providing a heat sink for energy from rapid nuclear decay during Noah's Flood. Finally, one hears that General Relativity is more open to the soul's action on the body than is earlier physics, because the soul's action violates energy conservation, but General Relativity already discards energy conservation anyway. That last claim is almost backwards, because Einstein's equations are logically equivalent to energy-momentum conservation laws (Anderson, 1967). (If one wants souls to act on bodies, souls had better couple to gravity also.) The question whether vanishing total energy of the universe (given certain topologies) would permit it to pop into being spontaneously is also implicated.

Given that Noether's theorems – the first, not just the second – apply to GR, can one interpret the continuity equations sensibly and block the unfortunate inferences? The Noether operator generalizes canonical stress-energy tensor to give conserved quantities due to symmetry vector fields ξ^μ (Bergmann, 1958; Goldberg, 1980; Sorkin, 1977; Szabados, 1991; Trautman, 1962). For simpler theories than GR, the Noether operator is a weight 1 tangent vector density $\mathfrak{T}^\mu_\nu \xi^\nu$, so the divergence of the current $\partial_\mu(\mathfrak{T}^\mu_\nu \xi^\nu)$ is tensorial (equivalent in all coordinate systems) and, for symmetries ξ^ν, there is conservation $\partial_\mu(\mathfrak{T}^\mu_\nu \xi^\nu) = 0$. GR (the Lagrangian density, not the metric!) has *uncountably many* 'rigid' translation symmetries $x^\mu \to x^\mu + c^\mu$, where $c^\mu{}_{,\nu} = 0$, for any coordinate system, preserving the action $S = \int d^4 x \mathcal{L}$. These uncountably many symmetries yield uncountable conserved energy-momentum currents. Why can they not all be real? The lore holds that because there are infinitely many currents, really there are not any. But just because it is infinite does not mean it is 0 (to recall an old phrase). Getting $\infty = 0$ requires an extra premise, to be uncovered shortly. For GR, the Noether operator is a conserved but *non-tensorial differential* operator on ξ, depending on $\partial \xi$ also. Hence one obtains coordinate-dependent results, with energy density vanishing at an arbitrary point, etc., the usual supposed vices of pseudotensors. If one expects only one energy-momentum (or rather, four), it should be tensorial, with the transformation law relating faces in different coordinates. But Noether tells us that there are *uncountably many* rigid translation symmetries.

If one simply "takes Noether's theorem literally" (Pitts, 2010) (apparently novelly, although Einstein and Tolman (Tolman, 1930) said nice things about pseudotensors), then uncountably many symmetries imply uncountably many conserved quantities. How does one get $\infty = 0$? By assuming that the infinity of conserved energies are all supposed to be faces of the same conserved entity with a handful of components – the key tacit premise of uniqueness. Suppose that one is told in Tenerife that "George is healthy" and "Jorge está enfermo" (is sick). If one expects the two sentences to be equivalent under translation (analogous to a coordinate transformation), then one faces a contradiction: George is healthy and unhealthy. But if George and Jorge then walk into the room together, there is no tension: *George* $\neq$ *Jorge*. An expectation of uniqueness underlies most objections to pseudotensors, but it is unclear what justifies that expectation. Making more sense of energy conservation makes its appearance in Einstein's physical strategy in finding his field equations less ironic. Indeed, conservation due to gauge invariance is a key step in spin-2 derivations, which improve on Einstein's physical strategy (Einstein and Grossmann, 1996; Deser, 1970; Pitts and Schieve, 2001). Noether commented on *converses* to her theorems (Noether, 1918); one should be able to derive Einstein's equations from the conservation laws, much as the spin-2 derivations do using symmetric gravitational stress-energy (hence perhaps needing Belinfante–Rosenfeld technology).

But what is the point of believing in gravitational energy unless it does energetic things? Can it heat up a cup of coffee? Where is the physical interaction? Fortunately these questions have decent answers: gravitational energy is roughly the non-linearity of Einstein's equations, so it mediates the gravitational self-interaction.

Why did Hermann Bondi change from a skeptic to a believer in energy-carrying gravitational waves (Bondi, 1957)?[1] Given a novel plane wave solution of Einstein's equations in vacuum, his equation (2), he wrote:

there is a non-flat region of space between two flat ones, that is, we have a plane-wave zone of finite extent in a non-singular metric satisfying Lichnerowicz's criteria [reference suppressed]. Consider now a set of test particles at rest in metric (2) before the arrival of the wave. (Bondi, 1957)

After the passage of the wave, there is relative motion.

Clearly, this system of test particles in relative motion contains energy that could be used, for example, by letting them rub against a rigid friction disk carried by one of them. (Bondi, 1957)

This argument has carried the day with most people since that time: gravitational energy-transporting waves exist and do energetic things.

This argument has roots in Feynman (Anonymous, 2015) (DeWitt, 1957, p. 143) (Feynman *et al.*, 1995, xxv, xxvi) Kennefick (2007). John Preskill and Kip Thorne, drawing partly on unpublished sources, elaborate:

At Chapel Hill, Feynman addressed this issue in a pragmatic way, describing how a gravitational wave antenna could in principle be designed that would absorb the energy "carried" by the wave [DeWi 57, Feyn 57]. In Lecture 16, he is clearly leading up to a description of a variant of this device, when the notes abruptly end: "We shall therefore show that they can indeed heat up a wall, so there is no question as to their energy content." A variant of Feynman's antenna was published by Bondi [Bond 57] shortly after Chapel Hill (ironically, as Bondi had once been skeptical about the reality of gravitational waves), but Feynman never published anything about it. The best surviving description of this work is in a letter to Victor Weisskopf completed in February, 1961 [Feyn 61]. (Feynman *et al.*, 1995, p. xxv)

Gravitational energy in waves exists in GR, and one of the main objections to localization can be managed by taking Noether's theorem seriously: there are infinitely many symmetries and energies. Another problem is the non-uniqueness of the pseudotensor, which one might address with either a best candidate (as in Joseph Katz's work) or a physical meaning for the diversity of them in relation to boundary conditions (James Nester

---

[1] I thank Carlo Rovelli for mentioning Bondi.

et al.). Even scalar fields have an analogous problem Callan *et al.* (1970). With hope there as well, energy in GR, although still in need of investigation, is not clearly a serious conceptual problem anymore. That is scientific progress à la Laudan.

13.5 Change in Hamiltonian General Relativity

Supposedly, change is missing in Hamiltonian General Relativity (Earman, 2002). That seems problematic for two reasons: change is evident in the world, and change is evident in Lagrangian GR in that most solutions of Einstein's equations lack a time-like Killing vector field (Ohanian and Ruffini, 1994, p. 352). A conceptual problem straddling the internal *vs.* external categories is "empirical incoherence", being self-undermining. According to Richard Healey,

[t]here can be no reason whatever to accept any theory of gravity... which entails that there can be no observers, or that observers can have no experiences, some occurring later than others, or that there can be no change in the mental state of observers, or that observers cannot perform different acts at different times. It follows that there can be no reason to accept any theory of gravity... which entails that there is no time, or no change. (Healey, 2002, p. 300)

Hence accepting the no-change conclusion about Hamiltonian GR would undermine reasons to accept Hamiltonian GR. Change in the world is safe. But what about the surprising failure of Hamiltonian–Lagrangian equivalence?

A key issue involves where one looks for change, and relatedly, one what means by "observables". According to Earman (who would not dispute the point about the scarcity of solutions with time-like Killing vectors), "[n]o genuine physical magnitude countenanced in GTR changes over time" (Earman, 2002). Since the lack of time-like Killing vectors implies that the metric does change, clearly genuine physical magnitudes must be scarce, rarer than tensors. Tim Maudlin appeals to change in solutions to Einstein's equations: "stars collapse, perihelions precess, binary star systems radiate gravitational waves..." but "a sprinkling of the magic powder of the constrained Hamiltonian formalism has been employed to resurrect the decomposing flesh of McTaggart..." (Maudlin, 2002). Maudlin's appeal to common sense and Einstein's equations is helpful, as is Karel Kuchař's (Kuchař, 1993), but one needs more detail, motivation and (in light of Kuchař's disparate treatments of time and space) consistency.

Fortunately the physics reveals a relevant controversy, with reformers recovering Hamiltonian–Lagrangian equivalence (Castellani, 1982; Gràcia and Pons, 1988; Mukunda, 1980; Pons and Salisbury, 2005; Pons *et al.*, 1997; Pons and Shepley, 1998; Pons *et al.*, 2010; Sugano *et al.*, 1986). Hamiltonian–Lagrangian equivalence was manifest originally (Anderson and Bergmann, 1951; Rosenfeld, 1930; Salisbury, 2010); its loss needs study. In constrained Hamiltonian theories (Sundermeyer, 1982), some canonical momenta are (in simpler cases) just 0 due to independence of $\mathcal{L}$ from some $\dot{q}_i$; these are "primary constraints". In many cases of interest (including electromagnetism, Yang–Mills fields, and

General Relativity), some functions of p, q, $\partial_i p$, $\partial_i q$, $\partial_j \partial_i q$ are also 0 in order to preserve the primary constraints over time. Often these "secondary" (or higher) constraints are familiar, such as the phase space analog $\partial_i p^i = 0$ of Gauss's law $\nabla \cdot \vec{E} = 0$, Gauss–Codazzi equations embedding space into space-time in General Relativity, etc. Some constraints have something to do with gauge freedom (time-dependent redescriptions leaving the state or history alone). One takes Poisson brackets (q, p derivatives) of all constraints pairwise. If the result is in every case 0 (perhaps using the constraints themselves), then all constraints are "first-class", as in Clerk Maxwell's electromagnetism, Yang–Mills, and GR in their most common formulations. In General Relativity, the Hamiltonian, which determines time evolution, is nothing but a sum of first-class constraints (and boundary terms). Given that first-class constraints are related to gauge transformations, the key question is how they are related. Does each do so by itself, or do they rather work as a team? There is a widespread belief that each does so individually (Dirac, 1964). Then the Hamiltonian generates a sum of redescriptions leaving everything as it was, hence there is no real change. This is a classical aspect of the "problem of time." Some try to accept this conclusion, but recall Healey's critique.

Because Einstein's equations and common sense agree on real change, something must have gone wrong in Hamiltonian GR or the common interpretive glosses thereon, but what? Here the Lagrangian-equivalent reforming party has given most of the answer, namely, that what generates gauge transformations is not each first-class constraint separately, but the gauge generator G, a specially tuned sum of first-class constraints, secondary *and primary* (Anderson and Bergmann, 1951; Castellani, 1982; Pons, 2005; Pons *et al.*, 1997, 2010). Thus electromagnetism has two constraints at each point but only one arbitrary function; GR has eight constraints at each point but only four arbitrary functions. Indeed one can show that an isolated first-class constraint makes a mess (Pitts (2014b,a), such as spoiling the relation expected relation $\dot{q} = \frac{\delta H}{\delta p}$ making the canonical momentum equal to the electric field or the extrinsic curvature of space within space-time. These canonical momenta are auxiliary fields in the canonical action $\int dt d^3 x (p\dot{q} - \mathcal{H})$, and hence get their physical meaning from $\dot{q}$. Because each first-class constraint makes a physical difference by itself (albeit a bad one), the GR Hamiltonian no longer is forced to generate a gauge transformation by being a sum of them. There is change in the Hamiltonian formalism whenever there is no time-like Killing vector, just as one would expect from Lagrangian equivalence.

We have been guided by the principle that the Lagrangian and Hamiltonian formalisms should be equivalent ... in coming to the conclusion that they in fact are. (Pons and Shepley, 1998, p. 17)

By the same token, separate first-class constraints do not change $p\dot{q} - \mathcal{H}$ by (at most) a total derivative, but G *does* (Pitts 2014a, 2014b).

To get changing observables in GR, one should recall the distinction between internal and external symmetries. Requiring that observables have 0 Poisson bracket with the electromagnetic gauge symmetry generator is just to say that things that we cannot observe (in the ordinary sense) are unobservable (in the technical sense). By contrast, requiring that observables have 0 Poisson bracket with the gauge generator in GR implies that the Lie derivative of an observable is 0 in every direction. Thus anything that varies spatiotemporally is "unobservable" – a result that cannot be taken seriously. The problem is generated by hastily generalizing the definition from internal to external symmetries. Instead one should permit observables to have Lie derivatives that are not 0 but just the Lie derivative of a geometric object – an infinitesimal Hamiltonian form of the identification of observables with geometric objects in the classical sense (Nijenhuis, 1952), *viz.*, set of components in each coordinate system and a transformation law.

13.6 Einstein's Real Λ Blunder in 1917

One tends to regard perturbative expansions and geometry as unrelated at best, if not negatively related.

The advent of supergravity [footnote suppressed] made relativists and particle physicists meet. For many this was quite a new experience since very different languages were used in the two communities. Only Stanley Deser was part of both camps. The particle physicists had been brought up to consider perturbation series while relativists usually ignored such issues. They knew all about geometry instead, a subject particle physicists knew very little about. (Brink, 2006, p. 40)

But some examples will show how perturbative expansions can help to reveal the geometric content of a theory that is otherwise often misunderstood, can facilitate the conception of novel geometric objects that one might otherwise fail to conceive, and permit conceptual and ontological insight.

Perturbative expansions can help to reveal the geometric content of a theory that one might well miss otherwise. Einstein in his 1917 cosmological constant paper first reinvented a long-range modification of Newtonian gravity (Einstein, 1923) – one might call it (anachronistically) non-relativistic massive scalar gravity – previously proposed in the nineteenth century by Hugo von Seeliger and Carl Neumann. But he then made a false analogy to his new cosmological constant Λ, a mistake never detected till the 1940s (Heckmann, 1942), not widely discussed till the 1960s, and still committed at times today. According to Einstein, Λ was "completely analogous to the extension of the Poisson equation to $\Delta\phi - \lambda\phi = 4\pi K\rho$ " (Einstein, 1923). Engelbert Schücking, a former student of Heckmann, provided a firm evaluation. "This remark was the opening line in a bizarre comedy of errors" (Schucking, 1991). The problem is that Λ is predominantly 0th order in ϕ (having a leading constant term), whereas the modified Poisson is 1st order in ϕ. Λ gives a weird quadratic potential for a point source, but the modified Poisson equation gives a massive graviton with plausible Neumann–Yukawa exponential fall-off (Freund *et al.*, 1969; Schucking, 1991). "However generations of physicists have parroted this nonsense" (Schucking, 1991). Massive theories of gravity generically involve two metrics, whereas Λ involves only one. Understanding geometric content sometimes is facilitated by a perturbative expansion.

13.7 Series, Non-linear Geometric Objects, and Atlases

Perturbative series expansions can also be useful for conceptual innovations. For example, non-linear realizations of the "group" of arbitrary coordinate transformations have tended to be invented with the help of a binomial series expansion for taking the symmetric square root of the metric tensor (DeWitt and DeWitt, 1952; Ogievetskiĭ and Polubarinov, 1965; Ogievetsky and Polubarinov, 1965). The exponentiating technology of non-linear group realizations (Isham *et al.*, 1971) is also at least implicitly perturbative. While classical differential geometers defined non-linear geometric objects (basically the same as particle physicists' non-linear group realizations as applied to coordinate transformations) (Aczél and Gołab, 1960; Szybiak, 1966; Tashiro, 1952), they generally provided no examples.

Perhaps the most interesting example involves the square root of the (inverse) metric tensor, or rather a slight generalization for indefinite metrics. The result is strictly a square root and strictly symmetric using $x^4 = ict$; otherwise it is a generalized square root using the signature matrix $\eta_{\alpha\beta} = diag(-1, 1, 1, 1)$. One has $r^{\mu\alpha}\eta_{\alpha\beta}r^{\beta\nu} = g^{\mu\nu}$ and $r^{[\mu\nu]} = 0$. Under coordinate transformations, the new components $r^{\mu'\nu'}$ are non-linear in the old ones (Ogievetsky and Polubarinov, 1965; Pitts, 2012). These entities augment tensor calculus and have covariant and Lie derivatives (Szybiak, 1963; Tashiro, 1952).

Defining the symmetric square root of a metric tensor might seem more of a curiosity for geometric completists than an important insight – but the symmetric square root of the metric makes an important conceptual difference with spinor fields used to represent fermions. Spinors in GR are widely believed to require an orthonormal basis (Cartan and Mercier, 1966; Lawson and Michelsohn, 1989; Weyl, 1929). But they do not, using $r^{\mu\nu}$ (Bilyalov, 2002; DeWitt and DeWitt, 1952; Ogievetskiĭ and Polubarinov, 1965; Ogievetsky and Polubarinov, 1965). One can have spinors in coordinates, but with metric-dependent transformations beyond 15-parameter conformal group (Borisov and Ogievetskii, 1974; Isham *et al.*, 1971; Ogievetskiĭ and Polubarinov, 1965; Pitts, 2012), the conformal Killing vectors for the unimodular metric density $\hat{g}_{\mu\nu} = (-g)^{-\frac{1}{4}}g_{\mu\nu}$. Such spinors have Lie derivatives beyond conformal Killing vectors – often considered the frontier for Lie differentiation of spinors (Penrose and Rindler, 1986, p. 101) – but they sprout new terms in $\mathcal{L}_\xi \hat{g}_{\mu\nu}$. One can treat symmetries without surplus structure and an extra local $O(1, 3)$ gauge group to gauge it away.

The (signature-generalized) square root of a metric, although not very familiar, fits fairly nicely into the realm of non-linear geometric objects, yielding a set of components in every coordinate system (with a qualification) and a non-linear transformation law. The entity is useful especially if one wants to know what sort of space-time structure is necessary for having spin-$\frac{1}{2}$ particles in curved space-time (Woodard, 1984). Must one introduce an orthonormal basis, then discard much of it from physical reality by taking an equivalence class under local Lorentz transformations? Or can one get by without introducing anything beyond the metric and then throwing (most of?) it away?

A curious and little known feature of this generalized square root touches on an assumption usually made in passing in differential geometry. Although one can (often) make a binomial series expansion in powers of the deviation of the metric from the signature matrix, and (more often) one can take a square root using generalized eigenvalues, there are exotic coordinate systems in which the generalized square root does not exist due to the indefinite signature (Bilyalov, 2002; Deffayet *et al.*, 2013; Pitts, 2012). This fact is trivial to show in two space-time dimensions (signature matrix *diag*$(-1, 1)$) using the quadratic formula: just look for negative eigenvalues. The fact generally has not been noticed previously because most treatments (a great many are cited in Pitts, 2012) worked near the identity. Such a point could have been noticed some time ago by Hoek, but a fateful innocent inequality was imposed that restricted the coordinates (with signature $+ - --$).

We shall assume that [the metric tensor $g_{\mu\nu}$] is pointwise continuously connected with the Minkowski metric (in the space of four-metrics of Minkowski signature) and has $g_{00} > 0$. (Hoek, 1982)

The lesson to learn is that there can be feedback from the fibers over space-time to the atlas of admissible coordinate systems for non-linear geometric objects given an indefinite signature. Naively assuming a maximal atlas causes interesting and quite robust entities not to exist. Such a result sounds rather dramatic when expressed in modern vocabulary. But coordinate inequalities are old (Hilbert, 2007), familiar (Møller, 1972), and not very dramatic classically; coordinates can have qualitative physical meaning while lacking a quantitative one. A principal square root is related to the avoidance of negative eigenvalues of $g^{\mu\nu}\eta_{\nu\rho}$ (Higham, 1987, 1997). Null coordinates are fine; the coordinate restriction is mild. Amusingly, coordinate *order* can be important: if (x, t, y, z) is bad, switching to (t, x, y, z) suffices (Bilyalov, 2002).

13.8 Massive Gravity: 1965–72 Discovery of 2010 Pure Spin-2

The recent (re)invention of pure spin-2 massive gravity (de Rham *et al.*, 2011; Hassan and Rosen, 2011) used the symmetric square root of the metric, as did the first invention (Ogievetsky and Polubarinov, 1965), though not the second (Pitts, 2011; Zumino, 1970). This problem has a curious history, from which Ogievetsky and Polubarinov (1965) have been unjustly neglected. That paper highly developed the symmetric square root of the metric perturbatively. It derived a two-parameter family of massive gravities, which, I note, includes two of the original three modern massive pure spin-2 gravities with a flat background metric. In light of the dependence of the space-time metric on the lapse function N in a $3 + 1$ ADM split, there were only two Ogievetsky–Polubarinov theories with any chance of being linear in the lapse (hence having pure spin-2 (Boulware and Deser, 1972)), although the naive cross-terms are rather discouraging. These are the $n = \frac{1}{2}, p = -2$ theory built around $\delta^\alpha_\mu (g^{\mu\nu}\eta_{\nu\alpha}\sqrt{-g}^2)^{\frac{1}{2}}$, a theory reinvented as equation (3.4) of Hassan and Rosen (2011), and the $n = -\frac{1}{2}, p = 0$ theory built around $\delta^\mu_\alpha (g_{\mu\nu}\eta^{\nu\alpha}\sqrt{-g}^0)^{\frac{1}{2}}$. A truly novel third theory is now known (Hassan and Rosen, 2011). A second novel modern result is the non-linear field redefinition of the shift vector (Hassan *et al.*, 2012), which allows the square root of the metric to be linear in the lapse.

More striking than the proposal of such theories long ago is the fact that in 1971–1972 Maheshwari already showed that one of the Ogievetsky–Polubarinov theories had pure spin-2 *non-linearly* (Maheshwari, 1972)! Thus the Boulware–Deser–Tyutin–Fradkin ghost (Boulware and Deser, 1972; Tyutin and Fradkin, 1972) (the negative energy sixth degree of freedom that is avoided by Fierz and Pauli to linear order but comes to life non-linearly) was *avoided before it was announced*. Unfortunately Maheshwari's paper made no impact, being cited only by Maheshwari in the mid-1980s. With Vainshtein's mechanism also suggested in 1972 (Vainshtein, 1972), there was no seemingly insoluble problem for massive spin-2 gravity in the literature. Massive spin-2 gravity was largely ignored from 1972 until *c.* 2000 largely because of failure to read Maheshwari's paper. This example illustrates the point (Chang, 2012) that the history of a science has resources for current science.

13.9 Conclusions

The considerations above support the idea that progress in knowledge about gravity can be made by overcoming various barriers, whether between general relativity and particle physics, or between physics and the history and philosophy of science. GR does not need to be treated *a priori* as exceptional, either in justifying choosing GR over rivals or in interpreting it. GR is well motivated non-mysteriously using particle physicists' arguments about the exclusion of negative-energy degrees of freedom, arguments that leave only a few options possible. To some degree the same holds even for the context of discovery of GR, given the renewed appreciation of Einstein's "physical strategy".

Because conceptual problems of GR often can be resolved, there is no need to treat it as *a priori* exceptional in matters of interpretation, either. Regarding gravitational radiation, Feynman reflected on the unhelpfulness of GR-exceptionalism:

What is the power radiated by such a wave? There are a great many people who worry needlessly at this question, because of a perennial prejudice that gravitation is somehow mysterious and different— they feel that it might be that gravity waves carry no energy at all. We can definitely show that they can indeed heat up a wall, so there is no question as to their energy content. (Feynman *et al.*, 1995, pp. 219, 220)

The conservation of energy and momentum – rather, energies and momenta – makes sense in relation to Noether's theorems. Change, even in local observables, is evident in the Hamiltonian formulation, just as in the Lagrangian/four-dimensional geometric form.

To say that GR should not be treated as *a priori* exceptional is not to endorse the strongest readings of the claim that GR is just another field theory, taking gauge-fixing and perturbative expansions as opening moves. The mathematics of GR logically entails some distinctiveness, such as the difference between external coordinate symmetries (with a transport term involving the derivative of the field) and internal symmetries as in electromagnetism and Yang–Mills. Identifying such distinctiveness requires reflecting on the mathematics and its meaning, as well as gross features of embodied experience, but it does not require conjectures about the trajectory of historical progress or divination of the spirit of GR.

Series expansions have their uses in GR. Einstein's failure to think perturbatively in 1917 about the cosmological constant generated lasting confusion and surely helped to obscure massive spin-2 gravity as an option. Many of the (re)inventions of the symmetric generalized square root of the metric began perturbatively. It permits spinors in coordinates, a fundamental geometric result, just as was Weyl's (1929) impossibility claim. Perturbative methods should not always be used or always avoided; they are one tool in the tool box for the foundations of gravity and space-time.

References

Aczél, J. and Gołab, S. 1960. *Funktionalgleichungen der Theorie der Geometrischen Objekte*. Warsaw: PWN.

Anderson, J. L. 1967. *Principles of Relativity Physics*. New York: Academic.

Anderson, J. L. and Bergmann, P. G. 1951. Constraints in Covariant Field Theories. *Physical Review*, **83**, 1018–25.

Anonymous. 2015. Sticky Bead Argument. In: *Wikipedia*. Last modified 28 February 2015.

Bergmann, P. G. 1958. Conservation Laws in General Relativity as the Generators of Coordinate Transformations. *Physical Review*, **112**, 287–9.

Bilyalov, R. F. 2002. Spinors on Riemannian Manifolds. *Russian Mathematics (Iz. VUZ)*, **46**(11), 6–23.

Bondi, H. 1957. Plane Gravitational Waves in General Relativity. *Nature*, **179**, 1072–3.

Borisov, A. B., and Ogievetskii, V. I. 1974. Theory of Dynamical Affine and Conformal Symmetries as the Theory of the Gravitational Field. *Theoretical and Mathematical Physics*, **21**, 1179–88.

Boulware, D. G. and Deser, S. 1972. Can Gravitation Have a Finite Range? *Physical Review, D*, **6**, 3368–82.

Brading, K. 2005. A Note on General Relativity, Energy Conservation, and Noether's Theorems. In Kox, A. J. and Eisenstaedt, J., eds. *The Universe of General Relativity*. Einstein Studies, volume 11. Boston: Birkhäuser, pp. 125–35.

Brink, L. 2006. A Non-Geometric Approach to 11-Dimensional Supergravity. In Liu, J. T., Duff, M. J., Stelle, K. S. and Woodard, R. P., eds. *DeserFest: A Celebration of the Life and Works of Stanley Deser*. Hackensack, NJ: World Scientific, pp. 40–54.

Brush, S. G. 1989. Prediction and Theory Evaluation: The Case of Light Bending. *Science*, **246**, 1124–29.

Bury, J. B. 1920. *The Idea of Progress: An Inquiry into Its Origin and Growth*. London: Macmillan.

Callan, Jr., C. G., Coleman, S. and Jackiw, R. 1970. A New Improved Energy-momentum Tensor. *Annals of Physics*, **59**, 42–73.

Cartan, É. and Mercier, A. 1966. *The Theory of Spinors*. Cambridge: Massachusetts Institute of Technology Press. French original 1937.

Castellani, L. 1982. Symmetries in Constrained Hamiltonian Systems. *Annals of Physics*, **143**, 357–71.

Chang, H. 2012. *Is Water H_2O? Evidence, Realism and Pluralism*. Boston Studies in the Philosophy and History of Science, vol. 293. Dordrecht: Springer.

Chisholm, R. M. 1973. *The Problem of the Criterion: The Aquinas Lecture, 1973*. Milwaukee, USA: Marquette University Press.

Deffayet, C., Mourad, J. and Zahariade, G. 2013. A Note on "Symmetric" Vielbeins in Bimetric, Massive, Perturbative and Non Perturbative Gravities. *Journal of High Energy Physics*, **1303**(086). arXiv:1208.4493 [gr-qc].

de Rham, C., Gabadadze, G. and Tolley, A. J. 2011. Resummation of Massive Gravity. *Physical Review Letters*, **106**, 231101. arXiv:1011.1232v2 [hep-th].

Deser, S. 1970. Self-Interaction and Gauge Invariance. *General Relativity and Gravitation*, **1**, 9–18. gr-qc/0411023v2.

DeWitt, Bryce S. and DeWitt, C. M. 1952. The Quantum Theory of Interacting Gravitational and Spinor Fields. *Physical Review*, **87**, 116–22.

DeWitt, C. M. 1957. *Conference on the Role of Gravitation in Physics at the University of North Carolina, Chapel Hill, March 1957, WADC Technical Report 57–216.* Wright Air Development Center, Air Research and Development Command, United States Air Force, Wright Patterson Air Force Base, Ohio. https://babel.hathitrust.org/shcgi/pt?id=mdp.39015060923078;view=1up;seq=7, scanned from the University of Michigan.

Dirac, P. A. M. 1964. *Lectures on Quantum Mechanics.* Belfer Graduate School of Science, Yeshiva University. Dover reprint, Mineola, New York, 2001.

Earman, J. 1992. *Bayes or Bust? A Critical Examination of Bayesian Confirmation Theory.* Cambridge: Massachusetts Institute of Technology Press.

Earman, J. 2002. Thoroughly Modern McTaggart: Or, What McTaggart Would Have Said if He Had Read the General Theory of Relativity. *Philosophers' Imprint*, **2**(3). www.philosophersimprint.org/.

Ehlers, J. 1973. The Nature and Structure of Spacetime. In Mehra, J., ed. *The Physicist's Conception of Nature.* Dordrecht: D. Reidel, pp. 71–91.

Einstein, A. 1916. Hamiltonsches Prinzip und allgemeine Relativitätstheorie. *Sitzungsberichte der Königlich Preussischen Akademie der Wissenschaften, Sitzung der physikalisch-mathematisch Klasse*, 1111–16. Translated as "Hamilton's Principle and the General Theory of Relativity" in H. A. Lorentz, A. Einstein, H. Minkowski, H. Weyl, A. Sommerfeld, W. Perrett and G. B. Jeffery, *The Principle of Relativity*, 1923; Dover reprint 1952, pp. 165–173.

Einstein, A. 1923. Cosmological Considerations on the General Theory of Relativity. In Lorentz, H. A., Einstein, A., Minkowski, H., Weyl, H., Sommerfeld, A., Perrett, W. and Jeffery, G. B., eds. *The Principle of Relativity*. London: Methuen. Dover reprint, New York, 1952. Translated from "Kosmologische Betrachtungen zur allgemeinen Relativitätstheorie," *Sitzungsberichte der Königlich Preussichen Akademie der Wissenschaften zu Berlin* (1917), pp. 142–52.

Einstein, A. and Grossmann, M. 1996. Outline of a Generalized Theory of Relativity and of a Theory of Gravitation. In Beck, A. and Howard, D., eds. *The Collected Papers of Albert Einstein, Volume 4, The Swiss Years: Writings, 1912–1914, English Translation.* Princeton: The Hebrew University of Jerusalem and Princeton University Press, pp. 151–88. Translated from *Entwurf einer verallgemeinerten Relativitätstheorie und einer Theorie der Gravitation*, Teubner, Leipzig, 1913.

Feynman, R. P., Morinigo, F. B., Wagner, W. G., Hatfield, B., Preskill, J. and Thorne, K. S. 1995. *Feynman Lectures on Gravitation.* Reading, MA: Addison-Wesley. Original by California Institute of Technology, 1963.

Freund, P. G. O., Maheshwari, A. and Schonberg, E. 1969. Finite-Range Gravitation. *Astrophysical Journal*, **157**, 857–67.

Giulini, D. 2008. What Is (Not) Wrong with Scalar Gravity? *Studies in History and Philosophy of Modern Physics*, **39**, 154–80. gr-qc/0611100v2.

Goldberg, J. N. 1980. Invariant Transformations, Conservation Laws, and Energy-Momentum. In Held, A., ed. *General Relativity and Gravitation: One Hundred Years After the Birth of Albert Einstein*, vol. 1. New York: Plenum Press, pp. 469–89.

Goodman, Nelson. 1983. *Fact, Fiction, and Forecast*. Fourth edn. Cambridge: Harvard University Press.

Gràcia, X. and Pons, J. M. 1988. Gauge Generators, Dirac's Conjecture, and Degrees of Freedom for Constrained Systems. *Annals of Physics*, **187**, 355–68.

Hassan, S. F. and Rosen, R. A. 2011. On Non-Linear Actions for Massive Gravity. *Journal of High Energy Physics*, **1107**(009). arXiv:1103.6055v3 [hep-th].

Hassan, S. F., Rosen, R. A., and Schmidt-May, A. 2012. Ghost-free Massive Gravity with a General Reference Metric. *Journal of High Energy Physics*, **12**(02), 26. arXiv:1109.3230 [hep-th].

Healey, R. 2002. Can Physics Coherently Deny the Reality of Time? In Callender, C. ed. *Time, Reality & Experience*. Cambridge: Cambridge University Press. Royal Institute of Philosophy Supplement 50, pp. 293–316.

Heckmann, O. 1942. *Theorien der Kosmologie*. Revised edn. Berlin: Springer. Reprinted 1968.

Higham, N. J. 1987. Computing Real Square Roots of a Real Matrix. *Linear Algebra and Its Applications*, **88**, 405–30.

Higham, N. J. 1997. Stable Iterations for the Matrix Square Root. *Numerical Algorithms*, **15**, 227–42.

Hilbert, D. 2007. The Foundations of Physics (Second Communication). In Renn, J. and Schemmel, M., eds. *The Genesis of General Relativity, Volume 4: Gravitation in the Twilight of Classical Physics: The Promise of Mathematics*, vol. 4. Dordrecht: Springer, pp. 1017–38. Translated from "Die Grundlagen der Physik. (Zweite Mitteilung)," *Nachrichten von der Königliche Gesellschaft der Wissenschaft zu Göttingen. Mathematisch-Physikalische Klasse* (1917), pp. 53-76.

Hoek, J. 1982. On the Deser–van Nieuwenhuizen Algebraic Vierbein Gauge. *Letters in Mathematical Physics*, **6**, 49–55.

Howson, C. 2000. *Hume's Problem: Induction and the Justification of Belief*. Oxford: Clarendon Press.

Isham, C. J., Salam, A. and Strathdee, J. 1971. Nonlinear Realizations of Space-Time Symmetries. Scalar and Tensor Gravity. *Annals of Physics*, **62**, 98–119.

Kennefick, D. 2007. *Traveling at the Speed of Thought: Einstein and the Quest for Gravitational Waves*. Princeton: Princeton University Press.

Kuchař, K. V. 1993. Canonical Quantum Gravity. In Gleiser, R. J., Kozameh, C. N. and Moreschi, O. M, eds. *General Relativity and Gravitation 1992: Proceedings of the Thirteenth International Conference on General Relativity and Gravitation held at Cordoba, Argentina, 28 June–4 July 1992*. Bristol: Institute of Physics Publishing. gr-qc/9304012, pp. 119–50.

Lakatos, I. 1971. History of Science and Its Rational Reconstruction. In Buck, R. C. and Cohen, R. S, eds. *PSA: Proceedings of the Biennial Meeting of the Philosophy of Science Association, 1970*. Boston Studies in the Philosophy of Science. Dordrecht: D. Reidel, pp. 91–136.

Laudan, L. 1977. *Progress and Its Problems: Towards a Theory of Scientific Growth*. Berkeley: University of California.

Lawson, Jr., H. B. and Michelsohn, M.-L. 1989. *Spin Geometry*. Princeton: Princeton University Press.

Logunov, A. A. and Folomeshkin, V. N. 1977. The Energy-momentum Problem and the Theory of Gravitation. *Theoretical and Mathematical Physics*, **32**, 749–71.

Maheshwari, A. 1972. Spin-2 Field Theories and the Tensor-Field Identity. *Il Nuovo Cimento*, **8A**, 319–30.

Maudlin, T. 2002. Thoroughly Muddled McTaggart: Or, How to Abuse Gauge Freedom to Generate Metaphysical Monstrosities. *Philosophers' Imprint*, **2**(4). www.philosophersimprint.org/.

Møller, C. 1972. *The Theory of Relativity*. Second edn. Oxford: Clarendon.

Morrison, P. J. 1998. Hamiltonian Description of the Ideal Fluid. *Reviews of Modern Physics*, **70**, 467–521.

Mukunda, N. 1980. Generators of Symmetry Transformations for Constrained Hamiltonian Systems. *Physica Scripta*, **21**, 783–91.

Nijenhuis, A. 1952. *Theory of the Geometric Object*. Ph.D. thesis, University of Amsterdam. Supervised by Jan A. Schouten.

Noether, E. 1918. Invariante Variationsprobleme. *Nachrichten der Königlichen Gesellschaft der Wissenschaften zu Göttingen, Mathematisch-Physikalische Klasse*, 235–257. Translated as "Invariant Variation Problems" by M. A. Tavel, *Transport Theory and Statistical Physics*, **1**, 183–207 (1971), LaTeXed by Frank Y. Wang, arXiv:physics/0503066 [physics.hist-ph].

Norton, J. D. 1993. General Covariance and the Foundations of General Relativity: Eight Decades of Dispute. *Reports on Progress in Physics*, **56**, 791–858.

Norton, J. D. 2011. Observationally Indistinguishable Spacetimes: A Challenge for Any Inductivist. In Morgan, G. J., ed. *Philosophy of Science Matters: The Philosophy of Peter Achinstein*. Oxford: Oxford University Press, pp. 164–76.

Ogievetsky, V. I. and Polubarinov, I. V. 1965a. Spinors in Gravitation Theory. *Soviet Physics JETP*, **21**, 1093–100.

Ogievetsky, V. I. and Polubarinov, I. V. 1965b. Interacting Field of Spin 2 and the Einstein Equations. *Annals of Physics*, **35**, 167–208.

Ohanian, H. C. 2008. *Einstein's Mistakes: The Human Failings of Genius*. New York: W. W. Norton & Company.

Ohanian, H., and Ruffini, R. 1994. *Gravitation and Spacetime*. Second edn. New York: Norton.

Penrose, R. and Rindler, W. 1986. *Spinors and Space-time, Volume 2: Spinor and Twistor Methods in Space-time Geometry*. Cambridge: Cambridge University Press.

Pitts, J. B. 2010. Gauge-Invariant Localization of Infinitely Many Gravitational Energies from All Possible Auxiliary Structures. *General Relativity and Gravitation*, **42**, 601–22. 0902.1288 [gr-qc].

Pitts, J. B. 2011. Universally Coupled Massive Gravity, II: Densitized Tetrad and Cotetrad Theories. *General Relativity and Gravitation*, **44**, 401–26. arXiv:1110.2077.

Pitts, J. B. 2012. The Nontriviality of Trivial General Covariance: How Electrons Restrict 'Time' Coordinates, Spinors (Almost) Fit into Tensor Calculus, and 7/16 of a Tetrad Is Surplus Structure. *Studies in History and Philosophy of Modern Physics*, **43**, 1–24. arXiv:1111.4586.

Pitts, J. B. 2014a. Change in Hamiltonian General Relativity from the Lack of a Time-like Killing Vector Field. *Studies in History and Philosophy of Modern Physics*, **47**, 68–89. http://arxiv.org/abs/1406.2665.

Pitts, J. B. 2014b. A First Class Constraint Generates Not a Gauge Transformation, But a Bad Physical Change: The Case of Electromagnetism. *Annals of Physics*, **351**, 382–406. Philsci-archive.pitt.edu; arxiv.org/abs/1310.2756.

Pitts, J. B. and Schieve, W. C. 2001. Slightly Bimetric Gravitation. *General Relativity and Gravitation*, **33**, 1319–50. gr-qc/0101058v3.

Pons, J. M. 2005. On Dirac's Incomplete Analysis of Gauge Transformations. *Studies in History and Philosophy of Modern Physics*, **36**, 491–518. arXiv:physics/0409076v2.

Pons, J. M. and Salisbury, D. C. 2005. Issue of Time in Generally Covariant Theories and the Komar-Bergmann Approach to Observables in General Relativity. *Physical Review D*, **71**, 124012. gr-qc/0503013.

Pons, J. M. and Shepley, L. C. 1998. Dimensional Reduction and Gauge Group Reduction in Bianchi-Type Cosmology. *Physical Review D*, **58**, 024001. gr-qc/9805030.

Pons, J. M., Salisbury, D. C. and Shepley, L. C. 1997. Gauge Transformations in the Lagrangian and Hamiltonian Formalisms of Generally Covariant Theories. *Physical Review D*, **55**, 658–68. gr-qc/9612037.

Pons, J. M., Salisbury, D. C. and Sundermeyer, K. A. 2010. Observables in Classical Canonical Gravity: Folklore Demystified. *Journal of Physics: Conference Series*, **222**, 012018. First Mediterranean Conference on Classical and Quantum Gravity (MCCQG 2009); arXiv:1001.2726v2 [gr-qc].

Reichenbach, H. 1938. *Experience and Prediction: An Analysis of the Foundations and the Structure of Knowledge*. Chicago: University of Chicago Press.

Renn, J. 2005. Before the Riemann Tensor: The Emergence of Einstein's Double Strategy. In Kox, A. J. and Eisenstaedt, J, eds, *The Universe of General Relativity*. Einstein Studies, volume 11. Boston: Birkhäuser, pp. 53–65.

Renn, J. and Sauer, T. 1999. Heuristics and Mathematical Representation in Einstein's Search for a Gravitational Field Equation. In Goenner, H., Renn, J., Ritter, J. and Sauer, T., eds. *The Expanding Worlds of General Relativity*. Einstein Studies, vol. 7. Boston: Birkhäuser, pp. 87–125.

Renn, J. and Sauer, T. 2007. Pathways Out of Classical Physics: Einstein's Double Strategy in his Seach for the Gravitational Field Equations. In Renn, J., ed. *The Genesis of General Relativity, Volume 1: Einstein's Zurich Notebook: Introduction and Source*. Dordrecht: Springer, pp. 113–312.

Rosenfeld, L. 1930. Zur Quantelung der Wellenfelder. *Annalen der Physik*, **397**, 113–52. Translation and commentary by Donald Salisbury, Max Planck Institute for the History of Science Preprint 381, www.mpiwg-berlin.mpg.de/en/resources/preprints.html, November 2009.

Roseveare, N. T. 1982. *Mercury's Perihelion from Le Verrier to Einstein*. Oxford: Clarendon Press.

Rovelli, C. 2002. Notes for a Brief History of Quantum Gravity. In Jantzen, R. T., Ruffini, R. and Gurzadyan, V. G., eds. *Proceedings of the Ninth Marcel Grossmann Meeting (held at the University of Rome "La Sapienza," 2–8 July 2000)*. River Edge, New Jersey: World Scientific. gr-qc/0006061, pp. 742–68.

Russell, B. 1927. *The Analysis of Matter*. London: Kegan Paul, Trench, Trubner & Co. New York: Harcourt, Brace and Company.

Salisbury, D. C. 2010. Léon Rosenfeld's Pioneering Steps toward a Quantum Theory of Gravity. *Journal of Physics: Conference Series*, **222**, 012052. First Mediterranean Conference on Classical and Quantum Gravity (MCCQG 2009).

Schucking, E. L. 1991. The Introduction of the Cosmological Constant. In Zichichi, A., de Sabbata, V. and Sánchez, N., eds. *Gravitation and Modern Cosmology: The Cosmological Constant Problem, Volume in honor of Peter Gabriel Bergmann's 75th birthday*. New York: Plenum, pp. 185–7.

Shimony, A. 1970. Scientific Inference. In Colodny, R. G., ed. *The Nature & Function of Scientific Theories*. University of Pittsburgh Series in the Philosophy of Science, vol. 4. Pittsburgh: University of Pittsburgh Press, pp. 79–172.

Sorkin, R. 1977. On Stress-Energy Tensors. *General Relativity and Gravitation*, **8**, 437–49.

Sugano, R., Saito, Y. and Kimura, T. 1986. Generator of Gauge Transformation in Phase Space and Velocity Phase Space. *Progress of Theoretical Physics*, **76**, 283–301.

Sundermeyer, K. 1982. *Constrained Dynamics: With Applications to Yang–Mills Theory, General Relativity, Classical Spin, Dual String Model*. Berlin: Springer. Lecture Notes in Physics, volume 169.

Szabados, L. B. 1991. *Canonical Pseudotensors, Sparling's Form and Noether Currents*. www.rmki.kfki.hu/ lbszab/doc/sparl11.pdf.

Szybiak, A. 1963. Covariant Derivative of Geometric Objects of the First Class. *Bulletin de l'Académie Polonaise des Sciences, Série des Sciences Mathématiques, Astronomiques et Physiques*, **11**, 687–90.

Szybiak, A. 1966. On the Lie Derivative of Geometric Objects from the Point of View of Functional Equations. *Prace Matematyczne=Schedae Mathematicae*, **11**, 85–8.

Tashiro, Y. 1952. Note sur la dérivée de Lie d'un être géométrique. *Mathematical Journal of Okayama University*, **1**, 125–8.

Tolman, R. C. 1930. On the Use of the Energy-Momentum Principle in General Relativity. *Physical Review*, **35**, 875–95.

Trautman, A. 1962. Conservation Laws in General Relativity. In Witten, L., ed. *Gravitation: An Introduction to Current Research*. New York: John Wiley and Sons, pp. 169–98.

Tyutin, I. V. and Fradkin, E. S. 1972. Quantization of Massive Gravitation. *Soviet Journal of Nuclear Physics*, **15**, 331–4.

Vainshtein, A. I. 1972. To the Problem of Nonvanishing Gravitation Mass. *Physics Letters B*, **39**, 393–4.

van Dongen, J. 2010. *Einstein's Unification*. Cambridge: Cambridge University Press.

Weyl, H. 1929. Elektron und Gravitation. *Zeitschrift für Physik*, **56**, 330–52. Translation in Lochlainn O'Raifeartaigh, *The Dawning of Gauge Theory*. Princeton: Princeton University Press (1997), pp. 121–44.

Woodard, R. P. 1984. The Vierbein Is Irrelevant in Perturbation Theory. *Physics Letters B*, **148**, 440–4.

Zumino, B. 1970. Effective Lagrangians and Broken Symmetries. In Deser, S., Grisaru, M. and Pendleton, H., eds. *Lectures on Elementary Particles and Quantum Field Theory: 1970 Brandeis University Summer Institute in Theoretical Physics*, vol. 2. Cambridge: M. I. T. Press, pp. 437–500.

Part IV

Quantum Foundations and Quantum Gravity

14

Is Time's Arrow Perspectival?

CARLO ROVELLI

14.1 Introduction

We observe entropy decrease towards the past. Does this imply that in the past the world was in a non-generic microstate? I point out an alternative. The subsystem to which we belong interacts with the universe via a relatively small number of quantities, which define a coarse graining. Entropy happens to depend on coarse graining. Therefore, the entropy we ascribe to the universe depends on the peculiar coupling between us and the rest of the universe. Low past entropy may be due to the fact that *this coupling* (rather than microstate of the universe) is non-generic. I argue that for *any* generic microstate of a sufficiently rich system there are *always* special subsystems defining a coarse graining for which the entropy of the rest is low in one time direction (the "past"). These are the subsystems allowing creatures that "live in time" – such as those in the biosphere – to exist. I reply to some objections raised to an earlier presentation of this idea, in particular by Bob Wald, David Albert and Jim Hartle.

14.2 The Problem

An imposing aspect of the Cosmos is the mighty daily rotation of Sun, Moon, planets, stars and all galaxies around us. Why does the Cosmos so rotate? Well, it is not really the Cosmos that rotates, it is us. The rotation of the sky is a *perspectival* phenomenon: we understand it better as due to the peculiarity of our own moving point of view, rather than a global feature of all celestial objects.

A vivid feature of the world is its being in color: each dot of each object has one of the colors out of a three-dimensional (3D) color-space. Why? Well, it is us that have *three* kinds of receptors in our eyes, giving the 3D color space. The 3D space of the world's colors is *perspectival*: we understand it better as a consequence of the peculiarity of our own physiology, rather than the Maxwell equations.

The list of conspicuous phenomena that have turned out to be perspectival is long; recognizing them has been a persistent aspect of the progress of science.

A vivid aspect of reality is the flow of time; more precisely: the fact that the past is different from the future. Most observed phenomena violate time reversal invariance strongly.

Could this be a perspectival phenomenon as well? Here I suggest that this is a likely possibility.

Boltzmann's *H*-theorem and its modern versions show that for most microstates away from equilibrium, entropy increases in *both* time directions. Why then do we observe lower entropy in the past? For this to be possible, most microstates around us appear to be very *non*-generic. This is the problem of the arrow of time, or the problem of the source of the second law of thermodynamics [1, 2]. The common solution is to believe that the universe was born in an extremely non-generic microstate [3]. Roger Penrose even considered the possibility of a fundamental cosmological law breaking time-reversal invariance, forcing initial singularities to be extremely special (vanishing Weil curvature) [4].

Here I point out that there is a different possibility: past low entropy might be a *perspectival* phenomenon, like the rotation of the sky.

This is possible because entropy depends on the system's microstate *but also* on the coarse graining under which the system is described. In turn, the relevant coarse graining is determined by the concrete existing interactions with the system. The entropy we assign to the systems around us depends on the way we interact with them – as the apparent motion of the sky depends on our own motion.

A subsystem of the universe that happens to couple to the rest of the universe via macroscopic variables determining an entropy that happens to be low in the past is a system to which the universe appears strongly time oriented; as it appears to us. Past entropy may appear low because of our own perspective on the universe.

Specifically, I argue below that the following conjecture is plausible:

Conjecture In a sufficiently complex system, there is always *some* subsystem whose interaction with the rest determines a coarse graining with respect to which the system satisfies the second law of thermodynamics (in some time direction).

An example where this is realized is given below.

If this is correct, we have a new way for facing the puzzle of the arrow of time: the universe is in a generic state, but is sufficiently rich to include subsystems whose coupling defines a coarse graining for which entropy increases monotonically. These subsystems are those where information can pile up and "information gathering creatures" such as those composing the biosphere can exist.

All phenomena related to time flow, all phenomena that distinguish the past from the future, can be traced to (or described in terms of) entropy increase. Therefore the difference between past and future may follow from the peculiarities of our coupling to the rest of the universe, rather than from a peculiarity of the microstate of the universe; like the rotation of the cosmos.

14.3 A Preliminary Conjecture

To start with, consider classical mechanics. Quantum theory is discussed in the last section. It is convenient to use Gibbs' formulation of statistical mechanics rather than Boltzmann's, because Boltzmann takes for granted the split of a system in a large number of equal subsystems (the individual molecules), and this may precisely obfuscate the key point in the context of general relativity and quantum field theory, as we shall see.

Consider a classical system with many degrees of freedom in a ("microscopic") state s, element of a phase space Γ, evolving in time as $s(t)$. Let $\{A_n\}$, be a set of ("macroscopic") observables – real functions on Γ, labeled by the index n. This set defines a coarse graining. That is, it partitions Γ in unequal regions where the A_n are constant. The largest of these regions is the equilibrium region. The entropy of a state s can be defined as the volume of the region where it is located. With a (suitably normalized and time invariant) measure ds, entropy is then

$$S_{A_n} = \log \int_{\Gamma} ds' \prod_n \delta(A_n(s') - A_n(s)), \tag{14.1}$$

where the family of macroscopic observables A_n is indicated in subscript to emphasize that the entropy depends on the choice of these observables. Notice that this definition applies to any microstate.[1]

As the microstate s evolves in time so does its entropy

$$S_{A_n}(t) = \log \int_{\Gamma} ds' \prod_n \delta(A_n(s') - A_n(s(t))). \tag{14.2}$$

Boltzmann's H-theorem and its modern versions imply that under suitable ergodic conditions *if we fix the choice of the macroscopic observables* A_n, for most microstates out of equilibrium at t_0, and for any finite Δt, we have $S_{A_n}(t_0 + \Delta t) > S_{A_n}(t_0)$ irrespectively of the sign of Δt.

I want to bring the attention, instead, to the dependence of entropy on the family of observables, and enunciate the following first conjecture. If the system is sufficiently complex and ergodic, for most paths $s(t)$ that satisfy the dynamics and for each orientation of t, there is a family of observables A_n such that

$$\frac{dS_{A_n}}{dt} \geq 0. \tag{14.3}$$

In other words, *any* motion appears to have initial low entropy (and non-decreasing entropy) under *some* coarse graining.

The conjecture becomes plausible with a concrete example. Consider a set Σ of N distinguishable balls that move in a box, governed by a time reversible ergodic dynamics. Let the box have an extension $x \in [-1, 1]$ in the direction of the x coordinate, and be ideally

---

[1] This equation defines entropy up to an an additive factor, because phase space volume has the dimension of $[Action]^N$, where N is the number of degrees of freedom. This is settled by quantum theory, which introduces a unit of action, the Planck constant, whose physical meaning is to determine the minimum empirically distinguishable phase space volume, namely the maximal amount of information in a state. See Haggard and Rovelli [5].

divided in two halves by $x=0$. For any given subset $\sigma \subset \Sigma$ of balls, define the observable A_σ to be the mean value of the x coordinate of the balls in σ. That is, if x_b is the x coordinate of the ball b, define

$$A_\sigma = \frac{\sum_{b \in \sigma} x_b}{\sum_{b \in \sigma} 1}. \tag{14.4}$$

Let $s(t)$ be a generic physical motion of the system, say going from $t=t_a$ to $t=t_b > t_a$. Let σ_a be the set of the balls that is at the right of $x=0$ at $t=t_a$. The macroscopic observable $A_a \equiv A_{\sigma_a}$ defines an entropy that is in the large N limit and for most motions $s(t)$ satisfies

$$\frac{S_{A_a}(t)}{dt} \geq 0. \tag{14.5}$$

This is the second law of thermodynamics.

But let us now fix the motion $s(t)$, and define a different observable as follows. Let σ_b be the set of the balls that is at the left of $x=0$ at $t=t_b$. The macroscopic observable $A_b \equiv A_{\sigma_b}$ defines an entropy that is easily seen to satisfy

$$\frac{S_{A_b}(t)}{dt} \leq 0. \tag{14.6}$$

This is again the second law of thermodynamics, but now in the reversed time $-t$. It holds for the generic motion $s(t)$, with a specific observable.

This is pretty obvious: if at time t_a we ideally color in white all the balls at the right of $x=0$ (see Figure 14.1), then the state at t_a is low entropy with respect to *this* coarse graining, and the motion mixes the balls and raises the entropy as t moves from t_a to t_b.

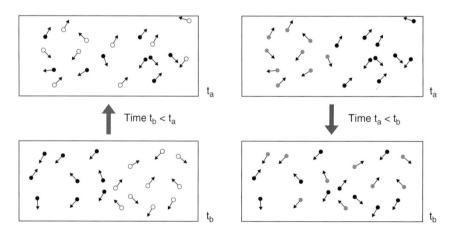

Figure 14.1 The same history, seen with different filters: for a filter seeing the white balls that are on the right at time t_a, entropy is low at t_a. A filter that sees the gray balls on the left at t_b defines an entropy low at t_b. Since the direction of time flow is determined by increasing entropy, time flows in a different direction with respect to the two different observables.

But if instead we color in gray the balls that are at the left of $x = 0$ *at the time t_b*, then the reverse is true and entropy increases in the reverse t direction.

The point is simple: for any motion there is a macroscopic family of observables with respect to which the state at a chosen end of the motion has low entropy: it suffices to choose observables that single out well the state at the chosen end. I call these observables "time oriented". They are determined by the state itself.

This simple example shows that, generically, past low entropy is not a feature of a special physical history of microstates of the system. Each such history may *appear* to be time oriented (that is: have increasing entropy) under a suitable choice of macroscopic observables.

Can this observation be related to the fact that we see entropy increase in the world? An objection to this idea is: how can a physical fact of nature, such as the second law, depend on a choice of coarse graining, which – so far – seems subjective and arbitrary? In the next section I argue that there is nothing arbitrary in the choice of the coarse graining and the macroscopic observables. These are fixed by the coupling between subsystems. Different choices of coarse graining represent different possible subsystems.

To pursue the analogy that opens this chapter, different reference systems from which the universe can be observed are concretely realized by different rotating bodies, such as the Earth.

14.4 Time-Oriented Subsystems

The fact that thermodynamics and statistical mechanics require a coarse graining, namely a "choice" of macroscopic observables, appears at first sight to introduce a curious element of subjectivity into physics, clashing with the objectivity of the predictions of science.

But of course there is nothing subjective in thermodynamics. A cup of hot tea does not cool down because of what I know or do not know about its molecules. The "choice" of macroscopic observables is dictated by the ways the system under consideration couples. The macroscopic observables of the system are those coupled to the exterior (in thermodynamics, those that can be manipulated and measured). The thermodynamics and the statistical mechanics of a system defined by a set of macroscopic observables A_n describe (objectively) the way the system interacts when coupled to another system via *these* observables, and the way *these* observables behave.

For instance, the behaviour of a box full of air is going to be described by a certain entropy function if the air is interacting with the exterior via a piston that changes its volume V. But the *same* air is going to be described by a *different* entropy function if it interacts with the exterior via *two* pistons with filters permeable to oxygen and nitrogen, respectively (see Figure 14.2). In this case, the macroscopic observables are others and chemical potentials enter the game. It is not our abstract "knowledge" on the relative abundance of oxygen and nitrogen that matters: it is the presence or not of a physical coupling of this quantity to the exterior, and the possibility of their independent variation, to determine

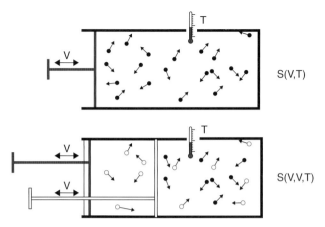

Figure 14.2 The same system – here a volume of air – is described by different entropy functions, describing different interactions it can have; here via its total volume or via the volume of its distinct chemical components.

which entropy describes the phenomena. Different statistics and thermodynamics of the same box of air describe different interactions of the box with the exterior.

In the light of this consideration, let us reconsider the box of the previous section replacing the abstract notion of "observable" by a concrete interaction between subsystems.

Say we have the N balls in a box as above, but now we add a new set of 2^N "small" balls[2], with negligible mass, that do not interact among themselves but interact with the previous ("large") balls as follows. Each small ball is labeled by a subset $\sigma \subset \Sigma$ and is attracted by the balls in σ and only these, via a force law such that the total attraction is in the direction of the center of mass A_σ of the balls in σ (see Figure 14.3).

Generically, a small ball interacts with a large number of large balls, but it does so only via a single variable: A_σ. Therefore it interacts with a statistical system, for which A_σ is the single macroscopic observable. For each small ball σ, the "rest of the universe" behaves as a thermal system with entropy S_{A_σ}.

It follows from the considerations of the previous sections that given a *generic* motion $s(t)$ there will generically be at least one small ball, the ball σ_a for which the entropy of the rest of the box is never decreasing in t, in the thermodynamical limit of large N. (There will also be another small ball, σ_b for which the entropy of the rest of the box is never *increasing* in t.)

Imagine that the box is the universe and each "small" ball σ is itself a large system with many degrees of freedom. Then generically there is at least one of these, namely σ_a (in fact many) for which the rest of the universe has a low-entropy initial state. In other words, it is plausible to expect the validity of the conjecture stated in the introduction, which I repeat here:

---

[2] 2^N is the number of subsets of Σ, namely the cardinality of its power set.

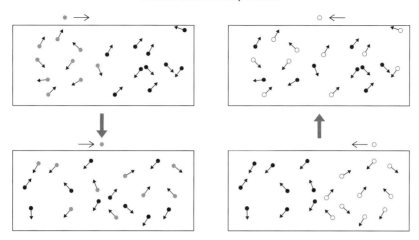

Figure 14.3 The "small" balls are represented on top of the box. The white and gray ones are attracted, respectively, by the large white and gray balls. Both interact with a statistical system where entropy changes, but entropy increases in the opposite direction with respect to each of them.

Conjecture In a sufficiently complex system, there is always *some* subsystem whose inter-action with the rest determines a coarse graining with respect to which the system satisfies the second law of thermodynamics (in some time direction).

That is: low past entropy can be fully perspectival.

Now, since σ_a interacts thermodynamically with a universe which was in a low-entropy state in the past, the world seen by σ_a appears organized in time: observed phenomena display marked and consistent arrows of time, which single out one direction. σ_a interacts with a world where entropy increases, hence "time flows" in one specific direction. The world seen by σ_a may include dissipation, memory, traces of the past, and all the many phenomena that characterize the universe as we see it. Within the subsystem σ_a, time ori-ented phenomena that require the growth of entropy, such as evolution or accumulation of knowledge, can take place. I call such a subsystem "time oriented".

Could this picture be related to the reason why we see the universe as a system with a low-entropy initial state? Could the low entropy of the universe characterize our own coupling with the universe, rather than a peculiarity of the microstate of the universe?

In the next section I answer some possible objections to this idea. Some of these were raised and discussed at the 2015 Tenerife Conference on the Philosophy of Cosmology.

14.5 Objections and Replies

1. *Isn't this just shifting the problem? Instead of the mystery of a strange (low past entropy) microstate of the universe, we have now the new problem of explaining why we belong to a peculiar system?*

Yes. But it is easier to explain why the Earth happens to rotate, rather than having to come up with a rationale for the full cosmos to rotate. The next question addresses the shifted problem.

2. *For most subsystems the couplings are such that entropy was not low at one end of time. Why then should we belong to a special subsystem that couples in such a peculiar manner?*

Because this is the condition for us to be what we are. We live in time. Our own existence depends on the fact of being in a situation of strong local entropy production: biological evolution, biochemistry, life itself, memory, knowledge acquisition, culture... As emphasized by David Albert, low past entropy is what allows us to reconstruct the past, and have memory and therefore gives us our sense of identity. Being part of a time-oriented subsystem is the condition for all this. This is a mild use of anthropic reasoning. It is analogous to asking why do we live on a planet's surface (non-generic place of the universe) and answering that this is simply what we are: creatures living on ground, needing water and so on. Our inhabiting these quarters of the universe is no more strange than me being born in a place where people happen to speak my own language.

3. *Assuming that we choose a coarse graining for which entropy is low at initial time t_a. Would not entropy then move very fast to a maximum, in the time scale of the molecular interactions, and then just fluctuate around the maximum? (Point raised by Bob Wald).*

There are different time scales. The thermalization time scale can be hugely different from the time scale of the molecular interactions, and in fact it is clearly so in our universe. Given a history of an isolated system, a situation where entropy increases can exist only for a time scale shorter than the thermalization time. This is precisely the situation in which we are in the universe: the Hubble time is much longer than the time scale of microphysics, but much shorter than the thermalization time of the visible universe. So, the time scales are fine.

4. *The interactions in the real universe are not as arbitrary as in the example of the heavy and small balls. In fact, in the universe there are no more than a small number of fundamental interactions. (Point raised by David Albert).*

The fundamental interactions are only a few, but the interaction channels they open are innumerable. The example of the colors makes this clear: the relevant elementary interaction is just the electromagnetic interaction. But our eyes pick up an incredibly tiny component of the electromagnetic waves. They pick up three variables out of an infinite number: they recognize a 3D space of colors (with some resolution) out of the virtually infinite dimensional space of waveforms. So we are *precisely* in the situation of the small balls, which only interact with a tiny fraction of the variables of the external world. It can be argued that these are the most relevant for us. This is precisely the point: it is by interacting with some specific variables that we may pick up time oriented features of the world. Another simple example is a normal radio: it can easily tune on a single band, out of the entire electromagnetic spectrum. We are in a similar situation. For instance, we observe a relatively tiny range of phenomena, among all those potentially existing in the full range of time scales existing in the 60 orders of magnitude between the Planck time and cosmological time.

5. *We see entropy increase in cosmology, which is the description of the whole, without coarse graining.*

Current scientific cosmology is not the dynamics of everything: it is the description of an extremely coarse grained picture of the universe. Cosmology is a feast of coarse graining.

6. *The observables that we use to describe the world are coarse grained but they are the natural ones.*

Too often "natural" is just what we are used to. Considering something "natural" is to be blind to subjectivity. For somebody it is natural to write from left to right. For others, the opposite.

7. *Our interactions pick up variables that are determined by the spatio-temporal structure of the world: spacetime integrals of conserved quantities. Quasi-classical domains are determined by these. Are these sufficiently generic for the mechanism you suggest? (Point raised by Jim Hartle).*

Yes they are, because spacetime averages of conserved quantities carry a very large amount of information, as our eyes testify. But this point is better raised in the context of quantum gravity, where the spacetime-regions structure is itself an emergent classical phenomenon that requires a quasi-classical domain. The emergence of a spatio-temporal structure from a quantum gravitational context may be related to the emergence of the second law in the sense I am describing here. In both cases there is a perspectival aspect of the emergence.

8. *Can the abstract picture of the coarse graining determining entropy be made concrete with an example?*

The biosphere is an oriented subsystem of the universe. Consider the thermodynamical framework of life on Earth. There is a constant flow of electromagnetic energy on Earth: incoming radiation from the Sun and outgoing radiation towards the sky. Microscopically, this is a certain complicated solution of the Maxwell equation. But as far as life is considered, most details of this solution (such as the precise phase of a single solar photon falling on the Pacific ocean) are irrelevant.

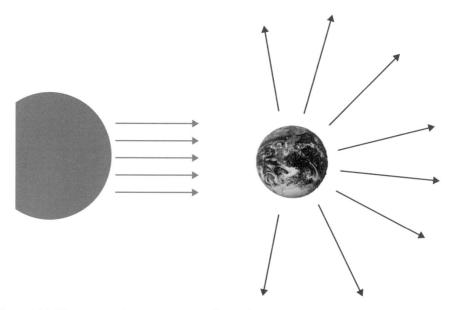

Figure 14.4 Electromagnetic energy enters and exits the Earth. The biosphere interacts with coarse grained aspects of it (frequency) with respect to which there is entropy production, and therefore time orientation, on the planet.

What matters to life on Earth are energy and a certain range of frequency, integrated over small regions. This determines a coarse graining and therefore a notion of entropy. Now the incoming energy is the same as the outgoing energy, but not so for the frequency. The Earth receives energy E (from the Sun) at higher frequency ν_a and emits energy (towards the sky) at lower frequency ν_b (see Figure 14.4). This is a fact about the actual solution of the Maxwell equations in which we happen to live. If we take energy and frequency as macroscopical observables, then an entropy is defined by such coarse graining. Roughly, this entropy counts the number of photons; at frequency ν the number of photons in a wave of energy E is $N = E/\hbar\nu$. If the received energy is emitted at lower frequency, the emitted entropy S_b is higher than the received entropy S_a. The process produces entropy: $S_b \gg S_a$. This entropy production is not a feature of the solution of the Maxwell equations alone: it is a feature of this solution *and* a set of macroscopic observables (integrated energy and frequency: oriented observables for this solution of the Maxwell equation) to which living systems couple.

Any system on Earth whose dynamics is governed by interactions with E and ν has a source of negative entropy at its disposal. This is what is exploited by the biosphere on Earth to build structure and organization. The point I am emphasizing is that what is relevant and peculiar here is not the individual solution of the Maxwell equation describing the incoming and outgoing waves: it is the peculiar manner in which the interaction with this energy is coarse grained by the biosphere.

14.6 Quantum Theory and General Relativity

Quantum phenomena provide a source of entropy distinct from the classical one gener-ated by coarse graining: entanglement entropy. The state space of any quantum system is described by a Hilbert space $\mathcal{H}$, with a linear structure that plays a major role for physics. If the system can be split into two components, its state space splits into the tensor product of two Hilbert spaces: $\mathcal{H} = \mathcal{H}_1 \otimes \mathcal{H}_2$, each carrying a subset of observables. Because of the linearity, a generic state is not a tensor product of component states; that is, in general $\psi \neq \psi_1 \otimes \psi_2$. This is entanglement. Restricting the observables to those of a subsystem, say system 1, determines a quantum entropy over and above classical statistical entropy. This is measured by the von Neumann entropy $S = -\mathrm{tr}[\rho \log \rho]$ of the density matrix $\rho = \mathrm{tr}_{\mathcal{H}_2} |\psi\rangle\langle\psi|$. Coarse graining is given by the restriction to the observables of a single subsystem.

The conjecture presented in this chapter can then be extended to the quantum context. Consider a "sufficiently complex" quantum system.[3] Then:

Conjecture Given a generic state evolving in time as $\psi(t)$, there exist splits of the system into subsystems such that the von Neumann entropy is low at initial time and increases in time.[4]

The point here is to avoid assuming a fixed tensorial structure of $\mathcal{H}$ *a priori*. Instead, given a generic state, we can find a tensorial split of $\mathcal{H}$ which sees von Neumann entropy grow in time.

This conjecture, in fact, is not hard to prove. A separable Hilbert space admits many discrete bases $|n\rangle$. Given any $\psi \in \mathcal{H}$, we can always choose a basis $|n\rangle$ where $\psi = |1\rangle$. Then we can consider two Hilbert spaces, $\mathcal{H}_1$ and $\mathcal{H}_2$, with bases $|k\rangle$ and $|m\rangle$, and map their tensor product to $\mathcal{H}$ by identifying $|k\rangle \otimes |m\rangle$ with the state $|n\rangle$ where (k, m) appear, say, in the n-th position of the Cantor ordering of the (n, m) couples $((1,1), (1,2), (2,1), (1,3), (2,2), (3,1), (1,4)\dots)$. Then, $\psi = |1\rangle \otimes |1\rangle$ is a tensor state and has vanishing von Neu-mann entropy. On the other hand, recent results show that entanglement entropy generically evolves towards maximizing entropy of a fixed tensor split (see Deutsch *et al.* [6]).

Therefore for any time evolution $\psi(t)$ there is a split of the system into subsystems such that the initial state has zero entropy and then entropy grows. Growing and decreasing of (entanglement) entropy is an issue about how we split the universe into subsystems, not a feature of the overall state of things (on this, see Tegmark [7]). Notice that in quantum field theory there is no single natural tensor decomposition of the Fock space.

Finally, let me get to general relativity. In all the examples above, I have considered non-relativistic systems where a notion of the single time variable is clearly defined. I have therefore discussed the *direction* of time, but not the *choice* of the time variable. In special relativity, there is a different time variable for each Lorentz frame. In general

---

[3] This means: with a sufficient complex algebra of observables and a Hamiltonian which is suitably "ergodic" with respect to it. A quantum system is not determined uniquely by its Hilbert space, Hamiltonian and state. All separable Hilbert spaces are isomorphic, and the spectrum of the Hamiltonian, which is the only remaining invariant quantity, is not sufficient to characterize the system.

[4] And others for which entropy is low at final time.

relativity, the notion of time further breaks into related but distinct notions, such as proper time along worldliness, coordinate time, clock time, asymptotic time, cosmological time... Entropy increase becomes a far more subtle notion, especially if we take into account the possibility that thermal energy leaks to the degrees of freedom of the gravitational field and therefore macrostates can includes microstates with different spacetime geometries. In this context, a formulation of the second law of thermodynamics requires us to identify not only a *direction* for the time variable, but also the *choice* of the time variable itself in terms of which the law can hold [8]. In this context, a split of the whole system into subsystems is even more essential than in the non-relativistic case, in order to understand thermodynamics [8]. The observations made in this chapter therefore apply naturally to the non-relativistic case.

The perspectival origin of many aspects of our physical world has been recently emphasized by some of the philosophers most sensible to modern physics [9, 10]. I believe that the arrow of time is not going to escape the same fate.

The reason for the entropic peculiarity of the past should not be sought in the cosmos at large. The place to look for them is in the split, and therefore in the macroscopic observables that are relevant to us. Time asymmetry, and therefore "time flow", might be features of a subsystem to which we belong, features needed for information gathering creatures like us to exist, not features of the universe at large.

Acknowledgments

I am indebted to Hal Haggard and Angelo Vulpiani for useful exchanges and to Kip Thorne, David Albert, Jim Hartle, Bob Wald and several other participants to the 2014 Tenerife Conference on the Philosophy of Cosmology for numerous inputs and conversations.

References

[1] J. Lebowitz, Boltzmann's entropy and time's arrow. *Phys. Today*, **46** (1993), 32.
[2] H. Price, *Time's Arrow*. (Oxford: Oxford University Press, 1996).
[3] S. M. Carroll and H. Tam, Unitary Evolution and Cosmological Fine-Tuning, (2010) arXiv:1007.1417.
[4] R. Penrose, Singularities and Time-Asymmetry. In S. W. Hawking and W. Israel, eds. *General Relativity: An Einstein Centenary Survey*, (Cambridge: Cambridge University Press, 1979), pp. 581–638.
[5] H. M. Haggard and C. Rovelli, Death and resurrection of the zeroth principle of thermodynamics, *Journal of Modern Physics D*, **22** (2013), 1342007, arXiv:1302.0724.
[6] J. M. Deutsch, H. Li and A. Sharma, Microscopic origin of thermodynamic entropy in isolated systems, *Physical Review E*, **87** (2013), no. 4, 042135, arXiv:1202.2403.
[7] M. Tegmark, How unitary cosmology generalizes thermodynamics and solves the inflationary entropy problem, *Physical Review D*, **85** (June, 2012), 123517, arXiv:1108.3080.
[8] T. Josset, G. Chirco and C. Rovelli, Statistical mechanics of reparametrization-invariant systems. It takes Three to Tango, (2016), arXiv:1503.08725.
[9] J. Ismael, *The Situated Self*. (Oxford: Oxford University Press, 2007).
[10] H. Price, *Naturalism without Mirrors*. (Oxford: Oxford University Press, 2011).

15

Relational Quantum Cosmology

FRANCESCA VIDOTTO

15.1 Conceptual Problems in Quantum Cosmology

The application of quantum theory to cosmology raises a number of conceptual questions, such as the role of the quantum-mechanical notion of "observer" or the absence of a time variable in the Wheeler–DeWitt equation. I point out that a relational formulation of quantum mechanics, and more in general the observation that evolution is always relational, provides a coherent solution to this tangle of problems.

A number of confusing issues appear when we try to apply quantum mechanics to cosmology. Quantum mechanics, for instance, is generally formulated in terms of an observer making measurements on a system. In a laboratory experiment, it is easy to identify the system and the observer: but what is the observer in quantum cosmology? Is it part of the same universe described by the cosmological theory, or should we think of it as external to the universe? Furthermore, the basic quantum dynamical equation describing a system including gravity is not the Schrödinger equation, but rather the Wheeler–DeWitt equation, which has no time parameter: is this related to the absence of an observer external to the universe? How do we describe the quantum dynamics of the universe without a time variable in the dynamical equation and without an observer external to the universe?

I suggest that clarity on these issues can be obtained by simply recognizing the relational nature of quantum mechanics and more in general the relational nature of physical evolution. In the context of quantum theory, this nature is emphasized by the so-called *relational* interpretation of quantum mechanics (Rovelli, 1996). The relational nature of evolution, on the other hand, has been pointed out by the partial-observable formulation of general covariant dynamics (Rovelli, 2002).

A central observation is that there is a common confusion between two different meanings of the expression "cosmology". This is discussed below, in Section 15.1.1. A considerable amount of the difficulties mentioned above stems, I argue, from this confusion. The discussion clarifies on the role of the quantum mechanical observer in cosmology, which is considered in Section 15.1.2. In Section 15.2, I briefly describe the relational interpretation of quantum mechanics and some related aspects of quantum theory. In Section 15.3.2, I discuss the timeless aspect of the Wheeler–DeWitt equation. In Section 15.3.5, I show how this perspective on quantum cosmology can be taken as an effective conceptual structure in the application of loop quantum gravity to cosmology.

15.1.1 Cosmology is not About Everything

The expression "cosmology" is utilized to denote two very different notions. The first is the subject of exploration of the cosmologists. The second is the "totality of things". These are two very different meanings.

To clarify why, consider a common physical pendulum. Its dynamics is described by the equation $\ddot{q} = -\omega q$, and we know well how to deal with the corresponding classical and quantum theory. Question: does this equation describe "everything" about the physical pendulum? The answer is obviously negative, because the pendulum has a complicated material structure, ultimately made by fast moving electrons, quarks, and whatever, not to mention the innumerable bacteria most presumably living on its surface and their rich bio-chemistry... The point is that the harmonic oscillator equation certainly does not describe the *totality* of the physical events on the real pendulum: it describes the behavior of one dynamical variable, neglecting everything else happening at smaller scales.

In a very similar manner, cosmology – in the sense of "what cosmologists actually study" – describes a number of large scale degrees of freedom in the universe. This number may be relatively large: it may include all the measured CMB modes, or the observed large scale structures. But it remains immensely smaller than the total number of degrees of freedom of the real universe: the details of you reading now these words do not appear in any of the equations written by cosmologists. In strict sense, "cosmology", defined as the object of study of the cosmologists, denote the *large scale* degrees of freedom of the universe. The fact that many shorter scale degrees of freedom are neglected is no different from what happens in any other science: a biologist studying a cat is not concerned with the forces binding the quarks in the nucleus of an atom in the cat's nose.

There is however a different utilization of the term "cosmology" that one may find in some physics articles: sometimes it is used to denote a hypothetical science dealing with "the totality of all degrees of freedom of Nature". For the reason explained above, the two meanings of "cosmology" are to be kept clearly distinct, and much of the conceptual confusion raised by quantum cosmology stems from confusing these two different meanings of the term. For clarity, I use here two distinct terms: I call "cosmology" the science of the large scale degrees of freedom of the Universe, namely the subject matter of the cosmologists. We can call "totology" (from the latin "totos", meaning "all") the science – if it exists – of *all* degrees of freedom existing in reality.

The important point is that cosmology and totology are two different sciences. If we are interested in the quantum dynamics of the scale factor, or the emergence of the cosmo logical structures from quantum fluctuations of the vacuum, or in the quantum nature of the Big Bang, or the absence of a time variable in the Wheeler–DeWitt equation, we are dealing with specific issues in cosmology, not in totology.

15.1.2 The Observer in Cosmology

The considerations above indicate that a notion of "observer" is viable in cosmology. In cosmology, the "observer" is formed by ourselves, our telescopes, the measurement apparatus on our spacecraft, and so on. The "system" is formed by the large scale degrees of freedom of the universe. The two are clearly dynamically distinct. The observer is not "part of the system", in the dynamical sense.

Of course the observer is "inside" the system in a spacial sense, because the scale factor describes the dimension of a universe within which the observer is situated. But this is no more disturbing than the fact that a scientist studying the large scale structure of the magnetic field of the Earth is situated within this same magnetic field. Being in the same region of space does not imply being in the same degrees of freedom.

The notion of a dynamically external observer may be problematic in totology, but it is not so in conventional cosmology, and therefore there is no reason for it to be problematic in quantum cosmology.

Actually, there is an aspect of the conventional presentation of quantum theory which becomes problematic: the idea that a system has an intrinsic physical "state" which jumps abruptly during a measurement. This aspect of quantum theory becomes implausible in cosmology, because the idea that when we look at the stars the entire universe could jump from one state to another is not very palatable. But whether or not we interpret this jump as a physical event happening to the state depends on the way we interpret the quantum theory. This is why the application of quantum theory to cosmology bears on the issue of the interpretation of quantum mechanics. There are some interpretations of quantum mechanics that become implausible when utilized for quantum cosmology. But not all of them. Below, I describe an interpretation which is particularly suitable for cosmology and which does not demand implausible assumptions such as the idea that our measurement could change the entire intrinsic state of the universe, as demanded by textbook Copenhagen interpretation. As we shall see, in the relational interpretation of quantum mechanics the "quantum state" is not interpreted as an intrinsic property of a system, but only as the information one system has about another. There is nothing implausible if this changes abruptly in a measurement, even when the observed system is formed by the large scale degrees of freedom of the universe.

15.2 Relational Quantum Mechanics

The relational interpretation of quantum mechanics was introduced in Rovelli (1996) and has attracted the interest of philosophers such as Michel Bitbol, Bas van Frassen and Mauro Dorato (see Dorato (2013) and references therein).

The point of departure of the relational interpretation is that the theory is about quantum events rather than about the wave function or the quantum state. The distinction can be traced to the very beginning of the history of quantum theory: Heisenberg's key idea was to replace the notion of an electron continuously existing in space with a lighter ontology, the one given just by discrete tables of numbers. The electron, in Heisenberg's vision, can be thought of as "jumping" from one interaction to another.[1] In contrast, one year later, Schrödinger was able to reproduce the technical results of Heisenberg and his collaborators using a wave in space. Schrödinger's wave evolved into the modern notion of quantum state. What is quantum mechanics about: an evolving quantum state, as in Schrödinger; or a discrete sequence of quantum events that materialise when systems interact?

Relational quantum mechanics is an interpretation of quantum mechanics based on the second option, namely on Heisenberg's original intuition. The advantage is that the quantum state is now interpreted as a mere theoretical booking device for the information about a system S that a *given* system O might have gathered in the course of its past interactions with S. Therefore, the quantum state of S is not intrinsic to S: it is the state of S *relative* to O. It describes, in a sense, the information that O may have about S. No surprise if it jumps abruptly at a new interaction, because at each new interaction O can gather new information about S.

Thus, if we adopt this reading of quantum theory, there is no meaning in "the wave function of the entire universe", or "the quantum state of everything", because these notions are extraneous to the relational interpretation. A quantum state, or a wave function,[2] can only refer to the interactions between *two* interacting subsystems of the universe. It has no more reality than the distributions of classical statistical mechanics: tools for computing. What is real is not the quantum state: what is real are the individual quantum events. For all these reasons, the relational interpretation of quantum mechanics is particularly suitable for quantum cosmology.

Of course there is a price to pay for this remarkable simplification, as always the case with quantum theory. The price to pay here is that we are forced to recognise that the fact itself that a quantum event has happened, or not, must be interpreted as relative to a given system. Quantum events cannot be considered absolute; their existence is relative to the physical systems involved in an interaction. Two interacting systems realise a quantum event relative to one another, but not necessarily relative to a third system. This caveat, discussed in detail in Rovelli's original article and in the numerous philosopher's articles on the relational interpretation, is necessary to account for interference and to accommodate the equality of all physical systems. In fact, this is the metaphysical core of the relational interpretation, where a naive realism is traded for mature realism, able to account for Heisenberg's lighter ontology.

---

[1] Heisenberg gives a telling story about how he got the idea. He was walking in a park in Copenhagen at night. All was dark except for a few islands of light under street lamps. He saw a man walking under one of those, then disappearing in the dark, then appearing again under the next lamp. Of course, he thought, man is big and heavy and does not "really" disappear: we can reconstruct his path through the dark. But what about a small particle? Maybe what quantum theory is telling us is precisely that we cannot use the same intuitions for small particles. There is no classical path between their appearance here and their appearance there. Particles are objects that manifest themselves only when there is an interaction, and we are not allowed to fill up the gap in between. The ontology that Heisenberg proposes does not increase on the ontology of classical mechanics: it reduces it. It is *less*, not *more*. Heisenberg removes excess baggage from classical ontology and is left with a minimum necessary to describe the world.

[2] The "wave function" is the representation of the quantum state by means of a function on the spectrum of a complete commuting family of observables; for instance, a wave on configuration space.

In cosmology, however, it is a small price to pay: the theory itself guarantees that as long as the quantum effect in the interactions between quantum systems can be disregarded, different observing systems give the same description of an observed system and we are not concerned about this lighter ontology, This is definitely the case in cosmology.[3]

Under the relational reading of quantum theory, the best description of reality we can give is the way things affect one another. Things are manifest in interactions (quantum relationalism). Quantum theory describes reality in terms of facts appearing in interactions, and relative to the systems that are interacting. Cosmology describes the dynamics – possibly including the quantum dynamics – of the large scale degrees of freedom of the universe and the way these are observed and measured by our instruments.

Before closing this section on the relational interpretation, I discuss below two important notions on which this interpretation is based: *discreteness* and quantum *information*. I give also, below, a discussion on the role of the *wave function* of quantum mechanics, and on its common overestimation.

15.2.1 Discreteness

Quantum mechanics is largely a discovery of a very peculiar form of discreteness in Nature. Its very name refers to the existence of peculiar discrete units: the "quanta". Many current interpretations of quantum mechanics underemphasise this discreteness at the core of quantum physics, and many current discussions on quantum theory neglect it entirely. Historically, discreteness played a pivotal role in the discovery of the theory:

- Planck (1900): finite size packets of energy $E = h\nu$
- Einstein (1905): discrete particles of light
- Bohr (1912): discrete energy levels in atomic orbits
- Heisenberg (1925): tables of numbers
- Schrödinger (1926): discrete stationary waves
- Dirac (1930): state space and operators with possibly discrete spectra
- Spin: discrete values of angular momentum
- QFT: particles as discrete quanta of a field.

---

[3] In spite of this lighter ontology, this interpretation of quantum theory takes fully the side of realism, in the sense that it assumes that the universe exists independently from any conscious observer actually observing it. Consciousness, mind, humans, or animals, have no special role. Nor has any role any particular structure of the world (macroscopic systems, records, information gathering devices...). Rather, the interpretation is democratic: the world is made of physical systems, which are on equal ground and interact with one another. Facts are realized in interactions and the theory describes the probabilities of the outcome of future interaction, given past ones. But this is a realism in the weak sense of Heisenberg. In the moment of the interaction there is a real fact. But this fact exists relatively to the interacting systems, not in the absolute. In the same manner, two objects have a well defined velocity with respect to one another, but we cannot say that a single object has an absolute velocity by itself, unless we implicitly refer to some other reference object. Importantly, the structure of quantum theory indicates that in this description of reality the manner in which we split the world in subsystems is largely arbitrary. This freedom is guaranteed by the tensorial structure of QM, which grants the arbitrariness of the positioning of the boundary between systems which has been studied by Wigner: the magic of the quantum mechanics formalism is that we can split up the Hilbert space into pieces, and the formalism keeps its consistency.

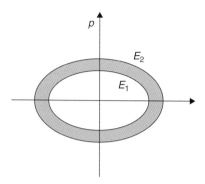

Figure 15.1 In the shaded region R there is an infinite number of classical states, but only a finite number of quantum ones.

As all these examples indicate, the scale of all these examples of discreteness is always set by $\hbar$. But what is actually discrete, in general, in quantum mechanics? It is easy to answer by noticing that $\hbar$ has the dimensions of an action and that the (Liouville) volume of the phase space of any system has always the dimension of an action (per degree of freedom). The constant $\hbar$ sets a unit of volume in the phase space of any system. The physical discovery at the basis of quantum theory is the impossibility of pinpointing the classical state of a system with a precision superior to such a unit volume.

If we have made some measurements on a system, with a given accuracy, and, consequently, we know that the system is in a certain region R of its phase space, then the quantum theory tells us that there is only a *finite* number of possible (orthogonal, namely distinguishable) states in which the system can be, given by

$$N = \frac{Volume(R)}{\hbar} \ . \tag{15.1}$$

Let us consider a simple example: say that measuring the energy E of a harmonic oscillator we learn that $E \in [E_1, E_2]$. This determines a region R of phase space, with volume $V = \int_{E\in[E_1,E_2]} dp\, dq = 2\pi \frac{E_2-E_1}{\omega}$ (see Figure 15.1). There is an infinite number of possible classical states, and an infinite number of possible values that the energy can actually have. In the quantum theory – that is, in Nature – the energy cannot take all these values, but only a finite number, which is $I = \frac{E_2-E_1}{\hbar\omega} = \frac{V}{\hbar}$. This is the maximum number of orthogonal states compatible with the previous measurement.

The example shows that a region of phase space is the specification of a certain amount of information we have about a system. This is general: quantum mechanics teaches us that in every finite region of phase space we can only accommodate a finite number of orthogonal states, namely that there is a finite information that we can extract from any finite region of phase space.

The same is true in quantum field theory. The quanta of a field are discrete particles (Dirac) and this is precisely a manifestation of this same discreteness; in particular, this is manifested in the discreteness of the spectrum of the energy of each mode.

Discreteness, in this sense, is the defining property of all quantum systems. The discreteness scale is set by $\hbar$, an action, or phase space volume.

15.2.2 Information

The notion of information useful in quantum theory is Shannon's *relative information*, which is defined as follows. Given two systems, such that we can find the first system in N_a states and the second in N_b, we say that the two have information about one another if, because of physical constraint, we can find the combined state in a number N_{ab} of states which is smaller than $N_a \times N_b$. The relative information is then defined by

$$I = \log_2(N_a \times N_b) - \log_2(N_{ab}) \tag{15.2}$$

The utility of this definition of information is that there is nothing mental or subjective about it: it is simply a measure that physics establishes between two degrees of freedom. For instance, as long as my pencil is not broken, each extreme of the pencil has information about the other, because the two can be in a smaller number of places than two separated objects. Knowledge of the position of one give some information about the position of the other.

The existence of this correlation is a measurable property of the combined system. If we have information about a system, we can make predictions about the outcome of future interactions with it. We can call "relevant" information that portion of the information that we have about a system which is not redundant in view of such predictions. In the relational interpretation, physics is the theory of the relevant relative information that systems can have about one another. In Rovelli (1996), two basic postulates were proposed, meant to capture the physical content of quantum theory:

Postulate I There is a maximum amount of relevant information that can be extracted from a system.

Postulate II It is always possible to acquire new relevant information about a system.

Remarkably, the entirety of the quantum mechanical formalism (Hilbert spaces, self adjoint operators, eigenvalues and eigenstates, projection postulate...) can be recovered on the basis of these two postulates, plus a few other more technical assumptions. The effort of understanding the physical meaning of these additional postulates, and to provide a mathematically rigorous reconstruction theorem has been developed by a number of authors and is still in progress (Grinbaum, 2003; Hoehn, 2014).

A given system S can be characterized by a set of variables. These take values in a set (the spectrum of the operators representing them, possibly discrete). In the course of an interaction with a second system O, the effect of the interaction on O depends on the value of one of these variables. We can express this by saying that the system O has then the information that this variable of the system S has the given value. This is one of the elementary quantum events that quantum theory is concerned with. The first postulate captures an essential aspect of quantum theory: the impossibility of associating a single point in phase space to the state of a system: to do so, we would need an infinite amount of information. The maximum localization in phase space is limited by $\hbar$, which determines the minimal physical cell in phase space. The information about the state of a system is therefore limited. The second postulate distinguishes quantum systems from classically discrete systems: even if finite, information is always incomplete, in the sense that new information can be gathered, at the price of making previous information irrelevant for predicting the future interactions: by measuring the L_x component of angular momentum we destroy information we previously had on the L_z component.

15.2.3 The Role of the Wave Function

If the real entities in quantum theory are discrete quantum events, what is the quantum state, or the wave function, which we commonly associate with an evolving system on many applications of quantum theory?

A good way to make sense of a quantity in a theory is to relate it to the corresponding quantity in an approximate theory, which we understand better for historical reasons. We can therefore investigate the meaning of the quantum mechanical wave function by studying what it becomes in the semiclassical limit. As is well known, if $\psi(x,t)$ is the wave function of a Newtonian particle:

$$\psi(x,t) \sim e^{\frac{i}{\hbar}S(x,t)} \tag{15.3}$$

where $S(x,t)$ is the Hamilton function, the Schrödinger equation becomes in the classical limit:

$$-i\hbar \frac{\partial}{\partial t}\psi(x,t) = -\frac{\hbar^2}{2m}\frac{\partial^2}{\partial x^2}\psi(x,t) + V(x)\psi(x,t) \tag{15.4}$$

$$\longrightarrow \quad \frac{\partial}{\partial t}S(x,t) = \frac{1}{2m}\frac{\partial^2}{\partial x^2}S(x,t) + V(x) \tag{15.5}$$

which is the Hamilton–Jacobi equation.

The Hamilton function is a theoretical device, not something existing for real in space and time. What exists for real in space and time is the particle, not its Hamilton function. I am not aware of anybody suggesting to endow the Hamilton function with a realistic interpretation.

So, why therefore would anybody think of doing so with its corresponding quantum object, the wave function? An interpretation of quantum theory that considers the wave function a real object sounds absurd: it is assigning a reality to a calculation tool. If I say: "Tomorrow we can go hiking, or we can go swimming", I am not making a statement implying that tomorrow the two things will both happen: I am only expressing my ignorance, my uncertainty, my lack of knowledge, my lack of information, about tomorrow. Perhaps tomorrow neither programme would realise, but not both. The wave function of a particle expresses probabilities, and these are related to our lack of knowledge. It does not mean that the wave function becomes a real entity spread in space, or, worse, in configuration space. The wave function (the state) is like the Hamilton function: a computational device, not a real object (Durr *et al.*, 1995).

In science, abstract concepts have sometimes been recognized for real entities. But often the opposite has happened: a misleading attitude has come from realism barking at the wrong tree: examples are the crystal spheres, the caloric, the ether... A realist should not be realist about the quantum state, if she wants to avoid the unpalatable alternative between the Scylla of the quantum collapse or the Charybdis of the branching many worlds. We can make sense of quantum theory with a softer realism, rather than with an inflated one. The wave function is about the relative information that systems have about each other: it is about information or lack of information, it is about what we can expect about the next real quantum event.

15.2.4 Relevance for Quantum Cosmology

As argued in the previous section, cosmology, in the conventional sense, is the study of the dynamics of these large scale degrees of freedom. It is not about everything, it is about a relatively small number of degrees of freedom. It is essentially based on an expansion in modes and neglects short wavelength modes, where "short" includes modes of millions of light years. It describes the way the large modes affect all the (classical) rest, including us. The dynamics of these degrees of freedom is mostly classical, but it can be influenced by quantum mechanics, which introduces discreteness and lack of determinism, in some regimes such as the early universe. This quantum aspect of the dynamics of the universe at large can be taken into account by standard quantum mechanics. The "observer" in this case is simply formed by ourselves and our instruments, which are not part of the large scale dynamics. The quantum state of the system represents the information we have gathered so far about the large scale degrees of freedom. As usual in quantum theory, we can often "guess" aspects of these states, completing our largely incomplete observations about them.

As far as the other interpretation of the term "cosmology" is concerned, namely as totology, in the light of the relational interpretation of quantum theory, it is perhaps questionable whether a coherent "quantum totology" makes any sense at all. If we define a system as the totality of anything existent, it is not clear what is the meaning of studying how this system appears to an observer, since by definition there is no observer. The problem is open, but it is not much related to the concerns of the real-life cosmologists, or to questions such as what happened at the big bang.

We can then use standard quantum theory, for instance in the form of a Hilbert space for those large degrees of freedom, observables that describe the physical interaction between the large degrees of freedom and our telescopes. For instance we can measure the temperature of the CMB. This can be done with the usual conceptual tools of quantum theory. The system does not include the observer. The observer in cosmology is indeed the majority of stuff in the universe: all the degrees of freedom of the universe except for the large scale one. This definition of cosmology eliminates any problem about the observer. A measurement does not need to "affect the large scale universe". It only affects our information about it.

15.3 Quantum Gravity and Quantum Cosmology

As best as we know, gravity is described by general relativity. The peculiar symmetries of general relativity add specific conceptual issues to the formulation of quantum cosmology. Among these is the fact that instead of a Schrödinger equation, evolving in a time variable t, we have a Wheeler–DeWitt equation, without any explicit time variable.

It is important to distinguish different issues. The discussion of the two previous sections would not change if gravity was described by Newton's theory and our main quantum dynamical equation was a Schrödinger equation. The distinction between cosmology and totology would not change. Contrary to what is often stated, the lack of a time variable in the Wheeler–DeWitt equation has no direct relation with the existence or the absence of an external quantum observer, because the problem given a quantum description of the totality of things would be the same also if gravity was Newtonian.

The additional, specific problem raised by general relativity is how to describe *evolution* in the presence of general covariance. The solution, on the other hand, is well known and is very similar to the solution of the problem of the observer discussed above: quantum theory describes the state of a system *relative* to another – arbitrary – system. General relativity describes the evolution of some observables *relative* to other – arbitrary – observables.

This solution is recalled below.

15.3.1 The Relational Nature of General Relativity

The celebrated idea of Einstein in 1915 is that spacetime *is* the gravitational field. This implies that spacetime is a dynamical field. General relativity does not imply that the gravitational field is particularly different from other fields. It implies that all physical fields do not live in spacetime: rather, the universe is made of several fields, interacting with one another. One of these is the gravitational field. The description of some specific aspect of a configuration of this field is what we call geometry.

In the region of the universe where we live, a good approximation is obtained by neglecting both the dynamics of the gravitational field and its local curvature, and using the gravitational field as a fixed background with respect to which we can define acceleration and write Newton's second law.

Localization is relative, to other dynamical objects, including the gravitational field. This is also true for temporal localization. In general relativity there is no fixed background structure, nor a preferred time variable with respect to which events are localized in time. In Newtonian physics there is a time along which things happen; in general relativity there is no preferred time variable. Physical variables evolve with respect to one another. For instance, if we keep a clock at a fixed altitude and we throw a second clock upward so that it raises and then falls back next to the first clock, the readings of the two clocks will then differ. Given sufficient data, general relativity allows us to compute the value of the first t_1 as a function of the second t_2, or vice versa. None of the two variables t_1 or t_2 is a more legitimate "time" than the other.

This observation can be formalised in terms of the notion of *partial observable* (Rovelli, 2013), which provides a clean way to deal with the peculiar gauge structure of general relativity.

In the example above, the two variables t_1 and t_2, representing the reading of the two clocks, are partial observables. Both quantities can be measured, but none of the two can be predicted, of course, because we do not know "when" an observation is made. But the value of each can be predicted once the other is known. The theory predicts (with sufficient information) the *relation* between them.

In the Hamiltonian formulation of the theory, the dynamics of general relativity is generated by constraints, as a direct consequence of the absence of background. The solution of those constraints codes the evolution of the system, without external time with respect to which evolution can be described (Rovelli, 1991). Partial observables are functions on the extended phase space where constraints are defined. On the constraint surface, the constraints generate orbits that determine relations between partial observables. These relations express the classical dynamical content of the theory.

The prototypical example of a partial observable which can be measured but not predicted is the conventional time variable of Newtonian physics: a quantity that we routinely determine (looking at a clock) but we can not predict from the dynamics of the system.

The physical phase space is the space of these orbits. A point of phase space cannot be interpreted as the characterization of the state of the system at a given time, because the theory has no notion of time. Rather, it can be interpreted as a way for designing a full solution of the equations of motion. But the dynamical information is not just in the physical phase space: it is in the relation between partial observables that each orbit determines.

In the quantum theory, the strict functional dependence between partial observables determined by the classical dynamics is replaced by transition amplitudes and transition probabilities. Thus, in a general covariant theory, physical observations are given by the transition amplitudes between eigenstates of partial observables. Formally, these are given by the matrix elements of the projector on the solution of the Wheeler–DeWitt operator between these states, which are defined on the same extended Hilbert space on which the Wheeler–DeWitt operator is defined.

15.3.2 Time in Quantum Cosmology

In cosmology it is often convenient to choose one of the variables and treat it as a "clock variable", that is, an independent variable with respect to which we study the evolution of the others. The choice is dictated by convenience, and has no fundamental significance whatsoever. For instance, there is no particular reason for choosing an independent variable that evolves monotonically along the orbits, or that defines a unitary evolution in the quantum theory.

In the case of a homogeneous and isotropic cosmology, where only one degree of freedom is considered, at least a second one is needed, to be used as a clock. A common choice is a massless scalar field. A more ingenious strategy can be implemented with enough degrees of freedom: for instance in the Bianchi I cosmology where the three spatial direction can evolve independently, one spatial direction can be taken to play the role of time. One should be careful to deal with regions where the chosen time fails to be monotonic. For instance the scale factor, that is implicitly taken as a clock in many models of quantum cosmology, is not monotonic if there is a recollapse.

Consider the study of the quantum mechanics of the scale factor a plus a single other degree of freedom, say a scalar field ϕ representing the average matter energy density. The two variables ϕ and a are partial observables. Predictions are extracted from their relative evolution of realizing Einstein's relationalism (Ashtekar, 2007).

In loop quantum cosmology, in particular, the dynamics studied include effects of the fundamental *spacetime discreteness* revealed by loop quantum gravity, using the technique of loop quantization. Among the results of the theory are the generic resolution of curvature singularities and the indication of the existence of a bounce replacing the initial singularity: a classical contracting solution of Einstein's equations can be connected to an expanding one via quantum tunneling. The bounce is a consequence of the Heisenberg relations for gravity, in the same way in which for an atomic nucleus those prevent the electrons from falling in (Ashtekar *et al.*, 2006; Bojowald, 2001). In the easiest case of a FLRW universe with no curvature, the effective equations provide a simple modification of the Friedmann equation that is:

$$\left(\frac{\dot{a}}{a}\right)^2 = \frac{8\pi G}{3}\rho\left(1 - \frac{\rho}{\rho_c}\right), \tag{15.6}$$

where ρ_c is the critical density at which the bounce is expected and that can be computed to correspond roughly speaking to the Planck density. The effects of the bounce on standard cosmological observables, such as CMB fluctuations, have been lengthily studied, see for instance (Ashtekar and Barrau, 2015).

These results are of course tentative and wait for an empirical confirmation, but the standard difficulty regarding time evolution does not plague them.

15.3.3 Covariant Loop Quantum Gravity

If we move to the quantum description of a small number of degrees of freedom, as in the last section, to a full quantum theory of gravity, which is ultimately needed in quantum cosmology, some interesting structures appear.

The main point is that in the absence of time we have to modify the notion of "physical system" used in Section 15.2. The reason is that the notion of quantum system implies a permanence in time which loses meaning in the fully covariant theory: how do we identify the "same system" at different times in a covariant field theory?

The solution is to restrict to local processes. The amplitudes of quantum gravity can be associated to finite spacetime regions, and the states of quantum gravity to the boundary of these regions. In fact, in quantum gravity we may even identify the notion of a spacetime region with the notion of *process*, for which we can compute transition amplitudes. The associated transition amplitudes depend on the eigenstates of partial observables that we can identify with spatial regions bounding the spacetime region of the process. In particular, loop quantum gravity gives a mathematically precise definition of the state of space, the boundary observables, and the amplitude of the process, in this framework. The possibility of this "boundary" formalism (Oeckl, 2003) stems from a surprising convergence between general relativity and quantum theory, which we have implicitly pointed out above.

We can call "Einstein's relationalism" the fact that in general relativity localization of an event is relative to other events. We can call "quantum relationalism" the fact that quantum theory is about the manner a system affects another system. In Bohr's quantum theory, the attention was always between the quantum system and the classical world, but we have seen that relational quantum theory allows us to democratize this split and describe the influence of any system on any other. These two relationalisms, however, appear to talk to one another, because of the locality of all interactions (Vidotto, 2013).

Indeed, one of the main discoveries in modern physics is locality: interactions at distance of Newton's kind do not seem to be part of our world. In the standard model things interact only when they "touch": all interactions are local. But this means that objects in interactions should be in the same place: interaction requires localization and localization requires interaction. To be in interaction corresponds to being adjacent in spacetime and vice versa: the two reduce to one another.

Quantum relationalism	$\longleftrightarrow$	Einstein's relationalism
Systems interact with other systems	$\longleftrightarrow$	Systems are located with regard to other systems
Interaction = Localization	$\longleftrightarrow$	Localization = Interaction

Bringing the two perspectives together, we get to the boundary formulation of quantum gravity: the theory describes processes and their interactions. The manner a process affects another is described by the Hilbert state associated to its boundary. The probabilities of one or another outcome are given by the transition amplitudes associated to the bulk, and

Francesca Vidotto

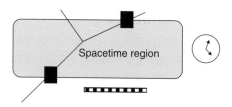

Figure 15.2 Boundary values of the gravitational field = geometry of box surface = distance and time separation of measurements.

are obtained from the matrix elements of the projector on the solutions of the Wheeler–De Witt equation.

Let us make this more concrete. Consider a process such as the scattering of some particles at CERN. If we want to take into account the gravitational field, we need to include it as part of the system. In doing quantum gravity, the gravitational field (or spacetime) is part of the system. Distance and time measurements are field measurements like the others in general relativity: they are part of the boundary data of the problem.

Thinking in terms of functional integrals, we have to sum over all possible histories, but also all possible geometries associated to a given finite spacetime region.

In the computation of a transition amplitude, we need to give the boundary data of the process that are for instance the position of a particle at an initial and a final time. We use rods and clocks to define them. But those measure geometrical information that is just the value of the gravitational field. Everything we have to give is the value of the fields on the boundary. This includes the gravitational fields from which we can say how much time has passed and the distance between the initial and the final point. Geometrical and temporal data are encoded in the boundary state (see Figure 15.2), because this is also the state of the gravitational field, which is the state of spacetime.

This clarifies that in quantum gravity a process is a spacetime region.

Now, we have seen that in relational quantum mechanics we need systems in interaction. What defines the system and when is it interacting? For spacetime, a process is simply a region of spacetime. Spacetime is a quantum mechanical process once we do quantum gravity. Vice versa, this now helps us to understand how to do quantum gravity.

Notice that from this perspective quantum gravitational processes are defined locally, without any need to invoke asymptotic regions. Summarizing:

Quantum dynamics of spacetime

Processes	$\rightarrow$	Spacetime regions
States	$\rightarrow$	Boundaries = spacial regions
Probability	$\rightarrow$	Transition amplitudes
Discreteness	$\rightarrow$	Quanta of space

15.3.4 Discreteness in Quantum Gravity

In Section 15.2.1, I have discussed how $\hbar$ gives us a unit of action in phase space, and a conversion factor between action and information. In gravity the phase space is the one of possible four-dimensional (4D) geometries, and there is the Newton constant G, which transforms regions of phase space in lengths. What kind of discreteness does this imply?

The answer is the well known Planck length, originally pointed out by Bronstein while debating a famous argument on field's measurability by Landau, in the case of the gravitational field (Rovelli and Vidotto, 2014).

The argument is simple: in order to check what happens in a small region of spacetime, we need a test particle. The smaller the region the more energetic the particle should be, until energy curves spacetime to form a black hole, whose horizon beyond which nothing can be seen is larger than the original region we wanted to prove. Because of this, it is not possible to probe scales smaller than the Planck length $\ell_{P\ell}$. This is the core of core quantum gravity: the discovery that there is a minimal length.

These handwaving semiclassical arguments can be made rigorous in the loop theory studying the phase space of general relativity and the corresponding operators. Geometrical quantities, such as area, volumes and angles, are functions of the gravitational field that are promoted to operators. Their discrete spectra describe a spacetime that is granular in the same sense in which the electric field is made of photons. For instance, the spectrum of the area can be computed and it results in being discrete (Rovelli and Smolin, 1995):

$$A = 8\pi \ell_{P\ell} \sqrt{j(j+1)}, \, j \in \frac{\mathbb{N}}{2} \tag{15.7}$$

where j is a half integer, similarly to what happens for the angular momentum. This has a minimal eigenvalue: a minimal value for the area.

Loop quantum gravity describes how these quanta of spacetime interact with one another. The notion of geometry emerges only from the semiclassical picture of these interactions. Formally, the theory is defined as follows. Every quantum field theory can be given in terms of a triple $(\mathcal{H}, \mathcal{A}, W)$: respectively a Hilbert space where the states live, an algebra of operators, and the dynamics defined in the covariant theory by a transition amplitude.

The interactions with the field manifest the discreteness of quantum mechanics: the fundamental discreteness appears in the presence of particles, that are just the quanta of a field, and in the spectrum of the energy of each mode of the field. The same structure applies to loop quantum gravity: states, operators and transition amplitudes can be properly defined (Rovelli and Vidotto, 2014) and there is a fundamental discreteness: the granularity of spacetime, yielded by the discreteness of the spectrum of geometrical operators. The geometry is quantized: eigenvalues are discrete and operators do not commute. Nodes carry discrete quanta of volume (*quanta of space*) and the links discrete quanta of area. Area and volume form a complete set of commuting observables and have discrete spectra.

States in loop quantum gravity are associated to graphs characterized by N nodes and L links. They can be thought as analogous to the N-particle states of standard quantum

(a) Spinnetwork   (b) Spinfoam

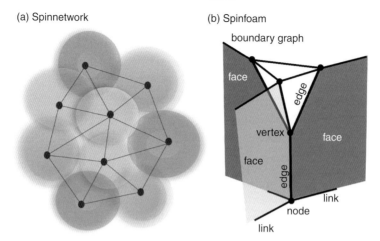

Figure 15.3 A spinnetwork provides a representation of the net of interactions between quanta of space building up the weave of space. A spinfoam represents the evolution of a spinnetwork. Links evolving create the faces of the foam. Vertices can create or annihilate nodes of the spinnetwork.

field theory, but with some further extra information given by the links, that turn out to be adjacency relations coding which "quanta of spacetime" are interacting with one another. These quanta *are* spacetime, they do not live in spacetime. The graphs are colored with quantum numbers, i.e. spins, forming the mathematical object called "spinnetwork" (Penrose, 1971). Penrose's "spin-geometry" theorem connects the graph Hilbert space with the description of the geometry of a cellular decomposition of spacetime (see Figure 15.3).

Notice that the full Hilbert space of the theory is formally defined in the limit of an infinite graph, but the physical theory is captured by a finite graph in the same way in which the Fock space is truncated to N particles. The truncation to a given finite graph captures the relevant degrees of freedom of the state we are interested to describe, disregarding those that need a "larger" graph to be defined.

A transition amplitude that represents a history of the geometry, in terms of graph states, becomes a history of the boundary graphs, or a "spinfoam". In a spinfoam quanta/nodes and links/relations get transformed into a new configuration by the action of interaction vertex in the bulk. A link spans a face through its history. This is the way of picturing a history of the quanta when these quanta make up spacetime themselves. This yields an ontological unification where all that exists are *covariant quantum fields* (Vidotto, 2014).

15.3.5 Spinfoam Cosmology

On a compact space we can expand the dynamical fields in discrete modes. The truncating of the theory to a finite number of modes defines an approximation to the full theory. This is neither a large scale approximation nor a short scale approximation, because the total space can still be very large or very small, as its scale is determined by the lowest modes of the gravitational field. Rather, the approximation is in the ratio between the largest and the smallest relevant wavelengths considered.

The graph expansion of the spin-network formulation of loop quantum gravity can be put in correspondence with this mode expansion of the fields on a compact space (Rovelli and Vidotto, 2008). A truncation on a fixed graph corresponds then to a truncation in the mode of expansion. The truncation provides a natural cut off of the infinite degrees of freedom of general relativity down to a finite number. Choosing a graph, we disregard the higher modes of this expansion. The truncation defines an approximation viable for gravitational phenomena where the ratio between the longest and the shortest wavelengths is bounded.

Since this is neither an ultraviolet nor an infrared truncation, what is lost are not wavelengths shorter than a given length, but rather wavelengths k times shorter than the full size of physical space, for some integer k.

This approximation is useful in cosmology. According to the cosmological principle, the dynamics of a homogeneous and isotropic space provides a good first order approximation to the dynamics of the real universe. Inhomogeneities can be disregarded in a first approximation. Notice that the approximation is not just a large scale approximation, because the universe may be small at some point of its evolution. Rather, the truncation is in the ratio between the scale of the inhomogeneities and the scale factor. At lowest order, we consider the dynamics of the whole universe as described solely by the scale factor; this ratio is a unit and a single degree of freedom is sufficient. We can then recover the rest of the theory adding degrees of freedom progressively. In the context of spinfoam cosmology, this can be obtained progressively refining the graph.

A graph with a single degree of freedom is just a single node: in a certain sense, this is the case of usual Loop Quantum Cosmology. To add degrees of freedom, we add nodes and links with a coloring (Borja *et al.*, 2012). These further degrees of freedom are a natural way to describe inhomogeneities and anisotropies.

Therefore, a single graph provides a useful calculation tool in cosmology. It is possible to generalize the spinfoam techniques for cosmology to large graphs. In a *regular* graph, which corresponds to a regular cellular decomposition, nodes and links become indistinguishable, and we obtain back the unique FLRW degrees of freedom (Vidotto, 2011). For an arbitrarily large regular graph, we can define coherent states and place them on a homogeneous and isotropic geometry, to represent macroscopic cosmological states.

Once we interpret the graph states as describing a cosmological evolution, we can compute cosmological transition amplitudes (Bianchi *et al.*, 2010, 2011; Vidotto, 2010). This transition amplitude makes concrete the notion of a sum over possible histories, namely all the possible 4-geometries compatible with the given three-dimensional (3D) states on the boundary.[4]

The advantage of this formalism is that it is fully Lorentzian, the amplitudes are infrared and ultraviolet fine, and they have a good classical behavior as they result in being peaked on solutions of classical general relativity. Since the theory is non-perturbative, we are allowed to use these equations in the deep quantum regime, where a perturbative calculation would exit its domain of validity. The hope is to obtain a full description of the quantum fluctuations at the bounce that replace the classical singularities.

Once again, the theory is tentative and may have technical difficulties, but there are no conceptual obstacles, if we adopt a fully relational perspective.

15.4 Conclusions

The world can be described in terms of facts. Facts happen at interactions, namely when a system affects another system, or, in a covariant theory, when a process affects another process. Relational quantum mechanics is the understanding of quantum theory in these terms. The resulting ontology is relational, characterizes quantum theory and is more subtle than that of classical mechanics.

Attributing ontological weight to the wave function is misleading. Quantum states are only the coding of past events that happened between two systems, or, in a generally covariant theory, between two processes. The way a system will affect another system in the future is probabilistically determined by the manner it has done so in the past, and physics is about the determination of such probabilistic relations. The quantum state codes the information relevant for this determination.

The amount of information is discrete in quantum theory. The minimal amount of information is determined by the Planck constant. The core of quantum theory is the discreteness of the information.

Cosmology is not about everything. It is about a few large scale degrees of freedom. It is based on an expansion in modes and neglects "short" wavelength modes, where "short" means millions of light years. Accordingly, in quantum cosmology the system does not include the observer. The observer is ourselves and our instruments, which are at a scale smaller than the cosmological scale.

In contrast, it is not clear whether a quantum theory of everything – a "totology" – makes sense, because quantum theory describes how a system affects another system. This question, however, has no bearing on standard quantum cosmology, which describes how the large scale degrees of freedom affect our instruments.

---

[4] In the lowest approximation, the classical theory expresses the dynamics as a relation between the scale factor and its momentum. Consequently, in quantum theory, at the first order in the vertex expansion the probability of measuring a certain "out" coherent state does not depend on the "in" coherent state. In other words, at the first order the probability is dominated by the product of the probabilities of each state to exist. Each term is given by a sum over all the possible 4D geometry compatible with the state representing a given 3D geometry. This is exactly the "spinfoam version" of the wave function of the universe (Hartle and Hawking, 1983).

Similarly, a preferred notion of time is not needed in quantum cosmology because the theory is about relations between partial observables, and dynamics is the study of these correlations.

Quantum gravity is the theory of the existence of a minimal length. Spacetime is (quantum) discrete. Since spacetime is dynamical, processes are spacetime regions. Neither space nor time are defined inside a process. Thanks to this, the application of quantum gravity to quantum cosmology cures the initial singularity and may lead to observable effects.

Using this relational understanding of quantum mechanics and of evolution, a coherent and consistent formulation of quantum cosmology is possible.

References

Ashtekar, A. 2007. An Introduction to Loop Quantum Gravity Through Cosmology. *Nuovo Cim.*, **122B**, 135–55.

Ashtekar, A. and Barrau, A. 2015. Loop quantum cosmology: From pre-inflationary dynamics to observations. *Classical and Quantum Gravity*, 32, 23, 234001.

Ashtekar, A., Pawlowski, T. and Singh, P. 2006. Quantum nature of the big bang. *Phys. Rev. Lett.*, **96**, 141301.

Bianchi, E., Rovelli, C. and Vidotto, F. 2010. Towards Spinfoam Cosmology. *Phys. Rev.*, **D82**, 084035.

Bianchi, E., Krajewski, T., Rovelli, C. and Vidotto, F. 2011. Cosmological constant in spinfoam cosmology. *Phys. Rev.*, **D83**, 104015.

Bojowald, M. 2001. Absence of singularity in loop quantum cosmology. *Phys. Rev. Lett.*, **86**, 5227–30.

Borja, E. F., Garay, I. and Vidotto, F. 2012. Learning about quantum gravity with a couple of nodes. *SIGMA*, **8**, 015.

Dorato, M. 2013. Rovelli's relational quantum mechanics, monism and quantum becoming. arXiv: 1309.0132.

Durr, D., Goldstein, S. and Zanghi, N. 1995. Bohmian mechanics and the meaning of the wave function. In *Foundations of Quantum Mechanics: A Symposium in Honor of Abner Shimony Boston, Massachusetts, 19–20 September, 1994*.

Grinbaum, A. (2003) Elements of information-theoretic derivation of the formalism of quantum theory. *Int. J. Quantum Inf.*, **1(3)** 289–300.

Hartle, J. B. and Hawking, S. W. 1983. Wave Function of the Universe. *Phys. Rev.*, **D28**, 2960–75.

Hoehn, P. A. 2014. Toolbox for reconstructing quantum theory from rules on information acquisition. arXiv: 1412.8323.

Oeckl, R. 2003. A 'general boundary' formulation for quantum mechanics and quantum gravity. *Phys. Lett.*, **B575**, 318–24.

Penrose, R. 1971. Angular momentum: An approach to combinatorial spacetime. In Bastin, T., ed. *Quantum Theory and Beyond*. Cambridge: Cambridge University Press.

Rovelli, C. 1991. Time in Quantum Gravity: Physics Beyond the Schrödinger Regime. *Phys. Rev.*, **D43**, 442–456.

Rovelli, Carlo. 1996. Relational quantum mechanics. *Int. J. Theor. Phys.*, **35**, 1637–78.

Rovelli, C. 2002. Partial observables. *Phys. Rev.*, **D65**, 124013.

Rovelli, C. 2013. Why Gauge? arXiv: 1308.5599 [hep-th].

Rovelli, C. and Smolin, L. 1995. Discreteness of area and volume in quantum gravity. *Nucl. Phys.*, **B442**, 593–622.

Rovelli, C. and Vidotto, F. 2008. Stepping out of Homogeneity in Loop Quantum Cosmology. *Class. Quant. Grav.*, **25**, 225024.

Rovelli, C. and Vidotto, F. 2014. *Covariant Loop Quantum Gravity*. Cambridge Monographs on Mathematical Physics. Cambridge: Cambridge University Press.

Vidotto, F. 2010. Spinfoam Cosmology: quantum cosmology from the full theory. *J. Phys.: Conf. Ser.*, **314**(012049).

Vidotto, F. 2011. Many-nodes/many-links spinfoam: the homogeneous and isotropic case. *Class. Quant Grav.*, **28**(245005).

Vidotto, F. 2013. Atomism and Relationalism as guiding principles for Quantum Gravity. Talk at the "Seminar on the Philosophical Foundations of Quantum Gravity", Chicago.

Vidotto, F. 2014. A relational ontology from General Relativity and Quantum Mechanics. Talk at the XIth International Ontology Congress, Barcelona, October.

16

Cosmological Ontology and Epistemology

DON N. PAGE

16.1 Introduction

In cosmology, we would like to explain our observations and predict future observations from theories of the entire universe. Such cosmological theories make ontological assumptions of what entities exist and what their properties and relationships are. One must also make epistemological assumptions or metatheories of how one can test cosmological theories. Here I shall propose a Bayesian analysis in which the likelihood of a complete theory is given by the normalized measure it assigns to the observation used to test the theory. In this context, a discussion is given of the trade-off between prior probabilities and likelihoods of the measure problem of cosmology, of the death of Born's rule, of the Boltzmann brain problem, of whether there is a better principle for prior probabilities than mathematical simplicity, and of an Optimal Argument for the Existence of God.

Cosmology is the study of the entire universe. Ideally in science, one would like the simplest possible theory from which one could logically deduce a complete description of the universe. Such theories must make implicit assumptions about the *ontology* of the universe, what are its existing entities and their nature. They may also make implicit assumptions about the ontology of the entire world (all that exists), particularly if other entities beyond the universe have relationships with the universe.

For example, I am a Christian who believes that the universe was created by an omnipotent, omniscient, omnibenevolent, personal God who exists outside it but who relates to it as His creation. Therefore, the ontology I assume for the world includes not only our universe but also God and other entities He may have created (such as new heavens and new earth for us after death). However, my assumption that God has created our universe as an entity essentially separate from Himself means that one can also look for a theory of the universe itself, without including its relationship to God, although there may be some aspects of such a theory that can only be explained by the assumption that the universe was created by God. (For example, even though a complete theory of our universe would necessarily imply the facts that there is life and consciousness within it, since such entities do exist within our universe, there may not be a good explanation of these aspects of the theory apart from the concept of creation by a personal God.)

Here I shall focus on theories of our universe as a separate entity, although near the end I shall speculate on some possible deeper explanations for why our universe is as it is.

I shall define a complete theory for a universe to be one that completely describes or specifies all properties of that universe. I shall assume that an observer who could observe all properties of a universe and who has unlimited reasoning power could deduce a unique complete theory for that universe. (Equivalent complete descriptions of the theory I regard as the same theory. Depending upon one's background knowledge and the effort needed to deduce consequences from logically equivalent assumptions, different equivalent complete descriptions may be assigned different levels of simplicity when one seeks the simplest complete description, but that involves subjective elements that I shall not get into here.)

Of course, we do not know for certain what is the unique complete theory of our universe. What we can know and how we can know it is the subject of *epistemology*.

Some of the complications of realistic epistemology arise from the limited reasoning powers of humans, such as the fact that we cannot deduce all the consequences of a set of axioms. (For example, the axioms of arithmetic and of complex numbers presumably imply whether the Riemann hypothesis is true or false, but so far humans have not been able to find a rigorous deduction of which is the case.) To avoid all the complications of limited reasoning power, for simplicity I shall just consider idealized cases in which observations are limited but reasoning powers are not.

Despite this idealization, we are also restricted by the fact that our observations are limited and do not show the entire universe. A limited observation does not imply a unique complete theory for a universe.

Therefore, beings like us with limited observations within the universe cannot hope to attain absolute certainty about a complete theory for the universe. The most we might hope for are posterior probabilities for different complete theories, probabilities taking into account our partial observations. But even the idealized observer cannot deduce the posterior probabilities of different theories fitting his or her partial observations without making subjective choices for the prior probabilities of these different theories.

Here I shall lump all of the information accessible to the idealized observer when he or she assigns posterior probabilities to different theories into a single observation O_k, which includes memory elements of what in ordinary language one might consider to be many previous observations. The observation O_k represents all that the idealized observer knows about the universe before coming up with various complete theories T_i to explain this observation and to which the idealized observer wishes to assign posterior probabilities $P(T_i|O_k)$ for the theory T_i that are conditional upon the observation O_k.

16.2 A Bayesian Analysis for the Probabilities of Theories

Consider theories T_i that for each possible observation O_k give a probability for that observation, $P(O_k|T_i)$. This of course is generically *not* the same as the probability $P(T_i|O_k)$ of the theory given the observation, which is a goal of science to calculate.

If we assign (subjective) prior probabilities $P(T_i)$ to the theories T_i (presumably higher for simpler theories, by Occam's razor) and use an observation O_k to test the theory, $P(O_k|T_i)$ is then the likelihood of the theory, and by Bayes' theorem we can calculate the posterior probability of the theory as

$$P(T_i|O_k) = \frac{P(T_i)P(O_k|T_i)}{\sum_j P(T_j)P(O_k|T_j)}. \tag{16.1}$$

We would like to get this posterior probability as high as possible by choosing a simple theory (high prior probability $P(T_i)$) that gives a good statistical fit to one's observation O_k (high likelihood $P(O_k|T_i)$).

Since a complete theory for a universe should completely specify all properties of that universe with certainty, one might think that it has no room for probabilities between 0 and 1, so that all likelihoods $P(O_k|T_i)$ would be 0 or 1. Then the posterior probabilities of these theories, given the observation O_k, would just be proportional to the prior probabilities of the theories T_i that give $P(O_k|T_i) = 1$.

For observations to play a bigger role in science, we would like a way of getting a range of likelihoods, $0 < P(O_k|T_i) < 1$. If the theories can give normalizable measures for the observations, then the normalized measures can be interpreted to be likelihoods $P(O_k|T_i)$ that can vary between 0 and 1.

In a classical theory, the measures could be functionals of the phase-space trajectory, such as the times that the system spends in different regions of the phase space. In a quantum theory, the measures could be certain functionals of the quantum state (maps from the quantum state to non-negative numbers).

Another argument against likelihoods all zero or one is the following: If a specific theory T_i leads to the definite existence of more than one observation O_j, one might be tempted to say that $P(O_j|T_i) = 1$ for all of these observations, giving unit likelihood for any theory T_i predicting the definite existence of the particular actual observation O_k used to test the theory. However, this procedure would mean that one could construct a theory of maximum (unit) likelihood, no matter what the observation turns out to be, just by having the theory predict that all possible observations definitely exist. But this seems to be too cheap, supposedly explaining everything but actually explaining nothing.

Therefore, I propose the normalization principle or metatheory that each theory T_i should give normalized probabilities for the different possible observations O_j that it predicts, so that for each T_i, the sum of $P(O_j|T_i)$ over all O_j is unity.

One could view theory construction as a contest, in which theorists are given only unit probabilistic resources to divide at will for each theory, and then others judge the theories by assigning prior probabilities to the theories based on their simplicity. Normalizing the products of these prior probabilities and of the probabilities the theories assign to the observation used to test the theory then gives the posterior probabilities of the theories.

For this to be a fair contest, each theory should be allowed a unit probability to distribute among all the observations predicted by the theory. Of course, if the theorists cheat and give unnormalized probabilities in their theories (such as assigning probabilities near one for more than one observation in the same theory), the judges can compensate for that by assigning correspondingly lower prior probabilities to such theories. However, it would give a more clear division between prior probabilities and likelihoods if theorists obeyed the proposed rules and normalized the observational probabilities for each theory so that they sum to one.

16.3 Trade-Off Between Prior Probabilities and Likelihoods

In a Bayesian analysis, we have (1) prior probabilities $P(T_i)$ for theories (intrinsic plausibilities), (2) conditional probabilities $P(O_k|T_i)$ for observations ('likelihoods' of the theories for a fixed observation), and (3) posterior probabilities $P(T_i|O_k) \propto P(T_i)P(O_k|T_i)$ for theories T_i, conditionalized upon the observation O_k.

Prior probabilities are subjective, usually higher for simpler theories. The highest prior probability might be for the theory T_1 that nothing concrete (contingent) exists, but then $P(O_k|T_1) = 0$. T_2 might be the theory that all observations exist equally: $P(O_k|T_2) = 1/\infty = 0$ when normalized (e.g. modal realism).

At the other extreme would be a maximal-likelihood theory giving $P(O_k|T_i) = 1$ for one's observation O_k (and zero probability for all other observations), but this seems to require a very complex theory T_i that might be assigned an extremely tiny prior probability $P(T_i)$, hence giving a very low posterior $P(T_i|O_k)$.

Thus there is a trade off between prior probabilities and likelihoods, that is, between intrinsic plausibility and fit to observations.

16.4 The Measure Problem of Cosmology

The *measure problem* of cosmology (see, e.g. Linde and Mezhlumian [1], Vilenkin [2], Freivogel [3] and Page [4]) is how to obtain probabilities of observations from the quantum state of the universe. This is particularly a problem when eternal inflation leads to a universe of unbounded size so that there are apparently infinitely many realizations or occurrences of observations of each of many different kinds or types, making the ratios ambiguous. There is also the danger of domination by Boltzmann Brains, observers produced by thermal and/or vacuum fluctuations [5–8].

The *measure problem* is related to the *measurement problem* of quantum theory, how to relate quantum reality to our observations that appear to be much more classical. An approach I shall take is to assume that observations are fundamentally conscious perceptions or sentient experiences (each perception being all that one is consciously aware of at once).

I shall take an Everettian view that the wave function never collapses, so in the Heisenberg picture, there is one single fixed quantum state for the universe (which could be a "multiverse"). I assume that instead of "many worlds", there are instead many different actually existing observations (sentient experiences) O_k, but that they have different positive measures, $\mu_k = \mu(O_k)$, which are in some sense how much the various observations occur, but they are *not* determined by just the contents of the observations.

For simplicity, I shall assume that there is a countable discrete set $\{O_k\}$ of observations, and that the total measure $\sum_k \mu_k$ is normalized to unity for each possible complete theory T_i that gives the normalized measures μ_k for all possible observations O_k. Then for a Bayesian analysis, I shall interpret the normalized measures μ_k of the observation O_k that each theory T_i gives as the probability of that observation given the theory, $P(O_k|T_i)$, which for one's observation O_k is the "likelihood" of the theory T_i.

16.5 Sensible Quantum Mechanics or Mindless Sensationalism

The map from the quantum state to the measures of observations could be non-linear. However, I assume a linear relationship in Sensible Quantum Mechanics [9–15] (which I have also called mindless sensationalism because it proposes that what is fundamental is not minds but conscious perceptions, which crudely might be called "sensations", although they include more of what one is consciously aware of, e.g. memories, than what is usually called "sensations"):

$$\mu(O_k) = \sigma[A(O_k)] \equiv \text{expectation value of the operator } A(O_k). \qquad (16.2)$$

Here σ is the quantum state of the universe (a positive linear functional of quantum operators), and $A(O_k)$ is a non-negative "awareness operator" corresponding to the observation or sentient experience (conscious perception) O_k. The quantum state σ (which could be a pure state, a mixed state given by a density matrix, or a C*-algebra state) and the awareness operators $\{A(O_k)\}$ (along with the linear relationship above and a description of the contents of each O_k) are all given by the theory T_i.

16.6 The Death of the Born Rule

Traditional quantum theory uses the Born rule with the probability of the observation O_k being the expectation value of $A(O_k) = \mathbf{P}_k$ that is a projection operator ($\mathbf{P}_j\mathbf{P}_k = \delta_{jk}\mathbf{P}_k$, no sum over k) corresponding to the observation O_k, so

$$P(O_k|T_i) = \sigma_i[\mathbf{P}_k] = \langle \mathbf{P}_k \rangle_i. \qquad (16.3)$$

Born's rule works when one knows where the observer is within the quantum state (e.g. in the quantum state of a single laboratory rather than of the universe), so that one has definite orthonormal projection operators. However, Born's rule does not work in a universe large enough that there may be identical copies of the observer at different locations, since then the observer does not know uniquely the location or what the projection operators are [16–19].

Why does the Born rule die? Suppose there are two identical copies of the observer, at locations B and C, that can each make the observations O_1 and O_2 (which do not reveal the location). Born's rule would give the probabilities $P_1^B = \sigma[\mathbf{P}_1^B]$ and $P_2^B = \sigma[\mathbf{P}_2^B]$ if the observer knew that it was at location B with the projection operators there being $\mathbf{P}_1^B$ and $\mathbf{P}_2^B$. Similarly, it would give the probabilities $P_1^C = \sigma[\mathbf{P}_1^C]$ and $P_2^C = \sigma[\mathbf{P}_2^C]$ if the observer knew that it was at location C with the projection operators there being $\mathbf{P}_1^C$ and $\mathbf{P}_2^C$.

However, if the observer is not certain to be at either B or C, and if $P_1^B < P_1^C$, then one should have $P_1^B < P_1 < P_1^C$. But there is no state-independent projection operator that gives an expectation value with this property for all possible quantum states. No matter what the orthonormal projection operators $\mathbf{P}_1$ and $\mathbf{P}_2$ are, there is an open set of states that gives expectation values that are not positively weighted means of the observational probabilities at the two locations. Thus the Born rule fails in cosmology.

In more detail, consider normalized pure quantum states of the form

$$|\psi\rangle = b_{12}|12\rangle + b_{21}|21\rangle, \tag{16.4}$$

with arbitrary normalized complex amplitudes b_{12} and b_{21}. The component $|12\rangle$ represents the observation 1 in the region B and the observation 2 in the region C; the component $|21\rangle$ represents the observation 2 in the region B and 1 in the region C. Therefore, $P_1^B = P_2^C = |b_{12}|^2$, and $P_2^B = P_1^C = |b_{21}|^2$.

For Born's rule to give the possibility of both observational probabilities being non-zero in the two-dimensional quantum state space being considered, the orthonormal projection operators must each be of rank one, of the form

$$\mathbf{P}_1 = |\psi_1\rangle\langle\psi_1|, \ \mathbf{P}_2 = |\psi_2\rangle\langle\psi_2|, \tag{16.5}$$

where $|\psi_1\rangle$ and $|\psi_2\rangle$ are two orthonormal pure states.

Once the state-independent projection operators are fixed, then if the quantum state is $|\psi\rangle = |\psi_1\rangle$, the expectation values of the two projection operators are $\langle\mathbf{P}_1\rangle \equiv \langle\psi|\mathbf{P}_1|\psi\rangle = \langle\psi_1|\psi_1\rangle\langle\psi_1|\psi_1\rangle = 1$ and $\langle\mathbf{P}_2\rangle \equiv \langle\psi|\mathbf{P}_2|\psi\rangle = \langle\psi_1|\psi_2\rangle\langle\psi_2|\psi_1\rangle = 0$. These extreme values of 1 and 0 are not positively weighted means of P_1^B and P_1^C and of P_2^B and P_2^C for any choice of $|\psi_1\rangle$ and $|\psi_2\rangle$ and any normalized choice of positive weights. Therefore, no matter what the orthonormal projection operators $\mathbf{P}_1$ and $\mathbf{P}_2$ are, there is at least one quantum state (and actually an open set of states) that gives expectation values that are not positively weighted means of the observational probabilities at the two locations. Thus Born's rule fails.

The failure of the Born rule means that in a theory T_i, the awareness operators $A_i(O_k)$, whose expectation values in the quantum state σ_i of the universe give the probabilities or normalized measures for the observations or sentient experiences O_k as $P(O_k|T_i) \equiv \mu_i(O_k) = \sigma_i[A_i(O_k)] \equiv \langle A_i(O_k)\rangle_i$, cannot be projection operators. However, the awareness operators could be weighted sums or integrals over spacetime of localized projection operators $\mathbf{P}_i(O_k, x)$ at locations denoted schematically by x, say onto brain states there that would produce the observations or sentient experiences.

16.7 The Boltzmann Brain Problem

In local quantum field theory with a definite globally hyperbolic spacetime, any positive localized operator (such as a localized projection operator) will have a strictly positive expectation value in any non-pathological quantum state. Therefore, if such a positive localized operator is integrated with uniform weight over a spacetime with infinite four-volume, it will give an awareness operator with an infinite expectation value.

If one takes the integral only up to some finite cutoff time t_c and normalizes the resulting awareness operators, then for a universe that continues forever with a three-volume bounded below by a positive value, the integrals will be dominated by times of the same order of magnitude as the cutoff time. If at late times the probability per four-volume drops very low for ordinary observers, then most of the measure for observations will be contributed by thermal or vacuum fluctuations, so-called Boltzmann brains. That is, Boltzmann brains will dominate the measure for observations.

If Boltzmann brains dominate the measure for observations, one might ask, "So what?" Could it not it be that our observations are those of ordinary observers? Or could it not it be that our observations really are those of Boltzmann brains? However, since Boltzmann brain observations are produced mainly by thermal or vacuum fluctuations, it would be expected that only a very tiny fraction of their measure would be for observations so ordered as our observations. This very tiny fraction, plus the even smaller fraction of ordered ordinary observer observations in comparison with the dominant disordered Boltzmann brain observations, would be only a very tiny fraction of the measure of all observations. Thus the normalized probability of one of our ordered observations (which we would use as the likelihood of the theory) would be highly diluted and hence much smaller than those of alternative theories in which Boltzmann brains do not dominate. If these alternative theories do not have prior probabilities that are too small, they would dominate the posterior probabilities.

In summary, Boltzmann brain domination, which is predicted by many simple extensions of current theories (e.g. with the awareness operators or their equivalent being obtained by a uniform integration over spacetime up to a cutoff that is then taken to infinity), gives a *reductio ad absurdum* for such theories, making their likelihoods very small. Surely there are alternative theories that avoid Boltzmann brain domination without such a cost of complexity that their prior probabilities would be decreased so much as the gain in likelihoods from not having the normalized probabilities of our ordered observations highly diluted by disordered Boltzmann brain observations.

The Boltzmann brain problem is analogous to the ultraviolet catastrophe of late nineteenth century classical physics: Physicists then did not believe that an ideal black body in thermal equilibrium would really emit infinite power, and physicists now do not believe that Boltzmann brains really dominate observations.

16.8 Volume Weighting Versus Volume Averaging

The approach that gives "awareness operators" as uniform integrals over spacetime of localized projection operators (or equivalently counts all observation occurrences equally, no matter when and where they occur in a spacetime) gives an especially severe Boltzmann brain problem in spacetimes with a positive cosmological constant (as ours seems to have) with the spatial hypersurfaces having three-volumes that asymptotically grow exponentially, as in the $k = 1$ slicing of the de Sitter spacetime. At each time, counting the number or measure of observations as growing with the volume is called "volume weighting".

In 2008 I proposed the alternative of volume averaging [20, 21], which gives a contribution to the measure for an observation from a hypersurface that is proportional to the *spatial density* of the occurrences of the observation on the hypersurface. This rewards the spatial frequency of observation occurrences rather than the total number that would diverge in eternal inflation as the hypersurface volume is taken to infinity.

Volume averaging ameliorates the Boltzmann brain problem by not giving more weight to individual spatial hypersurfaces at very late times when Boltzmann brains might be expected to dominate. However, when one integrates over a sequence of hypersurface with a measure uniform in the element of proper time dt, one gets a divergence if the time t goes to infinity. One needs some suppression at late times to avoid this divergence.

In 2010 I proposed Agnesi Weighting [22], replacing dt by $dt/(1 + t^2)$ where t is measured in Planck units. This year I have also proposed new measures depending on the spacetime average density (SAD) of observation occurrences within a proper time t from a big bang or bounce [4]. When these measures are combined with volume averaging and a suitable quantum state such as my symmetric-bounce one [23], they appear to be statistically consistent with all observations and seem to give much higher likelihoods than current measures using the extreme view that the measure is just given by the quantum state.

16.9 Is there a Better Principle than Mathematical Simplicity?

Mathematical simplicity seems to be a reasonably good guide in science for choosing prior probabilities. However, once observations are taken into account with the likelihoods, the highest posterior probabilities do not seem to go to the mathematically absolutely simplest theories (such as the theory that nothing concrete exists, which has zero likelihood given the fact that we observe something concrete, something not logically necessary).

Is there another principle that works better for predicting or explaining the properties of the actual world?

A conjectured principle for explaining the properties of the world is the following [24]: *The actual world is the best possible world.*

By the best possible world, I mean the one with maximum goodness. I take the goodness that is maximized to be the pleasantness (measure of pleasure) of conscious or sentient experiences, which is what I am calling their intrinsic goodness. Conscious experiences of pleasure (happiness, joy, contentment, satisfaction, etc.) would have positive goodness. Conscious experiences of displeasure (unhappiness, pain, agony, discontentment, dissatisfaction, etc.) would have negative goodness.

Our universe seems to have enormous positive goodness, but it also seems to have enormous negative goodness. Our universe also seems to have a very high degree of mathematical elegance and beauty. Humans can partially appreciate this, so that helps increase goodness. But it would seem that intrinsic goodness consciously experienced by humans would be higher if disasters, disease and cruelty were eliminated, even at the cost of less mathematical elegance and beauty for the laws of physics.

16.10 The Optimal Argument for the Existence of God

If the mathematical elegance of the universe were appreciated by a sentient Being outside the universe, that might increase the goodness of the world to a maximum, despite the sufferings within it. Goodness might be maximized if the Being had all possible knowledge, leading to the hypothesis of omniscience for full appreciation. Maximum goodness might also suggest the hypothesis that the Being has omnipotence for actualizing the best possible world. If the Being actualizes a world of maximum goodness, one might postulate that the Being is a Creator and has omnibenevolence. Such an omniscient, omnipotent, omnibenevolent Creator would fit the usual criteria for God.

Thus the assumption that the world has maximum goodness might suggest the conclusion of the existence of God:

Without God, it would seem that the goodness of the universe could be increased by violations of the laws of physics whenever such violations would lead to more pleasure within the universe. However, with God, such violations might decrease God's pleasure so much that total goodness would be decreased. Perhaps the actual world does maximize total goodness, despite suffering that is a consequence of elegant laws.

God may grieve over unpleasant consequences of elegant laws of physics and might even directly experience all of them Himself (as symbolized by the terrible suffering He experienced in the Crucifixion), but there may be that inevitable trade off that God takes into account in maximizing total goodness. If God really does maximize total goodness, He is doing what is best.

Let me give a draft syllogism for one form of this Optimal Argument for the Existence of God:

Assumptions:

1. The world is described by the simple hypothesis that it is the best, maximizing the pleasure within it.
2. It is most plausible that either (a) our universe exists in isolation, or (b) our universe is created by God whose pleasure is affected by the universe and who has a nature determining what gives Him pleasure.
3. Our universe could have had more pleasure.
4. If God exists, it is possible that the total pleasure of the world (including both that within our universe and within God) is maximized subject to the constraint of His nature.

Conclusions:

5. If our universe exists in isolation, 3 implies that it could have had more pleasure and hence the world could have been better, contradicting 1.
6. Therefore, 1 and 3 imply that option (a) of 2 is false.
7. Then 2 and 4 imply that it is most plausible that God exists and created our universe.

Of course, the assumptions of this argument are highly speculative, so it certainly does not give a proof of the existence of God from universally accepted axioms. It is merely suggestive, hopefully motivating further investigations of other evidence, such as of historical records about the founders and key events in the development of the world religions.

Assumption 1 is motivated by Occam's razor but is modified slightly from the typical scientific form that one seeks theories that have the simplest mathematical formulation. That form seems to work well when one constrains theories by observations, but it does not seem to give a fundamental explanation as to why our universe appears to be described by relatively simple mathematics but not the simplest possible mathematics. In making the assumption of the best world, I assume that goodness is fundamentally given by the measure of pleasure (the pleasantness) of conscious experiences (with displeasure counting negatively).

Assumption 2 is highly debatable, since there are many other possibilities that people have considered. Unless I made an expansion to include other possibilities, my argument might seem worthwhile mainly to those who are trying to choose between these two options.

Assumption 3 does seem rather obviously in agreement with our observations, at least if one considers all alternative logical possibilities for our universe.

Assumption 4 could be true in various different ways. A traditional free-will defense of the problem of evil might assume that God gets sufficient pleasure from having persons in His universe with libertarian free will (e.g. so that they can love Him freely), so that such a world maximizes the total pleasure of God and of His creatures despite the sufferings (displeasures, which I am counting as negative pleasures) within the universe. I am personally sceptical that it is logically possible for God to *create* totally from nothing creatures with libertarian free will, so I would instead postulate that God gets sufficient pleasure from a universe almost always obeying highly orderly and elegant laws of nature that He generally uses in His creation, so that the total pleasure of all conscience experiences (those of both God and His creatures) is maximized despite the sufferings that both God and His creatures also experience. But one might also consider other alternative ways to flesh out how assumption 4 could be true.

Once one makes these assumptions (and perhaps others that are implicit in my argument), then it does seem to me that the plausible existence of God as the Creator of our universe follows. Of course, the existence of God does not depend on these assumptions; God could exist even if one or more of these assumptions are false, just as I personally believe in His existence while being highly sceptical of one or more of the assumptions of nearly all of the classical arguments for the existence of God. But I do think that the

Optimal Argument for the Existence of God gives a somewhat new slant on why it might be plausible to believe in God. Furthermore, although one can do a Bayesian analysis for theories of our universe itself without reference to God, postulating the existence of God might help explain certain features of our best theories of our universe, such as why they are enormously more simple than they might have been but yet do not seem nearly so simple as what is purely logically possible.

16.11 Conclusions

I propose that in a Bayesian analysis in which the probability of a particular observation is used as the likelihood of the theory, the sum of the probabilities of all observations should be unity.

It seems plausible that one can find cosmological theories that avoid Boltzmann brain domination and explain the high order of our observations, although the *measure problem* is not yet solved.

The best theories do not seem to be the absolute simplest, so presumably something other than simplicity is maximized, such as goodness. The best explanation for the actual world may be that it is the *best possible world*. This assumption, with suitable assumptions about God's nature, such as love of mathematical elegance and of sentient beings, leads to the *Optimal Argument for the Existence of God*.

The world may maximize goodness only by including a God whose appreciation of the elegant universe and sentient beings He created overbalances the sufferings within the universe.

16.12 Acknowledgments

For my work on prior probabilities and likelihoods, on the measure problem of cosmology, on the death of Born's rule, and on the Boltzmann brain problem, I have benefited from discussions with many colleagues, including Andy Albrecht, Tom Banks, Raphael Bousso, Adam Brown, Steven Carlip, Sean Carroll, Brandon Carter, Willy Fischler, Ben Freivogel, Gary Gibbons, Steve Giddings, Daniel Harlow, Jim Hartle, Stephen Hawking, Simeon Hellerman, Thomas Hertog, Gary Horowitz, Ted Jacobson, Shamit Kachru, Matt Kleban, Stefan Leichenauer, Juan Maldacena, Don Marolf, Yasunori Nomura, Joe Polchinski, Steve Shenker, Eva Silverstein, Mark Srednicki, Rafael Sorkin, Douglas Stanford, Andy Strominger, Lenny Susskind, Bill Unruh, Erik Verlinde, Herman Verlinde, Aron Wall, Nick Warner and Edward Witten. Face-to-face conversations with Gary Gibbons, Jim Hartle, Stephen Hawking, Thomas Hertog and others were enabled by the gracious hospitality of the Mitchell family and Texas A & M University at a workshop at Great Brampton House, Herefordshire, England. My scientific research has been supported in part by the Natural Sciences and Engineering Research Council of Canada.

For my work on the Optimal Argument for the Existence of God and related philosophical and theological ideas, I am indebted to conversations with David Albert, Michael

Almeida, Luke Barnes, John Barrow, Nicholas Beale, Andrew Briggs, Peter Bussey, Bernard Carr, Sean Carroll, Brandon Carter, Khalil Chamcham, Robin Collins, Gary Colwell, William Lane Craig, Paul Davies, Stanley Deser, George Ellis, Peter Getzels, Shelly Goldstein, Stephen Hawking, Jim Holt, Colin Humphreys, John Leslie, Barry Loewer, Klaas Kraay, Robert Lawrence Kuhn, Robert Mann, David Marshall, Thomas Nagel, Elliot Nelson, Timothy O'Conner, Cathy Page, Jason Pollack, Joel Primack, Carlo Rovelli, Simon Saunders, Michael Schrynemakers, Richard Swinburne, Donald Turner, Aron Wall, Christopher Weaver, David Wilkinson, William Wootters, Dean Zimmerman, Henrik Zinkernagel and Anna Zytkow. This does not by any means imply that all or any of these people agree with my ideas; I have perhaps benefited most from those who have expressed sharp disagreement.

References

[1] A. D. Linde and A. Mezhlumian, Stationary Universe. *Phys. Lett. B*, **307**, 25 (1993) [gr-qc/9304015].

[2] A. Vilenkin, Predictions from Quantum Cosmology. *Phys. Rev. Lett.* **74**, 846 (1995) [gr-qc/9406010].

[3] B. Freivogel, Making Predictions in the Multiverse. *Class. Quant. Grav.* **28**, 204007 (2011) [arXiv:1105.0244 [hep-th]].

[4] D. N. Page, Spacetime Average Density (SAD) Cosmological Measures. *JCAP* **1411**, 038 (2014), arXiv:1406.0504 [hep-th].

[5] L. Dyson, M. Kleban and L. Susskind, Disturbing Implications of a Cosmological Constant. *JHEP* **0210**, 011 (2002) [hep-th/0208013].

[6] A. Albrecht, Cosmic Inflation and the Arrow of Time. In J. D. Barrow, P. C. W. Davies and C. L. Harper Jr, eds. *Science and Ultimate Reality: Quantum Theory, Cosmology, and Complexity.* (Cambridge: Cambridge University Press, 2004), pp. 363–401 [astro-ph/0210527].

[7] D. N. Page, Is Our Universe Likely to Decay within 20 Billion Years? *Phys. Rev. D*, **78**, 063535 (2008) [hep-th/0610079].

[8] D. N. Page, Is Our Universe Decaying at an Astronomical Rate? *Phys. Lett. B*, **669**, 197 (2008) [hep-th/0612137].

[9] D. N. Page, Sensible Quantum Mechanics: Are Only Perceptions Probabilistic? [quant-ph/9506010]. (1995) [arXiv:quant-ph/9506010].

[10] D. N. Page, Attaching Theories of Consciousness to Bohmian Quantum Mechanics. [quant-ph/9507006]. (1995) [arXiv:quant-ph/9507006].

[11] D. N. Page, Sensible Quantum Mechanics: Are Probabilities Only in the Mind? *Int. J. Mod. Phys. D*, **5**, 583 (1996) [gr-qc/9507024].

[12] D. N. Page, Quantum Cosmology Lectures. [gr-qc/9507028]. (1995) [arXiv:gr-qc/9507028].

[13] D. N. Page, Mindless Sensationalism: A Quantum Framework for Consciousness. In Q. Smith and A. Jokic, eds. *Consciousness: New Philosophical Perspectives*, (Oxford: Oxford University Press, 2003), pp. 468–506 [quant-ph/0108039].

[14] D. N. Page, Predictions and Tests of Multiverse Theories. In B. J. Carr, ed. *Universe or Multiverse?* (Cambridge: Cambridge University Press, 2007), pp. 411–29 [hep-th/0610101].

[15] D. N. Page, Consciousness and the Quantum. (2011) [arXiv:1102.5339 [quant-ph]].

[16] D. N. Page, Insufficiency of the Quantum State for Deducing Observational Probabilities. *Phys. Lett. B,* **678**, 41 (2009) [arXiv:0808.0722 [hep-th]].

[17] D. N. Page, The Born Rule Dies. *JCAP* **0907**, 008 (2009) [arXiv:0903.4888 [hep-th]].

[18] D. N. Page, Born Again. (2009) [arXiv:0907.4152 [hep-th]].

[19] D. N. Page, Born's Rule Is Insufficient in a Large Universe. (2010) [arXiv:1003.2419 [hep-th]].

[20] D. N. Page, Cosmological Measures without Volume Weighting. *JCAP,* **0810**, 025 (2008) [arXiv:0808.0351 [hep-th]].

[21] D. N. Page, Cosmological Measures with Volume Averaging. *Int. J. Mod. Phys. Conf. Ser.,* **01**, 80 (2011).

[22] D. N. Page, Agnesi Weighting for the Measure Problem of Cosmology. *JCAP,* **1103**, 031 (2011) [arXiv:1011.4932 [hep-th]].

[23] D. N. Page, Symmetric-Bounce Quantum State of the Universe. *JCAP,* **0909**, 026 (2009) [arXiv:0907.1893 [hep-th]].

[24] D. N. Page, The Everett Multiverse and God. In K. Kraay, ed. *God and the Multiverse: Scientific, Philosophical, and Theological Perspectives,* Routledge Studies in the Philosophy of Religion (Book 10). (London: Routledge, 2014), arXiv:1212.5608 [physics.gen-ph].

17

Quantum Origin of Cosmological Structure and Dynamical Reduction Theories

17.1 Introduction

Contemporary cosmology includes inflation as one of its central components. It corresponds to a period of accelerated expansion thought to have occurred very early in cosmic history, which takes the universe from relatively generic post Planckian era conditions to a stage where it is well described (with exponential accuracy in the number of e-folds) by a flat Robertson–Walker space-time (which describes a homogeneous and isotropic cosmology).

It was initially proposed to resolve various naturalness problems: flatness, horizons, and the excess of massive relic objects, such as topological defects, that were expected to populate the universe according to grand unified theories (GUT). Nowadays, it is lauded because it predicts that the universe's mean density should be essentially identical to the critical value, a very peculiar situation which corresponds to a spatially flat universe. The existing data support this prediction. However, its biggest success is claimed to be the natural account for the emergence of the seeds of cosmic structure in terms of primordial quantum fluctuations, and the correct estimate of the corresponding microwave spectrum.

This represents, thus, a situation which requires the combined application of general relativity (GR) and quantum theory, and that, moreover, is tied with observable imprints left from the early universe, in both the radiation in cosmic microwave background (CMB), and in the large scale cosmological structure we can observe today. It should not be surprising that, when dealing with attempts to apply our theories to the universe as a whole, we should come face to face with some of the most profound conceptual problems facing our current physical ideas.

In fact, J. Hartle had noted long ago [28] that serious difficulties must be faced when attempting to apply quantum theory to cosmology, although the situations that were envisioned in those early works corresponded to the even more daunting problem of full quantum cosmology, and not the limited use of quantum theory to consider perturbative aspects, which is what is required in the treatment of the inflationary origin of the seeds of cosmic structure. This led him and his collaborators to consider the Consistent Histories framework, which is claimed to offer a version of quantum theory that does not rely on any fundamental notion of measurements, as the favored approach to use in such contexts. We will have a bit more to say about this shortly.

As we will see, the treatment of this question forces us to address the so-called measurement problem of quantum mechanics, in a particularly clean situation where, not only the issues appear in a rather simple form, but also where most of the distracting complications that have served to elude dealing with the fundamental problems or to convince ourselves that we have a satisfactory resolution of them are simply absent. This chapter is based on previous works with various colleagues devoted to addressing the many issues that arise when confronting these questions. Specially significant in the following, are several articles [4, 6, 51].

17.2 Observations

The observational data providing support to our current views on the issue of structure in cosmology are the large scale galaxy surveys and the very precise temperature anisotropies of the CMB. We will focus here on the second, as they are observed now on the celestial two-sphere, which, in turn, are related to the inhomogeneities in the Newtonian potential Ψ[1] on the last scattering surface (the region from where the photons that reach us now were emitted just before the time in the universe's evolution when the plasma of electrons and protons became hydrogen atoms and thus transparent to photons),

$$\frac{\delta T}{T_0}(\theta, \varphi) = \frac{1}{3}\Psi(\eta_D, \vec{x}_D). \tag{17.1}$$

where $\delta T(\theta, \varphi)$ is the difference between the mean temperature $T_0 = 2.725K$ of the CMB, and that observed at the coordinates (θ, φ) on the celestial sphere, η_D is the (conformal) time of decoupling (i.e. the transformation of the plasma into hydrogen atoms), and $\vec{x}_D$ are spatial coordinates of the point, along that celestial direction, on the last scattering surface.

The variations of the Newtonian potential over the last scattering surface, in turn, are indicative of the departures from homogeneity and isotropy in the density of matter at that moment in the evolution of the universe, and the regions with higher densities represent the seeds of structure that will grow as a result of gravitational accretion.

The data are described in terms of the coefficients α_{lm} of the multipolar series expansion of the sky temperature map:

$$\frac{\delta T}{T_0}(\theta, \varphi) = \sum_{lm} \alpha_{lm} Y_{lm}(\theta, \varphi), \tag{17.2}$$

---

[1] The Newtonian potential, is, as we shall see shortly, one of the metric perturbations describing the departure of the space-time geometry of the universe from a state of perfect homogeneity and isotropy. It receives this name because, in the limit case of a static universe, this perturbation ends up playing the role of the gravitational potential in the Newtonian theory of gravity to which general relativity reduces, in the limit of slow particles, weak gravity, and sources with pressures that are much smaller than their densities.

where the coefficients are then given by

$$\alpha_{lm} = \int \frac{\delta T}{T_0}(\theta, \varphi) Y_{lm}^*(\theta, \varphi) d\Omega. \tag{17.3}$$

Here $Y_{lm}(\theta, \varphi)$ are the spherical harmonics. The different multipole numbers l correspond to different angular scales; low l to large scales and high l to small scales.

It is customary to present these data using the orientation averaged quantities given by

$$C_l \equiv \frac{1}{2l+1} \sum_l |\alpha_{lm}|^2 \tag{17.4}$$

At the theoretical level, the quantities α_{lm} are then given by expanding the Newtonian potential in its Fourier components $\Psi_{\vec{k}}$, and using Eq. (17.1) to obtain the expression:

$$\alpha_{lm} = \frac{4\pi i^l}{3} \int \frac{d^3k}{(2\pi)^3} j_l(kR_D) Y_{lm}^*(\hat{k}) \Delta(k) \Psi_{\vec{k}}(\eta_R), \tag{17.5}$$

with $j_l(kR_D)$ the spherical Bessel function of order l; η_R is the conformal time at the end of inflation when the inflation field is supposed to have decayed into ordinary matter, called the time of reheating, and R_D the co-moving radius of the last-scattering surface.

Between η_R and η_D, the inhomogeneities present at the end of inflation evolved by well known mechanisms. The resulting change is encoded in the so-called transfer functions $\Delta(k)$.

The data that are obtained by the modern observations culminating with the Planck satellite project, and that correspond to the radiation emitted at time η_D, are represented in Figure 17.1 below.

The anisotropies in the temperature of the CMB represent the first observable traces of the primordial inhomogeneities present at the time of decoupling (the time of the formation of the first atoms), and represent the seeds which eventually would evolve into all the structure in our Universe: galaxies, stars, planets, etc.

17.3 Brief Review of the Standard Treatments

We now offer a very brief review of the methods and ideas used in the standard accounts of inflation and the generation of the seeds of cosmic structure from primordial quantum fluctuations. Much more detailed accounts abound in the literature, and the reader is directed to those works, if interested. A highly recommended recent one is Weinberg [56].

The starting point of the analysis is a Robertson–Walker (RW) space-time background characterized by the space-time metric

$$dS^2 = a(\eta)^2 \{-d\eta^2 + d\vec{x}^2\} \tag{17.6}$$

expressed in terms of the so-called co-moving spatial coordinates $\vec{x}$ (where a fixed specific value of $\vec{x}$ is attached to each co-moving observer) and the conformal time is related to the proper time along the world lines of such observers by $dt = a(\eta)d\eta$. The scale factor a increases characterizing the universe's expansion, and initially does so at an accelerated rate, a situation usually described by saying it represents an inflating universe. This accelerated growth is the result of the gravitational dynamics under the influence of the potential associated with a scalar field which acts, for a while, like a cosmological constant. This scalar field is called the inflaton, and it is characterized by a classical background field $\phi = \phi_0(\eta)$, in a so-called slow roll condition so that the scale factor behaves approximately as $a(\eta) = \frac{-1}{\eta H_I}$, (during this period the "conformal time" takes negative values, i.e. the inflationary regime corresponds to $\eta \in (-\mathcal{T}, \eta_0), \eta_0 < 0$ where $-\mathcal{T}$ is the time when inflation began, and η_0, very close to 0, is the time inflation ends and reheating occurs).

On top of this background, one considers quantum fluctuations: that is, fluctuations $\delta\phi$ of the inflaton field, and metric fluctuations such as the so-called Newtonian potential Ψ, and the tensor or gravity wave terms δh_{ij} (representing the fluctuation of the spatial part of the metric tensor). These are treated quantum mechanically and assumed to be characterized by the "vacuum state" $|0\rangle$ (essentially the so-called Bunch–Davies vacuum)[2]. From these "fluctuations", one argues, the primordial inhomogeneities and anisotropies emerge.

The analysis leads to a remarkable agreement with observations. The actual analysis consists of the evaluation of the two-point correlation function of the inflaton field [3] in the the Bunch–Davies vacuum state $\langle 0|\delta\phi(x)\delta\phi(y)|0\rangle$ at the end of inflation, and the extraction of the power spectrum by writing

$$\langle 0|\delta\phi(x)\delta\phi(y)|0\rangle = \int d^3k \mathcal{P}(k)e^{i\vec{k}\cdot(\vec{x}-\vec{y})} \tag{17.7}$$

The result is an essentially flat spectrum $\mathcal{P}(k) = ak^{-3}$, which is then modified by the plasma-physics effects. We note that the characteristic oscillations we see in Figure 17.1 are the result of these late time plasma physics effects in the matter which originated from reheating, which are encoded in the transfer functions $\Delta(k)$, on top of the primordial flat spectrum. The physics of these acoustic oscillations is well understood, and will be mostly ignored in the rest of this chapter.

The remarkable fact is that the theory fits extremely well with the observations, a situation that can lead to an understandable and yet dangerous kind of complacency. "The theory works, that is it. What else do we want?" However, let us consider the following: according to this account, the Universe was homogeneous and isotropic, (both in the part that could be described at the "classical level", as well as that in quantum level) as

---

[2] This state is the one that would correspond to the standard Minkowski vacuum in the limit of vary large $\mathcal{T}$ in which the expansion rate would become negligible.

[3] More precisely of a linear combination of $\delta\phi$ and $\delta\Psi$, known as the Sasaki–Muchanov variable.

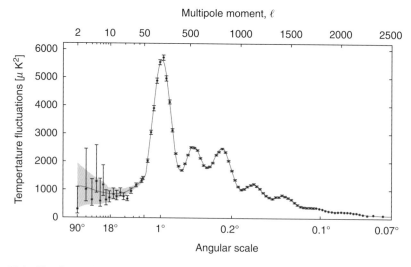

Figure 17.1   The figure shows the observed values of the C_l's together with the theoretical expectation obtained using the best fit of a handful of adjustable cosmological parameters. (Image: Courtesy of ESA/ Planck Collaboration).

a result of the early stages of inflationary process.[4] However, we end up with a situation which is not homogeneous and isotropic, as it includes the primordial inhomogeneities which will result in our Universe's structure and the conditions that permit our own existence.

How does this happen if the dynamics of the closed system does not break those symmetries? If we want to claim we have an explanation for the "origin" of cosmic structure we need to understand the corresponding transition, and answer the above question in a fully satisfactory way. If our currently accepted theories really do the job, then fine. However, if they do not, we can take this as a starting point for further inquiry.

In fact we can find this issue identified as an open problem in the recent book [56], which, on page 476 states: *"... the field configurations must become locked into one of an ensemble of classical configurations with ensemble averages given by quantum expectation values.... It is not apparent just how this happens ... "*.

A similar issue was considered by N. F. Mott in 1929 concerning the α decay of a radioactive nucleus [34]. He seemed to have thought that he resolved it and various researchers now seem to share that view. The issue is, what did he actually solve? He considered, within the context of non-relativistic quantum mechanics, the problem of a $J = 0$ nucleus placed at $(\vec{X} = \vec{0})$ initially in an excited state $\left|\Psi^+\right\rangle$ which is spherically symmetric, and ready to decay into an unexcited nucleus $\left|\Psi^0\right\rangle$, plus an α particle in state $\left|\Xi_\alpha\right\rangle$, which

---

[4] This is in fact true except from possible remnant features from the pre-inflationary regime which must be suppressed by e^{-N}, with N the number of e-folds of inflation generally expected to be at least of order $N \sim 70$, and which will be ignored from now on.

is also spherically symmetric. Next, he included into the problem two hydrogen atoms with their nuclei fixed at positions $\vec{a}_1$ and $\vec{a}_2$, and their corresponding electrons in the corresponding ground states. The analysis proceeded by computing the degree of alignment of the location of the atoms and the initial nucleus (i.e. $\vec{a}_2 = c\vec{a}_1$, with c real), if both atoms are to be excited by the outgoing α particle. The result is that the probability of both atoms getting excited is negligible unless there is a large degree of alignment, a result that according to Mott fully explains the fact that the α leaves straight line marks in a bubble chamber.

Is this therefore an example of a situation in which an initial state possessing spherical symmetry $|\Psi^+\rangle$ evolves into a final state lacking such symmetry, despite the assumption that the Hamiltonian (governing the decay $|\Psi^+\rangle \to |\Psi^0\rangle|\Xi_\alpha\rangle$ and the α particle evolution) is symmetric under rotations?

The answer of course is no. As we noted, the initial setting of the problem involves, besides the excited nucleus, two unexcited atoms, with localized nuclei, which break the rotational symmetry. In fact, the analysis that was carried out by Mott is not based on just the Hamiltonian for the evolution of the free α particle, but rather on the Hamiltonian for the joint evolution (including the interaction) of the α particle and the two electrons corresponding to the two localized hydrogen atoms. In fact, the projection postulate associated with a measurement is also playing a fundamental role in the evaluation of the probabilities, through the projection of the evolved state on the subspace corresponding to the two atoms being excited. In essence, the atoms, therefore, are taken to act as symmetry breaking detectors in the problem. Thus, if one replaces these atoms by some hypothetical detectors whose quantum description corresponded to spherically symmetric wave functions, each one with support, say, on a thin spherical shell with radius r_i, and then performs the analogous analysis one would be led to expect a spherical pattern of excitations rather than the symmetry breaking straight lines. That is, we would obtain a certain probability for finding the two detectors corresponding to the shells i^{th} and j^{th} in an excited state, and one could not have obtained any result that indicated a breakdown of the rotational symmetry.

The problem we face in the cosmological inflationary situation is thus clearly related to the "measurement problem" but in an aggravated form, because detectors, and beings capable of creating them, can only exist after the development of the inhomogeneities that generate galaxies, stars, and heavy chemical elements. Thus one cannot even rely on detectors or observers playing their standard role in the Copenhagen interpretation.

17.4 A Simplified Model: Mini-Mott

In order to help us consider conceptual issues, and the possible resolutions in a very clear and transparent setting, we find it useful to use a problem where the technicalities are all but absent. One should consider this example when contemplating any proposed solution to the question of the breaking of the quantum symmetries that one wishes to apply to the specific cosmological situation under consideration. In particular, one should use it as a test ground when considering the various proposals described in the section "The most common answers" below.

Consider thus a two level detector $|-\rangle$ (ground) and $|+\rangle$ (excited), and take two of them located at $x = x_1$ and $x = -x_1$. They are both initially in the ground state. Take a free particle with initial wave function $\psi(x,0)$ given by a simple Gaussian centered at $x = 0$ (so the whole set up is symmetric w.r.t $x \to -x$).

The particle's Hamiltonian: $\hat{H}_P = \hat{p}^2/2M$ while that of each detector is

$$\hat{H}_i = \epsilon \hat{I}_p \otimes \{|+\rangle^{(i)}\langle+|^{(i)} - |-\rangle^{(i)}\langle-|^{(i)}\}. \tag{17.8}$$

where $i = 1, 2$. The interaction of particle and detector 1 is

$$\hat{H}_{P1} = \frac{g}{\sqrt{2}}\delta(x - x_1\hat{I}_p) \otimes (|+\rangle^{(1)}\langle-|^{(1)} + |-\rangle^{(1)}\langle+|^{(1)}) \otimes I_2 \tag{17.9}$$

with a similar expression for the Hamiltonian term describing the particle's interaction with detector 2.

Next we consider the evolution of the system as given by Schrödinger's equation for the fully symmetric initial state:

$$\Psi(0) = \sum_x \psi(x,0)|x\rangle \otimes |-\rangle^{(1)} \otimes |-\rangle^{(2)} \tag{17.10}$$

and it is clear that after some time t we will have:

$$\Psi(t) = \sum_x \psi_1(x,t)|x\rangle \otimes |+\rangle^{(1)} \otimes |-\rangle^{(2)} + \sum_x \psi_2(x,t)|x\rangle \otimes |-\rangle^{(1)} \otimes |+\rangle^{(2)} \tag{17.11}$$

$$+ \sum_x \psi_0(x,t)|x\rangle \otimes |-\rangle^{(1)} \otimes |-\rangle^{(2)} + \sum_x \psi_D(x,t)|x\rangle \otimes |+\rangle^{(1)} \otimes |+\rangle^{(2)} \tag{17.12}$$

One might now interpret the last two terms easily: no detection and double detection (involving a bounce, and characterized by a small number $O(g^2)$). Also, we could think the first two terms indicate that the initial symmetry was broken with high probability: either detector 1 was excited or detector 2 was. That is, one might think that by adopting a standard interpretation (Copenhagen?) the puzzle is solved. Let us reconsider this. The symmetry was just broken and we can not pinpoint where? The problem can be seen by considering instead the alternative state basis for the detectors (or as it is often called an alternative "context")

$$|U\rangle \equiv |+\rangle^{(1)} \otimes |+\rangle^{(2)} \tag{17.13}$$

$$|D\rangle \equiv |-\rangle^{(1)} \otimes |-\rangle^{(2)} \tag{17.14}$$

$$|S\rangle \equiv \frac{1}{\sqrt{2}}[|+\rangle^{(1)} \otimes |-\rangle^{(2)} + |-\rangle^{(1)} \otimes |+\rangle^{(2)}] \tag{17.15}$$

$$|A\rangle \equiv |\frac{1}{\sqrt{2}}[|+\rangle^{(1)} \otimes |-\rangle^{(2)} - |-\rangle^{(1)} \otimes |+\rangle^{(2)}] \tag{17.16}$$

In fact this basis is more convenient for describing issues related to symmetries of the problem. It is then easy to see that the $x \to -x$ and $1 \to 2$ symmetry of the initial setting and of the dynamics prevents the excitation of an asymmetric term. The issue is then: can we or can we not describe things on this basis? And, if not, why not?

An experimental physicist in the laboratory would perhaps say he/she sees no problem, he/she has many things that in practice (for all practical purposes (FAPP) as J. Bell would say) indicate one should use the other basis (i.e. one knows that the detectors are always either excited or un-excited. One never perceives them in any state of superposition). However, the measurement problem is present here but takes the following form: exactly how does *our theory* account for that *experience* of our experimental colleague? The fact is that most often we just do not care.

However, if we now have to consider, as in cosmology, a situation where there are no experimentalists, and nothing else in the universe, we simply do not know what to do. In that situation, why would we believe the conclusions drawn in the first context (or choice of basis) but not those of the second? In other words, how do we account for the breakdown of the symmetry?

17.5 The Most Common Answers

The question we are considering can be restated as: How is it that the primordial inhomogeneities that act as the seeds of cosmic structure emerge from the quantum fluctuations in the inflaton field, if the state of that field is supposed to be a homogeneous and isotropic vacuum state, and the field dynamics does not break that symmetry? Here we reproduce some of the most common answers one tends to obtain when presenting colleagues with these issues, followed by a very brief characterization of what we see as their major shortcomings.

1. As in all QM situations, take into acount that "we make a measurement", and that breaks the symmetry
Even ignoring all the standard issues that come with the measurement problem in Quantum Theory, we must be aware that taking this view amounts to saying that the conditions that made possible our own existence have to be considered as being, at least in part, the result of our own actions. This would represent a problematic situation of circular causation, to say the least.

2. Environment-induced decoherence, possibly supplemented by a many worlds interpretation.
This approach requires, as a starting point, the identification of a collection of degrees of freedom (DoF) of the systems that are taken to act as an "environment" (and then the reduced density matrix is obtained by retracing over those DoF). In the standard presentation, such identification entails using our limitations to "measure things", as part of the argument that selects what will be considered as the environment. That, again, involves using the peculiarities of the human condition as part of the argument explaining the

emergence of the conditions that made possible our own existence. One way of characterizing the problem is that we need what is often called a "third person description" of
the emergence of primordial inhomogeneities. I think most people would object to any
explanation about the extinction of the dinosaurs that depended, even partially, on the
argument that relatively large mammals could not possibly survive in a planet dominated
by such efficient predators. That is, we would not want to explain the emergence of the
conditions that made possible our own existence (that is the emergence of structure, or the
extinction of the dinosaurs) based, even in part, on our specific weakness (be it the observational limitations of humans here and now to observe certain aspects of the cosmos,
or the lack of size and strength, or perhaps speed, of mammals to successfully confront
dinosaurs).

The fact that a successful decoherence indicates that the density matrix becomes diagonal, *does not* tell us that the situation is now described by one element of the diagonal
density matrix, but it is still described by all of them, and as such the situation is still
symmetric. One needs something like the many worlds interpretation (MWI) to say that
somehow one is dealing with a multiplicity of realized alternatives and that each one
corresponds to one element of the diagonal in the reduced density matrix.

However, MWI (which seems to have other drawbacks) requires some criteria to determine the alternatives, or basis, that characterize the word's multiplicity. *What would play
that role (i.e. selecting the preferred basis) in the situation at hand, if our limitations cannot
be used as part of the explanation?*

Even Zurek acknowledges: "The interpretation based on the ideas of decoherence and
ein-selection has not really been spelled out to date in any detail. I have made a few half-
hearted attempts in this direction, but, frankly, I was hoping to postpone this task, since the
ultimate questions tend to involve such anthropic attributes of the 'observership' as perception, awareness, or consciousness, which, at present, cannot be modeled with a desirable
degree of rigor." [59]. However in a recent article [58], Zurek seems to argue in a different direction based on a proposal called "Quantum Darwinism", which has been severely
criticized, as shown, for instance in Kastner [31].

In fact, in the cases of symmetric situations, one faces an extra specific problem with
the issue of basis selection: Consider a standard EPR situation, corresponding say to the
entangled two-particle state that results from the decay of a $J = 0$ particle into two identical
particles with spin 1/2 moving in opposite directions along the z axis of our laboratory's
coordinate system. Let $\{|+;x\rangle^{(1)}, |-;x\rangle^{(1)}\}$ be the basis if eigenstates of the component x
of the spin for particle 1 (and similarly for particle 2). The singlet state that results from
the decay of the system is:

$$|\Psi\rangle = (1/\sqrt{2})\{|+;x\rangle^{(1)} \otimes |-;x\rangle^{(2)} - |-;x\rangle^{(1)} \otimes |+;x\rangle^{(2)}\} \qquad (17.17)$$

This entangled state is clearly invariant under rotations about the z axis. Let us say we
consider now that the spin of particle 2 should be taken as an environment. We then evaluate

the reduced density matrix for particle 1, (by tracing over the DoF of particle 2). One then finds:

$$\rho^{(1)} = (1/2)\{|+;x\rangle\langle+;x| + |-;x\rangle\langle-;x|\} = (1/2)I \qquad (17.18)$$

This matrix is, however, diagonal in all possible bases for system 1. Thus, environmental decoherence does not offer a criterion for choosing the basis.

There is in fact a simple theorem, proved first in [6], which shows that this problem is generic:

Theorem: Consider a quantum system made of a subsystem S and an environment E, with corresponding Hilbert spaces H_S and H_E so that the complete system is described by states in the product Hilbert space $H_S \otimes H_E$. Let G be a symmetry group acting on the Hilbert space of the full system in a way that does not mix the system and environment. That is, the unitary representation O of G on $H_S \otimes H_E$ is such that $\forall g \in G$, $\hat{O}(g) = \hat{O}^S(g) \otimes \hat{O}^E(g)$, where $\hat{O}^S(g)$ and $\hat{O}^E(g)$ act on H_S and H_E, respectively. Let the system be characterized by a density matrix $\hat{\rho}$ which is invariant under G. Then the reduced density matrix of the subsystem is a multiple of the identity in each invariant subspace of H_S.

This result shows that in these cases decoherence is not even helpful in selecting a preferred (or pointer) basis.

3. Consistent (or de-cohering) histories

We believe the consistent histories approach has some serious problems in general, but in any event, in the particular case at hand the answer we obtain depends on the questions we ask. In particular, we can use the approach to conclude that, with probability 1, our universe today is homogeneous and isotropic (for details see [54]).

4. This is "just philosophy"

It seems clear that in this conference, in contrast with various others I have attended, I would not need to argue that conceptual clarity, as provided by philosophical considerations, is not something to be dismissed in the pursuit of a deeper understanding of the physical world.

17.6 Collapse Approach

We have argued above that the existing explanations for the generation of the primordial cosmic inhomogeneities as results from quantum fluctuations of a field in a completely homogeneous and isotropic state can not be considered fully satisfactory. On the other hand, it seems clear that overlooking that issue, inflationary account leads to a remarkable agreement with observations. It is thus hard to dismiss the proposal as a whole. We then take the conservative approach which is to argue that the overall scheme is correct but that there are some missing elements. As the issue seems to be connected with the measurement problem in general we feel it is natural to consider a modification of quantum theory which in principle seems capable of addressing the point. To many people any suggestion of modifying quantum mechanics seems something not far short of heresy. Of course the point is that it is perhaps the best tested theory in the history of physics, with an extraordinary number of applications in a large range of systems. How can we seriously consider such an adventurous proposal?

Well, to start with, we must recognize that the situation we face here is rather unique in that it involves i) quantum theory, ii) gravity and iii) actual observational data (as pointed out to me by J. Martin), and thus it would not be very surprising if some aspects of physics which have not been previously observed might show up within this context for the first time.

The issue is that we need to be able to point to some physical process, that occurs in time, when trying to explain the emergence of the primordial inhomogeneities and anisotropies; that is, the seeds of cosmic structure. After all emergence means precisely: *Something that was not there at a time, is there at a later time.* We need to explain the breakdown of the symmetry of the initial state, when the dynamics that results from the action of the scalar field in interaction with the space-time metric have no feature capable of doing so, just as in the mini-Mott example. The basis of our approach is thus the observation that theories incorporating something like spontaneous collapse of quantum states can do this.

Collapse theories: There is an important body of previous work on dynamical collapse theories starting with the early proposals by Pearle, Ghirardi, Rimini and Weber, and a long list of further developments [21–27, 38, 40–44, 46] including the ideas linking the issue with aspects of quantum gravity by Diosi and Penrose [9–18, 47–50]. There are also some recent advances to make it compatible with special relativity [2, 19, 45, 55]. A recent work by Weinberg indicates that the issues underlying such a proposal are starting to resonate with the larger physics community [57].

We propose to address the issue at hand by *adding*, to the standard inflationary paradigm, a quantum collapse of the wave function considered as a self induced process, such as that envisaged in those proposals. Here we will illustrate these ideas with one rather well developed proposal formulated in the context of ordinary quantum mechanics, known as CSL [42] in an "adaptation" to the situation at hand. This analysis was carried out in [4], while almost simultaneous treatments based on different adaptations of CSL to the problem were carried out in [32] and [7]. We must of course acknowledge that a truly viable resolution of the question at hand would need to be formulated within a fully generally covariant version of a collapse theory.

Let us first briefly consider how such a modification of quantum theory, involving dynamical collapses, would fit with our current views and physical ideas? This is a very delicate question and its careful exploration would no doubt require much more than what has been achieved so far. However, to see the possibilities let us recall that, in connection with quantum gravity, there are some serious issues and conceptual difficulties that are still outstanding, and that it would not be so surprising if their resolution would involve modifications that would allow, or even require, the incorporation of collapse-like effects. The two major issues in this regard are:

1. The problem of time: All canonical approaches to quantum gravity, including the Wheeler–de Witt proposals and the loop quantum gravity program lead to a timeless theory. The issue is closely tied to the diffeomorphism invariance of GR, which translates into the requirement that at the quantum level different slices of space-time be regarded as gauge equivalent.
2. Recovery of space-time: More generally, we do not know how to recover space-time from canonical approaches to quantum gravity.

Solutions to 1 usually start by using some dynamical variable as a physical clock and then considering relative probabilities (and wave functions). It seems that in those considerations one recovers only an approximate Schrödinger equation with corrections that violate unitarity [20]. Could something like this lie at the bottom of collapse theories? Regarding 2, we note that there are many suggestions indicating space-time might be an emergent phenomenon (see for instance [3], [30], and [52]).

How should we think about general relativity and its emergence from a deeper, quantum gravity theory? What is the nature of what in GR we describe as a classical space-time? These are very hard questions and I would not attempt anything close to a solid answer, in part because in any attempt to do so with some rigor one would need to have at hand a fully satisfactory and workable theory of quantum gravity, something we do not have at this time. However, it is often useful in physics to consider analogies. In this case I want to illustrate the nature of what we might be facing with the questions above, using a hydrodynamic analogy: A fluid is often described in terms of the Navier–Stokes (NS) equations characterizing the velocity field of the liquid elements. The fluids are also characterized by the densities and pressures satisfying some equation of state. We know, however, that such macroscopic characterization of the fluid, in which it is described by a simple and well formulated differential equation, is only an approximation, and that, at a more fundamental level, the fluid is made of individual molecules (which are themselves made of atoms, and so on). We know then that notions such as fluid density or fluid-element's velocity are emergent characterizations (whose regime of validity is limited), and that the NS equation is not only an approximation, but that it might be grossly violated in suitable circumstances (we can think of water passing through a very hot region and having part of it boil, or the breaking of waves in the sea).

It might well be, therefore, that the metric characterization of space-time emerges from a deeper quantum gravity regime in the same manner as the fluid description emerges from the atomic or molecular level characterization of its components. In that case the equations governing the macrodynamics of space-time, that is Einstein's equations, would be approximate characterizations of the bulk dynamics, but not a truly fundamental law of nature; just as the NS equations. Moreover, concepts like space and time would simply fail to exist at the fundamental level, just as fluid velocity or local density do not exist at the subatomic level, which we know characterizes more accurately than the hydrodynamical level, the fundamental nature of the fluid.

In that case, the level at which one can talk about space-time concepts is the classical description. However, some traces of the quantum aspects might appear as deviations from the simpler effective level characterization, just as the forming of the foam at the breaking of the sea waves reflects the underlying molecular constitution of the fluid. Those traces might, according to the ideas we want to study, include behaviors that would look like a quantum mechanical collapse of the wave function (which after all is based on a classical characterization of space-time, i.e. a single particle wave function depends on the space and time coordinates, and so does the operator describing a quantum field in relativistic quantum field theory).

17.6.1 General Setting

We now turn to describe the setting within which we will address the issue of the emergence of the primordial cosmological inhomogeneities from quantum fluctuations during inflation. The general idea, as we discussed above, is that at the quantum level, space-time and thus gravity, are very different from their classical characterization, and that at "large scales" those quantum aspects become manifest as small traces of its true nature, and that such traces include effects that look like a collapse of the quantum wave function matter fields.

Below we discuss a formal implementation of these ideas in the context of the relatively simpler schemes involving discrete collapses.

We take then the view that the inflationary regime is one where gravity already has a good classical description, and that Einstein's equation is generally satisfied (i.e. not always and not in general in an exact manner). However, the situation is such that matter fields still require a full quantum treatment. The setting will thus naturally be semi-classical Einstein's gravity (more precisely, we will rely on the notion of *semi-classical self-consistent configurations* (SSC) developed in [8]).

Definition: The set $\{g_{\mu\nu}(x), \hat{\varphi}(x), \hat{\pi}(x), \mathcal{H}, |\xi\rangle \in \mathcal{H}\}$ represents a SSC if and only if $\hat{\varphi}(x)$, $\hat{\pi}(x)$ and $\mathcal{H}$ correspond to a quantum field theory constructed over a space-time with metric $g_{\mu\nu}(x)$ and the state $|\xi\rangle$ in $\mathcal{H}$ is such that

$$G_{\mu\nu}[g(x)] = 8\pi G \langle\xi|\hat{T}_{\mu\nu}[g(x), \hat{\varphi}(x), \hat{\pi}(x)]|\xi\rangle. \tag{17.19}$$

This corresponds, in a sense, to the GR version of the Schrödinger–Newton equation [9].

To this setting we want to add, in order to describe the transition from a homogeneous and isotropic situation to one that is not, an extra element: THE COLLAPSE OF THE WAVE FUNCTION. That is, besides the unitary evolution describing the change in time of the state of a quantum field, there will be, sometimes, spontaneous jumps. For instance, the vacuum state of the quantum inflaton field, seen as the product of the individual harmonic oscillator states for each mode, could undergo a spontaneous transition to a state where one of such oscillators becomes excited:

$$\ldots.|0\rangle_{k_1} \otimes |0\rangle_{k_2} \otimes |0\rangle_{k_3} \otimes \ldots. \rightarrow \ldots.|\Xi\rangle_{k_1} \otimes |0\rangle_{k_2} \otimes |0\rangle_{k_3} \otimes \ldots. \qquad (17.20)$$

We must note here that semi-classical GR was deemed to be un-viable in the work [37] which, however, only considered the possibility of regarding the theory, and in particular the equation, as holding always in a 100 per cent precise manner, and thus precluded the consideration of spontaneous collapse of the wave function. The situation differs dramatically once we incorporate the possibility of self induced collapse of the wave function. A careful analysis of these issues can be found in [5], in which it was shown that the conclusions about the un-viability of semi-classical GR were rather premature.

Again, the view is that there is an underlying Quantum Theory of Gravity (probably with no notion of space or time). However, by the "time" we can recover space-time concepts, the semi-classical treatment is a very good one, and its regime of validity includes the inflationary regime as long as the curvature is not too large (i.e. $R << 1/l_{Planck}^2$, a condition that is supposed to hold in the inflationary regime). Moreover just as under some conditions we find some departures from the exact Navier–Stokes equation in the behavior of a fluid (breaking sea waves), under some conditions there are quantum collapses that might occur in association with some departures from Einstein's equation.

Now, we turn again to the SSC formalism and note that in order to describe the collapse we must go beyond the transition Eq. (17.20). That is, a collapse will be a transition from one SSC to another, not simply from one quantum state to another. The point is that a change in the quantum state, generically, implies a change in the expectation of the energy momentum tensor, and then, according to Eq. (17.19), a change in the space-time and therefore in the Hilbert space where the state "lives". Clearly the complete characterization of such a process becomes rather complex. An individual collapse then is associated with the "gluing"[5] of two space-times along the collapse hypersurface, while imposing certain continuity requirements. It is clear that Einstein's equations would not hold exactly at such junctures, and we take this as an indication that the quantum aspects of gravitation that have been ignored in arriving at the classical characterization of space-time, must be playing some role in the collapse process, just as the molecular dynamics of the fluid constituents play a role in the breaking of sea waves and the breakdown of the NS equation in association, say, with the generation of foam during such a process. The formalism is, in a sense, similar to the so called Israel's matching prescription developed to describe thin shells in general relativity [29]. For a detailed analysis of these issues see [8].

The discussion above refers to the discrete versions of spontaneous collapse theories, such as those embodied in the GRW proposal, and one would have to modify the treatment

---

[5] This is how one refers to the process of producing a new space-time by joining two space-times with boundary, along the corresponding boundaries. This might be done, provided that the induced metric on the corresponding boundaries of both manifolds are "the same" (i.e. there is an isometry between the boundaries).

in an appropriate manner, using adequate limiting procedures and integral representations of the dynamics, when dealing with continuous versions such as the CSL theory.

17.6.2 Practical Treatment of the Problem

In principle one would use the SSC formalism in treating the collapse of each individual mode of the inflation field. This would be extremely cumbersome and impractical. Furthermore, it is clear that things would become even more complex in the context of continuous collapse dynamics. We will thus use a simpler scheme that uses a single quantum field theory (QFT) construction rather than the multiplicity of constructions that are required following strictly the SSC prescription. We have checked that this is equivalent, at the lowest order in perturbation theory, to the one based on SSC [8].

We again split the treatment into that of a classical homogeneous ("background") part and an inhomogeneous part ("fluctuation"). That is, we write the physical metric as $g_{ab} = g^0_{ab} + \delta g_{ab}$, and similarly we write the scalar field as $\phi = \phi_0 + \delta\phi$. In both cases the first term represents the homogeneous and isotropic background, and the second the perturbation containing the deviation from such a symmetric state of affairs.[6] The background is taken again to be a flat Friedman–Robertson universe, and the homogeneous scalar field $\phi_0(\eta)$. In the strict SSC treatment this corresponds to the zero mode of the quantum field (i.e. the mode of the quantum field which has no dependence on the spatial coordinates).

The main difference in treatment between our approach and the standard ones will concern the treatment of the spatially dependent perturbations, which will be subject not only to a quantum treatment but also to a collapse dynamics. Furthermore, according to the ideas previously discussed, in our approach we quantize the scalar field but not the metric perturbations.

17.6.3 Continuous Spontaneous Localization

Here, we offer a very brief description of this theory, and refer the reader to Pearle [39] for a more detailed presentation. The simplest version of the theory is defined by two equations:
A modified Schrödinger equation, whose solution is:

$$|\psi, t\rangle_w = \hat{T} e^{-\int_0^t dt' \left[i\hat{H} + \frac{1}{4\lambda}[w(t') - 2\lambda\hat{A}]^2\right]} |\psi, 0\rangle. \tag{17.21}$$

where $\hat{H}$ is the usual Hamiltonian, $\mathcal{T}$ is the time-ordering operator, λ is a collapse rate parameter, and $\hat{A}$ is an operator to whose eigenstates the collapse tends. $w(t)$ is a random classical function of time, of white noise type, and where the probability of $w(t_i)$ taking on a particular value in an interval $dw(t_i)$, over the range $0 < t_i < t$ is given by the second

---

[6] This sort of split into background and perturbation is not trivial due to the "gauge freedom" intrinsic to diffeomorphism invariant theories such as GR, but we will mostly ignore such complications in this presentation.

equation, the probability rule:

$$PD_w(t) \equiv {}_w\langle\psi,t|\psi,t\rangle_w \prod_{t_i=0}^{t} \frac{dw(t_i)}{\sqrt{2\pi\lambda/dt}}. \tag{17.22}$$

The deterministic and unitary evolution, as described by the standard Schrödinger equation (i.e. the U – for unitary – process in Penrose's language), and the non-deterministic "jump" or collapse of the quantum state that in the textbook approach is associated with a measurement (i.e. the R – for reduction – process in Penrose's language) (corresponding to the observable $\hat{A}$) are unified. For non-relativistic QM, the regime for which the collapse theories were originally developed as a means to resolving the measurement problem, the proposal involves collapse to the joint eigenstates of a collection of commuting operators (essentially the mass density operators at all points of space) rather than the simpler collapse to the eigenstates of a single operator $\hat{A}$, as described above. This entails a classical field, a random function of time at each point of space, rather than the simpler single random function of time $w(t)$, as described above. In fact, one needs to use a kind of smeared operators, otherwise the localization that results from the collapse will involve a generation of infinite energy. See [1] for a discussion.

It is clear that the parameter λ must be small enough not to conflict with high precision tests of quantum theory in the domain of subatomic physics and big enough to result in rapid localization of "macroscopic objects". GRW suggested a range: $\lambda \sim 10^{-16}\text{s}^{-1}$ (i.e. it probably depends on the particle's mass).

The point is that collapse theories (such as CSL) can account for the breakdown of symmetries in the mini-Mott example, and in the cosmological setting. In particular, the original version of CSL, with the smeared position operator for individual particles as the universal collapse generating operator, clearly indicates that the first basis is the appropriate one to treat the mini-Mott problem, because the detectors are built from particles and their localization will lead to states in which the individual detectors will give a well defined level of excitation.

We will thus consider a version of CSL appropriate for the situation at hand. That involves considering a version of the theory adapted to quantum field theory, and which in principle, given the general relativistic setting of the problem, should also be fully covariant. We will ignore in this work this second requirement, in the clear understanding that a truly satisfactory resolution will need to address that aspect as well.

As we have already indicated, the space-time metric will be treated classically. We will be using a specific gauge (the so-called Newtonian gauge) and we will be ignoring tensor perturbations. The perturbed space-time will therefore be described by the metric:

$$ds^2 = a^2(\eta)\left[-(1+2\Psi)d\eta^2 + (1-2\Psi)\delta_{ij}dx^i dx^j\right], \Psi(\eta,\vec{x}) \ll 1 \tag{17.23}$$

where Ψ is just the so-called Newtonian potential introduced in Eq. (17.1). We will set $a = 1$ at the "present cosmological time", and write and assume that the inflationary regime takes place during the interval $-\mathcal{T} < \eta < \eta_0$, with $\eta = \eta_0$, negative and very small in absolute terms, and with the scale factor $a(\eta)$ given below Eq. (17.6). The inflaton scalar field $\phi(x)$ must be treated using QFT in curved space-time (using SSC). The quantum state of the scalar field and the space-time metric satisfy Einstein's semi-classical equation: $G_{\mu\nu} = 8\pi G \langle \xi | \hat{T}_{\mu\nu} | \xi \rangle$.

As we explained we will be concentrating on the modes other than the zero mode which we treat classically as an effective approximation. At the early stages of inflation, which we denote by $\eta = -\mathcal{T}$, the state of the scalar field perturbation is described by the Bunch–Davies vacuum, and the space-time is 100 per cent homogeneous and isotropic. In fact, in the vacuum state, the operators $\hat{\delta\phi}_k$ and $\hat{\pi}_k$ are characterized by Gaussian wave functions centered on 0 with uncertainties $\Delta\delta\phi_k$ and $\Delta\pi_k$, respectively, when expressed in the eigenbasis of either $\delta\phi_k$ or π_k.

The role of the collapse dynamics is to modify the quantum state, and generically will lead to modified expectation values of $\hat{\delta\phi}_k(\eta)$ and $\hat{\pi}_k(\eta)$. In order to proceed, we must now specify in some detail the rules governing the collapse. As we have indicated, this is supposed to be the result of some unknown aspect of physics, which we will here encode into an adapted version CSL theory. How can we deal with such a situation? Our approach is based on making an "educated guess", which can later be contrasted with observations. The collapse is assumed to take place for each mode independently according to the CSL dynamics and determined by independent stochastic functions. We must stress that in this approach our universe is viewed as corresponding to one specific realization of these stochastic functions (one for each $\vec{k}$).

The semi-classical Einstein equation we must focus on is (the $\eta-\eta$ component of $G_{\mu\nu} = 8\pi G \langle \xi | \hat{T}_{\mu\nu} | \xi \rangle$ at first order in the perturbations):

$$-k^2 \Psi(\eta, \vec{k}) = 4\pi G \phi_0'(\eta) \langle \hat{\phi}'(\vec{k}, \eta) \rangle = \frac{4\pi G \phi_0'(\eta)}{a} \langle \hat{\pi}(\vec{k}, \eta) \rangle \qquad (17.24)$$

where $\Psi(\eta, \vec{k})$ is the Fourier transform of the Newtonian potential, the prime represents the derivative with respect to conformal time η, and the expectation values are taken in the quantum state of matter field (in this case the inflaton field) that results from the CSL dynamics on the corresponding hypersurface.[7]

As we said, at the start of inflation ($\eta = -\mathcal{T}$) state is described by the Bunch–Davies vacuum, so the expectation $\langle \hat{\pi}(\vec{k}, \eta) \rangle$ vanishes, and thus, as long as the state of the field is that vacuum, the space-time will be 100 per cent homogeneous and isotropic. Of course the collapse dynamics will modify this unacceptable conclusion. The quantity of direct interest

---

[7] In the actual calculation, we found it convenient to work in the Schrödinger picture where the inflaton field is time-independent, and the state vector $|\psi, \eta\rangle$ undergoes evolution in conformal time under the joint influence of the Hamiltonian and collapse dynamics. In our treatment, it is most convenient to work with the individual modes of the inflaton field, that is, the Fourier components of $\phi(x)$, $\pi(x)$. These components turn out to satisfy essentially the position–momentum commutation relations, and their Hamiltonian turns out to be of a modified harmonic oscillator form. However, we proceed with the discussion in the interaction picture, where the state changes are due to the CSL dynamics while the field operators evolve with the free Hamiltonian.

to us is:

$$\frac{\Delta T(\theta,\varphi)}{\bar{T}} = (1/3)\Psi(\eta_D,\vec{x}) = c\int d^3k e^{i\vec{k}\cdot\vec{x}}\frac{1}{k^2}\langle\hat{\pi}(\vec{k},\eta_D)\rangle, \qquad (17.25)$$

where $c \equiv -\frac{4\pi G\phi_0'(\eta)}{3a}$. Here, $\vec{x}$ is a point on the intersection of our past light cone with the last scattering surface ($\eta = \eta_D$), and corresponds to the direction on the sky specified by θ,φ. Thus, as follows from Eq. (17.3), the expression for the quantity of direct observational interest is:

$$\alpha_{lm} = c\int d^2\Omega Y_{lm}^*(\theta,\varphi)\int d^3k e^{i\vec{k}\cdot\vec{x}}\frac{1}{k^2}\langle\hat{\pi}(\vec{k},\eta_D)\rangle. \qquad (17.26)$$

There is nothing analogous to this expression in the standard approaches! The expression above shows that the quantity of interest can be thought of as a result of a "random walk" on the complex plane. That is, the quantity of interest is the sum (represented by the second integral above) of many contributions (in this case infinite), one for each value of $\vec{k}$, where every individual contribution involves some random variable (the one determining the post-collapse value of $\langle\hat{\pi}(\mathbf{k},\eta)\rangle$). As is always the case one cannot predict the end point of such a random "walk"; however, one can focus on the most likely value of magnitude of the total displacement. It follows from Eq. (17.26) that the quantity corresponding to magnitude of the total displacement for one realization of the collapse dynamics (the one corresponding to that which took place in our universe) is:

$$|\alpha_{lm}|^2 = (4\pi c)^2\int d^3k d^3k' j_l(kR_D)j_l(k'R_D)Y_{lm}(\hat{k})Y_{lm}^*(\hat{k}')$$

$$\times\frac{1}{k^2 k'^2}\langle\hat{\pi}(\vec{k},\eta)\rangle\langle\hat{\pi}(\vec{k}',\eta)\rangle^*. \qquad (17.27)$$

Let us recall that we need the product of the expectation values and not the expectation value of the product!![8] Now we proceed to estimate the most likely value of the quantity above, through the use of an imaginary ensemble of realizations and identifying (for the sake of computational ease) the most likely value with the ensemble average. We need to compute the ensemble average at "late times". One can show that generically we will have: $\overline{(\langle\hat{\pi}(\mathbf{k},\eta)\rangle\langle\hat{\pi}(\mathbf{k}',\eta)\rangle^*)} = f(k)\delta(\mathbf{k}-\mathbf{k}')$. Therefore, from Eq. (17.27), we obtain,

$$\overline{|\alpha_{lm}|^2} = (4\pi c)^2\int_0^\infty dk j_l(kR_D)^2\frac{1}{k^2}f(k). \qquad (17.28)$$

At this point it should be noted that in the present analysis the essentially flat spectrum, which is known to fit well with observations, would correspond to $f(k)$ being proportional to k.

---

[8] The latter is what one finds in the standard treatments. The difference is very significant among other things because the latter is generically non-zero even when the state of the quantum field is homogeneous and isotropic, while the former will be non-vanishing only if the quantum state does not have those symmetries.

In order to proceed further, and study what our theoretical approach says regarding the function $f(k)$ above, we need to consider the theory controlling the dynamics of the collapse. As we said, we will here rely on CSL theory; however, in order to do that we still need to choose the operator $\hat{A}$ driving the collapse and the value of the parameter λ.

It is convenient to work with a re-scaled field $y(\eta, \vec{x}) \equiv a\delta\phi(\eta, \vec{x})$ and its momentum conjugate $\pi_y(\eta, \vec{x}) = a\delta\phi'(\eta, \vec{x})$. For simplicity, we put everything in a box of size L (to be removed at the end of the calculations), and focus on a single mode $\vec{k}$. We thus write:

$$Y \equiv (2\pi/L)^{3/2} y(\eta, \vec{k}), \qquad \Pi \equiv (2\pi/L)^{3/2} \pi_y(\eta, \vec{k}). \tag{17.29}$$

As we saw, in order to compare with the observations, we need to evaluate the ensemble average $\overline{\langle\hat{\Pi}\rangle^2}$, and determine under what circumstances, if any, this is $\sim k$. Recall that we must consider the quantity $\overline{\langle\hat{\Pi}\rangle^2}$ rather than the quantity $\langle\hat{\Pi^2}\rangle$ that is considered in standard treatments. This is on the one hand, what we need to compute is the most likely magnitude of the coefficients α_{lm} and not simply the quantum mechanical uncertainty of that quantity in the relevant quantum state. We can see that this is the case, among other reasons, because a non-vanishing value of the first clearly indicates a lack of isotropy in the CMB, while a non-vanishing value of the second implies no such thing. This is just as in the case of a harmonic oscillator (with potential $V = Kx^2$), where any state that is symmetric under $x \rightarrow -x$ has vanishing $\langle\hat{X}\rangle$, while even the symmetric ground state has non-vanishing $\langle\hat{X}^2\rangle$.

The explicit calculations are a little bit involved and the interested reader is directed to [4].

We start by considering the case where $\hat{\Pi}$ is taken as the collapse-generating operator toward one of whose eigenstates the state vector evolves (i.e. as the operator that acts as the generator of collapse). That is setting $\hat{A} = \hat{\Pi}$. In this case we obtain:

$$\overline{\langle\hat{\Pi}\rangle^2} = \frac{\lambda k^2 \mathcal{T}}{2} + \frac{k}{2} - \frac{k}{\sqrt{2}\sqrt{1 + \sqrt{1 + 4\lambda^2}}}. \tag{17.30}$$

We note that if we set the collapse rate parameter $\lambda = 0$ (turn off CSL), we would have the standard quantum mechanics result $\overline{\langle\hat{\Pi}\rangle^2} = 0$ since $\langle\hat{\Pi}\rangle = 0$. This is as it should be: If there is no mechanism for breaking the initial symmetry, then the value of a measure of the departure from such symmetry should be zero.

We further see that agreement with the observed scale-invariant spectrum (i.e. a result that to a good approximation is simply proportional to k) can be achieved if we assume the first term is dominant and we set $\lambda = \tilde{\lambda}/k$, where $\tilde{\lambda}$ is a constant (independent of k). We note that this replaces the dimensionless collapse rate parameter λ with parameter $\tilde{\lambda}$ of dimension time^{-1}.

In that case we obtain:

$$\overline{\langle \hat{\Pi} \rangle^2} = \frac{\tilde{\lambda} k \mathcal{T}}{2} + \frac{k}{2} - \frac{k}{\sqrt{2}\sqrt{1 + \sqrt{1 + 4(\tilde{\lambda}/k)^2}}}. \quad (17.31)$$

We note that although we found an expression corresponding almost to a scale invariant spectrum, our analysis indicates a particular type of modification that can in principle be searched for in the observational data.

Analogously, we have considered the case where $\hat{Y}$ is taken as generator of collapse and we obtained very similar, yet different, results. For details see [4].

Finally, the comparison and adjustment with the overall amplitude of the observed CMB anisotropies, using the GUT scale for the value of inflation potential, and other standard values for the slow-roll parameters (order of a few per cent), we are led to the estimate: $\tilde{\lambda} \sim 10^{-5} \mathrm{Mpc}^{-1} \approx 10^{-19} \mathrm{s}^{-1}$. It is noteworthy that this value is not very different from GRW suggestion of $10^{-16} \mathrm{s}^{-1}$, although these values apply in quite different settings: the former in applying CSL to everyday non-relativistic physics, the latter in applying CSL to the cosmological situation at hand. Moreover, we should note the relatively wide range of viable values of the parameter in the non-relativistic quantum settings and in particular its likely dependence on the particle's mass.

17.6.4 Collapse on Field Operators

One might be a bit concerned with the fact that in our treatment we have assumed that the collapse occurs mode by mode, given the fact that these quantum field modes are extended over the whole universe. The non-local aspect of this collapse suggests some possible conflict with relativistic causality. However, a moment's thought indicates that this is not very different from what occurs within any collapse theory treatment of something like an EPR situation. The collapse theories are known to work fine in these situations and imply no faster than light signaling despite their non-local aspects. After all, we all know that any successful description of the experimental test of Bell's inequalities implies some level of non-locality [53].

One way we can see that, in principle, this should be no serious obstacle for the approach we have taken is by expressing the collapse dynamics in terms of local operators. Thus, we next write the version of the CSL theory we have been led to, when described in terms of the space-time field operators. In one case we can start by defining

$$\tilde{y}(\vec{x}) \equiv \frac{1}{(2\pi)^{3/2}} \int d^3k\, e^{i\vec{k}\cdot\vec{x}} k^{1/2} y(\vec{k}) = (-\nabla^2)^{1/4} \hat{y}(\vec{x}), \quad (17.32)$$

where we recall that $y = a\delta\phi$. Then the state vector evolution is given by

$$|\psi, t\rangle = \mathcal{T} e^{-i\int_{-\mathcal{T}}^{\eta} d\eta' \hat{H} - \frac{1}{4\tilde{\lambda}} \int_{-\mathcal{T}}^{\eta} d\eta' \int d^3x [w(\vec{x},\eta') - 2\tilde{\lambda}\tilde{y}(\vec{x})]^2} |\psi, -\mathcal{T}\rangle. \quad (17.33)$$

This is just the standard CSL state-vector evolution, where the collapse-generating operators (toward whose joint eigenstates collapse tends) are $\tilde{y}(\vec{x})$ for all $\vec{x}$.

Similarly, in the case where we take $\hat{\Pi}$ as the generator of collapse we have

$$|\psi, \eta\rangle = \mathcal{T} e^{-i\int_{-\mathcal{T}}^{\eta} d\eta' \hat{H} - \frac{1}{4\lambda}\int_{-\mathcal{T}}^{\eta} d\eta' \int d^3x [w(\vec{x},\eta') - 2\tilde{\lambda}\tilde{\pi}(\vec{x})]^2} |\psi, -\mathcal{T}\rangle. \qquad (17.34)$$

where $\tilde{\pi}(\vec{x}) \equiv (-\nabla^2)^{-1/4}\hat{\pi}(\vec{x})$. The above equation describes just the standard CSL state-vector evolution, where the collapse-generating operators (toward whose joint eigenstates collapse drives all states) are $\tilde{\pi}(\vec{x})$ for all $\vec{x}$.

Thus we have seen that the distribution of matter in our actual universe and the observed CMB radiation which follows from it can be explained by the hypothesis of CSL dynamical collapse acting on the state vector describing inflation fluctuations in the early universe. Indeed, we have seen that can be done in two possible ways. However, it has turned out that the collapse-generating operator which gives such results is not the inflation fluctuation itself $\hat{y}(\mathbf{x})$ or its conjugate momentum $\hat{\pi}(\mathbf{x})$, but rather a peculiar differential operator acting on these, either $(-\nabla^2)^{-1/4}\hat{\pi}(\vec{x})$ or $(-\nabla^2)^{1/4}\hat{y}(\vec{x})$. A satisfactory explanation will have to wait for a general theory expressing, in all situations, from particle physics to cosmology, the exact form of the CSL-type of modification to the evolution of quantum states. Such generic theory would likely involve gravitation playing a fundamental role.

17.7 Discussion and Conclusions

We have reviewed the standard account of the generation of the seeds of cosmic structure from quantum fluctuations in inflationary cosmology. We have argued that that account lacks any element that could be used to explain the necessary breakdown in the homogeneity and isotropy of the vacuum state characterizing the quantum field. The issue is related to the interpretation of quantum mechanics, and, in particular, to the so-called "measurement problem". The discussion of the conceptual issues was facilitated here by considering, within the same conditions associated with the cosmological problem, a simplified version of an early analysis by Mott of the quantum decay of a nucleus in a spherically symmetric initial state leading to the well known straight traces in a bubble chamber: the mini-Mott problem. The idea was to use this model as a test ground to examine any proposal for the resolution of the main problem.

We have considered briefly the answers to the above conundrum that are most commonly offered by inflationary cosmologists. We have examined these answers and have argued that they are clearly unsatisfactory, basically from the start. That is simply because in the cosmological context we need to account for the emergence of the primordial inhomogeneities and anisotropies which are the seeds of all cosmic structure including galaxies, stars, planets, life, humans (or other sapient beings), and instruments. Therefore, we have to provide arguments that do not rely on instruments or observers at the stage where we want to understand the emergence of the primordial inhomogeneities.

We have noted that the application of ideas based on decoherence requires the prior identification of an environment, something that generically depends on our particular interests or measuring capabilities, and should therefore be off limits, for the use in an explanatory context, in the situation at hand. Moreover, we have argued that the symmetry of the situation ensures that the decoherence could only lead to reduced density matrices of a form that does not allow for the unique selection of the so-called pointer basis.

In the case of many worlds interpretations we argued that the various schemes possess no elements allowing one to select the basis, or context in which the ensemble of alternatives must be described, or in which the "splitting of the world" takes place (or any other characterization of the passing from the one world to the many worlds that characterize these proposals). That is, by making different choices of basis, either in the mini-Mott case or in the cosmological situation, one could be led to argue in favor or against the breakdown of the initial symmetry.

Something similar happens with the approaches such as consistent histories proposal. There is simply nothing like an unambiguous rule indicating which basis, context or "realm" to consider, and depending on the choice one could end assigning non-vanishing probabilities to non-symmetric histories, or an exactly vanishing probability for all except the symmetric ones.

One idea that we did not consider in this chapter is that provided by the de Broglie–Bohm (dBB) version of quantum theory. In that case, the physical situation is described not only by the wave function, but also by a set of configuration variables, and thus it is perfectly possible to have a system characterized by a symmetric wave function, whose state is nonetheless not fully symmetric. That is, the asymmetry can reside in the configuration variable. It is thus possible to consider dBB as a viable scheme to address our problem. In fact, various works based on the application of dBB theory to the inflationary situation have been published in recent years [60, 61]. However, when applying dBB to the cosmological case at hand one can not really argue that it describes in the cosmological context *the emergence* of the primordial inhomogeneities, simply because within this approach the symmetry was never there to start with. That is, even though the initial wave function is symmetric the initial values for the configuration variables were not, in the individual cases, initially symmetric (this despite the fact that one might consider an ensemble of realizations, and argue that at the statistical level the distribution of initial data was symmetric).

Thus, we conclude that none of the existing frameworks for quantum theory (with the possible exception of the dBB proposal, with the caveat explained above) seems to offer a satisfactory account leading to the desired breaking of the initial symmetry in the problems at hand, and leading to what we think are the appropriate characterizations of the late time situations where the symmetry is gone.

The analysis presented here indicates the attractiveness of incorporating something like the collapse of the wave function as a spontaneous dynamical aspect of nature, something that has been long advocated by some colleagues working on addressing the measurement problem in quantum theory. We have shown here how one can adapt one of the promising versions of such theories, the CSL proposal, to address the cosmological problem that motivated this research path. Of course, a completely satisfactory theory is still lacking, as that would need not only a general formulation which would work in situations ranging from the few particle non-relativistic quantum mechanics to which the initial GRW or CSL proposals were devoted, to the cosmological situations such as the the the one explored here, but will have to be a theory that is shown to be fully compatible with special and general relativity. In the meanwhile, we feel the analysis we have presented serves as a first set of steps in the explicit parametrization of the still mysterious aspect of physics that we have argued is missing from our understanding of the workings of nature and that is exhibited in a particularly blatant form in the context of cosmology. It is worthwhile mentioning that initial steps in the exploration of further applications of dynamical collapse theories indicate [33, 35, 36] they can be helpful in addressing other problems appearing at the interface of the quantum and gravitational realms, namely the so-called "black hole information paradox" and "the problem of time" in quantum gravity. On the other hand, we should accept the possibility that the approach based on non-unitary modifications of quantum theory that we have been exploring in this context, will be shown to be non-viable. However, even in that case, our analysis would have served to pinpoint a serious shortcoming in our understanding of the physical world. In any event, it seems quite clear that the research into these matters must continue.

Ackowledgments

I am thankful to Philip Pearle for reading this chapter and for the many useful comments he offered. This work was supported in part by the CONACYT grant no. 220738, and the joint CONACYT and MINCYT, grant no. 191302.

References

[1] Bassi, A., Ippoliti, E., and Vacchini, B. 2005. On the energy increase in space-collapse models. *Journal of Physics A: Mathematical and General*, **38**(37), 8017.

[2] Bedingham, D. J. 2011. Relativistic state reduction dynamics. *Found. Phys.*, **41**, 686–704.

[3] Bombelli, L., Lee, J., Meyer, D. and Sorkin, R. 1987. Space-Time as a Causal Set. *Phys. Rev. Lett.*, **59**, 521–4.

[4] Cañate, P., Pearle, P. and Sudarsky, D. 2013. Continuous spontaneous localization wave function collapse model as a mechanism for the emergence of cosmological asymmetries in inflation. *Phys. Rev.*, **D87**(10), 104024.

[5] Carlip, S. 2008. Is Quantum Gravity Necessary? *Class. Quant. Grav.*, **25**, 154010.

[6] Castagnino, M., Fortin, S., Laura, R. and Sudarsky, D. 2014. Interpretations of Quantum Theory in the Light of Modern Cosmology. (2014) arXiv:1412.75756.

[7] Das, S., Lochan, K., Sahu, S. and Singh, T. P. 2013. Quantum to classical transition of inflationary perturbations: Continuous spontaneous localization as a possible mechanism. *Phys. Rev.*, **D88**(8), 085020.

[8] Diez-Tejedor, A. and Sudarsky, D. 2012. Towards a formal description of the collapse approach to the inflationary origin of the seeds of cosmic structure. *JCAP*, **1207**, 045.

[9] Diosi, L. 1984. Gravitation and Quantum Mechanical Localization of Macro-Objects. *Phys. Lett. A*, **105**, 4–5, 199–202.

[10] Diosi, L. 1987. A Universal Master Equation for the Gravitational Violation of Quantum Mechanics. *Phys. Lett.*, **A120**, 377.

[11] Diosi, L. 1989. Models for universal reduction of macroscopic quantum fluctuations. *Phys. Rev.*, **A40**, 1165–1174.

[12] Diosi, L. and Lukacs, B. 1987. In Favor of a Newtonian Quantum Gravity. *Annalen Phys.*, **44**, 488.

[13] Diosi, L. and Lukacs, B. 1989. On the minimum uncertainty of space-time geodesics. *Phys. Lett.*, **A142**, 331.

[14] Diosi, L. 1997. Lorentz covariant stochastic wave function dynamics? arXiv:quant-ph/9704025.

[15] Diosi, L. 2000. Emergence of classicality: from collapse phenomenologies to hybrid dynamics. *Lect. Notes Phys.*, **538**, 243–50.

[16] Diosi, L. 2004. Probability of intrinsic time arrow from information loss. *Lect. Notes Phys.*, **633**, 125–35.

[17] Diosi, L. 2014. Gravity-related spontaneous wave function collapse in bulk matter. *New J. Phys.*, **16**(10), 105006.

[18] Diosi, L. and Papp, T. N. 2009. Schrödinger–Newton equation with complex Newton constant and induced gravity. *Phys. Lett.*, **A373**, 3244–7.

[19] Durr, D., Goldstein, S., Tumulka, R. and Zanghi, N. 2004. Bohmian mechanics and quantum field theory. *Phys. Rev. Lett.*, **93**, 090402.

[20] Gambini, R., Porto, R. A. and Pullin, J. 2004. Fundamental decoherence from relational time in discrete quantum gravity: Galilean covariance. *Phys. Rev.*, **D70**, 124001.

[21] Ghirardi, G. C., Rimini, A. and Weber, T. 1985. A model for a unified quantum description of macroscopic and microscopic systems. In *Quantum Probability and Applications II*, Lecture Notes in Mathematics, Vol. 1136. ISBN 978-3-540-15661-1. Springer Verlag, p. 223.

[22] Ghirardi, G. C., Rimini, A. and Weber, T. 1986. A Unified Dynamics for Micro and MACRO Systems. *Phys. Rev.*, **D34**, 470.

[23] Ghirardi, G. C., Nicrosini, O., Rimini, A. and Weber, T. 1988. Spontaneous Localization of a System of Identical Particles. *Nuovo Cim.*, **B102**, 383.

[24] Ghirardi, G. C., Grassi, R. and Pearle, Philip M. 1990a. Relativistic dynamical reduction models: General framework and examples. *Foundations of Physics*, **20**, 11, 1271–316.

[25] Ghirardi, G. C., Pearle, P. M. and Rimini, A. 1990b. Markov Processes in Hilbert Space and Continuous Spontaneous Localization of Systems of Identical Particles. *Phys. Rev.*, **A42**, 78–9.

[26] Ghirardi, G. C., Grassi, R. and Rimini, A. 1990c. A continuous spontaneous reduction model involving gravity. *Phys. Rev.*, **A42**, 1057–64.

[27] Ghirardi, G. C., Grassi, R. and Pearle, P. M. 1990d. Relativistic dynamical reduction models and nonlocality. *J. Found. Mod. Phys.*, 0109–123.

[28] Hartle, J. B. 2006. Generalizing quantum mechanics for quantum gravity. *Int. J. Theor. Phys.*, **45**, 1390–96.

[29] Israel, W. 1966. Singular hypersurfaces and thin shells in general relativity. *Nuovo Cim.*, **B44S10**, 1.

[30] Jacobson, T. 1995. Thermodynamics of space-time: The Einstein equation of state. *Phys. Rev. Lett.*, **75**, 1260–63.

[31] Kastner, R. E. 2014. Comment on "Quantum Darwinism, Decoherence, and the Randomness of Quantum Jumps," arxiv:1412.5206.

[32] Martin, J., Vennin, V. and Peter, P. 2012. Cosmological Inflation and the Quantum Measurement Problem. *Phys. Rev.*, **D86**, 103524.

[33] Modak, S. K., Ortz, L., Pea, I., and Sudarsky, D. 2014. Black Holes: Information Loss But No Paradox. arXiv:1406.4898 [gr-qc].

[34] Mott, N. F. 1929. The Wave Mechanics of α- Ray tracks. *Proc. of the Royal Soc. of London*, **126**, 79.

[35] Okon, E. and Sudarsky, D. 2014. Benefits of Objective Collapse Models for Cosmology and Quantum Gravity. *Foundations of Physics*, **44**(2), 114–3.

[36] Okon, E. and Sudarsky, D. 2015. The Black Hole Information Paradox and the Collapse of the Wave Function. *Foundations of Physics*, **45**(4), 461–70.

[37] Page, D. N. and Geilker, C. D. 1981. Indirect Evidence for Quantum Gravity. *Phys. Rev. Lett.*, **47**, 979–82.

[38] Pearle, P. M. 1984. Experimental tests of dynamical state-vector reduction. *Phys. Rev.*, **D29**, 235–40.

[39] Pearle, P. M. 2014a. Collapse Miscellany. In: Struppa, D. C. and Tollaksen, J. M., eds. *Quantum Theory: A Two-Time Success Story*. (Milan: Springer Milan), pp. 131–56.

[40] Pearle, P. M. 1976. Reduction of the State Vector by a Nonlinear Schrodinger Equation. *Phys.Rev.*, **D13**, 857–68.

[41] Pearle, P. M. 1979. Toward Explaining Why Events Occur. *Int. J. Theor. Phys.*, **18**, 489–518.

[42] Pearle, P. M. 1989. Combining Stochastic Dynamical State Vector Reduction With Spontaneous Localization. *Phys. Rev.*, **A39**, 2277–89.

[43] Pearle, P. M. 1999. Collapse models. *Lect. Notes Phys.*, **526**, 195.

[44] Pearle, P. M. 2000. Wave function collapse and conservation laws. *Found. Phys.*, **30**, 1145–60.

[45] Pearle, P. M. 2014b. A Relativistic Dynamical Collapse Model for a Scalar Field. arXiv:1404.5074.

[46] Pearle, P. M. and Squires, E. 1996. Gravity, energy conservation and parameter values in collapse models. *Found. Phys.*, **26**, 291.

[47] Penrose, R. 2000. Gravitational collapse of the wavefunction: An experimentally testable proposal. *Proceedings, 9th Marcel Grossman Meeting*, 3–6.

[48] Penrose, R. 2001. On gravity's role in quantum state reduction. In: Callender, C., ed. *Physics Meets Philosophy at the Planck Scale*, pp. 290–304.

[49] Penrose, R. 1996. On gravity's role in quantum state reduction. *Gen. Rel. Grav.*, **28**, 581–600.

[50] Penrose, R. 2014. On the Gravitization of Quantum Mechanics 1: Quantum State Reduction. *Found. Phys.*, **44**, 557–5.

[51] Perez, A., Sahlmann, H. and Sudarsky, D. 2006. On the quantum origin of the seeds of cosmic structure. *Class. Quant. Grav.*, **23**, 2317–54.

[52] Seiberg, N. 2007. Emergent spacetime. In The Quantum Structure of Space and Time. World Scientific. arXiv:hep-th/0601234.

[53] Shimony, A. 2013. Bell's Theorem. In Zalta, E. N., ed. *The Stanford Encyclopedia of Philosophy*, winter 2013 edn. http://plato.stanford.edu/entries/bell-theorem/.

[54] Sudarsky, D. 2011. Shortcomings in the Understanding of Why Cosmological Perturbations Look Classical. *Int. J. Mod. Phys.*, **D20**, 509–52.

[55] Tumulka, R. 2006. On spontaneous wave function collapse and quantum field theory. *Proc. Roy. Soc. Lond.*, **A462**, 1897–908.

[56] Weinberg, S. 2008. *Cosmology.* Oxford University Press.

[57] Weinberg, S. 2012. Collapse of the State Vector. *Phys. Rev.*, **A85**, 062116.

[58] Zurek, W. H. (2016) Quantum Darwinism, Decoherence, and the Randomness of Quantum Jumps. *Physics Today*, **67**, 10, 44 50

[59] Zurek, W. H. 1998. Decoherence, Einselection, and the existential interpretation: The Rough guide. *Phil. Trans. Roy. Soc. Lond.*, **A356**, 1793–820.

[60] Valentine, A. 2010. *Phys. Rev.*, **D82**, 063513, 43pp.

[61] Pinto-Neto, N., Santos, G. and Struyve, W. 2012. *Phys. Rev.*, **D85**, 083506, 4pp.

18

Towards a Novel Approach to Semi-Classical Gravity

WARD STRUYVE

18.1 Introduction

Quantum gravity is often considered to be the holy grail of theoretical physics. One approach is canonical quantum gravity, which concerns the Wheeler–DeWitt equation and which is obtained by applying the usual quantization methods (which were so successful in the case of high energy physics) to Einstein's field equations. However, this approach suffers from a host of problems, some of technical and some of conceptual nature (such as finding solutions to the Wheeler–DeWitt equation, the problem of time, . . .). For this reason one often resorts to a semi-classical approximation where gravity is treated classically and matter quantum mechanically [1, 2]. The hope is that such an approximation is easier to analyse and yet reveals some effects of quantum gravitational nature.

In the usual approach to semi-classical gravity, matter is described by quantum field theory on curved space-time. For example, in the case the matter is described by a quantized scalar field, the state vector can be considered to be a functional $\Psi(\phi)$ on the space of fields, which satisfies a particular Schrödinger equation

$$i\partial_t \Psi(\phi, t) = \widehat{H}(\phi, g)\Psi(\phi, t),$$

$$(18.1)$$

where the Hamiltonian operator $\widehat{H}$ depends on the space-time metric g. This metric satisfies Einstein's field equations

$$G_{\mu\nu}(g) = 8\pi G \langle \Psi | \widehat{T}_{\mu\nu}(\phi, g) | \Psi \rangle,$$

$$(18.2)$$

where the source term is given by the expectation value of the energy-momentum tensor operator.

This semi-classical approximation of course has limited validity. For example, it will form a good approximation when the matter state approximately corresponds to a classical state (i.e. a coherent state), but will fail to be so when the state is a macroscopic superposition of such states. Namely, for such a superposition $\Psi = (\Psi_1 + \Psi_2)/\sqrt{2}$, we have $\langle \Psi | \widehat{T}_{\mu\nu} | \Psi \rangle \approx (\langle \Psi_1 | \widehat{T}_{\mu\nu} | \Psi_1 \rangle + \langle \Psi_2 | \widehat{T}_{\mu\nu} | \Psi_2 \rangle)/2$, so that the gravitational field is affected by two matter sources, one coming from each term in the superposition. However, one expects that according to a full theory for quantum gravity, the states $|\Psi_1\rangle$ and $|\Psi_2\rangle$ each have their own gravitational field and that the total state is a superposition of those. And, indeed, Page and Geilker showed with an experiment that this semi-classical theory is not adequate [2, 3].

Of course, as already noted by Page and Geilker, it could be that this problem is not due to the fact that gravity is treated classically, but due to the choice of the version of quantum theory. Namely, Page and Geilker adopted the many worlds point of view, according to which the wave function never collapses. However, according to standard quantum theory the wave function is supposed to collapse during a measurement. Which physical processes act as measurements is of course rather vague and is the source of the measurement problem. But it could be that such collapses explain the outcome of their experiment. If an explanation of this type is sought, one should consider so-called spontaneous collapse theories, where collapses are objective, random processes that do not in a fundamental way depend on the notion of measurement. (See Diez-Tejedor and Sudarsky [4] and Derakhshani [5] for actual proposals combining such a spontaneous collapse approach with, respectively, Eq. (18.2) and its non-relativistic version.)

In this chapter, we consider an alternative to standard quantum mechanics, called Bohmian mechanics [6–9] and develop a semi-classical approximation based on it which is expected to improve upon the usual approach.

Bohmian mechanics solves the measurement problem by introducing an actual configuration (particle positions in the non-relativistic domain, particle positions or fields in the relativistic domain [11]) that evolves under the influence of the wave function. According to this approach, instead of coupling classical gravity to the wave function, it is natural to couple it to the actual matter configuration. For example, in the case of a scalar field there is an actual field ϕ_B whose time evolution is determined by the wave functional Ψ. There is an energy-momentum tensor $T_{\mu\nu}(\phi_B, g)$ corresponding to this scalar field and this tensor can be introduced as the source term in Einstein's field equations:

$$G_{\mu\nu}(g) = T_{\mu\nu}(\phi_B, g) \, . \tag{18.3}$$

This approach immediately solves the problem with the macroscopic superposition, since the energy-momentum tensor will correspond to just one of the macroscopic matter distributions.

However, there is an immediate problem with this approach, namely that Eq. (18.3) is not consistent. The Einstein tensor $G_{\mu\nu}$ is identically conserved, i.e. $\nabla^\mu G_{\mu\nu} \equiv 0$. So the Bohmian energy-momentum tensor $T_{\mu\nu}(\phi_B, g)$ must be conserved as well. However, the equation of motion for the scalar field does not guarantee this. (Similarly, in the Bohmian approach to non-relativistic systems, the energy is generically not conserved.)

As explained in Struyve [12], the root of the problem seems to be the gauge invariance, which in this case is the invariance under spatial diffeomorphisms. Because the scalar field and the space-time metric are connected by spatial diffeomorphisms, it seems that one can not just assume the metric to be classical without also assuming the scalar field ϕ_B to be classical (in which case the energy-momentum tensor is conserved).

A similar problem arises when we consider a Bohmian semi-classical approximation to scalar electrodynamics, which describes a scalar field interacting with an electromagnetic field. In this case, the wave equation for the scalar field is of the form

$$i\partial_t \Psi(\phi, t) = \widehat{H}(\phi, A)\Psi(\phi, t),$$ (18.4)

where A is the vector potential. There is also a Bohmian scalar field ϕ_B and a charge current $j^\nu(\phi_B, A)$ that could act as the source term in Maxwell's equations

$$\partial_\mu F^{\mu\nu}(A) = j^\nu(\phi_B, A),$$ (18.5)

where $F^{\mu\nu}$ is the electromagnetic field tensor. In this case, we have $\partial_\nu \partial_\mu F^{\mu\nu} \equiv 0$ due to the anti-symmetry of $F^{\mu\nu}$. As such, the charge current must be conserved. However, the Bohmian equation of motion for the scalar field does not imply conservation. Hence, just as in the case of gravity, a consistency problem arises. As explained in Struyve [12], this problem can be overcome by eliminating the gauge invariance, either by assuming some gauge fixing or (equivalently) by working with gauge-independent degrees of freedom. In this way, we can straightforwardly derive a semi-classical approximation starting from the full Bohmian approach. For example, in the Coulomb gauge, the result is that there is an extra current j_Q^ν which appears in addition to the usual charge current and which depends on the quantum potential, so that Maxwell's equations read

$$\partial_\mu F^{\mu\nu}(A) = j^\nu(\phi_B, A) + j_Q^\nu(\phi_B, A).$$ (18.6)

While it is easy to eliminate the gauge invariance in the case of electrodynamics, this is notoriously difficult in the case of general relativity. One can formulate a Bohmian approach for the Wheeler–DeWitt equation for a scalar matter field interacting with gravity, but the usual formulation does not explicitly eliminate the gauge freedom arising from spatial diffeomorphism invariance. Our expectation is that one could find a semi-classical approximation given such a formulation. At least we find our expectation confirmed in simplified models, called mini-superspace models, where this invariance is eliminated. We will illustrate this for the model described by the homogeneous and isotropic Friedmann–Lemaître–Robertson–Walker (FLRW) metric and a uniform scalar field.

In this chapter, we are merely concerned with the formulation of Bohmian semi-classical approximations. Practical applications will be studied elsewhere. Such applications already have been studied in the context of quantum chemistry, see Section 18.3. It appears that Bohmian semi-classical approximations yield better or equivalent results compared to the usual semi-classical approximation. (They are better in the sense that they are closer to the exact quantum results.) This provides good hope that also in other contexts, such as quantum gravity, the Bohmian approach also gives better results. Potential applications might be found in inflation theory, where the back-reaction from the quantum fluctuations onto the classical background can be studied, or in black hole physics, to study the back-reaction from the Hawking radiation onto space-time.

The outline of the chapter is as follows. After introducing Bohmian mechanics in Section 18.2, we will discuss how to derive a Bohmian semi-classical approximation in the context of non-relativistic quantum mechanics. Semi-classical approximations to other quantum theories can be derived in a similar way. We present such approximations for scalar quantum electrodynamics in Section 18.4 and for a mini-superspace model in Section 18.5. More examples and details can be found in Struyve [12].

18.2 Bohmian Mechanics

18.2.1 Non-Relativistic Quantum Mechanics

Non-relativistic Bohmian mechanics (also called pilot-wave theory or de Broglie–Bohm theory) is a theory about point-particles in physical space moving under the influence of the wave function [6–9]. The equation of motion for the configuration $X = (\mathbf{X}_1, \ldots, \mathbf{X}_n)$ of the particles is given by[1]

$$\dot{X}(t) = v^{\psi}(X(t), t), \tag{18.7}$$

where $v^{\psi} = (\mathbf{v}_1^{\psi}, \ldots, \mathbf{v}_n^{\psi})$, with

$$\mathbf{v}_k^{\psi} = \frac{1}{m_k} \mathrm{Im}\left(\frac{\nabla_k \psi}{\psi}\right) = \frac{1}{m_k} \nabla_k S \tag{18.8}$$

and $\psi = |\psi| e^{iS}$. The wave function $\psi(x, t) = \psi(\mathbf{x}_1, \ldots, \mathbf{x}_n)$ itself satisfies the non-relativistic Schrödinger equation

$$i\partial_t \psi(x, t) = \left(-\sum_{k=1}^{n} \frac{1}{2m_k} \nabla_k^2 + V(x)\right) \psi(x, t). \tag{18.9}$$

For an ensemble of systems all with the same wave function ψ, there is a distinguished distribution given by $|\psi|^2$, which is called the *quantum equilibrium distribution*. This distribution is *equivariant*. That is, it is preserved by the particles dynamics Eq. (18.7) in the sense that if the particle distribution is given by $|\psi(x, t_0)|^2$ at some time t_0, then it is given by $|\psi(x, t)|^2$ at all times t. This follows from the fact that any distribution ρ that is transported by the particle motion satisfies the continuity equation

$$\partial_t \rho + \sum_{k=1}^{n} \nabla_k \cdot (\mathbf{v}_k^{\psi} \rho) = 0 \tag{18.10}$$

and that $|\psi|^2$ satisfies the same equation, i.e.

$$\partial_t |\psi|^2 + \sum_{k=1}^{n} \nabla_k \cdot (\mathbf{v}_k^{\psi} |\psi|^2) = 0, \tag{18.11}$$

---

[1] Throughout the chapter we assume units in which $\hbar = c = 1$.

as a consequence of the Schrödinger equation. It can be shown that for a typical initial configuration of the universe, the (empirical) particle distribution for an actual ensemble of subsystems within the universe will be given by the quantum equilibrium distribution [8, 9, 10]. Therefore for such a configuration Bohmian mechanics reproduces the standard quantum predictions.

Note that the velocity field is of the form $j^{\psi}/|\psi|^2$, where $j^{\psi} = (j_1^{\psi}, \ldots, j_n^{\psi})$ with $j_k^{\psi} = \mathrm{Im}(\psi^* \nabla_k \psi)/m_k$ is the usual quantum current. In other quantum theories, such as for example quantum field theories, the velocity can be defined in a similar way by dividing the appropriate current by the density. In this way equivariance of the density will be ensured. (See Stuyve and Valentini [13] for a treatment of arbitrary Hamiltonians.)

This theory solves the measurement problem. Notions such as measurement or observer play no fundamental role. Instead measurement can be treated as any other physical process.

There are two aspects of the theory that are important for deriving the semi-classical approximation. Firstly, Bohmian mechanics allows for an unambiguous analysis of the classical limit. Namely, the classical limit is obtained whenever the particles (or at least the relevant macroscopic variables, such as the center of mass) move classically, i.e. satisfy Newton's equation. By taking the time derivative of Eq. (18.7), we find that

$$m_k \ddot{\mathbf{X}}_k(t) = -\nabla_k (V(x) + Q^{\psi}(x,t))\big|_{x=X(t)}, \tag{18.12}$$

where

$$Q^{\psi} = -\sum_{k=1}^{n} \frac{1}{2m_k} \frac{\nabla_k^2 |\psi|}{|\psi|} \tag{18.13}$$

is the quantum potential. Hence, if the quantum force $-\nabla_k Q^{\psi}$ is negligible compared to the classical force $-\nabla_k V$, then the kth particle approximately moves along a classical trajectory.

Another aspect of the theory is that it allows for a simple and natural definition for the wave function of a subsystem [8, 10]. Namely, consider a system with wave function $\psi(x,y)$ where x is the configuration variable of the subsystem and y is the configuration variable of its environment. The actual configuration is (X, Y), where X is the configuration of the subsystem and Y is the configuration of the other particles. The wave function of the subsystem $\chi(x,t)$, called the *conditional wave function*, is then defined as

$$\chi(x,t) = \psi(x, Y(t), t). \tag{18.14}$$

This is a natural definition since the trajectory $X(t)$ of the subsystem satisfies

$$\dot{X}(t) = v^{\psi}(X(t), Y(t), t) = v^{\chi}(X(t), t). \tag{18.15}$$

That is, for the evolution of the subsystem's configuration we can either consider the conditional wave function or the total wave function (keeping the initial positions fixed). (The conditional wave function is also the wave function that would be found by a natural opera-tionalist method for defining the wave function of a quantum mechanical subsystem [14].) The time evolution of the conditional wave function is completely determined by the time evolution of ψ and that of Y. This implies that the conditional wave function does not necessarily satisfy a Schrödinger equation, although in many cases it does. This wave function collapses according to the usual text book rules when an actual measurement is performed.

18.2.2 Quantum Field Theory

We will also consider semi-classical approximations to quantum field theories. More specifically, we will consider bosonic quantum field theories. In Bohmian approaches to such theories it is most easy to introduce actual field variables rather than particle positions [11, 15]. To illustrate how this works, let us consider the free massless real scalar field (for the treatment of other bosonic field theories see Struyve [15]). Working in the functional Schrödinger picture, the quantum state vector is a wave functional $\Psi(\phi)$ defined on a space of scalar fields in 3-space and it satisfies the functional Schrödinger equation

$$i\partial_t \Psi(\phi, t) = \frac{1}{2} \int d^3x \left(-\frac{\delta^2}{\delta\phi(\mathbf{x})^2} + \nabla\phi(\mathbf{x}) \cdot \nabla\phi(\mathbf{x}) \right) \Psi(\phi, t). \tag{18.16}$$

The associated continuity equation is

$$\partial_t |\Psi(\phi, t)|^2 + \int d^3x \frac{\delta}{\delta\phi(\mathbf{x})} \left(\frac{\delta S(\phi, t)}{\delta\phi(\mathbf{x})} |\Psi(\phi, t)|^2 \right) = 0, \tag{18.17}$$

where $\Psi = |\Psi|e^{iS}$. This suggests the guidance equation

$$\dot{\phi}(\mathbf{x}, t) = \frac{\delta S(\phi, t)}{\delta\phi(\mathbf{x})} \bigg|_{\phi(\mathbf{x})=\phi(\mathbf{x},t)}. \tag{18.18}$$

(Note that in this case we did not distinguish notationally the actual field variable from the argument of the wave functional.) Taking the time derivative of this equation results in

$$\Box\phi(\mathbf{x}, t) = -\frac{\delta Q^\Psi(\phi, t)}{\delta\phi(\mathbf{x})} \bigg|_{\phi(\mathbf{x})=\phi(\mathbf{x},t)}, \tag{18.19}$$

where

$$Q^\Psi = -\frac{1}{2|\Psi|} \int d^3x \frac{\delta^2 |\Psi|}{\delta\phi(\mathbf{x})^2}, \tag{18.20}$$

where Q^Ψ is the quantum potential. The classical limit is obtained whenever the quantum force, i.e. the right hand side of Eq. (18.19), is negligible. Then the field approximately satisfies the classical field equation $\Box\phi = 0$.

One can also consider the conditional wave functional of a subsystem. A subsystem can in this case be regarded as a system confined to a certain region in space. The conditional wave functional for the field confined to that region is then obtained from the total wave functional by conditioning over the actual field value on the complement of that region. However, in the following we will not consider this kind of conditional wave functional. Rather, there will be other degrees of freedom, like for example other fields, which will be conditioned over.

This Bohmian approach is not Lorentz invariant. The guidance Eq. (18.18) is formulated with respect to a preferred reference frame and as such violates Lorentz invariance. This violation does not show up in the statistical predictions given quantum equilibrium, since the theory makes the same predictions as standard quantum theory which are Lorentz invariant.[2] The difficulty in finding a Lorentz invariant theory resides in the fact that any adequate formulation of quantum theory must be non-local [16]. One approach to make the Bohmian theory Lorentz invariant is by introducing a foliation which is determined by the wave function in a covariant way [17]. In this chapter, we will not attempt to maintain Lorentz invariance. As such, the Bohmian semi-classical approximations will not be Lorentz invariant, (very likely) not even concerning the statistical predictions. This is in contrast with the usual approach like the one for gravity given by Eq. (18.1) and Eq. (18.2) which is fully Lorentz invariant. However, this does not take away the expectation that the Bohmian semi-classical approximation will give better or at least equivalent results compared to the usual approach.

18.2.3 Quantum Gravity

In canonical quantum gravity, the state vector is a functional $\Psi(h, \phi)$ on the space of 3-metrics $h_{ij}(\mathbf{x})$ on a three-dimensional (3D) manifold and fields $\phi(\mathbf{x})$ (in the case the matter is described by a quantized scalar field). The wave functional is static and merely satisfies the constraints [2]

$$\mathcal{H}\Psi(h, \phi) = 0, \tag{18.21}$$

$$\mathcal{H}_i\Psi(h, \phi) = 0. \tag{18.22}$$

Their explicit forms are not important here. The latter constraint expresses the fact that the wave functional is invariant under infinitesimal diffeomorphisms of 3-space. The former equation is the Wheeler–DeWitt equation. It is believed that this equation contains the dynamical content of the theory. However, it is as yet not clear how this dynamical content should be extracted. This is the problem of time [2, 18].

---

[2] Actually, this statement needs some qualifications since regulators need to be introduced to make the theory and its statistical predictions well defined [15].

In the Bohmian approach, there is an actual 3-metric and a scalar field, whose dynamics depends on the wave functional [19–21]. The dynamics expresses how the Bohmian configuration changes along a succession of 3D space-like surfaces.[3] Although the wave function is stationary, the Bohmian configuration will change along these surfaces for generic wave functions. This is how the Bohmian approach solves the problem of time.

Some cosmological applications of the Bohmian approach to quantum gravity are the explanation of the quantum-to-classical transition in inflation theory [22, 23] and the study of space-time singularities [24–26].

18.3 Non-Relativistic Quantum Mechanics

18.3.1 Usual Versus Bohmian Semi-Classical Approximation

Consider a composite system of just two particles. The usual semi-classical approach (also called the mean-field approach) goes as follows. Particle 1 is described quantum mechanically, by a wave function $\chi(\mathbf{x}_1, t)$, which satisfies the Schrödinger equation

$$i\partial_t \chi(\mathbf{x}_1, t) = \left[-\frac{1}{2m_1} \nabla_1^2 + V(\mathbf{x}_1, \mathbf{X}_2(t)) \right] \chi(\mathbf{x}_1, t), \tag{18.23}$$

where the potential is evaluated at the position of the second particle $\mathbf{X}_2$, which satisfies Newton's equation

$$
\begin{aligned}
m_2 \ddot{\mathbf{X}}_2(t) &= -\left\langle \chi \left| \nabla_2 V(\mathbf{x}_1, \mathbf{x}_2) \right|_{\mathbf{x}_2 = \mathbf{X}_2(t)} \right| \chi \right\rangle \\
&= \int d^3 x_1 |\chi(\mathbf{x}_1, t)|^2 [-\nabla_2 V(\mathbf{x}_1, \mathbf{x}_2)] \Big|_{\mathbf{x}_2 = \mathbf{X}_2(t)}.
\end{aligned} \tag{18.24}
$$

So the force on the right hand side is averaged over the quantum particle.

An alternative semi-classical approach based on Bohmian mechanics was proposed independently by Gindensperger *et al.* [27] and Prezhdo and Brooksby [28]. In this approach there is also an actual position for particle 1, denoted by $\mathbf{X}_1$, which satisfies the equation

$$\dot{\mathbf{X}}_1(t) = \mathbf{v}^\chi(\mathbf{X}_1(t), t), \tag{18.25}$$

where

$$\mathbf{v}^\chi = \frac{1}{m_1} \text{Im} \frac{\nabla \chi}{\chi}, \tag{18.26}$$

and where χ satisfies the Schrödinger equation (18.23). But instead of Eq. (18.24), the second particle now satisfies

$$m_2 \ddot{\mathbf{X}}_2(t) = -\nabla_2 V(\mathbf{X}_1(t), \mathbf{x}_2) \Big|_{\mathbf{x}_2 = \mathbf{X}_2(t)}, \tag{18.27}$$

---

[3] The succession of the surfaces is determined by the lapse function and different choices of lapse function lead to different Bohmian dynamics. This determines that the dynamics is not invariant under space-time diffeomorphisms. Again, the root of the problem is the non-locality of quantum theory.

where the force depends on the position of the first particle. So in this approximation the second particle is not acted upon by some average force, but rather by the actual particle of the quantum system. This approximation is therefore expected to yield a better approach than the usual approach, in the sense that it yields predictions closer to those predicted by full quantum theory, especially in the case where the wave function evolves into a superposition of non-overlapping packets. This is indeed confirmed by a number of studies, as we will discuss below.

Let us first mention some properties of this approximation and compare them to the usual approach. In the mean field approach, the specification of an initial wave function $\chi(\mathbf{X}_1, t_0)$, an initial position $\mathbf{X}_2(t_0)$ and velocity $\dot{\mathbf{X}}_2(t_0)$ determines a unique solution for the wave function and the trajectory of the classical particle. In the Bohmian approach also the initial position $\mathbf{X}_1(t_0)$ of the particle of the quantum system needs to be specified in order to uniquely determine a solution. Different initial positions $\mathbf{X}_1(t_0)$ yield different evolutions for the wave function and the classical particle. This is because the evolution of each of the variables $\mathbf{X}_1, \mathbf{X}_2, \chi$ depends on the others. Namely, the evolution of χ depends on $\mathbf{X}_2$ via Eq. (18.23), whose evolution in turn depends on $\mathbf{X}_1$ via Eq. (18.27), whose evolution in turn depends on χ via Eq. (18.25). (This should be contrasted with the full Bohmian theory, where the wave function acts on the particles, but there is no back-reaction from the particles onto the wave function.)

The initial configuration $\mathbf{X}_1(t_0)$ should be considered random with distribution $|\chi(\mathbf{X}_1, t_0)|^2$. However, this does not imply that $\mathbf{X}_1(t)$ is random with distribution $|\chi(\mathbf{X}_1, t)|^2$ for later times t. It is not even clear what the latter statement should mean, since different initial positions $\mathbf{X}_1(t_0)$ lead to different wave function evolution; so which wave function should $\chi(\mathbf{X}_1, t)$ be?

This semi-classical approximation has been applied to a number of systems. Prezhdo and Brookby studied the case of a light particle scattering off a heavy particle [28]. They considered the scattering probability over time and found that the Bohmian semi-classical approximation was in better agreement with the exact quantum mechanical prediction than the usual approximation. The Bohmian semi-classical approximation gives probability one for the scattering to have happened after some time, in agreement with the exact result, whereas the probability predicted by the usual approach does not reach one. The reported reason for the better results is that the wave function of the quantum particle evolves into a superposition of non-overlapping packets, which yields bad results for the usual approach (since the force on the classical particle contains contributions from both packets), but not for the Bohmian approach. These results were confirmed and further expanded by Gindensperger *et al.* [29]. Other examples have been considered [27, 30, 31]. In those cases, the Bohmian semi-classical approximation gave very good agreement with the exact quantum or experimental results. It was always either better or comparable to the usual approach. These results give good hope that the Bohmian semi-classical approximation will also give better results than the usual approximation in other domains such as quantum gravity.

18.3.2 Derivation of the Bohmian Semi-Classical Approximation

The Bohmian semi-classical approach can easily be derived from the full Bohmian theory.[4] Consider a system of two particles. In the Bohmian description of this system, we have a wave function $\psi(\mathbf{x}_1, \mathbf{x}_2, t)$ and positions $\mathbf{X}_1(t), \mathbf{X}_2(t)$, which respectively satisfy the Schrödinger equation

$$i\partial_t\psi = \left[-\frac{1}{2m_1}\nabla_1^2 - \frac{1}{2m_2}\nabla_2^2 + V(\mathbf{x}_1, \mathbf{x}_2)\right]\psi \qquad (18.28)$$

and the guidance equations

$$\dot{\mathbf{X}}_1(t) = \mathbf{v}_1^\psi(\mathbf{X}_1(t), \mathbf{X}_2(t), t), \qquad \dot{\mathbf{X}}_2(t) = \mathbf{v}_2^\psi(\mathbf{X}_1(t), \mathbf{X}_2(t), t). \qquad (18.29)$$

The conditional wave function $\chi(\mathbf{x}_1, t) = \psi(\mathbf{x}_1, \mathbf{X}_2(t), t)$ for particle 1 satisfies the equation

$$i\partial_t\chi(\mathbf{x}_1, t) = \left(-\frac{\nabla_1^2}{2m_1} + V(\mathbf{x}_1, \mathbf{X}_2(t))\right)\chi(\mathbf{x}_1, t) + I(\mathbf{x}_1, t), \qquad (18.30)$$

where

$$I(\mathbf{x}_1, t) = \left(-\frac{\nabla_2^2}{2m_2}\psi(\mathbf{x}_1, \mathbf{x}_2, t)\right)\bigg|_{\mathbf{x}_2 = \mathbf{X}_2(t)} + i\nabla_2\psi(\mathbf{x}_1, \mathbf{x}_2, t)\bigg|_{\mathbf{x}_2 = \mathbf{X}_2(t)} \cdot \mathbf{v}_2^\psi(\mathbf{X}_1(t), \mathbf{X}_2(t), t). \qquad (18.31)$$

So in case I is negligible in Eq. (18.30), up to a time-dependent factor times χ,[5] we are led to the Schrödinger equation (18.23). This will, for example, be the case if m_2 is much larger than m_1 (I is inversely proportional to m_2) and if the wave function slowly varies as a function of $\mathbf{x}_2$. We also have

$$m_2\ddot{\mathbf{X}}_2(t) = -\nabla_2\left[V(\mathbf{X}_1(t), \mathbf{x}_2) + Q^\psi(\mathbf{X}_1(t), \mathbf{x}_2, t)\right]\bigg|_{\mathbf{x}_2 = \mathbf{X}_2(t)}, \qquad (18.32)$$

with Q^ψ the quantum potential. We obtain the classical Eq. (18.27), if the quantum force is negligible compared to the classical force.

In this way we obtain the equations for a semi-classical formulation. In addition, we also have the conditions under which they will be valid. For other quantum theories, such as quantum gravity, we can follow a similar path to find a Bohmian semi-classical approximation.

---

[4] The derivation is very close to the one followed by Gindensperger *et al.* [27]. A difference is that they also let the wave function of the quantum system depend parametrically on the position of the classical particle. This leads to a quantum force term in the Eq. (18.27) for particle 2. However, this does not seem to lead to a useful set of equations. In particular, they can not be numerically integrated by simply specifying the initial wave function and particle positions. In any case, Gindensperger *et al.* drop this quantum force when considering examples [27, 29, 30], so that the resulting equations correspond to the ones presented above.

[5] If I contains a term of the form $f(t)\chi$, then it can be eliminated by changing the phase of χ by a time-dependent term.

18.4 Scalar Electrodynamics

We consider scalar electrodynamics to illustrate the issues with developing a Bohmian semi-classical approximation for a gauge theory. There are various *equivalent* ways of formulating the Bohmian approach [12]. These formulations can either be found by considering different gauges or by working with different choices of gauge-independent variables. Here, we will consider two examples of gauges, namely the temporal gauge, which is an incomplete gauge fixing, and the Coulomb gauge, which completely fixes the gauge symmetry. Using the former gauge, we are not immediately led to a semi-classical approximation, due to the remaining gauge freedom, while we *are* using the latter gauge.

In classical scalar electrodynamics, the equations of motion for the scalar field ϕ and the vector potential $A^\mu = (A_0, \mathbf{A})$ are

$$D_\mu D^\mu \phi + m^2 \phi = 0, \qquad \partial_\mu F^{\mu\nu} = j^\nu, \qquad (18.33)$$

where $D_\mu = \partial_\mu + ieA_\mu$ is the covariant derivative, $F^{\mu\nu} = \partial^\mu A^\nu - \partial^\nu A^\mu$ the electromagnetic field tensor and

$$j^\nu = ie \left(\phi^* D^\nu \phi - \phi D^{\nu *} \phi^* \right) \qquad (18.34)$$

is the charge current. The theory has a local gauge symmetry

$$\phi \to e^{ie\alpha} \phi, \quad A^\mu \to A^\mu - \partial^\mu \alpha. \qquad (18.35)$$

One possible choice of gauge is the temporal gauge $A_0 = 0$. It does not completely fix the gauge; there is still a residual gauge symmetry given by the time-independent transformations

$$\phi \to e^{ie\theta} \phi, \qquad \mathbf{A} \to \mathbf{A} + \nabla \theta, \qquad (18.36)$$

with $\dot{\theta} = 0$. Quantization in this gauge leads to the following functional Schrödinger equation for $\Psi(\phi, \mathbf{A}, t)$ [32]:[6]

$$i\partial_t \Psi = \int d^3 x \left(-\frac{\delta^2}{\delta \phi^* \delta \phi} + |\mathbf{D}\phi|^2 + m^2 |\phi|^2 - \frac{1}{2} \frac{\delta^2}{\delta \mathbf{A}^2} + \frac{1}{2} (\nabla \times \mathbf{A})^2 \right) \Psi, \qquad (18.37)$$

together with the constraint

$$\nabla \cdot \frac{\delta \Psi}{\delta \mathbf{A}} + ie \left(\phi^* \frac{\delta \Psi}{\delta \phi^*} - \phi \frac{\delta \Psi}{\delta \phi} \right) = 0. \qquad (18.38)$$

The constraint expresses the fact that the wave functional is invariant under time-independent gauge transformations, i.e. $\Psi(\phi, \mathbf{A}) = \Psi(e^{ie\theta} \phi, \mathbf{A} + \nabla \theta)$, with θ time-independent. The constraint is compatible with the Schrödinger equation: if it is satisfied at one time, it is satisfied at all times.

---

[6] The wave functional should be understood as a functional of the real and imaginary part of ϕ. In addition, writing $\phi = (\phi_r + i\phi_i)/\sqrt{2}$, we have that the functional derivatives are given by $\delta/\delta\phi = (\delta/\delta\phi_r - i\delta/\delta\phi_i)/\sqrt{2}$ and $\delta/\delta\phi^* = (\delta/\delta\phi_r + i\delta/\delta\phi_i)/\sqrt{2}$.

In the Bohmian approach [33], there are actual configurations ϕ and $\mathbf{A}$ that satisfy the guidance equations

$$\dot{\phi} = \frac{\delta S}{\delta \phi^*}, \qquad \dot{\mathbf{A}} = \frac{\delta S}{\delta \mathbf{A}}, \tag{18.39}$$

where $\Psi = |\Psi| e^{iS}$. These equations are invariant under the time-independent gauge transformations Eq. (18.36) because of the constraint Eq. (18.38).

In the framework of standard quantum theory, there is a natural semi-classical approximation that treats the vector potential classically and the scalar field quantum mechanically. The scalar field is described by a wave functional $\chi(\phi, t)$ which satisfies

$$i\partial_t \chi = \int d^3 x \left(-\frac{\delta^2}{\delta\phi^*\delta\phi} + |\mathbf{D}\phi|^2 + m^2|\phi|^2 \right) \chi \tag{18.40}$$

and the electromagnetic field satisfies Maxwell's equations (with $A_0 = 0$)

$$\partial_\mu F^{\mu\nu} = \langle \chi | \widehat{j}^\nu | \chi \rangle, \tag{18.41}$$

where

$$\langle \chi | \widehat{j}^0 | \chi \rangle = \int \mathcal{D}\phi \, \Psi^* \mathcal{C}\Psi = e \int \mathcal{D}\phi \, \Psi^* \left(\phi^* \frac{\delta\Psi}{\delta\phi^*} - \phi\frac{\delta\Psi}{\delta\phi} \right),$$

$$\langle \chi | \widehat{\mathbf{j}} | \chi \rangle = ie \int \mathcal{D}\phi |\Psi|^2 \left(\phi \mathbf{D}^* \phi^* - \phi^* \mathbf{D}\phi \right), \tag{18.42}$$

with

$$\mathcal{C}(\mathbf{x}) = e \left(\phi^*(\mathbf{x}) \frac{\delta}{\delta\phi^*(\mathbf{x})} - \phi(\mathbf{x})\frac{\delta}{\delta\phi(\mathbf{x})} \right) \tag{18.43}$$

the charge density operator in the functional Schrödinger picture. This theory is consistent since $\partial_\mu \langle \chi | \widehat{j}^\mu | \chi \rangle = 0$, as a consequence of the Schrödinger equation (18.40).

A natural guess for a Bohmian semi-classical approximation similar to the usual one is the following (and can be obtained from the full Bohmian approach by considering the conditional wave function $\chi(\phi, t) = \Psi(\phi, \mathbf{A}(t), t)$). An actual field ϕ is introduced that satisfies $\dot{\phi} = \delta S/\delta\phi^*$, where the wave functional satisfies Eq. (18.40), and Maxwell's equations read $\partial_\mu F^{\mu\nu} = j^\nu$, where j^μ is the classical expression for the charge current. However, the second order equation for the Bohmian field is

$$\ddot{\phi} - D^2\phi + m^2\phi = -\frac{\delta Q^\chi}{\delta\phi^*}, \tag{18.44}$$

where $Q^\chi = -\frac{1}{|\chi|} \int d^3 x \left(\frac{\delta^2 |\chi|}{\delta\phi^*\delta\phi} \right)$. As a consequence, $\partial_\mu j^\mu = -i\mathcal{C}Q^\chi$ and hence Maxwell's equations imply that $\mathcal{C}Q^\chi = 0$ or $Q^\chi = Q^\chi(|\phi|^2)$. This is a constraint on the wave functional that was absent in the usual semi-classical theory. It also seems to be a rather strong condition. It will, for example, be satisfied if the scalar field evolves classically (i.e. when the right-hand side of Eq. (18.44) is zero) but it is unclear whether there are other solutions.

So the conclusion seems to be that if we assume **A** classical, then ϕ should also behave classically. This is not surprising since the gauge symmetry implies that the physical (i.e. gauge invariant) degrees of freedom are some combination of the fields **A** and ϕ. So one can not just assume **A** classical and keep ϕ fully quantum.

In Struyve [12], we showed that the problem disappears if we eliminate the gauge freedom, for example, by using a gauge which completely fixes the gauge freedom. We discussed in detail the Coulomb and the unitary gauge and showed that a semi-classical approximation can easily be obtained by considering either the scalar or electromagnetic field classically.

Let us consider the Coulomb gauge $\boldsymbol{\nabla} \cdot \mathbf{A} = 0$ here. In the full Bohmian approach [12, 15] we have that there are actual fields[7] ϕ and $\mathbf{A}^T$ that are guided by a wave functional $\Psi(\phi, \mathbf{A}^T, t)$ which satisfies the functional Schrödinger equation

$$i\partial_t \Psi = \int d^3x \left(-\frac{\delta^2}{\delta\phi^*\delta\phi} + |(\boldsymbol{\nabla} - ie\mathbf{A}^T)\phi|^2 + m^2|\phi|^2 - \frac{1}{2}C\frac{1}{\nabla^2}C - \frac{1}{2}\frac{\delta^2}{\delta\mathbf{A}^{T2}} + \frac{1}{2}(\boldsymbol{\nabla}\times\mathbf{A}^T)^2 \right)\Psi.$$
$$(18.45)$$

The first three terms in the Hamiltonian correspond to the Hamiltonian of a scalar field minimally coupled to a transverse vector potential. The fourth term corresponds to the Coulomb potential and the remaining terms to the Hamiltonian of a free electromagnetic field. The guidance equations are

$$\dot{\phi} = \frac{\delta S}{\delta\phi^*} - e\phi\frac{1}{\nabla^2}CS, \qquad \dot{\mathbf{A}}^T = \frac{\delta S}{\delta\mathbf{A}^T}. \qquad (18.46)$$

Defining

$$A_0 = -i\frac{1}{\nabla^2}CS, \qquad (18.47)$$

we can rewrite the guidance equation for the scalar field as

$$D_0\phi = \frac{\delta S}{\delta\phi^*}. \qquad (18.48)$$

The definition of A_0 was motivated by analogy with the classical equations of motion [12].

While this Bohmian approach is equivalent to the one in the temporal gauge [12], it naturally leads to the following semi-classical approximation (by considering the conditional wave function for the scalar field). The wave functional $\chi(\phi, t)$ satisfies

$$i\partial_t\chi = \int d^3x \left(-\frac{\delta^2}{\delta\phi^*\delta\phi} + |(\boldsymbol{\nabla} - ie\mathbf{A}^T)\phi|^2 + m^2|\phi|^2 - \frac{1}{2}C\frac{1}{\nabla^2}C \right)\chi \qquad (18.49)$$

and guides the actual scalar field through

$$D_0\phi = \frac{\delta S}{\delta\phi^*}, \qquad (18.50)$$

---

[7] We have used the decomposition $\mathbf{A} = \mathbf{A}^T + \mathbf{A}^L$, where $\mathbf{A}^T$ and $\mathbf{A}^L$ are, respectively, the transverse and longitudinal part of the vector potential. The Coulomb gauge then corresponds to $\mathbf{A}^L = 0$.

where A_0 is defined as before and with S now the phase of χ. The vector potential $A^\mu = (A_0, \mathbf{A}^T)$ satisfies Maxwell's equations

$$\partial_\mu F^{\mu\nu} = j^\nu + j_Q^\nu, \tag{18.51}$$

where $j_Q^\nu = (0, \mathbf{j}_Q)$ is an additional "quantum" current, with

$$\mathbf{j}_Q = i\boldsymbol{\nabla}\frac{1}{\nabla^2}\mathcal{C}Q^\chi \tag{18.52}$$

and

$$Q^\chi = -\frac{1}{|\chi|}\int d^3x \left(\frac{\delta^2|\chi|}{\delta\phi^*\delta\phi} + \frac{1}{2}\mathcal{C}\frac{1}{\nabla^2}\mathcal{C}|\chi| \right) \tag{18.53}$$

the quantum potential. These equations are consistent in the sense that $\partial_\mu(j^\mu + j_Q^\mu) = 0$, as a consequence of the second order equation

$$D_\mu D^\mu \phi - m^2\phi = -\frac{\delta Q^\chi}{\delta\phi^*}, \tag{18.54}$$

which follows from taking the time derivative of Eq. (18.50).

18.5 Quantum Gravity: Mini-Superspace Model

The structure of the Bohmian approach to canonical gravity (outlined in Section 18.2.3) is similar to that of scalar electrodynamics in the temporal gauge. Namely, in both cases there is a constraint on the wave functional which expresses invariance under infinitesimal gauge transformations: spatial diffeomorphisms in the former case and phase transformations in the latter case. In both cases the gauge invariance seems to be the source of the problem in formulating a consistent Bohmian semi-classical approximation. In the case of quantum electrodynamics the problem was overcome by gauge fixing. Presumably one can find a similar solution in quantum gravity. However, finding a suitable gauge is a notoriously hard problem in this case. We can, however, consider a mini-superspace model which is a symmetry-reduced approach to quantum gravity where homogeneity and isotropy are assumed. In this case, the spatial diffeomorphism invariance is eliminated and we can straightforwardly develop a Bohmian semi-classical approximation, as we will now show.

In the classical mini-superspace model, the universe is described by the FLRW metric

$$ds^2 = N(t)^2 dt^2 - a(t)^2 d\Omega_3^2, \tag{18.55}$$

where N is the lapse function, $a = e^\alpha$ is the scale factor[8] and $d\Omega_3^2$ is the metric on 3-space with constant curvature k. Assuming matter that is described by a homogeneous scalar field

---

[8] The reason for introducing the variable α is that it is unbounded, unlike the scale factor, which satisfies $a > 0$.

ϕ, the equations of motion are [12, 34, 35]:

$$\frac{1}{2}\dot{\alpha}^2 = \frac{1}{\kappa}\left(\frac{1}{2}\dot{\phi}^2 + V_M\right) + V_G,$$ (18.56)

$$\ddot{\phi} + 3\dot{\alpha}\dot{\phi} + \partial_\phi V_M = 0,$$ (18.57)

where the gauge $N = 1$ is chosen,[9] $\kappa = 3/4\pi G$,

$$V_G = -\frac{1}{2}ke^{-2\alpha} + \frac{1}{6}\Lambda$$ (18.58)

is the gravitational potential, with Λ the cosmological constant, and V_M is the potential for the matter field.

Canonical quantization of the classical theory yields the Wheeler–DeWitt equation:

$$(\widehat{H}_G + \widehat{H}_M)\psi = 0,$$ (18.59)

where

$$\widehat{H}_G = \frac{1}{2\kappa e^{3\alpha}}\partial_\alpha^2 + \kappa e^{3\alpha}V_G, \qquad \widehat{H}_M = -\frac{1}{2e^{3\alpha}}\partial_\phi^2 + e^{3\alpha}V_M.$$ (18.60)

In the corresponding Bohmian approach [35], there is an actual FLRW metric of the form Eq. (18.55) and scalar field, whose time evolutions are determined by the guidance equations

$$\dot{\alpha} = -\frac{N}{\kappa e^{3\alpha}}\partial_\alpha S, \qquad \dot{\phi} = \frac{N}{e^{3\alpha}}\partial_\phi S,$$ (18.61)

where N is an arbitrary lapse function.[10] In the gauge $N = 1$, these equations imply

$$\frac{1}{2}\dot{\alpha}^2 = \frac{1}{\kappa}\left(\frac{1}{2}\dot{\phi}^2 + V_M + Q_M^\psi\right) + V_G + Q_G^\psi,$$ (18.62)

$$\ddot{\phi} + 3\dot{\alpha}\dot{\phi} + \partial_\phi(V_M + Q_M^\psi + \kappa Q_G^\psi) = 0,$$ (18.63)

where

$$Q_G^\psi = \frac{1}{2\kappa^2 e^{6\alpha}}\frac{\partial_\alpha^2|\psi|}{|\psi|}, \qquad Q_M^\psi = -\frac{1}{2e^{6\alpha}}\frac{\partial_\phi^2|\psi|}{|\psi|}.$$ (18.64)

We will now look for a semi-classical approximation where the scale factor behaves approximately classically. In order to do so, we assume again the gauge $N = 1$ and we

---

[9] The theory is time-reparameterization invariant. Solutions that differ only by a time-reparameterization are considered physically equivalent. Choosing the gauge $N = 1$ corresponds to a particular time-parameterization.

[10] Just as the classical theory, the Bohmian approach is time-reparameterization invariant. This is a special feature of mini-superspace models [36, 37]. As mentioned before, for the usual formulation of the Bohmian dynamics for the Wheeler–DeWitt theory of quantum gravity, a particular space-like foliation of space-time or, equivalently, a particular choice of "initial" space-like hypersurface and lapse function, needs to be introduced. Different foliations (or lapse functions) yield different Bohmian theories.

consider the conditional wave function $\chi(\phi, t) = \psi(\phi, \alpha(t))$, given a set of trajectories $(\alpha(t), \phi(t))$. Using

$$\partial_t \chi(\phi, t) = \partial_\alpha \psi(\phi, \alpha)\big|_{\alpha = \alpha(t)} \dot{\alpha}(t), \tag{18.65}$$

we can write

$$i \partial_t \chi = \widehat{H}_M \chi + I, \tag{18.66}$$

where[11]

$$I = \frac{1}{\dot{\alpha}} i \partial_t \chi \left(\dot{\alpha} + \frac{1}{\kappa e^{3\alpha}} \partial_\alpha S\big|_{\alpha(t)} \right) + \frac{1}{2\kappa e^{3\alpha}} \left[(\partial_\alpha S)^2 + i \partial_\alpha^2 S \right]\Big|_{\alpha(t)} \chi + \kappa e^{3\alpha} (V_G + Q_G^\psi)\big|_{\alpha = \alpha(t)} \chi. \tag{18.67}$$

When I is negligible (up to a real time-dependent function times χ), Eq. (18.66) becomes the Schrödinger equation for a homogeneous matter field in an external FLRW metric. We can further assume the quantum potential Q_G^ψ to be negligible compared to other terms in Eq. (18.62). As such, we are led to the semi-classical theory:

$$i \partial_t \chi = \widehat{H}_M \chi, \tag{18.68}$$

$$\dot{\phi} = \frac{1}{e^{3\alpha}} \partial_\phi S, \tag{18.69}$$

$$\frac{1}{2} \dot{\alpha}^2 = \frac{1}{\kappa} \left(\frac{1}{2} \dot{\phi}^2 + V_M + Q_M^\chi \right) + V_G \equiv -\frac{1}{\kappa e^{3\alpha}} \partial_t S + V_G. \tag{18.70}$$

Let us now consider when the term I will be negligible. The quantity in brackets in the first term would be zero when evaluated for the actual trajectory $\phi(t)$ (because of the guidance equation for α). As such, the first term will be negligible if the actual scale factor evolves approximately independently of the scalar field. The second term will be negligible if S varies slowly with respect to α or if the term in square brackets is approximately independent of ϕ. In the latter case, the second term becomes a time-dependent function times χ, which can be eliminated by changing the phase of χ. Similarly, if $Q_G^\psi \ll V_G$ then the third term also becomes a time-dependent function times χ.

In the usual semi-classical approximation, one has Eq. (18.68) and

$$\frac{1}{2} \dot{\alpha}^2 = \frac{1}{\kappa e^{3\alpha}} \langle \chi | \widehat{H}_M | \chi \rangle + V_G, \tag{18.71}$$

with χ normalized to one. These equations follow from Eqs. (18.1) and (18.2). In Struyve [12] an example is worked out for which the Bohmian semi-classical approximation gives better results than this approximation. (Note that Vink himself, in his seminal paper on applying the Bohmian approach to quantum gravity, considers a derivation of the usual semi-classical approximation, rather than the Bohmian one. But he hinted on the Bohmian semi-classical approximation in Kowalski-Gilman and Vink [38].)

---

[11] To obtain this equation, note that $\partial_\alpha^2 \psi = [(\partial_\alpha S)^2 + i \partial_\alpha^2 S + \partial_\alpha^2 |\psi|/|\psi|] \psi + 2 i \partial_\alpha S \partial_\alpha \psi$, so that $\partial_\alpha^2 \psi|_{\alpha = \alpha(t)} = [(\partial_\alpha S)^2 + i \partial_\alpha^2 + \partial_\alpha^2 |\psi|/|\psi|]_{\alpha = \alpha(t)} \chi + 2 i \partial_\alpha S \partial_t \chi / \dot{\alpha}$. Using this equation together with Eq. (18.59) we obtain Eq. (18.66).

18.6 Conclusion

We have shown how semi-classical approximations can be developed using Bohmian mechanics. We have obtained these approximations from the full Bohmian theory by assuming certain degrees of freedom to evolve approximately classically. This was illustrated for non-relativistic systems. If there is a gauge symmetry, like in electrodynamics or gravity, then extra care is required in order to obtain a consistent semi-classical theory. By eliminating the gauge symmetry, either by imposing a gauge or by working with gauge-independent degrees of freedom, we were able to find a semi-classical approximation in the case of scalar quantum electrodynamics. For quantum gravity, eliminating the gauge symmetry (more precisely the spatial diffeomorphism invariance) is notoriously hard. We have only considered the simplified mini-superspace approach to quantum gravity, which describes an isotropic and homogeneous universe, and where the diffeomorphism invariance is explicitly eliminated. More general cases in quantum gravity still need to be studied. For example, for the case of inflation theory, where one usually considers fluctuations around an isotropic and homogeneous universe, it should not be too difficult to develop a Bohmian semi-classical approximation.

Apart from possible applications in quantum cosmology, such as inflation theory, it might also be interesting to consider potential applications in quantum electrodynamics or quantum optics. In particular, since the results may be compared to the predictions of full quantum theory, this may give us a handle on where to expect better results for the Bohmian semi-classical approximation compared to the usual one in the case of quantum gravity where the full quantum theory is not known. That is, it might give us better insight into which effects are truly quantum and which effects are merely artifacts of the approximation.

Further developments may include higher order corrections to the semi-classical approximation. One way of doing this might be by following the ideas presented in Norsen [39] and Norsen *et al.* [40]. As explained there, one might introduce extra wave functions for a subsystem in addition to the conditional wave function. These wave functions interact with each other and the Bohmian configurations. By including more of those wave functions one presumably obtains better approximations to the full quantum result.

Finally, although we regard the Bohmian semi-classical approximation for quantum gravity as an approximation to some deeper quantum theory for gravity, one could also entertain the possibility that it is a fundamental theory on its own. At least, there is presumably as yet no experimental evidence against it.

18.7 Acknowledgments

This work was supported by the Actions de Recherches Concertées (ARC) of the Belgium Wallonia–Brussels Federation under contract no. 12-17/02.

References

[1] R. M. Wald, *Quantum Field Theory in Curved Spacetime and Black Hole Thermodynamics.* (Chicago: The University of Chicago Press, 1994).

[2] C. Kiefer, *Quantum Gravity.* International Series of Monographs on Physics **124**. (Oxford: Clarendon Press, 2004).

[3] D.N. Page and C.D. Geilker (1981) Indirect Evidence for Quantum Gravity, *Phys. Rev. Lett.,* **47**, 979–82.

[4] A. Diez-Tejedor and D. Sudarsky (2012) Towards a formal description of the collapse approach to the inflationary origin of the seeds of cosmic structure. *JCAP,* **07**, 045 and arXiv:1108.4928 [gr-qc].

[5] M. Derakhshani (2014) Newtonian semiclassical gravity in the Ghirardi–Rimini–Weber theory with matter density ontology. *Phys. Lett. A*, **378**, 990–8 and arXiv:1304.0471 [gr-qc].

[6] D. Bohm and B.J. Hiley. *The Undivided Universe.* (New York: Routledge, 1993).

[7] P.R. Holland. *The Quantum Theory of Motion.* (Cambridge: Cambridge University Press, 1993).

[8] D. Dürr and S. Teufel. *Bohmian Mechanics.* (Berlin: Springer-Verlag, 2009).

[9] D. Dürr, S. Goldstein and N. Zanghì. *Quantum Physics Without Quantum Philosophy.* (Berlin: Springer-Verlag, 2012).

[10] D. Dürr, S. Goldstein and N. Zanghì. (1992) Quantum Equilibrium and the Origin of Absolute Uncertainty. *J. Stat. Phys.*, **67**, 843–907 and arXiv:quant-ph/0308039.

[11] W. Struyve (2011) Pilot-wave approaches to quantum field theory, *J. Phys.: Conf. Ser.*, **306**, 012047 and arXiv:1101.5819v1 [quant-ph].

[12] W. Struyve (2015) Semi-classical approximations based on Bohmian mechanics. arXiv:1507.04771 [quant-ph].

[13] W. Struyve and A. Valentini. (2009) de Broglie–Bohm guidance equations for arbitrary Hamiltonians. *J. Phys. A*, **42**, 035301 and arXiv:0808.0290v3 [quant-ph].

[14] T. Norsen and W. Struyve. (2014) Weak Measurement and Bohmian Conditional Wave Functions. *Ann. Phys.*, **350**, 166–78 and arXiv:1305.2409 [quant-ph].

[15] W. Struyve. (2010) Pilot-wave theory and quantum fields. *Rep. Prog. Phys.*, **73**, 106001 and arXiv:0707.3685v4 [quant-ph].

[16] S. Goldstein, T. Norsen, D.V. Tausk and N. Zanghì. (2011) Bell's theorem. *Scholarpedia*, **6**(10), 8378.

[17] D. Dürr, S. Goldstein, T. Norsen, W. Struyve and N. Zanghì. (2014) Can Bohmian mechanics be made relativistic? *Proc. R. Soc. A*, **470**, 20130699 and arXiv:1307.1714 [quant-ph].

[18] K.V. Kuchař. (2011) Time and interpretations of quantum gravity. In G. Kunstatter, D. Vincent and J. Williams, eds. *Proceedings of the 4th Canadian Conference on General Relativity and Relativistic Astrophysics.* ed. (Singapore: World Scientific), 1992), reprinted in *Int. J. Mod. Phys. D*, **20**, 3–86.

[19] Y.V. Shtanov. (1996) Pilot wave quantum cosmology. *Phys. Rev. D*, **54**, 2564–70 and arXiv:gr-qc/9503005.

[20] S. Goldstein and S. Teufel. (2004) Quantum spacetime without observers: ontological clarity and the conceptual foundations of quantum gravity. In C. Callender and N. Huggett, eds. *Physics Meets Philosophy at the Planck Scale.* (Cambridge: Cambridge University Press), pp. 275–89 and arXiv:quant-ph/9902018.

[21] N. Pinto-Neto. (2005) The Bohm Interpretation of Quantum Cosmology. *Found. Phys.*, **35**, 577–603 and arXiv:gr-qc/0410117.

[22] B.J. Hiley and A.H. Aziz Mufti. (1995) In M. Ferrero and A. van der Merwe, eds. *Fundamental Theories of Physics*, **73**, (Dordrecht: Kluwer), pp. 141–56.

[23] N. Pinto-Neto, G. Santos and W. Struyve. (2012) The quantum-to-classical transition of primordial cosmological perturbations. *Phys. Rev. D*, **85**, 083506 and arXiv:1110.1339 [gr-qc].

[24] N. Pinto-Neto, F.T. Falciano, R. Pereira and E. Sergio Santini. (2012) Wheeler–DeWitt quantization can solve the singularity problem. *Phys. Rev. D*, **86**, 063504 and arXiv:1206.4021 [gr-qc].

[25] N. Pinto-Neto and J.C. Fabris. (2013) Quantum cosmology from the de Broglie–Bohm perspective. *Class. Quantum Grav.*, **30**, 143001 and arXiv:1306.0820 [gr-qc].

[26] F.T. Falciano, N. Pinto-Neto and W. Struyve. (2015) Wheeler–DeWitt quantization and singularities. *Phys. Rev. D*, **91**, 043524 and arXiv:1501.04181 [gr-qc].

[27] E. Gindensperger, C. Meier and J.A. Beswick. (2000) Mixing quantum and classical dynamics using Bohmian trajectories. *J. Chem. Phys.*, **113**, 9369–72.

[28] O.V. Prezhdo and C. Brooksby. (2001) Quantum Backreaction through the Bohmian Particle. *Phys. Rev. Lett.*, **86**, 3215–9.

[29] E. Gindensperger, C. Meier and J.A. Beswick. (2002) Quantum-classical dynamics including continuum states using quantum trajectories. *J. Chem. Phys.*, **116**, 8–13.

[30] E. Gindensperger, C. Meier, J.A. Beswick and M.-C. Heitz. (2002) Quantum-classical description of rotational diffractive scattering using Bohmian trajectories: Comparison with full quantum wave packet results. *J. Chem. Phys.*, **116**, 10051–9.

[31] C. Meier. (2014) Mixed Quantum-Classical Treatment of Vibrational Decoherence. *Phys. Rev. Lett.*, **93**, 173003.

[32] C. Kiefer. (1992) Functional Schrödinger equation for scalar QED. *Phys. Rev. D*, **45**, 2044–56.

[33] A. Valentini. (1992) *On the Pilot-Wave Theory of Classical, Quantum and Subquantum Physics*. PhD Thesis, International School for Advanced Studies, Trieste. www.sissa.it/ap/PhD/Theses/valentini.pdf.

[34] J.J. Halliwell. (1991) Introductory lectures on quantum cosmology. In S. Coleman, J. B. Hartle, T. Piran and S. Weinberg, eds. *Quantum Cosmology and Baby Universes*. (Singapore: World Scientific), pp. 159–243 and arXiv:0909.2566 [gr-qc].

[35] J.C. Vink. (1992) Quantumpotential interpretation of the wave function of the universe. *Nucl. Phys. B*, **369**, 707–28.

[36] J. Acacio de Barros and N. Pinto-Neto. (1998) The Causal Interpretation of Quantum Mechanics and the Singularity Problem and Time Issue in Quantum Cosmology. *Int. J. Mod. Phys. D*, **07**, 201–13.

[37] F.T. Falciano, N. Pinto-Neto and E. Sergio Santini. (2007) Inflationary nonsingular quantum cosmological model. *Phys. Rev. D*, **76**, 083521 and arXiv:0707.1088 [gr-qc].

[38] J. Kowalski-Glikman and J.C. Vink. (1990) Gravity-matter mini-superspace: quantum regime, classical regime and in between. *Class. Quantum Grav.*, **7**, 901–18.

[39] T. Norsen. (2010) The Theory of (Exclusively) Local Beables. *Found. of Phys.*, **40**, 1858–84 and arXiv:0909.4553v2 [quant-ph].

[40] T. Norsen, D. Marian and X. Oriols. (2015) Can the wave function in configuration space be replaced by single-particle wave functions in physical space? *Synthese*, **132**, 3125–51 and arXiv:1410.3676 [quant-ph].

Part V

Methodological and Philosophical Issues

19

Limits of Time in Cosmology

SVEND E. RUGH AND HENRIK ZINKERNAGEL

19.1 Introduction

What does time mean in cosmology? Are there any physical conditions which must be satisfied in order to speak about cosmic time? If so, how far back can time be extrapolated while still maintaining it as a well-defined physical concept? We have studied these questions in a series of papers over the last ten years. The present chapter is a summary of some main points from our investigations, as well as some further considerations regarding time in cosmology.

It is standard to assume that a number of important events took place in the first tiny fractions of a second "after" the big bang. For instance, the universe is thought to have been in a quark-gluon phase between $\sim 10^{-11}$–10^{-5} seconds, whereas the fundamental material constituents are massless due to the electroweak (Higgs) transition at times earlier than $\sim 10^{-11}$ seconds. A phase of inflation is envisaged (in some models) to have taken place around $\sim 10^{-34}$ seconds after the big bang. A rough summary of the phases of the early universe is given in Figure 19.1 (next page).[1]

What could be wrong (or at least problematic) with this backward extrapolation from now? A main point is that *physical* time in relativity theory, in contrast to a purely mathematical parameter with the label t, is bound up with the notion of proper time.[2] For example, Misner, Thorne and Wheeler write:

> ... proper time is the most physically significant, most physical real time we know. It corresponds to the ticking of physical clocks and measures the natural rhythms of actual events.
>
> Misner, Thorne and Wheeler (1973, p. 813)

The connection between physical time and proper time leads to two kinds of problems for the backward extrapolation. The first of these follows from the fact that proper time is closely related to physical clocks or processes. The nature (and availability) of such clocks or processes changes as we go back in time. The problem in this regard was hinted at

---

[1] For a detailed discussion of (the assumptions behind) this figure and the epochs indicated, see also Rugh and Zinkernagel (2009).

[2] Proper time along a (timelike or lightlike) world line (the path of a particle in four-dimensional spacetime) can be thought of as the time measured by a "standard" clock along that world line.

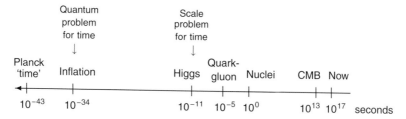

Figure 19.1 Contemplated phases of the early universe. The indicated quantum and scale problems for time are discussed in the text.

by Misner, Thorne and Wheeler in connection with a discussion of whether a singularity occurs at a finite past proper time. They note that no actual clock can adequately time the earliest moments of the universe:

Each actual clock has its "ticks" discounted by a suitable factor - $3 * 10^7$ seconds per orbit from the Earth-sun system, $1.1 * 10^{-10}$ seconds per oscillation for the Cesium transition, etc. Since no single clock (because of its finite size and strength) is conceivable all the way back to the singularity, a statement about the *proper time* since the singularity involves the concept of an infinite sequence of successively smaller and sturdier clocks with their ticks then discounted and added. [...] ... finiteness [of the age of the universe] would be judged by counting the number of discrete ticks on *realizable clocks*, not by accessing the weight of unrealizable mathematical abstractions. [Our emphasis]

Misner, Thorne and Wheeler (1973, p. 814)

The authors' discussion regarding this quote seems to imply that the progressively more extreme physical conditions, as we extrapolate the standard cosmological model back-wards, demand a succession of gradually more fine-grained clocks to give meaning to (or provide a physical basis of) the time of each of the epochs.[3] In this spirit, our view is that a minimal requirement for having a physical notion of time (with a scale) is that it must be possible to find physical processes (what we call "cores of clocks") with a sufficiently fine-grained duration in the physics envisaged in the various epochs of cosmic history. As we shall discuss below, this requirement of linking time to conceivable cores of clocks leads to a *scale problem* for time, since it becomes progressively more difficult to identify physical processes with a well-defined (and non-zero) duration in the very early universe.

A second kind of problem with the backward extrapolation follows since proper time is defined in terms of (possible) particle world lines or trajectories. Within the standard cosmological model, there is a privileged set of such world lines since matter on large scales is assumed to move in a highly ordered manner (allowing for the identification of a comoving reference frame and a global cosmic time equal to the proper time of any

---

[3] Regarding Misner, Thorne and Wheeler's examples in the quote, it is clear that one has to distinguish between *how fine-grained* a clock is (its precision) and *when* (in which cosmological epoch) such a clock could in principle be realized. For instance, no stable Cesium atoms – let alone real functioning Cesium clocks – can exist before the time of decoupling of radiation and matter, about 380,000 years after the big bang.

comoving observer). As we shall discuss, this implies that the notion of cosmic time is closely related to the so-called Weyl principle. Problems with the notion of a global cosmic time may arise if a privileged set of world lines becomes difficult to identify, e.g. in the very early universe above the electroweak (Higgs) phase transition or in a (complicated) inhomogeneous universe.

A more serious problem for time (which is a problem even for a local definition of time) arises if a point is reached in the backward extrapolation where the world lines themselves can no longer be identified. In particular, this appears to be the case if some point is contemplated, e.g. at the onset of inflation, where all constituents of the universe are of a quantum nature, leading to what can be called the *quantum problem* of time. Note that this problem arises roughly ten orders of magnitude "before" (in the backward extrapolation from now) reaching a possible quantum gravity epoch, and so before hitting the usual problem of time in quantum gravity models.

In the following, we first outline the scale problem for time and the close relation between time and clocks. We then address the relation between time and world lines in the set-up of the standard cosmological model. We briefly indicate how this relation may lead to problems for a global (cosmic) time concept in the very early universe above the electroweak phase transition or in a (complicated) inhomogeneous universe. We finally discuss the more serious local (quantum) problem for time in relation to the problem of identifying individual world lines.

19.2 Time and Clocks

The idea that time is dependent on change and/or motion is called relationism. It has been defended by classic thinkers like Aristotle and Leibniz, and in modern times by physicists like Barbour, Smolin and Rovelli. In our version of relationism, we argue in favour of a "time–clock" relation which asserts that time, in order to have a physical basis, must be understood in relation to physical processes which act as "cores" of clocks (Rugh and Zinkernagel, 2005, 2009; see also Zinkernagel, 2008). In the cosmological context, the time–clock relation implies that a necessary physical condition for *interpreting* the t parameter of the standard Friedmann–Lemaître–Robertson–Walker (FLRW) model as cosmic time in some "epoch" of the universe is the (at least possible) existence of a physical process which can function as a core of a clock in the "epoch" in question. In particular, we have suggested that in order to make the interpretation

$$t \leftrightarrow \text{time},$$

at a specific cosmological "epoch", the physical process acting as the core of a clock should: 1) have a well-defined duration which is sufficiently fine-grained to "time" the epoch in question; and 2) be a process which could conceivably take place among the material constituents available in the universe at this epoch.

The time–clock relation is in conformity with how time is employed in cosmology although cosmologists often formulate themselves in operationalist terms – that is, invoking observers measuring on factual clocks. For instance, Peacock writes concerning the FLRW model:

We can define a global time coordinate t, which is the time measured by clocks of these observers – i.e. t is the proper time measured by an observer at rest with respect to the local matter distribution.

Peacock (1999, p. 67)

While this reference to clocks (or "standard" clocks) carried by comoving observers is widely made in cosmology textbooks, there is usually no discussion concerning the origin and nature of these clocks. Part of the motivation for our investigations has been to provide a discussion of this kind.

The standard definition of the global time coordinate to which Peacock refers – and, in general, the question of how to make the t-time identification – can be read in at least two different ways: 1) Actual clocks should be available (operationalism); or 2) rudiments (cores) of clocks with a well-defined duration should, in principle, be present (time–clock relation). Clearly, the first possibility is not an option in the very early universe where no actual clocks, let alone observers and measurements, are available. As we shall see below, the viability of the second option depends upon the availability of physical processes with well-defined (and non-zero) duration.

We attempt to develop a position on the time concept which represents a departure from operationalism in several ways: (i) Time cannot be defined (reductively) in terms of clocks (since clocks and measurements depend on the time concept); (ii) no actual clocks are needed, we allow reference to possible (counterfactual) clocks, compatible with the physics of the epoch in question; (iii) we attempt to construct the *cores* of clocks out of available physics, but do not require that this core should be associated with a counter mechanism that could transform it into a real functioning clock; and (iv) we do not require the existence of observers and actual measurements. Nevertheless, the above formulated criterion for the $t \leftrightarrow$ time interpretation of being able to identify a process with a well-defined duration may still have an operationalist feel. For, as we shall see below, it means that there may be limits to time in cases where scales can be found in the physics, but where no physical process (core of a clock) can be identified which could in principle exemplify or realize the time scale in question.

Whereas cosmologists often refer to clocks as sketched above, they also define cosmic time "implicitly" by the specific cosmological model employed to describe the universe.[4] This can be done e.g. through the relation between time and the scale factor. If we for instance consider a radiation dominated epoch, the Einstein field equations may yield (see e.g. Rugh and Zinkernagel (2009), Section 4):

$$R(t) \propto \sqrt{t}$$

---

[4] This is related to a more general discussion of the implicit definition of time via natural laws, see Rugh and Zinkernagel (2009), Section 2.1.

In some sense, the scale factor here serves as a (core of a) clock. However, for this idea to work, one needs to have some bound system or a fixed physical length scale which does not expand (or which expands differently than the universe). Otherwise, there is no physical content of $R(t)$ and hence no physical content of "expansion". Eddington, for example, emphasized the importance of the expansion of the universe to be defined relative to some bound systems by turning things upside down: "The theory of the 'expanding universe' might also be called the theory of the 'shrinking atom'" (Eddington, 1933, quoted from Whitrow, 1980, p. 293).

The viewpoint presented here assumes that a physical foundation of time is closely related to *which* physical constituents are available (or at least possible) in the early universe. Such an assumption can be circumvented if one subscribes to some sort of Platonism (or mathematical foundationalism) according to which a purely mathematical definition of time, extracted e.g. from the formalism of general relativity (or, as simply the t parameter in some model), is sufficient. According to such a view, there would seem to be no problem in contemplating, say, periods like 10^{-100} seconds after the big bang. However, it is widely accepted that the standard cosmological model cannot be extrapolated below Planck scales and, accordingly, that the $t \leftrightarrow$ time interpretation cannot be made for t values below 10^{-43} seconds. This illustrates that a physical condition (namely that quantum gravity effects may be neglected) can imply a limitation for the $t \leftrightarrow$ time interpretation. But if it is accepted that there is at least *one* physical condition which must be satisfied in order to trust the backward extrapolation of the FLRW model and its time concept, it appears reasonable to require that also *other* physical conditions (which are necessary to set up the FLRW model) should be satisfied during this extrapolation.[5] Hence, we take it that Platonism is not a satisfactory position regarding time in cosmology. In our view, time has to have some physical basis (i.e. it must be embedded in the available physics) in order to be a well-defined physical concept.

19.2.1 The Scale Problem for Time

Let us now shortly review how the above considerations may lead to a scale problem for time in the early universe. The scale problem for time is related to two contemplated phase transitions at $\sim 10^{-5}s$ and $\sim 10^{-11}s$ in the early universe, where the notion of length and time scales (and their physical underpinning in terms of cores of clocks and rods) becomes progressively weaker and disappears at $\sim 10^{-11}s$ (if we consider well-known physics set by the standard model of particle physics).

$\sim 10^{-5}s$: No bound systems (the quark–hadron phase transition)

Physically based time and length scales are not independent notions. Einstein discussed an elementary clock system, "the light-clock", which involves the propagation of a light

---

[5] We shall return to this requirement in the final section. We also note that – except for the space-time singularity itself – there are no internal contradictions in the *mathematics* of the FLRW model (or classical general relativity) which suggests that this model should become invalid at some point, e.g. at the Planck scale.

signal across some physical length scale as when a light signal is being reflected back and forth between the ends of a rigid rod. If we ask whether we in principle may build up such Einstein light-clocks from the constituents of the early universe we note, as we extrapolate backwards in time (and the temperature rises), that it becomes progressively more difficult to find any spatially extended physical systems. In the so-called hadron era there are still bound (hadron) systems such as pions, neutrons and protons. At a transition temperature of $T \sim 10^{12}$ K ($\sim 10^{-5}s$ after the big bang) it is, however, believed that there is a quark–hadron phase transition, and above this transition point no bound states are left. The universe then consists of particles (like quarks, leptons, gluons and photons) which have no known spatial extension. If a rudiment (or core) of a rod has to be constructed from a bound physical system, we no longer have such rudiments of rods left in the universe, and we have to look elsewhere for physics which can set a physical length scale.

The quarks and leptons still possess the physical property of mass. Thus, one still has length scales if the Compton wavelength $\lambda = \lambda_C = \hbar/(mc)$ of these particles can be taken to set such a scale. However, a rod with spatial extension equal to the Compton wavelength leads to a "pair-production of rods" as a quantum effect (in general, the Compton wavelength is the length scale at which "pair production" of particle-antiparticle pairs occurs). It is thus difficult to imagine how the Compton wavelength divided by c corresponds to a physical process which could function as the core of a clock, e.g. in the above-mentioned light-clock.

Note that these considerations, and hence our first proposed time limit at 10^{-5} seconds, are based on the somewhat operationalist premise that one should in principle be able to identify a core of a clock (physical process) with a well-defined duration. The contemplated process is one in which a light signal travels a well-defined distance (namely a Compton wavelength of a quark or a lepton), but this process seems physically unrealizable insofar as the photon is converted to a particle-antiparticle pair during flight.[6]

$\sim 10^{-11}s$: Scale invariance (the electroweak phase transition)

According to the standard model of particle physics (which embodies a Higgs sector with a (set of) scalar field(s) ϕ) there is an electroweak phase transition at a transition temperature of $T \sim 300$ GeV $\sim 10^{15}$ K when the universe was $\sim 10^{-11}s$ old. Above this phase transition the Higgs field expectation value vanishes $< \phi > = 0$. This transition translates into *zero rest masses* of all the fundamental quarks and leptons (and massive force mediators) in the standard model. Without any masses in the theory it will exhibit a symmetry known as conformal invariance, and it will be impossible to find physical processes (among the microphysical constituents) with a well-defined (and non-zero) duration.[7]

---

[6] As different sorts of rudiments (cores) of clocks, we may consider the decay processes of unstable, massive particles such as the decay of the muons $\mu^- \rightarrow e^- \bar{\nu}_e \nu_\mu$, or the decay of the Z^0 particles, $Z^0 \rightarrow f\bar{f}$ (which can decay into any pair of fermions). But, as discussed in Rugh and Zinkernagel (2009), also these processes are difficult to conceive as functioning cores of clocks due to their quantum-mechanical and statistical nature.

[7] See however Rugh and Zinkernagel (2009, Section 5.3) for a brief discussion of some possible rudiments of mass which could remain above the Higgs transition (but which are, in our assessment, insufficient to ground e.g. a physical time scale).

Thus, not only can no core of a clock be identified. The relevant physics (the electroweak and strong sector) cannot set physical scales for time, scales for length and no scale for energy. If there is no scale for length and energy then there is no scale for temperature T. Metaphorically speaking, we may say that not only the property of mass of the particle constituents "melts away" above the electroweak phase transition but also the concept of temperature itself "melts" (i.e. T loses its physical foundation above this transition point).

In our assessment, therefore, the time scale assumed (e.g. in cosmology books) above the electroweak phase transition is purely speculative in the sense that it cannot be founded upon an extrapolation of well-known physics (due to conformal invariance) above the phase transition point. Thus, the time scale will have to be founded on the introduction of some new physics (beyond the standard model of particle physics), and is in this sense as speculative as the new (speculative) physics on which it is based.

It is of interest that Roger Penrose has recently attempted to turn the scale problem for time into a virtue in the construction of a new kind of cosmological scenario. Penrose cites our study[8] in connection with the following quote:

…close to the Big Bang, probably down to around 10^{-12} seconds after that moment, when temperatures exceed about 10^{16} K, the relevant physics is believed to become blind to the scale factor Ω, and *conformal* geometry becomes the space-time structure appropriate to the relevant physical processes. Thus, all this physical activity would, at that stage, have been insensitive to local scale changes. [Emphasis in original]

Penrose (2010, p. 142)

Our mentioning this point does not imply an endorsement of Penrose's proposal of an "Extraordinary New View of the Universe" (a conformal cyclic cosmology) in which approximate conformal invariance holds in both ends (the beginning and the remote future) of the universe. Nevertheless, there seems to be a certain agreement in philosophical outlook, also when Penrose mentions (2010, p. 93): "It is important for the *physical basis* of general relativity that extremely precise clocks actually exist in Nature, at a fundamental level, since the whole theory depends upon a naturally defined metric **g**" (our emphasis).

Why not refer to the Planck scales?

The combination of the constants $\hbar$ and c from relativistic quantum mechanics, and c and G from classical general relativity yields – as a mathematical combination of physical constants – the famous Planck scales. As concerns time, the Planck time scale $t_P = (\hbar G/c^5)^{1/2} \sim 10^{-43}$ seconds is immensely more fine-grained than time scales set by any physical process which we (in our investigations) have attempted to utilize as rudiments of clocks at various stages in cosmic history. *If* the Planck time scale were considered sufficient to provide a physical basis for the time scale in the early universe, then there

---

[8] Our study appeared as a handout in a first print in 2005 (Rugh and Zinkernagel, 2005) and was published in a revised version in 2009.

would be no scale problem for time anywhere along the extrapolation from now to the Planck times.

However, we see several related reasons to be suspicious that the Planck time scale does indeed provide a sufficient physical basis for the time scale in the early universe. First of all, the Planck scales are supposed to be the physical relevant scales of theories of quantum gravity, and such theories are still highly speculative. The Planck time scale is therefore at least as speculative as any other imagined time scale above the electroweak phase transition. Second, it is expected that quantum gravity effects are totally negligible at energy scales around the electroweak phase transition point (and negligible well into the "desert" above this phase transition). It appears dubious to ground time scales of Higgs-physics on quantum gravity effects which are irrelevant at Higgs-physics scales.[9] Third, even if we bypass the second problem, one may well question how physically reasonable the supposed physical processes would be for grounding the Planck scale (recall that, in our view, a physical basis for a time scale should be related to relevant physical processes). The Planck scales may be arrived at by setting the Compton wavelength equal to the Schwarschild radius of a black hole. It is thus a characteristic scale at which there is a pair production of black holes as a quantum effect. Consider this in the context of the discussion above on time and length scales in connection with the light-clock: At the Planck length scale there is a "pair production of rods" (the rod being the spatial extension of the quantum black hole), and the corresponding Planck time scale is the time it takes a light pulse to cross this length scale. This appears to be more of a mathematical construct than a conceivable physical process since the crossing of a light pulse is hardly a well-defined physical process within such violent fluctuations in the geometry. For instance, one may ask at which of the two pair produced quantum black holes the light pulse is supposed to end its "crossing"?[10]

In our assessment, then, if one wants to solve (or dissolve) the scale problem for time at and above the Higgs transition by referring to speculative processes in quantum gravity, such as "quantum pair production of black holes", then it should at least be admitted that the cosmic time scale constructed in this way is highly speculative.[11]

19.3 Time and World Lines

The above section has focused on the consequences of proper time being related to clocks. As we saw, this relation leads to the idea that time is related to physical processes – which

[9] Note that today we define a second as 9.192.631.770 vibrations of radiation caused by well-defined transitions in Cs-133 (see e.g. 't Hooft and Vandoren, 2014). This is a physical grounding (even operationally) of a time scale (a second) in terms of physical processes taking place at time scales substantially (ten orders of magnitude) smaller. It would be of interest if one could speculate any effect on Higgs-scale physics stemming from the quantum gravity scales 30 orders of magnitude below.
[10] Note that this third reason against using the Planck scale as a physical basis for the time scale in the early universe is, just like the 10^{-5} seconds limit discussed above, based on the somewhat operationalist premise that we should be able to point to a process (a core of a clock) which provides a definite time interval.
[11] As we shall discuss later, the *quantum* problem of time (Section 19.3.3) does not depend on whether we have a physically well-founded *scale* for time. This problem remains (e.g. at the onset of inflation) even if we base length and time scales (throughout cosmic history) on speculative Planck scale physics.

is a version of what is known as relationism. But there is a more direct route to relationism in cosmology which is independent of the mentioned time–clock relation (even if in conformity with it). This has to do with the fact that proper time is defined in terms of (possible) particle world lines. In the following we shall discuss how this implies a close relation between time and cosmic matter, both at the global and at the local level.

19.3.1 Setting up the FLRW Model with a Cosmic Time

In Rugh and Zinkernagel (2011) we discuss how the set-up of the FLRW model with a global time is closely linked to the motion, distribution and properties of cosmic matter. We now briefly review some key points of this discussion.

In relativity theory time depends on the choice of reference frame. For the universe, a reference frame cannot be given from the outside, so such a frame has to be "built up from within", that is, in terms of the (material) constituents within the universe. It is often assumed that the FLRW model may be derived just from the cosmological principle. This principle states that the universe is spatially homogeneous and isotropic (on large scales). It is much less well known that another assumption, called Weyl's principle, is necessary in order to arrive at the FLRW model and, in particular, its cosmic time parameter.[12] Whereas the cosmological principle imposes constraints on the *distribution* of the matter content of the universe, Weyl's principle imposes constraints on the *motion* of the matter content. Weyl's principle (from 1923) asserts that the matter content is so *well behaved* that a reference frame can be built up from it:

Weyl's principle (in a general form): The world lines of 'fundamental particles' form a spacetime-filling family of non-intersecting geodesics (a congruence of geodesic world lines).

The importance of Weyl's principle is that it provides a reference frame which is physically based on an expanding "substratum" of "fundamental particles" (e.g. galaxies or clusters of galaxies). In particular, if the (non-crossing) geodesic world lines are required to be orthogonal to a series of space-like hypersurfaces, a comoving reference frame is defined in which constant spatial coordinates are "carried by" the fundamental particles (see e.g. Figure 3.7 in Narlikar, 2002, p. 107). The time coordinate is a cosmic time which labels the series of hypersurfaces, and which may be taken as the proper time along any of the particle world lines. We note that the congruence of world lines is essential to the standard cosmological model since the symmetry constraints of homogeneity and isotropy are imposed with respect to such a congruence (see e.g. Ellis, 1999). Thus, Weyl's principle is *a precondition* for the cosmological principle; the former can be satisfied without the latter being satisfied but not vice versa.

---

12 In some cosmology textbooks – e.g. by Bondi, Raychaudhuri and Narlikar – the importance of Weyl's principle is emphasized and explicitly referred to. In other textbooks it appears, in our assessment (see Rugh and Zinkernagel, 2011), that the Weyl principle is implicitly assumed in the process of setting up the FLRW model.

In the early universe, problems may arise for the Weyl principle and thus for the possibility of identifying a reference frame and a global cosmic time parameter.[13] At present and for most of cosmic history, the comoving frame of reference can be identified as the frame in which the cosmic microwave background radiation (CMB) looks isotropic (see e.g. Peebles, 1993, p. 152), and cosmic matter is (above the homogeneity scale) assumed to be described as dust particles with zero pressure which fulfill Weyl's principle. But before the release of the CMB, the situation is less straightforward. For, as we go backwards in time, it may become increasingly more difficult to satisfy, or *even formulate*, the Weyl principle as a physical principle, since the nature of the physical constituents is changing from galaxies, to relativistic gas particles, and to entirely massless particles moving with velocity c.[14] Indeed, above the electroweak phase transition (before 10^{-11} seconds "after" the big bang), all constituents are massless and move with velocity c in any reference frame. There will thus be no constituents which are comoving (at rest). One might attempt to construct mathematical points (comoving with a reference frame) like a center of mass (or, in special relativity, center of energy) out of the massless, ultrarelativistic gas particles, but this procedure seems to require that length scales be available in order to e.g. specify how far the particles are apart (which is needed as input in the mathematical expression for the center of energy). As discussed earlier, the only option for specifying such length scales (above the electroweak phase transition) will be to appeal to speculative physics, and the prospects of satisfying Weyl's principle (and have a cosmic time) will therefore also rely on speculations beyond current well-established physics. The problem of building up the FLRW model with matter consisting entirely of consituents moving with velocity c may also be seen by noting that the set-up of the FLRW model requires matter (the energy-momentum tensor) to be in the form of a perfect fluid, as this is the only form compatible with the FLRW symmetries, see e.g. Weinberg (1972, p. 414). For this, a source consisting of pure radiation is not sufficient since one cannot effectively simulate a perfect fluid by "averaging over pure radiation".[15]

On top of this, the physical basis of the Weyl postulate (e.g. non-intersecting world lines of "fundamental particles"), and even that of proper time, appears questionable if some period in cosmic history is reached where the "fundamental particles" are described by wave-functions $\psi(x, t)$ referring to (entangled) quantum constituents. What is a "world line" or a "particle trajectory" then? (See also the section below on the quantum problem for time.)

---

[13] In Rugh and Zinkernagel (2013) we also argue that there is no approximate fulfillment of a Weyl principle and no well-defined global (multiverse) cosmic time concept in the eternal inflationary multiverse model outlined e.g. by Linde and Guth.

[14] In the early radiation phase, matter is highly relativistic (moving with random velocities close to c), and the Weyl principle is not satisfied for a typical particle but one may still introduce fictitious averaging volumes in order to create substitutes for "galaxies which are at rest"; see e.g. Narlikar (2002, p. 131).

[15] Krasinski (1997, pp. 5–9) notes that the energy-momentum tensor in cosmological models may contain many different contributions, e.g. a perfect fluid, a null fluid, a scalar field, and an electromagnetic field. He also emphasizes that a source of a pure null fluid or a pure electromagnetic field is not compatible with the FLRW geometry, and that solutions with such energy-momentum sources have no FLRW limit (Krasinski 1997, p. 13).

In the following we shall briefly question what happens to cosmic time if/when we cannot assume the validity of the standard FLRW model (next Subsection), before we turn to the question of what happens if/when the cosmic constituents become quantum (Subsection 19.3.3).

19.3.2 Cosmic Time in an Inhomogeneous Universe

The cosmological standard model is highly idealized and it is therefore of interest to inquire about cosmic time when the model's idealizing assumptions are relaxed. In particular, one may ask whether we still have a good cosmic time concept in our actual – at least up to very large scales – inhomogeneous universe? It is well known that close to massive objects time runs differently than in more "void like" segments of spacetime away from any such massive objects. The complexity of constructing a privileged time notion in such situations has been illustrated e.g. in the following example by Feynman.

How old is our earth? Since clocks (time) run differently in different gravitational potentials (time dilation in a gravitational field), time will run at a different rate in the center of the earth than on the surface of the earth. Feynman remarks:

… we might have to be more careful in the future in speaking of the ages of objects such as the earth, since the center of the earth should be a day or two younger than the surface!

Feynman (1995, p. 69)

In fact, the situation is slightly worse, for integrating up a relative time dilation factor of $\Delta\tau/\tau = \Delta\Phi/c^2$ over a coarse estimate of the (not precisely defined) lifespan of our earth ($\sim 5 \times 10^9$ years) yields some years in time difference.[16]

In a universe with an inhomogeneous distribution of the material constituents, the situation is less clear than in the Feynman example of a slightly inhomogeneous gravitational field throughout our earth. In some mathematically simplified spatially inhomogeneous models, it may be possible to maintain a Weyl principle and a notion of a global cosmic time (cf. e.g. Krasinski (1997) and references therein). However, if our universe exhibited fractal behavior and collisions on all scales it would be difficult to uphold a Weyl principle (even in an "average sense" where small scale collisions and inhomogeneities are averaged out). We may add that such fractal behavior and collisions on all scales appear to be a characteristic of envisaged multiverse inflationary scenarios like chaotic inflation, see e.g. discussion and references in Rugh and Zinkernagel (2013).

Not least due to the observed microwave background isotropy (and the remarkable isotropy of X-ray counts, radio source counts, and γ-ray bursts) it is generally expected (yet

---

[16] In an order of magnitude estimate we may assume that our earth is homogeneous and the potential difference between the center and the surface of our earth is then integrated up to $\Delta\Phi = GM/2R$ which translates into a relative time dilation effect $\Delta\tau/\tau = \Delta\Phi/c^2 = 1/4 \times (R_{Schw}/R) \sim 1/3 \times 10^{-9}$ (here $R_{Schw} = 2GM/c^2$ is the Schwarzschild radius of our earth with mass M and radius R). Integrating this relative time dilation over $\sim 5 \times 10^9$ years yields an order of magnitude estimate of ~ 2 years for the age difference (as measured by counterfactual clocks located in the center and at the surface of our earth over the lifespan of our earth).

debated) among cosmologists that there will be a transition from small-scale fractal behavior to large-scale homogeneity.[17] A recent study arguing this case is e.g. Scrimgeour *et al.* (2012).[18] Nevertheless, even if our universe is not fractal at the largest – but only at intermediate – distance scales, it is an interesting question how significantly this may change the cosmic time concept of the resulting cosmological model. Indeed, inhomogeneous models with a fractal matter distribution at intermediate scales will presumably exhibit more complicated conceptions of cosmic time than in the highly symmetric, idealized FLRW model universes.

One way to address what happens in an inhomogeneous universe is to attempt to construct a notion of cosmic time associated with an event (here, now) by looking at the proper time (e.g. Misner, Thorne and Wheeler (1973), §13.4 §27.4)

$$\tau = \tau(\gamma) = \int_\gamma \sqrt{g_{\mu\nu}dx^\mu dx^\nu} \ , \tag{19.1}$$

along (particle) timelike world lines (indicated with subscript γ and with 4-velocities $u^\mu(v) = dx^\mu(v)/dv$), which starts at the beginning of space time, and ends in the event (here, now).[19] But along which world lines γ should the proper time integral be taken?

Ellis (2012, pp. 9, 10) proposes to take the proper time integral along a specific set of preferred fundamental world lines, which (for realistic matter) are uniquely geometrically determined. This construction does not invalidate the Weyl principle but rather builds on it and develops it (Ellis, private communication).[20] The "present" is in this construction defined as the surface $\{\tau = constant\}$ determined by taking the proper time integral Eq. (19.1) over the family of fundamental world lines starting at the "big bang". However, according to Ellis, the equal time hypersurfaces can in generic situations be much more complicated (see discussion in Ellis, 2012, p. 10) than the simple equal (cosmic) time

[17] Moreover, it has been emphasized, e.g. by Barrow (2005), that large contrasts in density $\delta\rho/\rho$ are not necessarily mirrored in similar inhomogeneities in the gravitational potential Φ since the equation of the relative perturbation $\delta\Phi/\Phi$ of the gravitational potential has in it a huge suppression factor ($\delta\Phi/\Phi \sim \delta\rho/\rho \times (L/(c/H))^2$) if the size L of the density irregularity is small relative to the Hubble radius c/H.
[18] If we want to observationally test the expected homogeneity at large scales, one should pay attention to the danger of vicious circularities ("catch-22"). Distance measures like redshift-distance measures (at large distances) should not have built in the assumptions we want to test (the FLRW model as space-time metric, etc.). The analysis provided in e.g. Scrimgeour *et al.* (2012) is very elaborate but it is of interest that they note (p. 4): "To do this, we assume the FRW metric and ΛCDM. This is necessary for any homogeneity measurement, since we must always assume a metric in order to interpret redshifts. Therefore in the strictest sense this can *only* be used as a consistency test of ΛCDM. However, if we find the trend towards homogeneity matches the trend predicted by ΛCDM, then this is a strong consistency check for the model and one that an inhomogeneous distribution would find difficult to mimic" (our emphasis).
[19] Such a definition is only well-defined, i.e. the proper time is only finite, if there is a beginning of space-time e.g. in a "big bang" (see also Lachièze-Rey 2014, Section 5.3), or if we chose some arbitrary starting point (assumed to exist) from which we can integrate (Ellis 2012, Section 3).
[20] We note that Ellis assumes that there is a uniquely defined vector field of 4-velocities $u^\mu(v) = dx^\mu(v)/dv$ (if such 4-velocities are uniquely defined in each spacetime point on the manifold; this is equivalent to assuming the existence of a congruence of world lines which are non-crossing). According to Ellis' proposal, these 4-velocities (in order to be preferred fundamental world lines) should satisfy that they are timelike eigenlines of the Ricci tensor, $R_{\mu\nu}v^\nu = \lambda u_\mu$.

hypersurfaces in the FLRW model universes. In particular, Ellis remarks that the equal time hypersurfaces may not even necessarily be spacelike in an inhomogeneous spacetime.

It therefore appears to be a complicated – and to our knowledge still open – question whether the resulting concept of cosmic time exhibits the properties which allow for a "backward" extrapolation into an "early" inhomogeneous universe.

19.3.3 The Quantum Problem for Time

We have seen above that it may be difficult to identify a *global* cosmic time (without the Weyl principle), and in earlier sections also that there may not be a *scale* for time (before the Higgs transition). Even if this is so, it might still be possible to maintain a local time *order*, i.e. to ask about the past of some particular event – for instance, the past of the onset of inflation. However, as we shall indicate below, it may well be that not even a *local* (and scale-free) time order is available as time is extrapolated backwards in the very early universe.

The origin of this local (or quantum) problem for time is due to the widely assumed "quantum fundamentalist" view according to which the material constituents of the universe could be described *exclusively* in terms of quantum theory at some early stage of the universe. Such a perspective is natural in quantum cosmology (and quantum gravity), in which spacetime itself is treated quantum mechanically (see also Hartle 1991). From the point of view of such theories, it has been argued that a quantum problem of time appears already (in the backward extrapolation from now) at the onset of inflation. Thus, Kiefer affirms that:

The Universe was essentially "quantum" at the onset of inflation. Mainly due to bosonic fields, decoherence set in and led to the emergence of many "quasi-classical branches" which are dynamically independent of each other. Strictly speaking, the very concept of time makes sense only after decoherence has occurred. In addition to the horizon problem etc., inflation also solves the "classicality problem". [...] Looking back from our Universe (our semiclassical branch) to the past, one would notice that at the time of the onset of inflation our component would interfere with other components to form a timeless quantum-gravitational state. The Universe would thus cease to be transparent to earlier times (because there was no time).

Kiefer (2003, p. 208)

The problem here seems to be that our spacetime (and therefore time) "dissolves" into a superposition of spacetimes at the onset of inflation, and in this sense Kiefer acknowledges a quantum problem of time at this point. The situation, however, might be worse (i.e. the quantum problem may appear earlier in a backward extrapolation from now), since the appeal to decoherence is questionable. To see this, consider what one might call the cosmic measurement problem, which addresses the quantum mechanical measurement problem in a cosmological context:

The cosmic measurement problem: If the universe, either its content or in its entirety, was once (and still is) quantum, how can there be (apparently) classical structures now?

While many aspects of the cosmic measurement problem have been addressed in the literature, the perspective which we have tried to add is that the problem is closely related to providing a physical basis for the (classical) FLRW model with a (classical) cosmic time parameter. As illustrated in the Kiefer quote above, an often attempted response to the cosmic measurement problem is to proceed via the idea of decoherence. According to this idea, some degrees of freedom are regarded as irrelevant (they are deemed inaccessible to measurements and are traced out), and they are therefore taken to act as an environment for the relevant variables. The picture is that the environment in a sense "observes" the system in a continuous measurement process and thus suppresses superpositions of the system (see e.g. Kiefer, 1989).

However, as is widely known, decoherence cannot by itself solve the measurement problem and explain the emergence of a classical world.[21] For, if both environment and system are quantum, the total state of the system (relevant plus irrelevant degrees of freedom) is still a superposition. According to quantum mechanics, no definite (classical) state can therefore be attributed to any of the components. As argued by Sudarsky (2011, Section 4.1), this problem is only aggravated in the cosmological context since one cannot here appeal to the usual pragmatic considerations regarding what classical observers and their measurement apparatus would register.[22] In spite of such worries, Kiefer (2003) contemplates that decoherence successively classicalizes different constituents of the universe: At the onset of inflation, the inflaton field itself is classicalized and, at the end of inflation, decoherence converts the quantum fluctuations of the inflaton field into classical density perturbations (seeds of structure).[23]

But even if one were to bypass the strong arguments against decoherence as a solution to the cosmic measurement problem, a potentially more serious problem is lurking: If decoherence is to explain the emergence of classical structures, it cannot – as in environmentally induced decoherence – be a process in (cosmic) time, insofar as classical structures (particle world lines) are needed from the start to define time both locally and globally! There thus seems to be a *vicious circularity* if one invokes decoherence to explain the "emergence" of time, which we can formulate in slogan form:

Decoherence takes time and cannot therefore provide time.

This implies that several of the temporal expressions in the quote by Kiefer given above ("decoherence *sets in*", "*after* decoherence *has occurred*", etc.) are strictly speaking without meaning.

---

[21] For a simple explanation of this, and some references to the relevant literature, see e.g. Zinkernagel (2011, Section 2.1).

[22] From a pragmatic point of view, quantum mechanics may be seen as a theory of expected outcomes of measurements, in which both apparatus and observers are kept outside the quantum description. We have pointed out elsewhere (Rugh and Zinkernagel, 2005; Zinkernagel, 2016) that Bohr went beyond this pragmatic (or instrumental) interpretation. His view was rather a contextual one according to which any system can be treated quantum mechanically but not all systems can be treated this way at the same time.

[23] In this regard, Anastopoulos (2002) mentions a worry about decoherence closely related to the ones already noted: "...a sufficiently classical behavior for the environment seems to be necessary if it is to act as a decohering agent and we can ask what has brought the environment into such a state ad-infinitum".

Although the discussion above has focused on decoherence, we note that the quantum problem of time seems to be shared by other "quantum fundamentalist" views even when these do not rely essentially on decoherence (e.g. the spontaneous collapse model described in Sudarsky, 2011). Our point is that any interpretation of quantum mechanics will need a time concept – which is bound up with the notion of possible (classical) particle world lines – in order to address the early universe. The assumption of a quantum nature of the material (or otherwise) constituents of the universe makes it hard (or impossible) to associate these with well-defined particle trajectories. During inflation the only relevant constituent of the universe is taken to be the inflaton field φ which – in the last analysis – is a quantum field. And just as wave functions in non-relativistic quantum theory do not give rise to physical motion (of a particle or wave) in space and time – without assumptions solving the measurement problem – so quantum fields do not describe moving elementary particles in space with well-defined trajectories.

Up to this point we have discussed the quantum problem of time from a quantum fundamentalist point of view based on quantum cosmology or quantum gravity. Let us now proceed from the present (and more cautious) perspective, in which we start from a classical point of view and attempt to extrapolate proper time backwards. More specifically, consider the past of some event by extrapolating backwards the proper time integral along a world line with 4-velocity $u^\mu(v) = dx^\mu(v)/dv$, which ends in the event (formula as in Eq. (19.1) in Section 19.3.2). This approach assumes that we know the metric and that there are well-defined 4-velocities. The question then becomes whether such 4-velocities (or, equivalently, world lines) can always be constructed, i.e. physically realized as opposed to merely mathematically defined, from the available constituents (e.g. from a scalar field φ).[24]

In the inflationary scenario, the relevant candidate for constructing sensible notions of particle world lines and classical trajectories will have to come from the φ field. And even if we were allowed to take this field as effectively classical (described by the lowest order approximation in quantum field theory) during inflation (e.g. during a slow-roll evolution), the quantum problem of time will be faced at the on-set of inflation. At this point (supposed to be the "birth" of our bubble universe in a multiverse setting), the inflaton field is strongly quantum: Quantum fluctuations with amplitudes (within a factor of ten) of the order of the Planck scale are necessary to reset or lift the scalar field to a value where a new bubble (our universe) is born and becomes dominated by inflation (see e.g. Linde, 2004, Section 4). Thus, at the beginning of inflation (or the "birth" of our universe), the φ field is nowhere close to being a classical field on top of which we have small quantum fluctuations. Rather, it is entirely dominated by Planck scale quantum fluctuations.

In summary, to use the local time concept for contemplating times before inflation (or, indeed, earlier bubble universes in the multiverse), it must be possible to identify (or, at

---

[24] One idea here would be to equate the energy momentum tensor of the perfect fluid form with the energy momentum tensor for the scalar field. This results in the 4-velocity $u_\mu = A \cdot \partial_\mu \varphi$ where $A = (\partial^\nu \varphi \, \partial_\nu \varphi)^{-1/2}$, see e.g. Krasinski (1997, p. 8) and Hobson *et al.* (2006, p. 432).

least, to speculate) a particle world line along which proper time can be extrapolated backwards.[25] But, as we have seen, it is unclear how one would go about constructing any individual classical particle world line from the inflationary scalar field φ in a regime where its quantum behavior is dominant (at the onset of inflation). If such world lines (classical trajectories) cannot be constructed from the underlying physics (the φ field), it seems that the very conditions for speaking about the past of an event in general relativity are not fulfilled. Hence, in our assessment, a pure quantum phase in the early universe implies that proper time (and even its order aspect, that is, its ability to distinguish before and after) is no longer a well-defined concept.

19.4 Summary and Discussion

It is common practice to extrapolate the standard cosmological model back to at least the Planck time. In this chapter, we have tried to insist that this is problematic. The underlying philosophical reason is that the extrapolation of the FLRW model and its time concept requires, in our view, that the physical basis of time in the model and, more generally, the physical conditions needed to set up the model, are not invalidated along this extrapolation. This situation gives rise to a number of possible limits of time, respectively, at $\sim 10^{-5}s$, $10^{-11}s$, $10^{-34}s$, and $10^{-43}s$ "after" the mathematical point $t = 0$ in the FLRW model.

As briefly hinted in Section 19.2, we are aware that we are here making a philosophical choice – at least concerning the two first limits. For we are assuming that the natural laws need a physical basis at all points along the extrapolation, as opposed to just having a basis at the present epoch (when it is easy to identify not only length and time scales, but also physical processes with well-defined durations). The difference between the first two limits and the Planck time (and possibly the time of onset of inflation) is that the former two (phase transitions) do not mark events where the natural laws are expected to break down. Rather, the two phase transitions are predictions of the natural laws themselves (by contrast, classical gravity is expected to break down at the Planck scale). Hence, in the case of the first two time limits, the problem concerns the *interpretation* of the natural laws; i.e. whether we are entitled to interpret the laws as physical laws throughout the backward extrapolation, if the foundation for this interpretation (like the existence of cores of rods and clocks) disappears at some point along the extrapolation. Given our view of the interpretation of natural laws, the time concept in the early universe becomes speculative before the electroweak phase transition. As we have seen, before this point ($\sim 10^{-11}$ seconds), known physics becomes scale invariant and so one loses any (non-speculative) handle on how close we are to the singularity. We believe, but it should be further examined, that our position is a reasonable compromise between Platonism (mathematical foundationalism) and operationalism (which requires a method for actually measuring cosmic time).

---

[25] From our relationist point of view – in which time is necessarily related to physical processes – the time-like curves can only be identified (they only have a physical basis) if the motion of objects or test particles along these curves is at least in principle realizable given the available physics.

In Sections 19.3.1 and 19.3.2 we have seen that a *global* concept of cosmic time (with or without a scale) may become problematic in the early universe if the Weyl principle cannot be satisfied (e.g. if everything moves with the speed of light and no comoving reference frame can be constructed). Moreover, the discussion in Section 19.3.3 showed that not even a *local* concept of time, which could be used to address the past of some local event, may be available as time is extrapolated backwards in the very early universe. In particular, this seems to be the case if one assumes "quantum fundamentalism"; the idea that everything is quantum, and even if something looks classical *now*, there was an early time, e.g. 10^{-34} seconds, when nothing did. Thus, if all constituents are quantum at the onset of inflation $10^{-34}s$, it seems difficult (or impossible) to even construct a physical notion of proper (local) time along individual world lines which could order events in the very early universe. The upshot of our discussion on these points was that classical systems appear to be necessary throughout cosmic history (to have a reasonable time concept). It is standard to hold that quantum gravity sets in at $10^{-43}s$, i.e. that there is no time concept "before" this Planck time. But our discussion indicates that *if* one believes that everything is quantum, then one has a problem with time in general (and not only in quantum gravity)!

Let us finally briefly consider whether the possible limits to time are a misfortune for cosmology. We think not. Limits in science are good for at least two reasons. First, they should not be seen as stumbling blocks for research but rather as invitations to keep asking questions, e.g. as to which theories might describe what lies beyond the present temporal limits (or how the limits might be circumvented, e.g. by introducing speculative new physics). Such invitations can be expected to remain open since for any postulated theory describing earlier times, it will probably always be possible to ask: what lies beyond *that* theory? This leads to the second reason: The fact, if it is a fact, that there will always be something beyond our (current?) scientific understanding may be aesthetically attractive, if not also comforting.[26] Both of these reasons for endorsing limits are connected to that feeling of wonder which has been an important driving force throughout the history of cosmology.

Acknowledgments

We are grateful for discussions on the above topics with many people over the years (see acknowledgments in our previous manuscripts cited below). We also thank the organizers of the Philosophy of Cosmology conference at Tenerife for the opportunity to present this work. HZ thanks the Spanish Ministry of Science and Innovation (Project FFI2011-29834-C03-02) for financial support.

References

Anastopoulos, C. 2002. Frequently asked questions about decoherence. *International Journal of Theoretical Physics*, **41**, 1573–90.

---

[26] See Zinkernagel (2014) for some brief remarks on aesthetics in cosmology.

Barrow, J. D. 2005. Worlds without end or beginning. In D. Gough, ed. *The Scientific Legacy of Fred Hoyle*. Cambridge: Cambridge University Press, pp. 93–101.

Bondi, H. 1960. *Cosmology* (2nd edition). Cambridge: Cambridge University Press.

Ellis, G. F. R. 1999. 83 Years of general relativity and cosmology: progress and problems. *Classical and Quantum Gravity*, **16**, A37–75.

Ellis, G. F. R. 2012. Space time and the passage of time, gr-qc arXiV: 1208.2611.

Feynman, R. P. 1995. *Lectures on Gravitation* (edited by B. Hatfield). Reading, MA: Addison-Wesley.

Hartle, J. B. 1991. The Quantum Mechanics of Cosmology. In S. Coleman *et al., Quantum Cosmology and Baby Universes*. Singapore: World Scientific.

Hobson, M. P., Efstathiou, G. P. and Lasenby, A. N. 2006. *General Relativity*. Cambridge: Cambridge University Press.

Kiefer, C. 1989. Continuous measurement of intrinsic time by fermions. *Classical and Quantum Gravity*, **6**, 561–6.

Kiefer, C. 2003. Decoherence in Quantum Field Theory and Quantum Gravity. In E. Joos *et al.*, eds. *Decoherence and the Appearance of a Classical World in Quantum Theory*, 2nd edition. Berlin: Springer, pp. 181–225.

Krasinski, A. 1997. *Inhomogeneous Cosmological Models*. Cambridge: Cambridge University Press.

Lacheize-Rey, M. 2014. In search of relativistic time. *Studies in History and Philosophy of Modern Physics*, **46**, 38–47.

Linde, A. 2004. Inflation, quantum cosmology, and the anthropic principle. In J. D. Barrow, P. C. W. Davies and C. L. Harper, eds. *Science and Ultimate Reality*. Cambridge: Cambridge University Press, pp. 426–458.

Misner, C. W., Thorne, K. and Wheeler, J. A. 1973. *Gravitation*. New York: W. H. Freeman.

Narlikar, J. V. 2002. *An Introduction to Cosmology* (third edition). Cambridge: Cambridge University Press.

Peacock, J. A. 1999. *Cosmological Physics*. Cambridge: Cambridge University Press.

Peebles, P. J. 1993. *Principles of Physical Cosmology*. Princeton: Princeton University Press.

Penrose, R. 2010. *Cycles of Time – An Extraordinary New View of the Universe*. London: Random House.

Raychaudhuri, A. K. 1979. *Theoretical Cosmology*. Oxford: Clarendon Press.

Rugh, S. E. and Zinkernagel, H. 2005. Cosmology and the Meaning of Time, 76 pp. Distributed manuscript. Available upon request.

Rugh, S. E. and Zinkernagel, H. 2009. On the physical basis of cosmic time. *Studies in History and Philosophy of Modern Physics*, **40**, 1–19.

Rugh, S. E and Zinkernagel, H. 2011. Weyl's principle, Cosmic Time and Quantum Fundamentalism. In D. Dieks *et al.*, eds. *Explanation, Prediction and Confirmation. The Philosophy of Science in a European Perspective*. Berlin: Springer, pp. 411–424.

Rugh, S. E. and Zinkernagel, H. 2013. A Critical Note on Time in the Multiverse. In V. Karakostas and D. Dieks, eds. *Recent Progress in Philosophy of Science: Perspectives and Foundational Problems*. Berlin: Springer, pp. 267–279.

Scrimgeour, M. I. *et al.* 2012. The WiggleZ Dark Energy Survey: The transition to large-scale cosmic homogeneity, arXiv: 1205.6812v2.

Sudarsky, D. 2011. Shortcomings in the Understanding of Why Cosmological Perturbations Look Classical. *International Journal of Modern Physics D*, **20**, 509–52.

't Hooft, G. and Vandoren, S. 2014. *Time in Powers of Ten*. Singapore: World Scientific.

Weinberg, S. 1972. *Gravitation and Cosmology.* New York: Wiley and Sons.

Whitrow, G. J. 1980. *The natural philosophy of time.* Oxford: Clarendon Press.

Zinkernagel, H. 2008. Did Time have a Beginning? *International Studies in the Philosophy of Science*, **22**(3), 237–58.

Zinkernagel, H. 2011. Some trends in the philosophy of physics. *Theoria*, **26**(2), 215–41.

Zinkernagel, H. 2014. Introduction: Philosophical aspects of modern cosmology. *Studies in History and Philosophy of Modern Physics*, **46**, 1–4.

Zinkernagel, H. 2016. Niels Bohr on the wave function and the classical/quantum divide. *Studies in History and Philosophy of Modern Physics*, **53**, 9–19.

20

Self-Locating Priors and Cosmological Measures

CIAN DORR AND FRANK ARNTZENIUS

20.1 Introduction

It seems like bad news for a theory if it entails that almost all of those who perform a certain experiment get a certain result, and we actually perform that experiment and get a different result. But it is not immediately obvious why this should be bad news, or what kind of bad news it is, when the theory in question is logically consistent with the fact that we got the result we actually got. This is something we need to understand better. It is not enough just to say that other things being equal, a theory's having this feature is a reason not to believe it, since other things are never equal. In all the interesting cases, the theories in question will have a great many other features which are, *ceteris paribus*, reasons to believe them – they may be attractively strong and simple; they may accurately predict the values of certain measured parameters, and so forth. We need a framework of thinking about the bearing of evidence on theories that can give us some guidance about how these factors trade off against one another.

Indeed, we can see that it is not always bad news for a theory when we make observations that are atypical according to it. For consider the following theory (if you want to call it that): We will perform experiment E and get result A, although almost everyone else who performs experiment E will get result B. Since our getting any result other than A would refute this theory, this result looks like good news rather than bad news. So, this is another place where we need some guidance.

The need for such a framework is especially pressing when we turn our attention from the elaborate experiments physicists are paid to perform to the 'experiments' we perform all the time whether we like it or not – for example, the experiment of standing in front of a mirror and seeing what you look like. There might be gazillions of non-human observers in the universe, most of whom see completely different things when they look into mirrors. It seems foolish to reject a serious theory that posits a multitude of disparately shaped aliens just on the grounds that you see a human-like form (two arms, two legs, . . .) when you look in a mirror. Surely the place to look if you want to investigate a theory like that is a telescope, not a mirror! But it is unclear how this piece of common sense is to be reconciled with the idea that it is bad when a theory says that our observations are atypical. Small wonder that so many practising physicists are suspicious of considerations having to do with the typicality of our observations – 'anthropic' considerations – and would prefer to be able to make comparisons between theories without ever having to think about such matters.

Unfortunately for them, it is hard to see how one could possibly avoid the need to take such considerations into account. Since the work of Boltzmann – if not that of Democritus – physics has thrown up a succession of theories which seem to have the problematic feature that they make our actual observations excessively atypical, while being simple and attractive in other respects, and also logically consistent with our evidence, so that it is not clear what can be said against them without appealing to considerations of atypicality. In Boltzmann's picture, a fixed finite stock of particles continue to exist and to interact eternally, with the result that every possible dynamical state of those particles (with a fixed total energy) eventually comes arbitrarily close to being realised, including all those possible dynamical states that subserve the existence of observers making observations of any humanly possible kind. Nothing that we have observed is inconsistent with this theory. Its only obvious defect is that according to it, the vast majority of observations are utterly unlike our actual observations. They are, rather, the kinds of observations one would expect to be made by 'Boltzmann Brains' (Albrecht and Sorbo, 2004) – short-lived, isolated observers who came into existence as part of a recent, localized fluctuation from equilibrium. If one denies that this is any kind of problem for Boltzmann's theory, one seems forced into the position that observations could never bear in any way on a theory that entails that every possible observation is made at least once in the history of the universe. But this conclusion would be a disaster, since cosmologists have considered a multitude of serious hypotheses with exactly this feature. If empirical investigation can never favour some of these hypotheses over others, we are doomed to a paralysing level of scepticism.

These examples also remind us – if it was not clear enough already – that we need to think seriously about what it even means to say that our observations are 'atypical'. For given Boltzmann's theory, every possible observation is made not just once, but infinitely many times. The cardinalities are the same: there is a countable infinity of observations just like ours, and a countable infinity of observations unlike ours. So in what sense is it true to say that 'most' or 'almost all' observations are unlike ours? One could try to make sense of the 'most' claim by taking some kind of limit using a sequence of longer and longer finite temporal intervals. But what could make this the right way to compare the infinite sets? And how are we supposed to generalise it to more recent infinite-population theories which are set in relativistic spacetimes whose extent may be infinite in both temporal and spatial directions?

Quite apart from problems associated with infinity, the typicality or otherwise of our observations clearly depends on the class of objects you are considering: a feature that is typical among primates need not be typical among all animals. But what is the relevant class of things when we are trying to figure out whether our observations are, or are not, problematically atypical according to a certain theory?

In what follows we will develop a Bayesian framework for answering all these questions.

20.2 Bayesian Background

We will assume that an ideally rational person at any time t has degrees of belief ('credences') which can be represented by a probability function $\mathbf{C}(\cdot|E_t)$: the result of conditionalising a certain other probability function $\mathbf{C}$ – her 'priors' – on the total evidence E_t that she has at t. In this paper, 'probability functions' will always be functions P from *propositions* – things that can be true or false, and believed to various degrees – to real numbers, such that

  (i) $P(A) = 1$ whenever A is logically necessary
 (ii) $P(A) \leq P(B)$ whenever B is a logical consequence of A; and
(iii) $P(A \vee B) = P(A) + P(B)$ whenever A and B are logically incompatible.[1]

We take the set of propositions on which rational peoples' credence functions are defined to include both eternal, qualitative propositions (like *There exists or will exist or has existed at least one physicist*) and self-locating propositions (like *There is a physicist in front of me*) which attribute a qualitative property to the particular agent at the particular time in question.[2]

    Note that we are merely assuming that rational people can be so represented, not that they need to have priors in their heads in any psychologically realistic sense – let alone that they need to have had them in their heads temporally prior to any given episode of rational belief-formation. We are also not assuming any particular account of evidence, e.g. that only propositions about one's conscious experience at t can be part of one's evidence at t. Everything we say will be compatible with externalistic views on which a rich body of truths about one's surroundings and history count as part of one's evidence (e.g. Williamson, 2000). While the difference between these conceptions of evidence can make a big difference in some cases, we do not think it will matter to any of the theoretical comparisons we will be concerned with in this paper.

    One advantage of characterising a rational person's credences as the result of conditionalising her priors on her total evidence at the relevant time, rather than as the result of conditionalising her previous credence function on the evidence that she just received, is that it determines, in a prima facie plausible way, how a rational person's credences evolve when she forgets things, and how her credences evolve in response to changing self-locating evidence.[3] Most importantly, it gives us a setting in which we can pose questions not just about how *new* evidence should modify our credences, but also about how the

---

[1] Note that these conditions entail that $P(A)$ is never greater than 1 (since in that case $P(A\vee \geq A)$ would also have to be greater than 1 by (ii), contradicting (i)), and that $P(A)$ is never less than 0 (since in that case $P(\geq A)$ would have to be greater than 1 by (i) and (iii)). Note too that (iii) immediately extends to all finite disjunctions: if $A_1 \ldots A_n$ are pairwise logically incompatible, $P(A_1 \vee \ldots \vee A_n) = P(A_1) + \ldots + P(A_n)$. By contrast with the standard mathematical definition of 'probability function', we do not require probability functions to satisfy countable additivity, the analogue of this for countably infinite disjunctions.

[2] Our purposes in this chapter will not require taking any particular view as to the nature of propositions in general, or of self-locating propositions in particular. Perhaps the qualitative and the self-locating propositions are just two special subclasses of the class of all propositions, the former being those that are not 'directly about' any particular objects at all and the latter being those that are 'directly about' the person and time in question but nothing else. Or perhaps self-locating propositions should be treated as *sui generis*, as in the influential approach of Lewis (1979).

[3] See Arntzenius (2003) for further discussion of these issues.

evidence that we already have bears on a given theory. It is vital to be able to make sense of this, since in many cases it is quite an achievement to extract any observational predictions at all from a theory, and often the predictions we manage to extract will concern observations that we have already made. Such 'old evidence', whose relevance we want to be able to discuss, includes not just facts about experiments that have been done by physicists, but familiar facts of everyday life that are well known to everyone. In the present framework, even when a person has already updated her credences on certain evidence E, we can still say that E confirms H 'for that person if her prior $\mathbf{C}$ is such that $\mathbf{C}(H|E) > \mathbf{C}(H)$, and that E confirms H *simpliciter* if $\mathbf{C}(H|E) > \mathbf{C}(H)$ for any reasonable $\mathbf{C}$.

We stated above that the credences of a rational person are defined on self-locating as well as qualitative propositions. In recent years the epistemology of self-locating belief has been the focus of a substantial body of literature (for a survey, see Titelbaum, 2013). That literature tends to focus on thought experiments involving perfect duplication of experience between different people, or between the same person at different times. This might suggest that the inclusion of self-locating propositions in the framework is a technical innovation driven primarily by such thought experiments. But the idea that there is a crucial difference between learning that *you now* have a certain property and merely learning that *someone, sometime* has that property is really a completely intuitive one, which can be illustrated by any number of everyday cases. For instance, suppose that you and 20 friends have booked all the rooms in a 21-room hotel. You remember either reading in the hotel brochure that all but one room in the hotel is red, or reading that only one room is red. You just do not remember which, and you are initially about 50-50 as to the type of hotel that you have booked. Upon arrival you and your friends randomly pick rooms to go to. You then find that your room is red. If you took your evidence to be the qualitative proposition *Someone is in a red room* you would have no reason to modify your credences regarding the type of hotel that you are in, since that proposition had to be true either way. But if your evidence is the self-locating proposition *I am in a red room*, it strongly supports the hypothesis that all but one room in your hotel is red. And it seems obvious that this is the correct way to reason. This is presumably what you would conclude if you believed that you were the only person in the hotel, and it surely makes no relevant difference whether you believe that you have 20 friends with you or not.

Of course, we have claimed that one should form credences by conditionalising one's priors on one's *total* evidence at any given time, and it is unrealistic to think that *I am in a red room* is your total evidence. However, it is not unrealistic to think that this is the only part of your evidence that we need to take into account in order to assess how your evidence bears on the comparison between the two live hypotheses about what kind of hotel you are in. Formally, when E^- is a consequence of your total evidence E, E^- will exhaust the bearing of E on two hypotheses H_1 and H_2 whenever $\mathbf{C}(E|E^-H_1) = \mathbf{C}(E|E^-H_2)$. For in that case,

$$\frac{\mathbf{C}(H_1|E)}{\mathbf{C}(H_2|E)} = \frac{\mathbf{C}(EH_1)}{\mathbf{C}(EH_2)} = \frac{\mathbf{C}(E|E^-H_1)\mathbf{C}(E^-H_1)}{\mathbf{C}(E|E^-H_2)\mathbf{C}(E^-H_2)} = \frac{\mathbf{C}(E^-H_1)}{\mathbf{C}(E^-H_2)} = \frac{\mathbf{C}(H_1|E^-)}{\mathbf{C}(H_2|E^-)}$$

so that E^- can serve as a proxy for your total evidence E when assessing its bearing on these two hypotheses. In the present case, it is plausible that according to your priors, your total evidence in all its detail is just as likely on the assumption that you are in the only red room in the hotel as it is on the assumption that you are in one of 20 red rooms in the hotel, so that *My room is red* can serve as a proxy for your total evidence.

In this example, the hypothesis *All but one room in my hotel is red* is not a qualitative proposition, and it would obviously be crazy to think that your total qualitative evidence can serve as a proxy for your total evidence when the relevant hypotheses are self-locating propositions. But we could instead have focused on the qualitative hypothesis *There exists a 21-room hotel with 20 red rooms*. Given what you remember about the booking, you should clearly become more confident in this when you find yourself in a red room.

It is clear that your total evidence is not a qualitative proposition. We need not assume that it is a self-locating proposition either (i.e. one that attributes a qualitative property to the agent): perhaps we should instead take it to be a '*de re*' proposition like *I am in this particular red room*, which attributes to the agent a non-qualitative property involving a particular object. We will however be assuming that for the purposes of reasoning about qualitative and self-locating propositions, one's total self-locating evidence can serve as a proxy (in the sense explained above) for one's total evidence. Having absorbed the lesson that conditionalising on an existential generalisation often has very different effects from conditionalising on an instance of that generalisation, this assumption might seem implausible given the *de re* view of evidence – why would *I am in a red room* be any better as a proxy for *I am in this particular red room* than *Someone is in a red room*? But when we bear in mind that one's self-locating evidence might include propositions like *I am in a room that looks this highly distinctive way*, or *I am in a room that I remember having been in 20 years ago*, it becomes hard to imagine a plausible view on which the mere identity of the particular objects in one's environment would have any further capacity to discriminate among qualitative or self-locating hypotheses.

Some theorists have taken seriously the 'relevance-limiting thesis' according to which only qualitative evidence needs to be taken into account when we are reasoning about qualitative hypotheses.[4] Their idea for dealing with apparent counterexamples like our hotel case is to say that our total qualitative evidence is quite rich – not just *Someone is in a red room*, but *Someone is in a red room experiencing such-and-such detailed pattern of light and shade, hearing such-and-such sounds, having such-and-such memories*... When we are reasoning about hypotheses according to which the population of the universe is small

---

4 The label is due to Titelbaum (2013); defenders include Halpern and Tuttle (1993), Halpern (2004), Meacham (2008), and Neal (2006) (approvingly cited in Carroll, 2010, p. 401). In a somewhat similar vein, Hartle and Srednicki (2007, p. 1) claim that 'Cosmological models that predict that at least one instance of our data exists (with probability one) somewhere in spacetime are indistinguishable no matter how many other exact copies of these data exist', although their later work (Srednicki and Hartle, 2013, 2010) suggests that their view may be a 'permissivist' one on which ideal reasonableness permits, but does not require, the disposition to reason in such a way that one's credences concerning the accuracy of such models never evolve under the impact of new evidence.

enough that it is vanishingly unlikely that there would be more than one person satisfying such a rich description, this lets us recover the reasonable-looking patterns of reasoning described above, at the cost of having to count all sorts of intuitively irrelevant aspects of one's evidence as crucially relevant. For example, it is plausible that a reasonable prior credence function will count it as approximately 20 times more likely that someone will satisfy the above rich description conditional on there being 20 people in red rooms than conditional on there being only one person in a red room. But as soon as we start thinking about scenarios in which we can no longer reasonably neglect the possibility that there is more than one witness to the existential quantification that is our total qualitative evidence, the approach that looks only at qualitative evidence will start to generate distinctive and implausible results. For example, consider the following case:

Measuring a Parameter: Two qualitative theories T1 and T2 both entail that the population of the universe is vast but finite. They differ with regard to the value of a certain cosmological parameter α. T1 says that the true value of α is 34.31, and T2 says that it is 34.59. Because of this, T1 and T2 also differ as regards the distribution of results among the many repetitions in the history of the universe of a certain experiment E which fairly reliability measures the value of α. Conditional on T1, the expected proportion of those who get the result 34.31 among those who do E is approximately 1/20, while conditional on T2, the expected proportion is approximately 1/1000.

Intuitively, doing E and getting 34.31 very strongly favours T1 over T2. But if the populations are sufficiently large, our *qualitative* evidence will deserve high prior credence conditional on both T1 and T2, even if we are careful to include all manner of apparently irrelevant background details. Thus the view that only qualitative evidence matters in reasoning about qualitative hypotheses leads to a disastrous scepticism about our ability to bring empirical evidence to bear in distinguishing different large-population hypotheses.[5]

As if this were not bad enough, the relevance-limiting thesis faces the further problem that it requires a counterintuitive boost in the posterior probabilities of theories according to which the population is large relative to their prior probabilities, simply because the more people there are, the less unlikely it will be (according to a reasonable prior) that any given very detailed qualitative property has at least one instance. When combined with the approach's inability to allow for evidence-based discrimination between these large-universe hypotheses, this threatens to lead to a truly paralysing sceptical collapse.

One might still try to make a last gasp effort to make do only with purely qualitative evidence by adopting an *ultra*-fine-grained conception of evidence, on which your total qualitative evidence is the existential generalisation of a property so specific that you can

---

5 Neal (2006) attempts to address this problem by (i) adopting a very fine-grained conception of evidence, on which the population of the universe would have to be quite large (he suggests something in the order of $10^{10^{10}}$) for there to be a substantial chance that the qualitative property attributed by our total evidence has multiple instances; and (ii) proposing that we should simply ignore the possibility that the population is this large. This 'ignoring' strikes us as patently unreasonable, in spite of the dubious verificationist and ethical considerations which Neal offers in its favour. (On the ethical ramifications of infinite populations, see Arntzenius, 2014.)

legitimately neglect the possibility that more than one person has it, no matter how many people there are. For example, the property might specify the exact values of certain continuous parameters (perhaps having to do with one's mental state), in which case it is plausible that reasonable priors will assign its existential generalisation probability zero, even conditional on number of people being (countably) infinite. Implementing this strategy would require an elaboration of the framework in which conditional priors are taken as primitive rather than defined by $\mathbf{C}(A|B) = \mathbf{C}(AB)/\mathbf{C}(B)$, so that one can meaningfully conditionalise even on propositions whose prior credence is zero (see Hájek, 2003). The question how one's unconditional priors should be extended to allow for conditionalisation on probability-zero propositions raises tricky issues (see Dorr, 2010; Myrvold, 2015). Indeed, the project of formulating rules for extending unconditional priors to conditional priors involves puzzles that are in many ways analogous to those that arise for the project of formulating rules for extending qualitative priors to self-locating priors. Given that the ultra-fine-grained conception of evidence has severe foundational problems – intuitively, it vastly overstates the extent to which beings like us could ever hope to get their beliefs to correlate with the exact value of any continuous parameter – we will set it aside, while noting that the ideas about self-locating priors which we will discuss in this paper will have analogues within the ultra-fine-grained framework.

One final note: the Bayesian framework has often been combined with the idea that there are no rational constraints on priors beyond the probability axioms. This would make the present enquiry trivial. We will be taking it for granted that there are better and worse priors to have, and that factors such as simplicity can legitimately be appealed to in saying what makes the better ones better. This means that we accept that in certain senses of 'simple', you should be pretty confident that the world is 'simple' in the absence of relevant evidence. Some will find this suspiciously rationalistic. They will be tempted to think that reasonable priors should instead be 'unbiased' or 'uniform'. But making sense of a relevant notion of lack of bias or uniformity is extremely difficult (especially in the infinite case). And in the limited range of cases where one can make sense of it (e.g. when there are only finitely many possibilities or there is a unique measure having certain natural symmetries), such 'uniform' or 'unbiased' priors turn out to have the feature that you typically do not learn anything interesting about the world given any finite amount of evidence. In any case, legislating some notion of uniformity or lack of bias seems equally a prioristic to us.

20.3 A Principle About Finite Worlds

In our present state of understanding, it would be foolish to try to codify the features that make some priors more reasonable than others – simplicity and so forth – in the form of some precise collection of axioms. In general, we just have to muddle along as best we can by trusting our judgements about particular cases. However, there are a few special domains where we have the resources to formulate principles about what reasonable prior credence functions are like which go beyond the probability axioms, are not obviously false, and are precise enough to be worth arguing about. One of these domains is the epistemology of self-locating belief conditional on there being only finitely many observers. In this domain, we can formulate a general principle which, if true, will allow one completely to specify a unique reasonable assignment of prior credences to self-locating propositions (conditional on the population being finite), given as input a reasonable assignment of prior credences to qualitative propositions. This principle can be seen as a very limited principle of indifference: the intuition is that your self-locating priors, conditional on a certain qualitative state of affairs, should be indifferent among all the observers who exist in that state of affairs.[6] Or better – since self-locating propositions may address the question *when it is* as well as *who you are* – your conditional self-locating priors should be indifferent among all the portions of the lives of observers in the relevant state of affairs whose duration is the same.

To state this principle more rigorously, we will need to introduce some notation. Where P is a probability function and R is a real-valued random variable – which we can identify with a function that maps each real number x to a proposition '$R = x$' in such a way that the resulting propositions are pairwise inconsistent and jointly exhaustive – we will use '$\hat{P}(R)$' to denote the expectation value of R in P. Similarly we write '$\hat{P}(R|H)$' for the expectation value of R in $P(\cdot|H)$. When F and G are any properties, we write $\langle F : G \rangle$ for the random variable whose value is the ratio between the total, for all the things that are ever F, of the duration for which they are F and the total, for all the things that are ever G, of the duration for which they are G. For example, $\langle \text{physicist} : \text{philosopher} \rangle = 10$ is the proposition that the total duration of all philosophers' careers is positive and ten times smaller than the total duration across history of all physicists' careers.[7] Then our principle can be stated as follows:

PROPORTION Where H is any qualitative hypothesis which entails that the total duration of the lives of all observers is positive and finite, and F is any qualitative property, and $\mathbf{C}$ is any reasonable prior credence function such that $\mathbf{C}(H)$ is positive:

$$\mathbf{C}(I \text{ } am \text{ } F|H) = \hat{\mathbf{C}}(\langle F \text{ observer} : \text{observer}\rangle|H)$$

In words: your prior probability that you are F given H should equal your prior expectation, given H, for the proportion of all observer time taken up by F. Note that the expectation value of this random variable depends only on how $\mathbf{C}$ treats qualitative propositions. Thus PROPORTION fully determines one's self-locating priors (conditional on the total duration of the lives of all observers being positive and finite) as a function of one's qualitative priors.

To get a sense for the appeal of PROPORTION, let us return to the case of *Measuring a Parameter*. We can take it that T1 and T2 do not differ as regards the proportion of

---

[6] This basic thought is what Bostrom (2002) calls 'The Self-Sampling Assumption': 'One should reason as if one were a random sample from the set of all observers in one's reference class'. PROPORTION below is intended as a precisification of this vague formulation. Our principle does not talk about 'reference classes': where Bostrom would ask 'What is the right way of specifying the reference class?', we will simply ask 'What is the right way of defining 'observer'?'

[7] We leave it open what the relevant notion of 'duration' is: it might be taken to be the physical notion of proper time; subjective (psychological) time; some measure of complexity of evolution in the relevant system's physical state; or something else again.

observers who ever do experiment E, or as regards how far along they are in their lives (on average) when they do it. In that case, the expected ratio between the total amounts of observer time occupied by the properties *having done E and got 34.31* and *having done E* will be equal to the expected proportion of trials of E that yield the result 34.31 according to that theory. So according to PROPORTION, $\mathbf{C}$(*I did E and got 34.31*|T1) will be (approximately) 20 times greater than $\mathbf{C}$(*I did E and got 34.31*|T2) for any reasonable $\mathbf{C}$. Thus, assuming that none of the other details of your total evidence beyond the fact that you did E and got 34.31 are relevant to the comparison between T1 and T2, we can draw a conclusion about how the experiment should affect your credences: the ratio of your credence in T1 to your credence in T2 should be 20 times greater than it would have been with no relevant evidence.

Ideas in the vicinity of PROPORTION are often summed up with slogans like 'We should expect ourselves to be typical'. But such slogans need to be treated with care. If being a typical observer means having only those properties that most observers have, we should obviously expect *not* to be typical; indeed we should be confident that everyone has some properties that most observers lack. For the same reason, if being a typical observer means something like being an *average* observer, we should also expect not to be typical. To extract from PROPORTION the claim that we should be confident that we are typical observers, we need to devise an interpretation for 'typical observer' on which it is trivially true that if there are finitely many observers, most of them are typical. If we were concerned only with typicality in one particular quantitative *respect*, we could simply define 'typical' to mean 'having a value for the relevant quantity that is within so many standard deviations of the mean value among all observers'. But making sense of an 'all things considered' notion of typicality is a much harder task, and not one that we have any need to take on.

Let us turn next to a more controversial application of PROPORTION.

Brains or No Brains? Two theories *Brains* and *No Brains* are generally similar except that *Brains* predicts that, after the heat death of the universe, an enormous (but still finite) number of observers come into existence because of random vacuum fluctuations, while *No Brains* includes some mechanism that prevents this from happening. Most of the randomly produced observers that exist according to *Brains* are 'Boltzmann Brains' – i.e. things that just barely qualify as 'observers' in the relevant sense, however we end up cashing it out. Although almost all the Boltzmann Brains are short-lived, *Brains* predicts that there are so many of them that the total duration of their lives is much larger than the total duration of all the ordinary observers' lives. And while a few of the Boltzmann Brains will, by chance, have misleading experiences as of living in a world like ours, fake memories, etc., almost all of them will spend their entire lives enduring the rather unpleasant sorts of experiences one would expect given the inhospitable conditions in which they have come into existence.

Consider our current evidence – evidence, perhaps, as of sitting in a comfortable room, drinking a cup of tea while typing on a computer keyboard. *Brains* and *No Brains* differ radically with regard to the proportion of observer time occupied by this qualitative property. So according to PROPORTION, a reasonable prior credence function will assign this evidence much higher probability conditional on *No Brains* than conditional on *Brains*, so that the evidence counts heavily in favour of *No Brains*. Of course, this does not yet mean that we should be much more confident in *No Brains* than in *Brains*. This confidence depends on the priors for the two theories as well as on the evidence, and *No Brains* might have features in virtue of which it deserves a much lower prior credence – e.g. if the mechanism that prevents Boltzmann Brains from forming is an *ad hoc* postulate without any further motivation, it might detract greatly from *No Brains*'s simplicity. However, the larger the ratio of Boltzmann Brains to normal observers is according to *Brains*, the harder it will be to justify an asymmetry in the priors large enough to compensate for the force of the evidence.[8]

Brains or No Brains? shows that PROPORTION has some distinctive and controversial implications when combined with other plausible claims about reasonable prior credences. In that example, the required additional claim was to the effect that if theories are roughly similar in respect of simplicity, etc., they should not be assigned very different prior credences. In other examples, the additional claim that combines with PROPORTION to generate controversial implications is one that is often taken for granted in applications of Bayesian methods: namely, that conditional on the hypothesis that a certain function is the one that maps each proposition to its *objective chance* of being true – its physical probability – our prior credences should agree with that function.

(PP) $\mathbf{C}(A|P \text{ is the objective chance function}) = P(A)$ whenever defined.

(This is one version of the 'Principal Principle' from Lewis, 1980.) The combination of (PP) with PROPORTION makes for distinctive consequences when we are dealing with theories according to which are significant objective chances for substantially different total numbers of observers. For example, consider the following case from Bostrom (2002):

The Incubator Stage (a): The world consists of a dungeon with one hundred cells. The outside of each cell has a unique number painted on it (which cannot be seen from the inside); the numbers being the integers from 1 to 100. The world also contains a mechanism which we can term the *incubator*. The incubator first creates one observer in cell #1. It then... flips a fair coin. If the coin falls tails, the incubator does nothing more. If the coin falls heads, the incubator creates one observer in each of the cells ##2–100. Apart from this, the world is empty. It is now a time well after the coin has been tossed and any resulting observers have been created. Everyone knows all the above.

Stage (b): A little later, you have just stepped out of your cell and discovered that it is #1. (Bostrom, 2001, 363)

Let *S* be the qualitative description of this set-up, supplemented with the stipulation that all observers created by the incubator live equally long lives; let *H* and *T*, respectively, be the

---

[8] Of course, we do not have to rest content with the evidence we can get by sitting in our armchairs: we can also go out and do some experiments. However, unless these experiments have an incredibly strong ability to discriminate the theories, they will not change the epistemic situation very much.

conjunctions of S with the propositions that the coin lands Heads and that the coin lands Tails. Since S specifies that the chances of H and T are equal, (PP) says that $\mathbf{C}(H|S) = \mathbf{C}(T|S) = 1/2$ for any reasonable $\mathbf{C}$. Since H entails that there are exactly 100 equally long-lived observers of whom one is in a cell numbered #1, while T entails that all observer time is spent in a cell numbered #1, we have $\hat{\mathbf{C}}(\langle\text{observer in cell \#1} : \text{observer}\rangle|H) = 1/100$ and $\hat{\mathbf{C}}(\langle\text{observer in cell \#1} : \text{observer}\rangle|T) = 1$. So by PROPORTION, $\mathbf{C}(I\ am\ in\ cell\ \#1|H) = 1/100$ while $\mathbf{C}(I\ am\ in\ cell\ \#1|T) = 1$. We can thus apply Bayes's theorem to the probability function $\mathbf{C}(\cdot|S)$ to get

$$
\begin{aligned}
\mathbf{C}(H|I\ am\ in\ cell\ \#1 \wedge S) &= \frac{\mathbf{C}(I\ am\ in\ cell\ \#1|H)\mathbf{C}(H|S)}{\mathbf{C}(I\ am\ in\ cell\ \#1|S)} \\
&= \frac{\mathbf{C}(I\ am\ in\ cell\ \#1|H)\mathbf{C}(H|S)}{\mathbf{C}(I\ am\ in\ cell\ \#1|H)\mathbf{C}(H|S) + \mathbf{C}(I\ am\ in\ cell\ \#1|T)\mathbf{C}(T|S)} \\
&= \frac{0.01 \times 0.5}{0.01 \times 0.5 + 1 \times 0.5} = \frac{1}{101}
\end{aligned}
$$

So if we assume that at stage (a) you have no relevant evidence beyond S, and that you gain no relevant evidence at stage (b) beyond the proposition that you are in cell 1, your credence in Heads will decrease from 1/2 at stage (a) to very low at stage (b).

In thinking about cases like *The Incubator*, some have been attracted to an alternative view according to which your credence in Heads should be 1/2 at stage (b). On the less plausible version of this view, your credence should *also* be 1/2 at stage (a). But this is hard to take seriously, since the discovery that you are in cell #1 looks like strong evidence in favour of Tails (which entails it).[9] On the more plausible version of the view, your credence in Heads should be high at stage (a), so that it can be 1/2 even after the impact of this strong evidence.

Given our Bayesian framework, there are two ways to generate this high credence in Heads at stage (a): we could either revise PROPORTION in such a way that your evidence at stage (a) will count as heavily favouring Heads, or we could revise (PP) in such a way as to require reasonable priors to favour Heads over Tails. One natural thought that would motivate the relevant sort of revision to PROPORTION is the idea that the more observers there are, the less surprising it is that *you* are one of them, so that the self-locating proposition that you *exist* (or that you are an observer) should count as strong evidence for Heads (Bartha and Hitchcock, 1999). This is inconsistent with PROPORTION, which entails that your prior credence that you exist and are an observer (conditional on the total duration of observer time being positive and finite) should be 1. If we wanted existence or observer-hood to have evidential force, we could easily modify PROPORTION so as to concern not your prior unconditional credences, but your prior credences conditional on your existence

---

[9] If we add the stipulation that all the observers have exactly the same evidence at stage (a), the claim that your credence should not change between stage (a) and stage (b) follows from the relevance-limitation thesis discussed in Section 20.2. There are also some, such as Bostrom (2002), who are sympathetic to this claim but not to the relevance-limitation thesis.

or observerhood. The problem with this approach is that it is not really general enough. Deriving the judgement that your credence in Heads at stage (b) should be 1/2 in all variants of *The Incubator* that differ just with regard to the two population numbers associated with Heads and Tails would require a prior credence function in which the probability of *I am an observer* conditional on *There are n observers* increases *linearly* in proportion to *n*. And of course this is impossible, since conditional probabilities cannot exceed 1. The better option, then, is to keep PROPORTION while revising (PP), by building into the priors a proportional bias towards hypotheses according to which there are many observers, even when chances are equal.[10]

Those who favour a high credence in Heads at stage (a) in *The Incubator* will presumably take an analogous view about other comparisons between hypotheses that disagree about the total number of observers. For example, in *Brains or No Brains?*, they will hold that your prior credence in *Brains* conditional on *Either* Brains *or* No Brains *is true and I am an observer* should be very high, so that the posterior credences in *Brains* and *No Brains* given normal evidence (e.g. as of sitting drinking tea and typing) end up close. But considered as a general model for good reasoning about the population of the universe, this seems quite crazy. Consider our current state of ignorance as regards how hard it is for intelligent life to evolve in an arbitrary solar system. When combined with a cosmological theory according to which spacetime as a whole is finite (but very large), different answers to this question will generate radically different expected numbers of total observers. The 'bias towards high populations' idea will thus lead us to the absurd result that we should right now be confident, conditional on the universe being finite, that it is *very easy* for intelligent life to evolve – probably easy enough that even when we conditionalise on the (by the lights of this approach surprising) fact that we have not yet encountered any alien life, we should still be confident that the average galaxy contains many inhabited solar systems. While this 'abundant life' hypothesis is not itself unreasonable, it seems clear that there are also perfectly reasonable hypotheses on which life is far rarer than this, and that in our current state of ignorance, it would be quite unreasonable to assign a very low credence to these hypotheses.[11]

Some have argued that PROPORTION itself should be rejected on the grounds that it makes it easier than it should be to do astrobiology from the armchair. For example, Sean Carroll argues as follows against the claim that 'we should make predictions by asking what most observers would see':

Imagine we have two theories of the universe that are identical in every way, except that one predicts that an Earth-like planet orbiting the star Tau Ceti is home to a race of 10 trillion intelligent lizard beings, while the other theory predicts there are no intelligent beings of any kind in the Tau Ceti system. Most of us would say that we do not currently have enough information to decide between

---

[10] For further discussion of this strategy, including the details of the required modification of (PP), see Arntzenius and Dorr (MS).

[11] Our argument here echoes Bostrom's 'Presumptuous Philosopher' argument against what he calls the 'Self-Indication Assumption' (Bostrom, 2002).

these two theories. But if we are truly typical observers in the universe, the first theory strongly predicts that we are more likely to be lizards on the planet orbiting Tau Ceti, not humans here on Earth, just because there are so many more lizards than humans. But that prediction is not right, so we have apparently ruled out the existence of many observers without collecting any data at all about what is actually going on in the Tau Ceti system. (Carroll, 2010, p. 225)

Carroll does not tell us what the two theories in his example say about life other than on Tau Ceti and Earth. This matters when we apply PROPORTION: if both theories entail there being quadrillions of observers elsewhere, and say the same things about how many of those observers are likely to have the qualitative property ascribed by our total self-locating evidence, the lizards will make no appreciable difference. But perhaps Carroll was taking it for granted that both theories entail that every observer is either on Earth or on Tau Ceti. In that case, PROPORTION does entail that our armchair evidence counts strongly against the lizard theory. If our priors are not biased towards high populations, and the theories really are on a par in the other relevant respects, we should be confident that the lizard theory is false, just as in *The Incubator* we should be confident that we are alone in the universe when we know that we are in cell #1.[12]

If, like Carroll, you think this result is wrong, it is worth trying to get clear on what it is about the lizards that is driving your judgment. Consider a range of theories:

T0 In any given solar system there is a certain tiny chance ϵ for life to evolve at all, but if observers do come into existence they are likely to be human-like creatures whose DNA uses two base pairs, having five toes on each foot.
T1 In any given solar system, the chance of human-like life evolving is ϵ, but the chance of lizard-like life evolving is $10,000\epsilon$.
T2 In any given solar system, the chance of human-like creatures with five toes on each foot evolving is ϵ, while the chance of human-like creatures with six toes on each foot evolving is $10,000\epsilon$.
T3 In any given solar system, the chance of human-like creatures whose DNA uses only two base pairs is ϵ, while the chance of human-like creatures whose DNA uses three or more base pairs is $10,000\epsilon$.

Suppose that all the theories are on a par as regards simplicity, and agree that the number of solar systems is finite but far greater than $1/\epsilon$.

According to PROPORTION, our actual evidence (as of being human-like, with five toes and two base pairs) strongly supports T0 over all of T1–T3. In the case of T3, this result is quite intuitive. Suppose we had known about T0 and T3 and their predictions before investigating our DNA: then, surely, the discovery that we have two base pairs would have strongly favoured T0. The situation seems similar in every relevant respect to *Measuring a*

[12] Hartle and Srednicki (2007) make a similar argument about aliens living in the atmosphere of Jupiter. They argue that it would be unreasonable to reject a theory according to which there are many such aliens 'solely because humans would not then be typical of intelligent beings in our solar system'. Of course we agree with this, but we note that the theories in their example are not described in enough detail – in particular, with regard to what they say about life outside the solar system – to determine what PROPORTION says about them.

Parameter. Perhaps there is some temptation to think that the situation with T2 is different, given that we have always known how many toes we have. But how could that difference matter? The order in which we get evidence does not normally have much significance as regards what we should believe once we have the evidence; the mere fact that the toe-counting experiment was performed long ago is not in itself any kind of reason to discount its significance. This takes us back to T1 and the lizard beings. Carroll does not tell us very much about what is driving his judgement about them. Is it their scaly skin? Their cold blood? Their bizarre social structures? Their alien sensory experiences? We have the feeling that as the scenario is fleshed out in more detail, the sense that there could be any important difference between T1 and T2 will start to fade away.

The ways of fleshing out T1 that make it most plausible that we should think about it differently from T2 and T3 involve the lizard beings having a mental life that is in some deep respect very different from ours. But in these versions of the case it is no longer obvious what PROPORTION says, because it is no longer clear whether the lizards count as 'observers'. So far we have been treating PROPORTION as a single univocal principle; but we should now admit that as we are conceiving it, it is really a schema that can be filled in in many different ways depending on how one interprets 'observer'. Some of the instances of the schema have crazily counterintuitive consequences. For example, if we plug in 'five-toed biped' for 'observer', we will get the absurd result that no amount of evidence should shake your confidence that you are a five-toed biped, conditional on there being a positive finite number of five-toed bipeds. If we understand 'observer' as 'living being or rock', we will end up with the absurd result that our actual evidence heavily favours hypotheses according to which the ratio of rocks to living beings is low over hypotheses that according to which it is high. We can thus refine our understanding of how the schematic notion of an 'observer' should be understood by considering our judgements about such cases. But there will still, inevitably, be hard cases where the intuitions are unclear.

Some friends of PROPORTION might hope to draw some principled, sharp line between observers and non-observers.[13] We do not think this can be done, but we also do not think that this is a problem for the basic thought underlying PROPORTION. Given that all the relevant factors are continuous, we should probably allow reasonable prior credence functions that blur the line in one way or another – for example, by assigning real-valued 'degrees of observerhood' which one integrates over time, instead of simply looking at the duration for which a single property is instantiated, and/or by taking a weighted average of many different probability functions each of which obeys PROPORTION on a different conception of observerhood. Probably, too, we should be somewhat permissive, allowing different reasonable prior credence functions corresponding to different ways of cashing out observerhood. Perhaps, if Carroll's lizards are sufficiently mentally alien, they may be among the things that can reasonably be excluded altogether, or assigned an observerhood

---

[13] A natural thought for those inclined towards some kind of mind-body dualism is that the sharp line in question is the one between consciousness and lack of consciousness (see, e.g. Page, this volume).

score much less than that of humans, or excluded by most of the probability functions that enter into the final weighted average.

The general dialectical situation here is quite similar to the situation we are in in connection with the idea that reasonable prior credence functions favour simpler theories over more complex ones: there are many different ways of making the notion of simplicity precise, and we doubt that there is any uniquely natural, rationally compulsory way of measuring simplicity and taking it into account. Things are messy, but this should not stop us from trusting our judgements about particular cases where it is relatively obvious how the simplicity comparisons pan out. The appropriate methodology for working out how simplicity relates to reasonableness seems to be one of 'reflective equilibrium'; insofar as we are drawn to something like PROPORTION, our attitude about what to count as an observer should be worked out in the same way.

Fortunately, the details about what counts as an observer for the purposes of PROPOR-TION do not seem to be relevant to any of the theoretical comparisons that have arisen in physics. For example, in realistic ways of filling in the details of *Brains or No Brains?*, it will not actually matter whether we count disembodied brains as observers when calculating the relevant proportions. We will get almost the same results if we only count fully embodied creatures, or creatures that live lives long enough to realise certain cognitive capacities, or creatures that achieve certain kinds of interaction with their environments. In each case, *Brains* still will entail (a) that the total duration of the lives of the post-heat-death 'observers' is far greater than the total duration of the pre-heat-death 'observers', and (b) that the qualitative property ascribed by our total evidence takes up a much lower proportion of post-heat-death observer time than of pre-heat-death observer time.

One further point: if we were only concerned with the question how we should respond to some change in our evidence, taking for granted that our credences *before* the change are reasonable, there would be no reason to concern ourselves with the question what to count as an observer. For given certain very weak assumptions, we can use PROPORTION to derive a formula which specifies our new credences conditional on some H in terms of our old credences conditional on H, in which the notion of observerhood does not appear at all except in delimiting the possible values of H:

PROPORTIONAL UPDATE: When a reasonable person has evidence *I am E* and credence function $\mathbf{C}_t$ at t, and evidence *I am E^+* and credence function $\mathbf{C}_{t+}$ at t^+, and H is a qualitative proposition that entails that the total duration of observer time is finite and that the total duration of E is positive if the total duration of E^+ is,

$$\mathbf{C}_{t+}(I\ am\ F|H) = \frac{\hat{\mathbf{C}}_t(\langle FE^+ : E\rangle|H)}{\hat{\mathbf{C}}_t(\langle E^+ : E\rangle|H)}$$

whenever the right-hand side is defined.

To derive PROPORTIONAL UPDATE from PROPORTION, the only assumption we need to make about the property of observerhood is that it is entailed by both E and E^+. Begin with a definition and two preliminary observations. *Definition:* when R and S are two random variables, let $R \times S$ be the random variable such that for any non-zero x, $R \times S = x$ is the disjunction of all conjunctions $R = y \wedge S = z$ where $yz = x$, while $R \times S = 0$ is just $R = 0 \vee S = 0$ (and thus can be true even when one of R and S is undefined). *First observation:* given that E and E^+ entail observerhood and that H entails that F has positive duration whenever E^+ does, H entails that $\langle FE^+ : \text{observer} \rangle = \langle FE^+ : E \rangle \times \langle E : \text{observer} \rangle$. *Second observation:* PROPORTION yields the following expression for the posterior expectation of any qualitative random variable R conditional on H:

$$\hat{C}(R | I \text{ am } E \wedge H) = \frac{\hat{C}(R \times \langle E : \text{observer} \rangle | H)}{\hat{C}(\langle E : \text{observer} \rangle | H)}$$

This gives us what we need to establish PROPORTIONAL UPDATE:

$$C_{t+}(I \text{ am } F | H) = \frac{C(I \text{ am } F \text{ and } E^+ | H)}{C(I \text{ am } E^+ | H)} = \frac{\hat{C}(\langle FE^+ : \text{observer} \rangle | H)}{\hat{C}(\langle E^+ : \text{observer} \rangle | H)}$$

$$= \frac{\hat{C}(\langle FE^+ : E \rangle \times \langle E : \text{observer} \rangle | H)}{\hat{C}(\langle E^+ : E \rangle \times \langle E : \text{observer} \rangle | H)} = \frac{\hat{C}(\langle FE^+ : E \rangle | I \text{ am } E \wedge H)}{\hat{C}(\langle E^+ : E \rangle | I \text{ am } E \wedge H)} = \frac{\hat{C}_t(\langle FE^+ : E \rangle | H)}{\hat{C}_t(\langle E^+ : E \rangle | H)}$$

where the equalities are justified respectively by the fact that E^+ is your evidence at t^+, by PROPORTION, by the first observation, by the second observation, and by the fact that E is your evidence at t.

PROPORTIONAL UPDATE: in turn yields a rule that we can use in the same way that people standardly use Bayes's rule, to express how much a change in evidence favours one qualitative hypothesis over another (when the conditions of PROPORTIONAL UPDATE are met):[14]

$$\frac{C(H_1)}{C_{t+}(H_2)} = \frac{\hat{C}_t(\langle E^+ : E \rangle | H_1)}{\hat{C}_t(\langle E^+ : E \rangle | H_2)} \frac{C_t(H_1)}{C_t(H_2)}$$

The question what counts as an observer for the purposes of PROPORTION can thus be bracketed when we are only concerned with assessing the impact of new evidence. However, unlike many Bayesians, we are interested in questions about synchronic rationality (what credences are reasonable given certain evidence) as well as diachronic rationality (how one's credences should evolve given certain changes in one's evidence, assuming they were reasonable to begin with). So we do not take this as a dissolution of the question what counts as an observer.[15]

---

[14] *Proof:* $C_{t+}(H_1)/C_{t+}(H_2) = C_{t+}(I \text{ am such that } H_1 | H_1 \vee H_2)/C_{t+}(I \text{ am such that } H_2 | H_1 \vee H_2) = \hat{C}_t(\langle E^+\text{-such-that-}H_1 : E \rangle | H_1 \vee H_2)/\hat{C}_t(\langle E^+\text{-such-that-}H_2 : E \rangle | H_1 \vee H_2)$ (by PROPORTIONAL UPDATE) $= (\hat{C}_t(\langle E^+ : E \rangle | H_1)C_t(H_1 | H_1 \vee H_2))/(\hat{C}_t(\langle E^+ : E \rangle | H_2)C_t(H_2 | H_1 \vee H_2)) = \hat{C}_t(\langle E^+ : E \rangle | H_1)C_t(H_1)/\hat{C}_t(\langle E^+ : E \rangle | H_2)C_t(H_2)$.

[15] Garriga and Vilenkin (2008) also note that for the purposes of assessing the impact of new evidence, there is no need to talk about any 'reference class' other than the one given by the old evidence. They seem to be

In conclusion, it seems to us that once we modify PROPORTION so as to remove the suggestion that there is a unique, binary notion of observerhood that all reasonable prior credence functions have to respect, the result is an attractive principle that yields defensible results across a wide range of cases. If we only ever had to think about finite populations, the simplicity and strength of this principle combined with the plausibility of its consequences would constitute good grounds for accepting it. However, we cannot reasonably assign a credence of zero to the hypothesis that there are infinitely many observers. And given this, we should surely want whatever we say about the epistemology of self-locating belief in finite worlds to emerge as a special case of some more general epistemological theory that also has something to say about infinite worlds. Thus, we will have to investigate the infinite case before forming a final view about PROPORTION. In the remaining sections we will make a start on this project.

20.4 Infinite Populations

Let us begin with some especially straightforward infinite-world hypothesis, where there is an obvious, uniquely natural way of generalising PROPORTION-style reasoning by taking limits.

Chessboard: Black and white houses are arranged on a two-dimensional plane, in a chessboard pattern. At a certain time, one person is born in each house, and lives the next 20 years inside that house. At that point all the doors of the houses are unlocked, and the people get to leave their houses, see what colour they are, and explore their immediate neighbourhood. Sixty years later, everyone dies. No other living creatures ever exist.

What prior credence should you have that you were born in a black house, conditional on *Chessboard*? Or equivalently: if you are still locked inside your house, how confident should you be, conditional on *Chessboard*, that you will find it to be black when you get let out? The natural answer is 1/2.

It is sometimes suggested that there is a deep conceptual problem about endorsing this natural answer. Perhaps the thought is that in claiming that your credence that you are in a black house conditional on *Chessboard* should be 1/2, we are somehow forgetting about the fact that it is not true to say that the *proportion* of observers who are in black houses if *Chessboard* is true is 1/2 (or any other number), since there is no such thing as the ratio of infinity to infinity. But this assumes that claims about proportions provide the only possible basis for favouring some credences over others in this case. We see no grounds for any such assumption.

Finding a fully general principle that entails the natural answer concerning *Chessboard* is a very tall order. But we can take a step in that direction by formulating a principle that

interested only in what we called 'diachronic rationality', and thus take their method to be a full solution to the problem of defining observerhood. PROPORTIONAL UPDATE is an improvement on the updating method they describe, which does not take account of the time-relativity of evidence.

tells us how to assign prior credences to self-locating propositions conditional on hypotheses like *Chessboard*, which describe approximately static arrangements of observers in a fixed background space.

LIMITING PROPORTION: Suppose that F is some qualitative property, and H is a qualitative proposition that entails that every finite region only ever contains finitely many observers each of whom has a finite life, and that

(i) There is a certain real number x such that, for any all-encompassing nested sequence of concentric spheres $\sigma_1, \sigma_2, \sigma_3 \ldots$, x is the limit of the sequence $x_1, x_2, x_3 \ldots$, where x_i is the proportion of the total duration of the lives of observers whose lives are confined to σ_i during which they are F.

(ii) Observers do not move around too much: there is a finite upper bound to the lengths of the journeys they take over the course of their lives.[16]

Then $\mathbf{C}(I\ am\ F|\mathrm{H}) = x$ for any reasonable prior credence function $\mathbf{C}$ for which $\mathbf{C}(\mathrm{H})$ is positive.

(We call a sequence of regions 'all-encompassing' just in case its union is the entire space.)

One might worry that there is something objectionably arbitrary about using the family of orderings of observers generated by the nested spheres to set a constraint on priors. After all, so long as the cardinality of F observers is the same as the cardinality of non-F observers, one can find orderings of the observers in which the limiting proportion of F observers takes any value one pleases. However, in general, the definition of one of these competitor orderings – or of a family of such orderings that agree on the limiting proportion of F observers – will be far more complicated than the definition of the family of orderings generated by nested sequences of spheres. For example, in the case of *Chessboard* the sequences of observers in which the limiting frequency of observers in black houses is anything other than 1/2 are, intuitively, quite crazy, jumping around in ever-larger leaps with no discernible logic beyond the imperative to make the limiting frequency come out at a specified value. Thus, insofar as one is comfortable with the idea that considerations of simplicity can play a legitimate role in making a difference between reasonable and unreasonable priors, it is hard to see how there could be any deep problem with the thesis that prior credences based on taking limits in nested spheres are more reasonable than prior credences based on taking limits using some ordering that gives a limiting proportion other than 1/2.

Someone might object that our judgement that the nested-sphere-based orderings are simpler than the jumpy orderings that give different limiting frequencies is a merely 'relative' one. The notion of a sphere is defined in terms of a certain metric; but given any

16 Given this condition, we will get the same limiting proportion whether we look, for each sphere, at the observers whose lives are confined to that sphere; or at the observers whose lives overlap that sphere; or at the portions of the lives of observers spent in that sphere. These limits can come apart in far-fetched possibilities where the observers move about at unbounded speeds. We consider the puzzles raised by such possibilities in Arntzenius and Dorr (MS).

ordering of the observers, one could always define a new metric according to which that
ordering counts as being derived from a sequence of nested 'spheres'. Relative to the new
metric, the sequences that looked well behaved relative to the old metric will make arbitrarily large jumps. But we do not see any problem here. There are very important differences
between the real metric – the one that matters in physics – and the cooked-up quantities
relative to which the crazy orderings look simple, and these seem like exactly the sorts
of differences we should expect reasonable people to be sensitive to. Moreover, since just
about any theory can be made to look simple by expressing it in a language with appropriately cooked-up vocabulary, it is hard to see how considerations of simplicity could play
any substantive role in an epistemology that gave no role to the contrast between natural
quantities and cooked-up ones.[17]

A different way of motivating the claim that there is a special conceptual problem about
infinite populations comes from the idea that reasonable prior credence functions should
be *permutation-invariant*, in the following sense:

PERMUTATION-INVARIANCE: When H entails that every observer bears R to exactly one
observer and that exactly one observer bears R to every observer, $\mathbf{C}(I\ am\ F|H) = \mathbf{C}(I\ bear\ R$
to someone who is F|H) for any reasonable prior credence function $\mathbf{C}$ for which $\mathbf{C}(H) > 0$.

PERMUTATION-INVARIANCE is inconsistent with LIMITING PROPORTION. Suppose we
define an ordering of all the observers with no first or last member, in which every third
observer is in a black house and the rest are in white houses. If we let R be the relation
every observer bears to the next observer in this ordering, PERMUTATION-INVARIANCE
entails that $\mathbf{C}(I\ am\ in\ a\ black\ house|Chessboard) = \mathbf{C}(I\ bear\ R\ to\ someone\ in\ a\ black$
house|Chessboard) = $\mathbf{C}(I\ bear\ R\ to\ someone\ who\ bears\ R\ to\ someone\ in\ a\ black\ house|$
Chessboard). Since *Chessboard* entails that every observer falls into exactly one of these
categories, $\mathbf{C}(I\ am\ in\ a\ black\ house|I\ am\ an\ observer\ and\ Chessboard\ is\ true)$ must equal
1/3 (if it is defined at all). But for the same reason, since we can define periodic orderings
of the observers corresponding to any rational number between 0 and 1, PERMUTATION-
INVARIANCE also entails that $\mathbf{C}(I\ am\ in\ a\ black\ house|I\ am\ an\ observer\ and\ Chessboard$
is true) is either ill defined or equal to x for every other rational $x \in (0, 1)$. This requires
that either $\mathbf{C}(Chessboard)$ is zero, or $\mathbf{C}(I\ am\ an\ observer|Chessboard)$ is zero. Since the
same reasoning will apply to other infinite-world hypotheses, the upshot is that we should
be sure that the number of observers is not infinite. We take this to be a decisive reason to
give up PERMUTATION-INVARIANCE.

20.5 Infinite Worlds with Multiple Simple Measures

LIMITING PROPORTION is not applicable to most of the infinite-population hypotheses that
arise in the context of cosmology. The reason for this is that in relativistic spacetimes there

---

[17] For more on naturalness and simplicity, see Lewis (1983) and Dorr and Hawthorne (2013).

is no useful notion of a 'four-dimensional sphere' – the closest analogues of spheres are regions bounded by hyperboloids, but these regions will in general contain infinite numbers of observers and hence be useless for the purpose of taking limits. One possible response to this limitation would be to embark on a quest for a generalisation of LIMITING PRO-PORTION: a single natural rule which prescribes reasonable prior self-locating credences conditional on any infinite-world hypothesis that physicists are likely to take seriously. But merely formulating a principle at this level of generality, let alone arguing for its truth, would be a very difficult task.

We do not think this is the right way to go. The moral we want to draw from the previous section's qualified defence of LIMITING PROPORTION is not even that LIMITING PRO-PORTION is true without exception, but that those who are comfortable with the idea that considerations of simplicity play a role in making the difference between reasonable and unreasonable priors face no special conceptual problem when it comes to infinite worlds. In typical cases where the method of taking limits in nested sequences of spheres yields well-defined self-locating priors, it is also far simpler than any method yielding different results. But this is not always the case; and in cases where there are simple self-locating probability functions that disagree with those yielded by the method of nested spheres, the claim that reasonable priors must accord with that method is much more tendentious. Consider:

Uneven Road: Inhabited houses are arranged along an infinite road running east–west. At one point there is a wall across the road. To the west of the wall, the houses are 100 metres apart; to the east, they are 10 km apart.

How confident should you be, conditional on *Uneven Road*, that you are in the western (thickly settled) part of the road? LIMITING PROPORTION entails that your credence should be 100/101, since this is the limiting proportion of observers to the west of the wall in any all-encompassing sequence of nested concentric spheres. But this is not the only principled answer that could be given in this case: there is also some temptation to think that your credence should be 1/2. This will seem natural insofar as one is gripped by the thought that the spacing of the houses is rationally irrelevant. And this answer can also be generated by a reasonably simple (albeit rather less general) method of assigning probabilities to self-locating propositions, namely a method which looks, not at nested sequences of concentric spheres, but at nested sequences of segments of the road in which each member of the sequence expands on its predecessor by adding the same number of houses in both directions.

The prior credences prescribed by LIMITING PROPORTION in this case strike us as somewhat dogmatic: finding that you do not live in a densely packed region does not seem like *very* strong evidence against *Uneven Road*. But the claim that you should be equally confident that you are in the western and eastern regions also seems unpromising – it is hard to see what plausible general principle could underlie such a prescription. We suggest that in cases like this, where there are multiple simple recipes for assigning credences to self-locating propositions conditional on some qualitative hypothesis, the most reasonable

approach is to split the difference. Conditional on such a hypothesis, reasonable prior credences will be generated by taking a *weighted average* of the credences that result from
the different simple methods, in which the simpler ones get weighted more heavily.

    To make this talk of 'recipes' more precise: let a *cosmological measure* be a function
μ from qualitative properties to qualitative random variables, such that conditional on the
proposition that $\mu(G)$ is defined for at least one property G:

(i) $\mu(F) = 1$ follows from *Always, everything is F, and*
(ii) $\mu(F) \leq \mu(F')$ follows from *Always, everything F is F'*
(iii) $\mu(F) + \mu(F') = \mu(F'')$ follows from *Always, everything F'' is either F or F'' but not
      both*

Given a measure μ and any probability function P on qualitative propositions which
assigns probability 1 to μ being defined, we can extend P to a self-locating probability
function $P^{[\mu]}$ simply by taking $P^{[\mu]}(I\ am\ F)$ to be $\hat{P}(\mu(F))$. A self-locating probability
function thus corresponds to the combination of a qualitative probability function and
a cosmological measure. In these terms, our proposed 'compromising' approach can be
stated as follows:

COMPROMISE: For any reasonable prior **C** and sufficiently specific qualitative H,

$$\mathbf{C}(I\ am\ F|H) = \sum_i w_i \hat{\mathbf{C}}(\mu_i(F)|H)$$

where μ_i are simple measures which are well-defined according to H, and w_i are weights
summing to 1, generally higher for simpler μ_i.

The 'sufficiently specific' H should be, at the minimum, specific enough to settle, of each
simple measure, whether it is well defined or not. More generally, the point is to focus on
hypotheses that pin things down in enough detail that there is no controversy about what
the *qualitative* priors should be conditional on them, so that all of the debate pertains to the
self-locating priors.

    (We can be more precise about the restriction to 'sufficiently specific' H if we derive
COMPROMISE from the following attractive account of the role of simplicity in reasonable
priors:

SIMPLICITY: A reasonable prior credence function **C** is a weighted average $\sum_i w_i P_i$ of
self-locating probability functions P_i, where w_i is generally higher for simpler P_i.

Plausibly all or almost all of the weights should go to probability functions P_i that are of
the form $Q_i^{[\mu_i]}$ for some qualitative probability function Q_i and measure μ_i. In this setting,
the 'sufficiently specific' H mentioned by COMPROMISE can be characterised as those
for which $Q_i(\cdot|H)$ and $Q_j(\cdot|H)$ are everywhere approximately identical whenever both are
defined and Q_i and Q_j are simple enough to receive significant weight. If this condition
is met, $\hat{\mathbf{C}}(R|H)$ will be approximately the same as $\hat{Q}_i(R|H)$ for any qualitative random

variable R; and so we have $\mathbf{C}(I\ am\ F|H) = \mathbf{C}(I\ am\ F \wedge H)/\mathbf{C}(H) = \sum_i w_i P_i(I\ am\ F \wedge H)/\mathbf{C}(H) = \sum_i w_i \hat{Q}_i(\mu_i(F\ such\ that\ H))/\mathbf{C}(H) \approx \sum_i w_i \hat{\mathbf{C}}(\mu_i(F\ such\ that\ H))/\mathbf{C}(H) = \sum_i w_i \hat{\mathbf{C}}(\mu_i(F)|H).)$

COMPROMISE is especially plausible when we turn to hypotheses in which LIMITING PROPORTION does not apply, but where there are multiple other simple methods of assigning probabilities to self-locating propositions. In many such cases, the quest for a general principle which would privilege a particular simple method as the one corresponding to a reasonable self-locating prior credence function seems misguided. Consider:

The Cliff: There is an infinite half-plane dotted with black, grey and white houses, terminated to the north by a straight, infinite cliff edge. The houses and their inhabitants get exponentially smaller and more tightly packed as we approach the cliff. The distance between the centre of a house and the cliff is always a power of two (in metres). The black houses on the line 2^n metres from the cliff are distributed randomly, in such a way that the expected number of houses in each segment of length l is equal to $l/2^n$: thus the average spacing between black houses on a given line is equal to the distance between that line and the cliff. The distribution of grey and white houses is determined by the distribution of black houses, as follows: for each black house, there is a grey house exactly halfway between it and the cliff, and for each grey house, there is a white house exactly halfway between it and the cliff. These are the only grey and white houses (see Figure 20.1).

There are two simple ways of assigning probabilities to self-locating propositions conditional on *The Cliff*, which assign different probabilities to *I am in a black house*. One

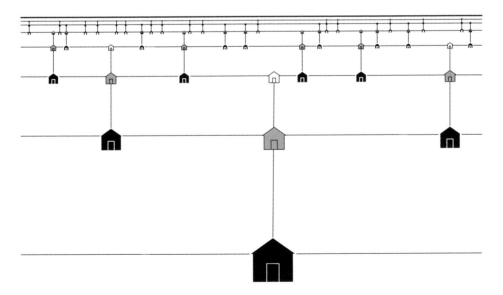

Figure 20.1 The Cliff

approach assigns a probability of 1/3, based on the fact that each north–south line of houses contains three houses of which one is black. The other approach assigns a probability of 4/7, based on the fact that on any given *east–west* line of houses, black houses are twice as common as grey houses, and grey houses are twice as common as white houses, so that (with chance 1) the limiting proportion of black houses along any east–west line is 4/7. For the same reason, the limiting proportion of black houses in any all-encompassing nested sequence of *rectangular* regions (all with finite populations) will also be 4/7.[18]

Instead of the compromising approach, one might suggest a permissivist approach according to which *both* of the simple self-locating probability functions correspond to optimally reasonable prior credence functions conditional on *The Cliff*. More generally, the thought would be each sufficiently simple measure can be used to generate a maximally reasonable prior credence function conditional on a given detailed qualitative hypothesis. (There is some pressure for those who take this permissivist view to allow that weighted averages of maximally reasonable prior credence functions are also maximally reasonable.) But this view is implausibly liberal. By modifying the details of *The Cliff*, one can make the two candidate credences as close as one pleases to 1 and 0 – just let each black house be the southernmost member of a very long series of non-black houses, and allow the density of black houses to increase by an arbitrarily large factor with each step towards the cliff.[19] The extreme credence functions that assign *I am in a black house* probabilities close to 0 or 1 in these cases seem clearly less reasonable than the weighted-average credence functions that assign intermediate credences – it seems absurdly dogmatic to treat either the discovery that one is in a black house, or the discovery that one is not, as incredibly weighty evidence against *The Cliff*.

In endorsing COMPROMISE, we do not mean to commit ourselves to the strong claim that simplicity is the *only* factor relevant to setting the weights assigned in a reasonable prior. There may be some further conditions that measures need to meet in order to deserve any weight at all. Note that all the measures we have considered make reference to a notion of observerhood, something which – as we discussed in connection with PROPORTION – could be understood in many different ways. A flat-footed extension of the compromising approach to the question what should count as an observer, according to which we simply take a weighted average of all probability functions which can be defined by appealing to simple criteria of observerhood – even crazy ones that count rocks as observers! – would have quite implausible consequences. Thus, we can already see that a fuller articulation of the compromising approach will need to appeal to some considerations other than simplicity to keep the final weighted average from being dominated by probability functions

---

[18] In this case there are no all-encompassing, nested sequences of *concentric* circles each of which has a finite population: since circles that extend past the cliff edge have infinite populations, any all-encompassing sequence of nested circles in which each circle has a finite population must have centres that get further and further away from the cliff edge. The limiting proportion of black houses in any such sequence is also 4/7.

[19] Indeed, if we modify the case so that each black house is the southernmost member of an *infinite* series of non-black houses, the north–south way of assigning probabilities will require assigning probability zero to being in a black house, while we can still make the east–west proportion of black houses arbitrarily close to 1.

defined using crazy (but simple) criteria of observerhood. The question what these considerations should look like is closely bound up with the question whether PROPORTION is true. It could easily turn out that the best way of excluding the crazy probability functions has as a consequence that the only probability functions that should receive any positive weight are functions that agree with PROPORTION in finite worlds (as do all the limiting procedures that we have considered). If this proved to be so, it would be good news for PROPORTION. If not, PROPORTION might start to look like an ugly and *ad hoc* addition, out of keeping with the spirit of the compromising approach. We leave further investigation of this question as a topic for future research.

20.6 The Compromising Approach and the Measure Problem in Cosmology

Let us consider one way in which modern cosmology prompts us to take seriously the possibility that there are infinitely many observers. According to the theory of inflation, if you follow the geodesic paths that are currently occupied by observable galaxies back far enough, you eventually – after 14 billion years or so – reach an 'inflationary' era during which the paths (as we follow them backward) approach one another at an exponential rate. This theory is fantastically successful by normal scientific standards. But models of the universe as a whole which provide a mechanism for such inflation typically feature *eternal* inflation – a kind of universe in which pockets of ordinary, non-inflating space keep forming, but in such a way that the inflating portion of space is never completely filled, but keeps expanding and giving rise to new non-inflating pockets. In the most plausible such models, there are many different kinds of pockets, only a few of which are hospitable to life. Nevertheless there is plenty of life: in fact there will be infinitely many life-friendly pockets as well as infinitely many life-unfriendly ones, and the life-friendly pockets will typically contain infinitely many observers each. We attempt to illustrate the general picture in Figure 20.2.

Since these hypotheses are set in relativistic spacetime, the proposal to assign probabilities by taking limiting relative frequencies in sequences of nested spheres does not even make sense. Nor does any other way of assigning self-locating probabilities look to be overwhelmingly simpler than all the others. Instead, what we generally find is a multiplicity of non-equivalent, reasonably simple cosmological measures. Here are a just a few examples. The 'proper time measure' (see Linde, 1986; Garcia-Bellido and Linde, 1995) is defined by starting with an (almost) arbitrary bounded, smooth, spacelike hypersurface; using it to construct a nested sequence of four-dimensional regions, each of which is the union of all timelike geodesic segments of a particular finite length, perpendicular to and extending futurewards from the chosen hypersurface; and assigning self-locating probabilities by taking limits within this sequence of regions, as in LIMITING PROPORTION. The 'scale factor measure' (De Simone *et al.*, 2010) is similar, except that instead of following the geodesics along by constant amounts of proper time, we use a time co-ordinate given by the scale factor – essentially, we follow each geodesic as far as we need to to reach a hypersurface on which the distances between nearby geodesics are a constant multiple of

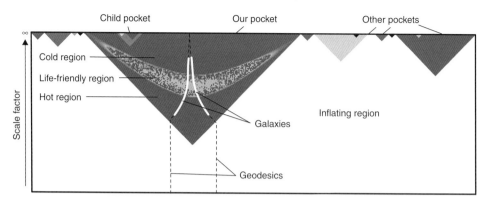

Figure 20.2 Eternal inflation

the distances between them on the initial hypersurface. For the 'causal diamond measure' (Bousso, 2006), we choose a timelike geodesic and take limits in a nested sequence of four-dimensional regions sandwiched between the forward light cones of points increasingly early on that geodesic and the backward light cones of points increasingly late on that geodesic. (Unlike the sequences of nested spheres considered in LIMITING PROPORTION, the sequences of regions considered by these three measures need not be all-encompassing; nevertheless, if the definitions of the measures are filled in properly, all the sequences of regions meeting the relevant criteria should yield the same limiting proportions.) There are also measures that are not based on taking limits in sequences of bounded regions at all: for example, the 'pocket-based measure' of Garriga *et al.*, (2006) works in two stages, first generating a separate measure for each vacuum state, and then aggregating these using a separate recipe for assigning weights to the different vacuum states.

Some of these measures have turned out to be 'pathological' in that they assign vanishingly low probability to our actual evidence – according to them, 'almost all' observers in the relevant models are, in some way or other, drastically and manifestly unlike us. For example, the proper time measure suffers from what is sometimes called the 'youngness paradox' (see Bousso, Freivogel and Yang, 2008; Tegmark, 2005). Since new pocket universes are being created in the inflating region at such a high rate, at each region in one of the relevant nested sequences, the pocket universes that have only just been added constitute a high proportion of all the pocket universes in that region. Because there are so many 'new' pockets, the observers in the new pockets who have come into existence quite soon after their local Big Bang (the initial boundary of their pocket) outnumber all the observers in the older pockets. As a result, when we take limits using these sequences, observations like ours – e.g. of measuring the temperature of the microwave background to be 2.7 K – will be assigned far lower probabilities than the kinds of observations that would be expected closer to a Big Bang, e.g. measurements of higher background temperatures. Indeed, the measure is so drastically skewed towards early observers that by its lights, 'almost all' observers are 'Boltzmann Babies' – freak observers who fluctuate into existence at times when the universe was so hot as to be extremely inhospitable to life, and spend the entirety of their short lives being roasted to death by their infernally hot surroundings.[20] Some other measures suffer from a more familiar kind of pathology: they are dominated not by observers who live much *closer* than us to their local Big Bang, but by observers who live much *further* – Boltzmann Brains who have fluctuated into existence in the endless freezing vacuum. At least some versions of the pocket-based measure suffer from this pathology (Page, 2008). The problem here is that there is a sense in which each and every inhabited pocket universe with a positive cosmological constant is dominated by Boltzmann Brains: if you begin with a finite-volume portion of the initial boundary of such a pocket universe, and evolve this region further and further forwards into the future (excluding any 'child' pockets that might form inside the initial one), you will continue to add Boltzmann Brains without bound, but you will only ever have come across finitely many ordinary observers.[21] This makes it quite challenging to devise a simple measure on observers in any one vacuum state that is not dominated by Boltzmann Brains (and, as usual, also dominated by Boltzmann Brains of the usual sort, living brief and bizarre lives).[22]

There is a rough analogy here with *The Cliff*. The spacelike surfaces represented by horizontal lines in Figure 20.2 are dominated by Boltzmann Babies, just as the east–west lines in *The Cliff* are dominated by black houses, whereas timelike paths (ignoring child pockets) are dominated by Boltzmann Brains, just as the north–south lines in *The Cliff* are dominated by non-black houses. Fortunately, whereas in the case of *The Cliff* there do not seem to be any other comparably simple measures, in the case of eternally inflating spacetime there is a wider array of alternatives that have not been shown to be in any way pathological.

In introducing the multiplicity of measures, cosmologists often characterise it as a deep foundational problem – the 'measure problem'. Tegmark (2014, p. 314) goes so far as to call it 'the greatest crisis facing physics today'. Some regard this problem as a weighty reason to reject inflation in favour of some rival theory: for instance, Steinhardt (2011, pp. 42–3) says that 'The notion of a measure, an *ad hoc* addition, is an open admission that inflationary theory on its own does not explain or predict anything', and rhetorically asks 'If inflationary theory makes no firm predictions, what is its point?'. Many others

---

[20] It is also worth noting that, even conditional on our actual evidence about the present and past, a probability function generated by the proper time measure will assign high probability to the proposition that we are about to *start* getting roasted to death.

[21] In Figure 20.2, imagine that the two dotted geodesics are the boundaries of a region that is open to the future but bounded in spacelike directions. Then the intersections of this region with the 'hot' and 'life-friendly' parts of our pocket universe have finite spacetime volume and contain finitely many observers, whereas the intersection of this region with the 'cold' part has infinite spacetime volume, and contains infinitely many Boltzmann Brains.

[22] These claims about the existence of Boltzmann Brains in the very late parts of pockets with positive cosmological constants are standard, but are sensitive to issues about the interpretation of quantum theory. Boddy, Carroll, and Pollack (this volume) and Goldstein, Struyve, and Tumulka (n.d.) point out ways in which the question of Boltzmann Brains may require rethinking on Everettian and pilot-wave interpretations, respectively. By contrast, the earlier remark about the domination of the proper time measure by Boltzmann Babies seems much less interpretation sensitive.

regard the problem as analogous to the divergences in quantum field theory: not a reason to reject the relevant hypotheses altogether, but a reason to believe that they are mere approximations to, or characterisations of, some emergent behaviour in some yet-to-be-discovered underlying theory that does not suffer from the problem. For instance, Tegmark (2014, p. 316) thinks that the measure problem is 'telling us' that we will have to give up on the idea that 'space can have an infinite volume, that time can continue forever, and that there can be infinitely many physical objects'. Similarly, Bousso and Freivogel (2007) treat the pathological features of certain measures as a reason to treat the infinite population of observers as 'figments of our imagination', instead favouring the minimalistic (and, *prima facie*, objectionably anthropocentric) view that a causal diamond that includes everything causally accessible from our worldline is 'all there is, as far as the semiclassical description of the universe goes'.[23]

While we have no objections to physicists following their hunches, one moral we want to draw from our discussion in previous sections is that the 'measure problem' does not constitute a *reason to disbelieve* the infinite-population hypotheses that give rise to it. There is no good *a priori* or empirical reason to be confident that our universe is one of the 'well-behaved' infinite ones with a unique simple measure, let alone that it is finite. The fact that there are *some* simple self-locating probability functions that give high probability to a certain hypothesis about the structure of the world as a whole while giving a vanishingly small probability to our evidence does not constitute any kind of reason to reject that hypothesis. What would constitute a reason to reject the hypothesis would be the discovery that *every* simple self-locating probability function that assigns it substantial probability assigns vanishingly small probability to our evidence.[24] And we have not discovered anything like this in the case of eternal inflation.

Despite the hand-wringing about foundational problems, the actual practice that cosmologists have adopted in reasoning scientifically about infinite-population hypotheses looks very much like what would be recommended by our 'compromising' account of the role of simplicity considerations in reasonable priors. A theorist will come up with a definition of a measure that works for a certain model of the cosmos, and try to figure out what kinds of observations are probable according to that measure (often a difficult task). When a particular measure is shown to have some pathological feature such as assigning high probability to being a Boltzmann Baby, the theorist's reaction is not to give up straight away on the relevant cosmological model, or suddenly to start treating sceptical scenarios in which we actually *are* Boltzmann Babies (or freak observers of some other sort) as live options. Rather, the theorist will start looking for some alternative simple measure which does not suffer from any such pathology. The goal is to find a pair of a simple cosmological hypothesis and a simple measure that together give reasonably high probability to certain

---

[23] Bousso and Freivogel seem to think that this eliminativist attitude is required by the use of their causal diamond measure, but we do not see why this should be so.

[24] 'Vanishingly small' here really means: small by comparison with the probability assigned to our evidence by other simple self-locating probability functions that do not assign high probability to the given cosmological hypothesis.

characteristic properties attributed to us by our actual evidence, such as observing a background temperature not too far from what we actually observe. And the cosmologists seem to be making considerable progress towards this goal – indeed the whole enterprise looks like science at its best.

All of this is exactly as it should be on the compromising approach. When generating reasonable self-locating priors from reasonable qualitative priors by setting $\mathbf{C}(I\ am\ F|H) = \sum_i w_i \hat{\mathbf{C}}(\mu_i(F)|H)$, it will often happen that some of the μ_i simple enough for the weight factor w_i to be substantial will be 'pathological' in the sense that they give a very low measure even to the barest outlines of our actual evidence. For example, there may be some simple μ_i such that H entails that μ_i (being a Boltzmann Baby in the midst of being roasted to death) is close to 1. But this is not a problem: so long as there are *any* reasonably simple measures that are not pathological in this way, the final weighted average will be dominated by those terms. That is: our posteriors will be approximately as they would be if we had started, not with our actual priors, but with a weighted average that excluded the pathological measures. Indeed, it is reasonable to hope that by doing the right experiments, we can enrich our total evidence to the point where one simple measure μ_k will assign it a vastly higher probability than any other comparably simple measure. In that case, our posterior self-locating credence $\mathbf{C}_t(I\ am\ F) = \mathbf{C}(I\ am\ FE)/\mathbf{C}(I\ am\ E)$ will be approximately equal to $\hat{\mathbf{C}}(\mu_k(FE))/\hat{\mathbf{C}}(\mu_k(E))$. For the purposes of making predictions, we will not have to think about anything other than the particular simple measure μ_k, and we will not need to concern ourselves with detailed questions about how exactly simplicity should be measured and weighted.

20.7 'Which Measure does Nature Subscribe to?'

Cosmologists sometimes say rather mysterious things in describing what we are learning about the world when we engage in this process of defining different measures and trying to find out which ones assign high probability to our evidence. For example, Max Tegmark describes the enterprise as follows:

There is some correct measure that nature subscribes to, and we need to figure out which one it is, just as was successfully done in the past for the measures allowing us to compute probabilities in statistical mechanics and quantum physics. (Tegmark, 2005, p. 2)

This remark suggests the following picture. In addition to facts of the familiar sort studied by physics (facts about the disposition of fields in spacetime, the wave function, and so on), there are facts about *which measure nature subscribes to*. We can investigate such facts empirically, since our evidence that we have a certain property F counts in favour of the hypothesis that nature subscribes to a measure μ for which $\mu(F)$ is high. More generally: for any measure μ, when we conditionalise any reasonable prior credence function $\mathbf{C}$ on the proposition that nature subscribes to μ, the result (if well defined) is just given by μ

itself:

DEFER TO NATURE $\mathbf{C}(I \text{ am } F|\text{Nature subscribes to } \mu) = \hat{\mathbf{C}}(\mu(F))$

Given this, all that remains to be done to pin down a particular reasonable prior credence function $\mathbf{C}$ is to specify $\mathbf{C}(\text{Nature subscribes to } \mu)$ for every μ. Presumably this should be higher for simpler μ.

Although Tegmark's remark is unusually explicit, the picture is not unique to him. It is also suggested by a certain way of using the word 'theory' that some cosmologists favour in this context, on which a theory is something that 'builds in' or 'comes with' a particular measure or self-locating probability function. As Linde (2007, p. 32) puts it: 'the probability measure becomes a part of the theory, and we test both the theory and the measure by comparing them with observations'. By itself this is a harmless terminological choice; but it becomes consequential when it is combined with the natural assumption that theories are propositions, things capable of being true or false.[25] For clearly it makes no sense to say that a measure – which is just a function from properties to real-valued random variables obeying certain axioms – is true or false. The question has to be whether the measure enjoys some special status that plays something like the epistemological role characterised by DEFER TO NATURE. And the choice becomes a serious metaphysical commitment if it is combined with the further assumption that the truth or falsity of the relevant theories remains a qualitative question, so that the relevant distinguished status is not merely a distinguished relation that the relevant measure stands in to *us*.[26]

We find these putative facts about what measure nature subscribes to obscure and problematic. We see no good reason to posit them at all, rather than taking seriously a more economical picture of reality as fully characterised by facts of a more familiar physical kind (such as facts about fields in spacetime).

You may have noticed that DEFER TO NATURE is structurally parallel to (PP), the standard 'Principal Principle' that specifies how reasonable prior credence functions behave conditional on a hypotheses about what probability function enjoys a different special status, that of being the true *objective chance* function. This might suggest that those who do not regard facts about objective chance as objectionably spooky should have no metaphysical objection to facts about what measure nature subscribes to. But there is at least the following disanalogy between the two cases. One widely held view about objective chance is Humean reductivism, which in its best-developed form (Elga, 2004; Lewis, 1994), identifies the proposition that a probability function P is the objective chance function with the proposition that P optimally balances the desiderata of simplicity and 'fit' (assigning high probability to truths, especially simple truths). It is reasonable to hope

[25] We do not attribute this assumption to Linde.

[26] For example, Page (this volume) proposes that each 'complete theory for a universe' should 'give normalized probabilities for the different possible observations O_j that it predicts, so that for each T_i, the sum of $P(O_j|T_i)$ over all O_j is unity', while also defining a 'complete theory for a universe' as one that 'completely describes or specifies all properties of that universe', in a context where it it is clear that the 'properties' in question are *qualitative* properties.

that by finding the right weighting of these factors, we could define a property F, necessarily instantiated by at most one probability function, such that $P(P \text{ is } F)$ will be close to 1 for any P simple enough to deserve substantial weight in a reasonable prior credence function. In that case, $\mathbf{C}(\cdot | P \text{ is } F)$ will be well approximated by P, so that (PP) will be approximately true (see Lewis, 1994, §10). We could certainly adopt a similar, reductivist Humean account of 'being the measure to which nature subscribes'. But on this understanding, the talk of 'nature' is deeply misleading, since the proposition that nature subscribes to μ will be a self-locating proposition rather than a qualitative one. If there is a simple measure μ that assigns a much higher probability to the properties that truly characterise *me* than any other simple measure, then I will speak truly when I say 'Nature subscribes to μ'. But someone else (some Boltzmann Brain, say), whose qualitative properties are atypical according to μ but typical according to some other simple measure μ^*, will say something false in uttering the same sentence. If we want to conceive of facts about what measure nature subscribes to as qualitative, Humean reductivism is not an option.

Our metaphysical objection to positing irreducible facts about what measure nature subscribes to will, of course, not weigh so heavily with those who already endorse an anti-Humean realist account of objective chance. But even they should want some argument for positing such facts. In the case of facts about objective chance, one can point to the pervasive role the concept seems to play in our ordinary thought and in the sciences. By contrast, the concept 'measure to which nature subscribes' seems like a recent innovation; so far, no-one seems to have given anything that looks like an argument for positing a domain of facts answering to this concept.

One argument that might be given for positing such facts is that there is no way to understand the rationality and cognitive significance of the cosmologists' enterprise of looking for simple cosmological models and simple measures on those models which assign reasonably high probability to our evidence – without making such a posit. But one of the morals of our discussion in the previous sections is that this argument is unsuccessful. Take a reasonable prior credence function as conceived by someone who posits facts about what nature subscribes to, and simply delete all the probabilities assigned to propositions about that peculiar subject matter, leaving the probabilities of ordinary qualitative and self-locating propositions unchanged. So long as the original probability function gave simpler measures a higher probability of being subscribed to, the output of this procedure will fit our compromising picture of reasonable priors: it will be a weighted average of self-locating probability functions in which simple ones receive higher weight. Thus, insofar as we are interested in how our evidence bears on these ordinary questions, there is no need to take seriously the idea that we are uncovering hidden facts about what nature subscribes to. We can treat this as nothing more than a colourful manner of speaking: for example, we can say 'We have learnt that nature subscribes to μ' when what we really mean is something like 'We have received evidence such that the result of conditionalising our priors on it is approximately equal as the result of conditionalising a self-locating probability function corresponding to μ on it'.

20.8 Conclusion

To sum up: for completely ordinary reasons, we should realise that the propositions we believe and that constitute our evidence are not always qualitative, and we should implement the familiar Bayesian idea that a reasonable credence function is derived from reasonable priors by conditionalisation on evidence in a way that treats self-locating propositions and qualitative propositions as being on a par. In such a framework there is nothing mysterious or surprising about the idea that our total self-locating evidence *I am E* might support one qualitative proposition over another, even when both propositions entail that *someone* is *E*, or entail that there are infinitely many people. The question *when* this happens reduces to the question what priors are reasonable. We should not, in general, expect to be able to establish our claims about reasonable priors by deducing them from precisely stated, uncontroversial principles, or regard it as a deep problem for some hypothesis about the world if we cannot find principles of this sort which completely pin down the result of conditionalising any reasonable prior credence function on that hypothesis. (If this is what it means for a hypothesis to make 'firm predictions', firm predictions are overrated.) Once we give up on the misguided hope for a knockdown demonstration that reasonable priors have to work in a certain way, we can get a long way with the familiar, vague idea that reasonable priors should be influenced by considerations of simplicity. In particular, the idea that reasonable priors are weighted averages of simple probability functions, in which simpler probability functions are weighted more heavily, yields prescriptions for reasoning about infinite-population hypotheses that are both intuitively plausible, and a good fit for the methodology that cosmologists have actually adopted in practice.

References

Albrecht, A. and Sorbo, L. (2004) Can the Universe Afford Inflation? *Physical Review D*, **70**.6, 063528 arXiv: hep-th/0405270.

Arntzenius, F. (2003) Some Problems for Conditionalization and Reflection. *Journal of Philosophy*, **100**, 356–70.

Arntzenius, F. (2014) Utilitarianism, Decision Theory and Eternity. *Philosophical Perspectives*, **28**.1, 31–58.

Arntzenius, F. and Dorr, C. (n.d.) What to Expect in an Infinite World.

Bartha, P. and Hitchcock, C. (1999) No-one Knows the Date or the Hour: An Unorthodox Application of Rev. Bayes's Theorem. *Proceedings of the Biennial Meeting of the Philosophy of Science Association* 66, S339–53.

Boddy, K., Carroll, M. and Pollack, J. (this volume) Why Boltzmann Brains Do Not Fluctuate into Existence from the De Sitter Vacuum. arXiv: 150502780 [hep-th].

Bostrom, N. (2001) The Doomsday Argument, Adam and Eve, UN++, and Quantum Joe. *Synthese*, **127**, 359–87.

Bostrom, N. (2002) *Anthropic Bias: Observation Selection Effects in Science and Philosophy*. New York: Routledge.

Bousso, R. (2006) Holographic Probabilities in Eternal Inflation. *Phys. Rev. Lett.*, **97**, 191302 arXiv: hep-th/0605263.

Bousso, R. and Freivogel, B. (2007) A Paradox in the Global Description of the Multiverse. *Journal of High Energy Physics 06* 18. arXiv: hep-th/0610132.

Bousso, R., Freivogel, B. and Yang, I.-S. (2008) Boltzmann Babies in the Proper Time Measure. *Physical Review D.* **77**.10, 103514. arXiv: 0712.3324 [hep-th].

Carroll, S. M (2010) *From Eternity to Here.* New York: Dutton.

De Simone, A. *et al.* (2010) Boltzmann Brains and the Scale-Factor Cutoff Measure of the Multiverse. *Physical Review D*, **82**.6, 063520. arXiv: 0808.3778 [hep-th]

Dorr, C. (2010) The Eternal Coin: A Puzzle About Self-Locating Conditional Credence. *Philosophical Perspectives*, **24**.1, 189–205.

Dorr, C. and Hawthorne, J. (2013) Naturalness. In Bennett, K. and Zimmerman, D, eds. *Oxford Studies in Metaphysics* vol. 8, Oxford: Oxford University Press, pp. 3–77.

Elga, A. (2004) Infinitesimal Chances and the Laws of Nature. *Australasian Journal of Philosophy*, **82**.1, 67–76.

Garcia-Bellido, J. and Linde, A. (1995) Stationarity of Inflation and Predictions of Quantum Cosmology. *Physical Review D*, **51**.2, 429. arXiv: hep-th/9408023.

Garriga, J., Schwartz-Perlov, D. *et al.* (2006) Probabilities in the Inflationary Multiverse. *JCAP*, 0601, 017. arXiv: hep-th/0509184v3 [hep-th].

Garriga, J. and Vilenkin, A. (2008) Prediction and Explanation in the Multiverse. *Physical Review D*, **77**, 043526. arXiv: 0711.2559v3 [hep-th].

Goldstein, S. Struyve, W. and Tumulka, R. (n.d.) The Bohmian Approach to the Problems of Cosmological Quantum Fluctuations. Forthcoming in Ijjas, A. and Loewer, B., eds. *Introduction to the Philosophy of Cosmology,* Oxford: Oxford University Press. arXiv: 1508.01017 [gr-qc].

Hájek, A. (2003) What Conditional Probability Could Not Be. *Synthese*, **137**, 273–323.

Halpern, J. (2004) Sleeping Beauty Reconsidered: Conditioning and Reflection in Asynchronous Systems In Gendler, T. S. and Hawthorne, J., eds. *Oxford Studies in Epistemology* vol. 1, Oxford University Press, pp. 111–42.

Halpern, J. and Tuttle, M. (1993) Knowledge, Probability, and Adversaries, *Journal of the Association for Computing Machinery*, **40**, 917–62.

Hartle, J. and Srednicki, M. (2007) Are We Typical? *Physical Review D*, **75**.12, 123523. arXiv: 0704.2630v3 [hep-th].

Lewis, D. (1979) Attitudes *De Dicto* and *De Se*. *Philosophical Review*, **88**, 513–43.

Lewis, D. (1980) A Subjectivist's Guide to Objective Chance. In Jeffrey, R. C., ed. *Studies in Inductive Logic and Probability*, vol. 2, Berkeley: University of California Press, pp. 263–93.

Lewis, D. (1983) New Work for a Theory of Universals. *Australasian Journal of Philosophy*, **61**, 343–77.

Lewis, D. (1994) Humean Supervenience Debugged. *Mind*, **103**, 473–90.

Linde, A. (1986) Eternally Existing Self-reproducing Chaotic Inflanationary Universe. *Physics Letters B*, **175**.4, 395–400.

Linde, A. (2007) Sinks in the Landscape, Boltzmann Brains and the Cosmological Constant Problem. *Journal of Cosmology and Astroparticle Physics*, 022. arXiv: hep-th/0611043.

Meacham, C. (2008) Sleeping Beauty and the Dynamics of *De Se* Beliefs. *Philosophical Studies*, **138**, 245–69.

Myrvold, W. C. (2015). You Can't Always Get What You Want: Some Considerations Regarding Conditional Probabilities. *Erkenntnis*, **80**, 573–603.

Neal, R. M. (2006) *Puzzles of Anthropic Reasoning Resolved Using Full Non-indexical Conditioning*. Technical Report 0607, Department of Statistics, University of Toronto. arXiv: math/0608592v1.

Page, D. (2008) Is Our Universe Likely to Decay Within 20 Billion Years? *Phys. Rev. D*, **78**, 063535.

Page, D. (this volume) Cosmological Ontology and Epistemology. arXiv: 1412.7544 [hep-th].

Srednicki, M. and Hartle, J. B. (2010) Science In a Very Large Universe. *Physical Review D*, **81**, 123524. arXiv: 0906.0042v3 [hep-th].

Srednicki, M. and Hartle, J. B. (2013) The Xerographic Distribution: Scientific Reasoning in a Large Universe. In: *Journal of Physics: Conference Series* Vol. 462. 1. IOP Publishing, p. 012050. arXiv: 1004.3816v1 [hep-th].

Steinhardt, P. (2011) The Inflation Debate. *Scientific American*, 36–43.

Tegmark, M. (2005) What Does Inflation Really Predict? *Journal of Cosmology and Astroparticle Physics*.

Tegmark, M. (2014) *Our Mathematical Universe*. New York: Knopf.

Titelbaum, M. G. (2013) Ten Reasons to Care About the Sleeping Beauty Problem. *Philosophy Compass*, **8**.11, 1003–17.

Williamson, T. (2000) *Knowledge and Its Limits*, Oxford: Oxford University Press.

21

On Probability and Cosmology: Inference Beyond Data?

MARTIN SAHLÉN

21.1 Cosmological Model Inference with Finite Data

In physical cosmology we are faced with an empirical context of gradually diminishing returns from new observations. This is true in a fundamental sense, since the amount of information we can expect to collect through astronomical observations is finite, owing to the fact that we occupy a particular vantage point in the history and spatial extent of the Universe. Arguably, we may approach the observational limit in the foreseeable future, at least in relation to some scientific hypotheses (Ellis, 2014). There is no guarantee that the amount and types of information we are able to collect will be sufficient to statistically test all reasonable hypotheses that may be posed. There is under-determination both in principle and in practice (Butterfield, 2014; Ellis, 2014; Zinkernagel, 2011). These circumstances are not new, indeed cosmology has had to contend with this problem throughout history. For example, Whitrow (1949) relates the same concerns, and points back to remarks by Blaise Pascal in the seventeenth century: 'But if our view be arrested there let our imagination pass beyond; ... We may enlarge our conceptions beyond all imaginable space; we only produce atoms in comparison with the reality of things'. Already with Thales, epistemological principles of uniformity and consistency have been used to structure the locally imaginable into something considered globally plausible. The primary example in contemporary cosmology is the Cosmological Principle of large-scale isotropy and homogeneity. In the following, the aim will be to apply such epistemological principles to the procedure of cosmological model inference itself.

The state of affairs described above naturally leads to a view of model inference as inference to the best explanation/model (e.g. Lipton, 2004; Maher, 1993), since some degree of explanatory ambiguity appears unavoidable in principle. This is consistent with a Bayesian interpretation of probability which includes *a priori* assumptions explicitly. As in science generally, inference in cosmology is based on statistical testing of models in light of empirical data. A large body of literature has built up in recent years discussing various aspects of these methods, with Bayesian statistics becoming a standard framework (Hobson, 2010; Jaynes, 2003; von Toussaint, 2011). The necessary foundations of Bayesian inference will be presented in the next section.

Turning to the current observational and theoretical status of cosmology, a fundamental understanding of the dark energy phenomenon is largely lacking. Hence we would like to collect more data. Yet all data collected so far point consistently to the simplest model of a cosmological constant, which is not well understood in any fundamental sense. Many of the other theoretical models of dark energy are also such that they may be observationally indistinguishable from a cosmological constant (Efstathiou, 2008). Another important area is empirical tests of the inflationary paradigm, the leading explanation of the initial conditions for structure in the Universe (e.g. Smeenk, 2014). Testing such models necessitates, in principle, the specification or derivation of an *a priori* probability of inflation occurring (with particular duration and other relevant properties). The *measure problem* is the question of how this specification is to be made, and will be the departure point and central concern in the following sections.

We will argue that the measure problem, and hence model inference, is ill defined due to ambiguity in the concepts of probability, global properties, and explanation, in the situation where additional empirical observations cannot add any significant new information about some relevant global property of interest. We then turn to the question of how model inference can be be made conceptually well defined in this context, by extending the concept of probability to general valuations (under a few basic restrictions) on partially ordered sets known as lattices. On this basis, an extended *axiological Bayesianism* for model inference is then outlined. The main purpose here is to propose a well-motivated, systematic formalisation of the various model assessments routinely, but informally, performed by practising scientists.

21.2 Bayesian Inference

Inference can be performed on different levels. An important distinction is that between parameter and model inference: the first assumes that a particular model is true and derives the most likely model parameter values on the basis of observations, whereas the latter compares the relative probability that different models are true on the basis of observations. The two inferential procedures can be regarded as corresponding epistemically to description (parameter inference) and explanation (model inference), respectively. This chapter will focus on model inference, which becomes particularly troublesome in the global cosmological context. We present both cases below for completeness. For more on Bayesian inference, see e.g. Jaynes (2003).

21.2.1 Parameter Inference

Bayesian parameter inference is performed by computing the *posterior probability*

$$p(\theta|D;M) = \frac{\mathcal{L}(D|\theta;M)\Pi(\theta;M)}{P(D;M)} \, , \tag{21.1}$$

where D is some collection of data, M the model under consideration and θ the model parameters. The *likelihood* of the data is given by $\mathcal{L}(D|\theta;M)$ and the *prior* probability distribution is $\Pi(\theta;M)$. The normalisation constant $P(D;M)$ is irrelevant for parameter inference, but central to model inference, and will be discussed next. The expression above is known as *Bayes' theorem*.

21.2.2 Model Inference

The *Bayesian evidence* for a model M given data D can be written

$$P(D; M) = \int \mathcal{L}(D|\theta; M) \Pi(\theta; M) d\theta , \qquad (21.2)$$

where symbols are defined as above. The Bayesian evidence is also called the *marginalised likelihood*, reflecting the fact that it measures the average likelihood across the prior distribution, and is thus a measure of overall model goodness in light of data and pre-knowledge. It is used in inference to compare models, with a higher evidence indicating a better model. Conventional reference scales (e.g. the Jeffreys scale) exist to suggest when a difference in evidence is large enough to prefer one model over another (Hobson, 2010).

Looking at Eq. (21.2), the Bayesian evidence is clearly sensitive to the specification of the prior distribution. A prior is usually specified based on *previous* empirical knowledge from parameter estimation, or may be predicted by the theoretical model, or given by some aesthetic principle. This highlights two things: without empirical pre-knowledge, a prior is entirely based on theoretical or philosophical assumption, and a prior is also not cleanly separable from the model likelihood and can therefore to a degree be regarded as part of the model. As increasing amounts of data are collected, the influence of the initial prior is gradually diminished through the process of *Bayesian updating*, i.e. the current posterior probability becomes the (new) prior probability for a future data analysis. Through this process, the posterior eventually converges to a distribution essentially only dependent on the total numbers of and accuracies of measurements. Increasingly numerous and precise measurements make the initial prior insignificant for the posterior. When data are extremely limited relative to the quantity/model of interest, this process stops short and the initial prior can then play a significant role in the evidence calculation. This will be of importance in the discussion that follows.

21.3 Global Model Inference in Cosmology

Cosmology, by its nature, seeks to describe and explain the large-scale and global properties of the Universe. There is also, by the nature of the field, a problem of finite data and underdetermination that becomes particularly poignant for measuring and explaining some global properties of the Universe. This will typically be associated with features on physical scales corresponding to the size of the observable Universe or larger, or features in the very early Universe. On the one hand, there is an epistemological question of knowledge based on one observation (i.e. the one realisation of a universe we can observe): how accurate/representative is our measurement? On the other hand, there is an ontological question of whether a property is truly global: if not, how might it co-depend on other properties, with possible implications for the evaluation of probabilities and inference? We shall therefore distinguish epistemically and ontically global properties in the following. In general, a global property will be defined here as some feature of the Universe which remains constant across the relevant domain (e.g. observable Universe, totality of existence).

A conventional approach to global properties is to treat separate regions of the Universe as effectively separate universes, such that sampling across regions in the Universe corresponds to sampling across different realisations of a universe. While this approach is useful for understanding the statistics of our Universe on scales smaller than the observable Universe, when approaching this scale the uncertainty becomes increasingly large and eventually dominates. This uncertainty is commonly called *cosmic variance* (see e.g. Lyth and Liddle, 2009). **We will explicitly only be concerned with the case when this cosmic variance cannot be further reduced by additional empirical observations to any significant degree, for some global property of interest.**

A case in point of particular contemporary relevance concerns the initial conditions of the Universe – what is the statistical distribution of such initial conditions? This is often described as 'the probability of the Universe', or 'the probability of inflation' since the inflationary paradigm is the leading explanation for producing a large, geometrically flat universe and its initial density fluctuations. More formally, it is known as the *measure problem*: what is the probability measure on the space of possible universes (known as *multiverse*)? The measure problem is important, because parameter inference might non-negligibly depend on this measure, and model inference should non-negligibly depend on this measure. Meaningfully performing inference at this level of global properties therefore depends on finding some resolution for how to approach the measure problem. In recent years, this has led to intense debate on the scientific status of inflation theory, string theory, and other multiverse proposals (Carr, 2007; Dawid, 2013, 2015; Ellis and Silk, 2014; Kragh, 2014; Smeenk, 2014; Steinhardt, 2011).

This is not the place for addressing the range of approaches to this problem in the literature. Proposals commonly rely on some relative spatial volume, aesthetic/theoretical principle (e.g. Jeffreys prior, maximum entropy), or dynamical principle (e.g. energy conservation, Liouville measure). The reader is referred to Carr (2007); Smeenk (2014) and references therein for more details.

21.4 The Measure Problem: A Critical Analysis

21.4.1 Preliminaries

It is helpful to recognise that the measure problem is a sub-problem, arising in a particular context, related to the broader question of how to perform model inference in relation to global properties of the Universe. It arises as a problem from the desire to provide explanation for some certain global properties of the Universe, and so depends on a view of what requires explanation and what provides suitable explanation. In pursuing statistical explanation, the problem naturally presents itself through the application of conventional Bayesian statistical inference as we have seen above, and particularly in the calculation of Bayesian evidence, where the assignment of a prior probability distribution for parameter values is essential. The model and/or prior will also explicitly or implicitly describe how different global properties co-depend, and more generally prescribe some particular structure for the unobserved ensemble of universes (hence, the multiverse). This ensemble may or may not correspond to a physically real such structure.

In addressing the measure problem, one might therefore explore the implications of varying the conditions, assumptions, and approaches described above. To what extent is the measure problem a product thereof? We will consider this question in the following. The analysis echoes issues raised in e.g. Ellis' and Aguirre's contributions in Carr (2007), Ellis (2014) and Smeenk (2014), while providing a new context and synthesis. In the following, Kolmogorov probability will be contrasted with Bayesian probability for illustration and motivation. We note that other foundations and definitions of probability, which we do not consider, also exist (e.g. de Finetti's approach) – see Jaynes (2003).

21.4.2 Analysis

Structure of Global Properties

Statistical analysis for global properties in cosmology typically relies on certain unspoken assumptions. For example, it is commonly assumed that the constants of Nature are statistically independent, or that a multiverse can be meaningfully described by slight variations of the laws of physics as we know them – for example as in Tegmark *et al.* (2006). For many practical purposes, these assumptions are reasonable or irrelevant. However, in some cosmological contexts, especially in relation to global properties and questions of typicality/fine-tuning, such assumptions can impact on the conclusions drawn from observations. For example, it has been argued that fine-tuning arguments rely on untestable theoretical assumptions about the structure of the space of possible universes (Ellis, 2014).

A distinction was made in the preceding section between epistemically global and ontically global properties of the Universe. A central point is that it is impossible to make this distinction purely observationally: the set of ontically global properties will intersect the set of epistemically global properties, but which properties belong to this intersection set cannot be determined observationally. Hence, it is possible that some ontically global properties remain unknown, and that some epistemically global properties are not ontically global. This implies that in general, global properties will be subject to an uncertainty associated with these sources of epistemic indeterminacy.

In consequence, when seeking to determine and explain some certain global properties through analysis of observational data, the possibility that the values of these global properties could depend on some other global properties – known or unknown – cannot be excluded empirically. One example of this possibility concerns the constants of Nature, whose values may be interdependent (as also predicted by some theories). Another example is the distinction between physical (global) law and initial conditions of the Universe: in what sense are these concepts different? They are both epistemically global properties, and from an epistemological point of view clearly interdependent to some extent (e.g.

'observable history = initial conditions + evolution laws'). Yet their epistemic status is often considered to be categorically different, on the basis of extrapolated theoretical structure. These issues are discussed further in e.g. Ellis (2014) and Smeenk (2014).

To explore these ideas further, let us consider a cosmological model described by some global parameters θ_p (e.g. constants of Nature or initial conditions). This model also contains a specification of physical laws, which normally are considered fixed in mathematical form. For the sake of argument, let us now assume that deviations from these laws can be meaningfully described by some additional set of parameters $\delta\theta_l$. Such a $\delta\theta_l$ will give rise to a shift $\delta\mathcal{L}(\theta_p, \delta\theta_l)$ in the data likelihood for our assumed exhaustive observations, relative to the same likelihood assuming standard physical laws.

The function $\delta\mathcal{L}$ will in principle depend on the relationship between θ_p and $\delta\theta_l$ in some more general model picture (we have no reason *a priori* to exclude such a more general picture). The shift $\delta\theta_l$ should also affect the prior $\Pi(\theta_p)$. While this may not be explicitly stated, a parameter prior for θ_p is always specified under the assumption of certain physical laws. For the case of global parameters, the distinction between parameters and laws becomes blurred, as they are both global properties: they may in fact be co-dependent in some more general model. However, the correlation matrix (or other dependence) between θ_p and $\delta\theta_l$ cannot be independently determined from observations, since only one observable realisation of parameter values is available (i.e. our one observable Universe). Hence, the shift $\delta\theta_l$ should in general induce a shift $\Pi \rightarrow \Pi + \delta\Pi$ in the assumed prior, due to the empirically allowed and theoretically plausible dependencies between θ_p and $\delta\theta_l$. On this basis, there will in general be some function $\delta\Pi$ that should be included for probabilistic completeness. But this function is essentially unconstrained since it cannot be independently verified or falsified. Hence, we are in principle always free to renormalise the Bayesian evidence by an arbitrary (non-negative) amount without violating any empirical constraints. Model inference in the conventional sense therefore becomes ill defined/meaningless in this context.

The problem here is that while we know that there in general should be co-dependencies/correlations between laws/parameters, we are unable to account for them. This means that Kolmogorov's third axiom (the measure evaluated on the 'whole set' equals the sum of the measures on the disjoint subsets) is generically violated due to unaccounted-for correlations between known and unknown global properties (see Jaynes, 2003; Kolmogorov, 1933, for details on Kolmogorov's probability theory). The axiom could be regarded as unphysical in this context. This may also lead to problems satisfying Kolmogorov's first (non-negative probabilities) and second axiom (unitarity) for the prior. While Bayesian probability based on Cox's theorem (Cox, 1946, 1961) does not require Kolmogorov's third axiom (discussed further in the following subsection), potential problems equally arise in relation to negative probabilities and non-unitarity. Therefore, we find that probability in this context is better thought of as *quasiprobability* (occurring also e.g. in the phase space formulation of quantum mechanics, Wigner, 1932).

Foundations for Inference

Statistical analysis of empirical measurements is the paradigm within which inference usually takes place in science, including cosmology. It is therefore important for us to consider the foundations of probability, as applicable to this process. A central distinction as regards probability is that between *physical probability* and *inductive probability*. The first is some putative ontic probability, the second corresponds to the epistemological evaluation of empirical data. Albrecht and Phillips (2014) claim that there is no physically verified classical theory of probability, and that physical probability rather appears to be a fundamentally quantum phenomenon. They argue that this undermines the validity of certain questions/statements based on non-quantum probability and makes them ill-defined, and that a 'quantum-consistent' approach appears able to provide a resolution of the measure problem. The precise relationship between physical and inductive probability is a long-standing topic of debate, which we will not go into great detail on here (for a review, see Jaynes, 2003). The essential point for our discussion is the possible distinction between the two, and the idea that inductive probability can be calibrated to physical probability through repeated observations, e.g. in a frequency-of-outcome specification, or a Bayesian posterior updating process. In that way, credence can be calibrated to empirical evidence.

This procedure fails when considering the Universe as a whole, for which only one observable realisation exists. A conventional inductive approach to probability therefore becomes inadequate (unless one assumes that the observable Universe defines the totality of existence, but this is a rather absurd and arbitrary notion which inflationary theory also contradicts). This can also be regarded as a failure to satisfy the second axiom in Kolmogorov's definition of probability: that there are no elementary events outside the sample space (Jaynes, 2003; Kolmogorov, 1933). Without a well-defined empirical calibration of sample space and evidence, one is at risk of circular reasoning where inductive probability is simply calibrated to itself (and hence, the *a priori* assumptions made in the analysis). Related situations in cosmology have been discussed in the context of the inductive disjunctive fallacy by Norton (2010).

Bayesian statistics has become the standard approach in cosmology, due to the limitations of 'frequentist' methods when data are scarce in relation to the tested hypotheses (the whole-Universe and Multiverse cases being the extreme end of the spectrum). The formalism of Bayesian statistics can be motivated from rather general assumptions. For example, Cox's theorem shows that Bayesian statistics is the unique generalisation of Boolean logic in the presence of uncertainty (Cox, 1946, 1961). This provides a logical foundation for Bayesian statistics. In recent years, it has further been shown that the Bayesian statistical formalism follows from even more general assumptions (the lattice probability construction, see Knuth and Skilling, 2012; Skilling, 2010), which we will return to in Section 21.5.1.

There is a difference between the definition of probability based on Kolmogorov's three axioms (Jaynes, 2003; Kolmogorov, 1933), and Bayesian probability based on Cox's theorem (Cox, 1946, 1961; Van Horn, 2003). Both constructions are measure-theoretic in nature, but Kolmogorov probability places a more restrictive requirement on valid measures. In Bayesian probability, measures are finitely additive, whereas Kolmogorov measures are countably additive (a subset of finitely additive measures). This means that in Bayesian probability (based on Cox's theorem), in principle the measure evaluated on the full sample space need not equal the sum of the measures of all disjoint subsets of the sample space. This can be understood to mean that integrated regions under such a measure do not represent probabilities of mutually exclusive states. In the preceding subsection, we discussed this possibility in the context of unaccounted-for correlations in the structure of global properties. In the Bayesian statistical set-up, this is thus not a problem in principle, although problematic negative probabilities and non-unitarity could also occur in this case.

We can thus see certain benefits with a modern Bayesian statistical framework, relative to the Kolmogorov definition of probability, even though some issues also present themselves. This leaves, at least, an overall indeterminacy in 'total probability' (through $\delta\Pi$). More broadly, it also remains an open question what the status of the logical foundations is. Is there a unique logical foundation? What is its relation to the physical Universe/Multiverse – is it a physical property? (cf. Putnam, 1969).

Modes of Explanation

In addition to the above, a fundamental question is which phenomena, or findings otherwise, are thought to require explanation (on the basis of current theoretical understanding), and what provides that explanation. In the case of the measure problem, it is often considered that the initial conditions of the Universe appear to have been very special, i.e. in some sense very unlikely, and therefore need to be explained (Ellis, 2014; Smeenk, 2014). This proposition clearly rests on some *a priori* notion of probable universes, often based on extrapolations of known physics (e.g. fixed global laws).

The main mode of explanation in cosmology is based on statistical evaluation of observational data. This is usually done using the formalism of Bayesian statistics. The conventional approaches for providing a solution to the measure problem are also statistical/probabilistic in nature, and can be regarded as picking some 'target' probability regime that is to be reached for the posterior probability, for a proposed measure to be considered explanatory (another alternative is some strong theoretical/structural motivation). There are broadly speaking two modes of explanation in this vein: 'chance' and 'necessity'. For example, the measured values of physical constants are to be highly probable (fine-tuning/anthropics), average (principle of mediocrity), or perhaps necessarily realised at least 'somewhere' (string-theory landscape). However, in view of the discussion in the preceding subsections, it appears impossible to independently establish such probabilities, and hence the notions of and distinctions between chance and necessity become blurred. Statistical explanation therefore suffers from the same problems as detailed in the preceding subsections, with the risk for circular confirmatory reasoning that in reality is not actually explanatory. This critique echoes common objections to the epistemic theories of justification called *foundationalism* and *coherentism*. Epistemic justification based on coherentism can provide support for almost anything through circularity, and foundationalism can become arbitrary through the assertion of otherwise unjustified foundational beliefs. Neither of these two epistemological approaches appear able to provide satisfactory epistemic justification in response to the ambiguity in the effective Bayesian prior that we are discussing.

In terms of model structure, the typical form of explanation takes the shape of evolutionary, universal laws combined with some set of initial conditions at the 'earliest' time. Some additional aesthetic/structural criteria such as simplicity, symmetry, and conserved quantities may also be implicitly invoked. These are usually introduced as part of the theoretical modelling, rather than as intrinsically integrated with the inferential procedure. Therefore, possible co-dependencies with other explanatory criteria (which may be thought of as a type of global properties!) are usually not considered or explored.

In conclusion, statistical explanation is ill defined in the context of the measure problem in Bayesian statistics, and the relation to other possible explanatory principles typically neglected. How to interpret a Bayesian evidence value, or differences between such values for different models, is therefore here not clear.

21.4.3 A Synthesis Proposal

We thus find that the measure problem is ill-defined, in principle, in conventional Bayesian statistical inference in the measure problem context. This is due to a compound ambiguity in the definition of probability, prior specification, and evidence interpretation – based on the observations above that the concepts of

- laws and global parameters/initial conditions
- probability
- explanation

are ambiguous when considering a global, whole-Universe (or Multiverse) context. Hence, measure problem solution proposals in the Bayesian statistical context ultimately are subjective statements about how 'strange' or 'reasonable' one considers the observable Universe to be from some particular point of view. This suggests that conventional Bayesian model inference concerning global properties in the circumstances we consider is non-explanatory, and at best only self-confirmatory. Additional information is needed to provide meaningful epistemic justification.

The ambiguities listed above can be regarded, in the language of conventional Bayesian inference, as each introducing some effective renormalisation in the expression for Bayesian evidence. Can these ambiguities be accommodated in such a way that inference becomes conceptually well defined? There appear to be three possible, consistent, responses to this:

1. Declare model inference meaningless in this context and in the spirit of Wittgenstein stay silent about it.
2. Introduce additional philosophical/explanatory principles that effectively restrict the prior (e.g. an anthropic principle, a dynamical or aesthetic principle on model space).
3. Introduce novel empirical/epistemic domains (e.g. mathematical-structure space of accepted scientific theory, possibility space of accepted scientific theory).

This author argues that the natural and consistent approach involves a combination of 2. and 3. To pursue this, we need to explicitly address the ambiguous concepts and the nature of the ambiguity. The lattice conception of probability (reviewed in Section 21.5.1), including the measure-theoretic Bayesian probability as a special case, provides a way to do this. It allows the generalisation of Bayesian probability and evidence to include more general valuation functions which can encompass the ambiguity in model structure, probability and explanation in a consistent way. It also demonstrates that combining valuation functions is done uniquely by multiplication, suggesting that the identified ambiguities can be decomposed in such a way.

In conclusion, the measure problem has an irreducible axiological component. While there may be some hope of falsifying certain extreme measures, it should however be clear that any successful resolution would need to address this axiological component. The proposed construction constitutes a natural extension of the Bayesian statistical framework, and can be motivated and naturally implemented starting from the notions of partially ordered sets and measure theory. It aims to explicitly include valuations which are conventionally left implicit, to provide a conceptually and theoretically consistent framework. The workings of such an *axiological Bayesianism* will now be presented.

21.5 Axiological Bayesianism

21.5.1 Lattice Probability

Kevin H. Knuth and John Skilling have developed a novel approach to the foundations of Bayesian probability, based on partially ordered sets and associated valuations (Knuth and Skilling, 2012). It generalises Kolmogorov's and Cox's earlier work. An overview is given in Skilling (2010), but the main features will be outlined here. The construction starts off from a general set of possibilities, for example a set of different models or parameter values, but where our ultimate purpose is to quantitatively constrain to a subset of inferred preferable possibilities. The set is built up by a 'null' element, a set of basis elements (e.g. {Model A, Model B}), and the set of all combinations of logical disjunctions ('OR') between the basis elements (e.g. {Model A-OR-Model B}).

On this set, *partial ordering* is defined, denoted here by '$<$'. For elements x and y, we have that $x < y$ means that y includes x. The ordering is required to be transitive, i.e.

$$x < y \text{ and } y < z \Longrightarrow x < z. \tag{21.3}$$

The concept of *least upper bound* is introduced separately. The least upper bound to x and y, if it exists, is the least element at or including both x and y. We denote it by $x \vee y$. The *greatest lower bound* of x and y is defined analogously, and denoted $x \wedge y$.

A *lattice* is a partially ordered set with a well-defined least upper bound, reflecting the idea that the ordering induces a structure on the set. It also obeys (among other conventional axioms) associativity, $(x \vee y) \vee z = x \vee (y \vee z)$. This property is central to the probability construction based on valuations that now follows.

On the lattice, a function prescribing a quantitative valuation to each lattice element is then introduced. The purpose of this valuation is to rank elements. Requiring that such a valuation respects the ordering and the lattice structure, it can be shown that any valuation *m* must satisfy

$$m(x \vee y) = m(x) + m(y), \tag{21.4}$$

without loss of generality. This is essentially what defines a mathematical *measure*. From this it also follows that the valuation of general lattice elements (constructed via use of '$\vee$') can be built up by addition from valuation prescriptions for the basis elements.

One can also consider a direct product '$\times$' of lattices, which under similar assumptions on *m* as above leads to the requirement

$$m(x \times y) = m(x)m(y) \tag{21.5}$$

on the valuation *m*. Combinations are thus always multiplicative. This will be of particular importance in the following.

Turning to the question of how to define probability, it can be shown that under preservation of lattice structure, associativity and unitarity, conventional Bayesian probability calculus follows, with a probability *p* defined by

$$p(x|t) \equiv \frac{m(x \wedge t)}{m(t)}, \tag{21.6}$$

where *t* is some lattice context that one considers *a priori*. This expression generalises the conventional probability concept and calculus to any valuation concordant with ordering, lattice structure and associativity. It provides therefore a basis for a generalised inference procedure. A prescription for how to reason rationally also within the ambiguous context we consider.

21.5.2 Gevidence: Generalisation of Bayesian Evidence

The concept of explanation is intrinsically tied to the concept of probability in the context of statistical explanation. In generalising the concept of probability to general valuations (which, as shown above, must also be mathematical measures in the Knuth–Skilling construction), statistical explanation can therefore in that process also be generalised. Such a generalisation will then involve an evaluation of how well a model corresponds to some set of explanatory principles encoded in valuations (e.g. predicting empirical observations, satisfying aesthetic criteria, etc.). We therefore turn to the question of how Bayesian evidence can be generalised on the basis of the lattice probability construction, to provide a resolution of the conceptual problems associated with the measure problem. A key question for implementing a generalisation of Bayesian evidence for model inference, is how to combine several valuations corresponding to different explanatory criteria and empirical/epistemic domains, into a compound 'net' valuation. The preceding subsection gives the unique answer: by multiplication. Given a set of valuations/lattice representations, there is thus a unique way to define a probability based on these through multiplication.

In analogy with the way in which different physical measurements can be combined to form joint likelihoods, other explanatory criteria can be combined in different ways using the multiplicative prescription. We therefore define a generalised Bayesian evidence – let us call it *gevidence* for short – by

$$P(a,b,c,...;M) \equiv \int p(a|\theta;M)p(b|\theta;M)p(c|\theta;M)...\Pi(\theta;M)d\theta , \qquad (21.7)$$

where the letters *a*, *b*, *c*, ... refer to different valuation prescriptions corresponding to explanatory criteria. It is also useful to define the log-quantity

$$L_{abc...} \equiv \log P(a,b,c,...;M) , \qquad (21.8)$$

which is more useful when performing model comparison, since the log-quantities add/subtract between models.

While any valuation measure in itself cannot be 'proven', just like for a parameter prior it can be founded on theoretical principle and experience. The novel measures are fundamentally no different from conventionally used priors on model parameter space – they simply generalise to higher orders of model characteristics. The proposed construction provides a prescription for how to carry out model comparison on the basis of such measures. It is not primarily intended as a tool to exclude models, but rather a means of maintaining a principled and systematic approach to comparing models, given certain assumptions. Note, however, that it is possible to perform inference on the explanatory criteria through re-conditionalising, as discussed in Section 21.5.4.

21.5.3 Evidence, Elegance, Beneficence

While the proposed construction does not remove epistemic ambiguity, it provides a rational, natural and well-defined framework for examining this ambiguity in a systematic, rational and explicit way to determine relative model fitness. A rough way to categorise the possibilities is on the basis of empirical/epistemic domain. A practical way to classify models is then by theoretical/mathematical structure and physical possibility space. This can roughly be translated to correspond to aesthetic and ethical principles. While this terminology may appear unorthodox, it emphasises the axiological element which is present regardless, but does not therefore imply the presence of any aesthetic or moral agent. Model inference thus divides into interlinking empirical, aesthetic and ethical comparisons. This classification may also be useful in that it reflects how scientists intuitively tend to approach informal model assessment. To structure the problem, let us therefore represent the gevidence in the specific form

$$P(D,A,E;M) = \int \mathcal{L}(D|\theta;M)p(A|\theta;M)p(E|\theta;M)\Pi(\theta;M)d\theta , \qquad (21.9)$$

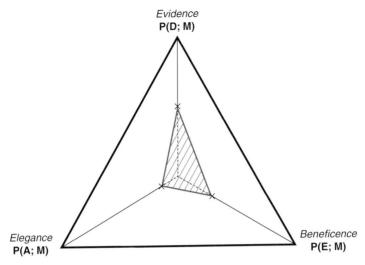

Figure 21.1 A schematic axiological Bayesian triangle representation of model evidence $P(D; M)$, elegance $P(A; M)$, and beneficence $P(E; M)$.

where D denotes 'data', A denotes 'aesthetics', and E denotes 'ethics'. We thus represent models on a direct product set of empirical observables, model structure and model possibility space. Each of these probabilities may in themselves be subdivided by multiplication into any number of component valuations.

We may also form the partial gevidences $P(A, D; M)$, $P(A, E; M)$, and $P(D, E; M)$, which provide additional information about the ways in which the different explanatory criteria corroborate or contradict each other. We shall refer to the individual gevidences as *evidence* $[P(D; M)]$, *elegance* $[P(A; M)]$, and *beneficence* $[P(E; M)]$. The log-quantities L_{ADE}, L_{AD}, L_{AE}, L_{DE}, L_D, L_A, and L_E will also be of interest for model comparison.

This quantification offers a formalisation of model comparison and explanation across the categories, and hence also of problem formulation: just as an unexpectedly rare state in model phase space (fine-tuning) may prompt explanation, e.g. an unexpectedly un-aesthetic/aesthetic model structure may prompt explanation.

21.5.4 Implementation and Application

Let us now turn to how, in practice, the type of framework proposed could be implemented and used. A few essential elements are needed:

- Basis elements for lattice (representation of models);
- Aesthetic measure/s on model-structure space;
- Ethical measure/s on model possibility space;
- Computational capability to evaluate on model-structure space and possibility space.

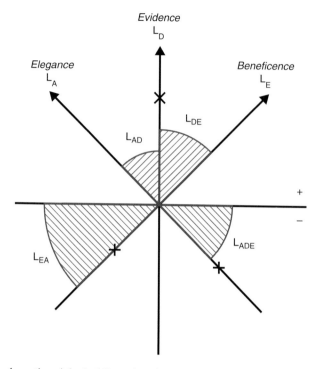

Figure 21.2 A schematic axiological Bayesian circle representation of model gevidence, useful for model comparison. The implicit model conditioning has been dropped in the figure. The figure is divided into a positive and negative half-plane. There are three axes on which are crosses to indicate the values of L_D, L_A and L_E. These axes and the axis separating the half-planes divide the figure into eight equal segments. Within these areas are plotted shaded circle segments with areas corresponding to the values of L_{AD}, L_{DE}, L_{AE} and L_{ADE}. The shaded circle segments are placed in the segment delineated by the axes corresponding to the two criteria in question, e.g. L_{AD} is placed between the axes for L_D and L_A. In the case of AE, the half-plane axis forms one of the axes. L_{ADE} is plotted in the remaining segment. Positive values are indicated by placing the shaded circle segment in the positive half-plane, and negative values in the negative half-plane. (This is reinforced by choosing the orientation of the shading lines to distinguish positive and negative values.) The areas can be directly compared within figures for a given model, and between figures for different models, to indicate relative model fitness (a model difference figure can also be constructed in the same way). Recall that the larger the values are, the stronger the support is for a model.

Out of these, the first three appear straightforward to achieve. Some aesthetic measures of e.g. simplicity are naturally incorporated in the conventional Bayesian statistical framework (Ockham's razor), others may conventionally be used more informally. The proposed framework formalises model structure considerations beyond 'parameter shaving' with Ockham's razor. A list of some possible aesthetic criteria (after Ellis, 2014) are shown in Figure 21.3. As an example, 'connectedness to the rest of science' might be quantified on the basis of the different physical constants that appear in a model. Ellis and Uzan

Satisfactory structure
(a) internal consistency, (b) simplicity (Ockham's razor),
(c) 'beauty' or 'elegance';

Intrinsic explanatory power
(a) logical tightness,
(b) scope of the theory – unifying otherwise separate phenomena;

Extrinsic explanatory power
(a) connectedness to the rest of science,
(b) extendability – a basis for further development.

Figure 21.3 Example aesthetic model comparison criteria, after Ellis (2014).

(2014) argue that Higgs inflation could be regarded as a preferred inflation model on such grounds. Ethical measures are not commonly discussed, although scientists (as everyone) will at some level be influenced by such considerations, for example due to their philosophical position along the axis from materialism to idealism (ethics defined on exclusively materialistic or idealistic grounds can clearly differ significantly). Explicit consideration of ethics in relation to cosmology is given by Knobe *et al.* (2006), who discuss the ethical implications of inflationary cosmology, and in Murphy and Ellis (1996) where it is argued that scientific cosmology points toward a kenotic ethic. Computational capability on model structure space should be reasonably adequate with today's technology. However, computations of/on possibility spaces may present serious challenges especially for complex theories and measures.

In practice, the framework can be used much the way models are compared in the light of different data sets both separately and jointly to test consistency and combined inferential power. Figure 21.1 shows schematically a simple way to illustrate the evidence $P(D; M)$, elegance $P(A; M)$ and beneficence $P(E; M)$. A fuller representation is shown schematically in Figure 21.2, where the log-gevidences are shown for all possible combinations of criteria. This figure gives a complete picture of the model gevidence, and can be directly compared between models to understand which model is best overall, or with respect to combinations of only some of the criteria. This offers possibilities to explore multi-factor explanations, i.e. part empirical, part aesthetic, part ethical. It gives a clear picture to what extent empirical data and axiological criteria are consistent with each other.

One may also examine which aesthetic and ethical criteria are 'preferred' with the help of the framework. By re-conditionalising the probability, one can compute e.g. $P(D|A; M) = P(D, A; M)/P(A; M)$ which quantifies the empirical support for the elegance principle A (under model M). One could also separately study the support for particular explanatory criteria across some range of models of relevance by comparing $P(C) = \Sigma_M P(C; M)\Pi(M)$ for different criteria C, which thus effectively extends the empirical/epistemic domain to model structure and model possibility space.

21.6 Concluding Discussion

In this chapter, it has been argued – in concordance with earlier observations by Ellis (2014), Smeenk (2014), and others – that

1. Model inference in cosmology involves both evaluation of empirical statistical evidence and application of other interpretative principles.
2. The Bayesian statistical framework, particularly, suffers from the measure problem in relation to explanation of global properties in a whole-Universe and Multiverse context (notably, inflationary initial conditions).
3. Some interpretative principles are not in themselves empirically testable by conventional Bayesian statistical tests.
4. Bayesian statistical explanation therefore is effectively qualitative in the whole-Universe and Multiverse context, such that Bayesian probability becomes ambiguous quasiprobability and the measure problem ill defined.
5. It is possible to extend Bayesian statistical inference in a natural and well-defined way to explicitly account for non-observational explanatory principles and provide a conceptually well-defined inferential procedure.

These considerations lead to the following conclusion. If we accept probability calculus as founded on the lattice construction, then the conventional scientific method can be regarded as a special case of a more general part-subjective, *but uniquely rational*, framework for reasoning we have termed 'axiological Bayesianism'. This framework generalises Bayesian statistics to define a more general version of Bayesian evidence for model inference. We have called this 'gevidence' and divided it into three main sub-components: evidence, elegance, and beneficence. This enables the inclusion of probabilities based on valuation measures on model structure and possibility space, that combine in a unique way. The framework appears to have overlap with Dawid's concept of *non-empirical theory evaluation* (Dawid, 2013), and to lend itself to the epistemic theory of justification called *foundherentism* (Haack, 1993), a synthesis of foundationalism and coherentism. The framework can be further justified by appealing to epistemological principles of uniformity/unity and consistency/coherence to be extended to new domains, i.e. model inference and comparison.

Potential problems with the proposed inference approach arise if the rules of probability are themselves global empirical properties of the Universe, just like a physical law, since such a 'probability law' could be different in other parts of a multiverse. The quasi-probability nature of Bayesian statistics in our analysis suggests that the framework may be extended to alternative logical foundations for probability. It remains to be seen how the framework of axiological Bayesianism might be developed and applied in practice. The details of model representation, associated product-space construction and measures, as well as computational techniques, need to be worked out. A reference scale for gevidence differences would be desirable (perhaps based on the concept of *information*, see Skilling, 2010).

When there are very limited data, it is inevitable in principle that some type of hermeneutic process comes into play when engaging in inference. We must then either accept additional explanatory criteria (i.e. not based on data likelihood) as valid 'scientific method', or appeal to some additional principle that invalidates such criteria. One possible such principle might be that the type of subjectivity inherent in axiological Bayesianism is outside the realms of science, and hence the framework is to be rejected. This is a perfectly valid position to take. However, since it was shown above that the subjective ambiguity is also present implicitly in the conventional Bayesian framework, this principle also excludes making statements about the measure problem using that same framework. One should then choose to stay silent on the matter, to remain consistent. Hence, the proposed framework appears to be an in principle necessary, conceptually consistent, and theoretically natural (though not necessarily unique) generalisation of the Bayesian statistical framework for addressing the measure problem and similar questions, for those who wish not to stay silent.

Acknowledgements

Thanks are due to Jeremy Butterfield, Khalil Chamcham, Daniel Darg, George Ellis, Hans Halvorson, Andrew Liddle, Laura Mersini-Houghton, Brian Pitts, Joseph Silk, David Sloan, Chris Smeenk, and Henrik Zinkernagel, for useful conversations in this area.

References

Albrecht, A. and Phillips, D. 2014. Origin of probabilities and their application to the multiverse. *Physical Review D*, **90**(12).

Butterfield, J. 2014. On under-determination in cosmology. *Studies in History and Philosophy of Modern Physics*, **46**, 57–69.

Carr, B. (ed.) 2007. *Universe or multiverse?* Cambridge, UK: Cambridge University Press.

Cox, R. T. 1946. Probability, Frequency and Reasonable Expectation. *American Journal of Physics*, **14**, 1–13.

Cox, R. T. 1961. *The Algebra of Probable Inference*. Baltimore, USA: Johns Hopkins University Press.

Dawid, R. 2013. *String Theory and the Scientific Method*. Cambridge, UK: Cambridge University Press.

Dawid, R. 2015. Physics theory: 'Simple' or 'elegant' criteria are not valid. *Nature*, **518**(7539), 303.

Efstathiou, G. 2008. Limitations of Bayesian evidence applied to cosmology. *Monthly Notices of the Royal Astronomical Society*, **388**(3), 1314–20.

Ellis, G. F. R. 2014. On the philosophy of cosmology. *Studies in History and Philosophy of Modern Physics*, **46**, 5–23.

Ellis, G. F. R. and Silk, J. 2014. Defend the integrity of physics. *Nature*, **516**(7531), 321–3.

Ellis, G. F. R. and Uzan, J-P. 2014. Inflation and the Higgs particle. *Astronomy & Geophysics*, **55**(1), 19–20.

Haack, S. 1993. *Evidence and Inquiry*. Oxford, UK: Blackwell.

Hobson, M. P. *et al.* (eds.) 2010. *Bayesian methods in cosmology*. Cambridge, UK: Cambridge University Press.

Jaynes, E. T. 2003. *Probability Theory*. Cambridge, UK: Cambridge University Press.

Knobe, J., Olum, K. D. and Vilenkin, A. 2006. Philosophical implications of inflationary cosmology. *British Journal for the Philosophy of Science*, **57**(1), 47–67.

Knuth, K. H. and Skilling, J. 2012. Foundations of Inference. *Axioms*, **1**(1), 38.

Kolmogorov, A. N. 1933. *Grundbegriffe der Wahrscheinlichkeitrechnung*. Ergebnisse der Mathematik und Ihrer Grenzgebiete, vol. 2. Berlin-Heidelberg: Springer.

Kragh, H. 2014. Testability and epistemic shifts in modern cosmology. *Studies in History and Philosophy of Modern Physics*, **46**, 48–56.

Lipton, P. 2004. *Inference to the best explanation*, 2nd edn. London, UK: Routledge.

Lyth, D. H. and Liddle, A. R. 2009. *The Primordial Density Perturbation*. Cambridge, UK: Cambridge University Press.

Maher, P. 1993. *Betting on theories*. Cambridge, UK: Cambridge University Press.

Murphy, N. and Ellis, G. F. R. 1996. *On the Moral Nature of the Universe. Theology & the Sciences*. Minneapolis, USA: Fortress Press.

Norton, J. D. 2010. Cosmic Confusions: Not Supporting versus Supporting Not. *Philosophy of Science*, **77**(4), 501–23.

Putnam, H. 1969. Is Logic Empirical? In Cohen, R. S., and Wartofsky, M. W. (eds). *Proceedings of the Boston Colloquium for the Philosophy of Science 1966/1968. Boston Studies in the Philosophy of Science,* vol. 5. Dordrecht: Springer Netherlands.

Skilling, J. 2010. Foundations and algorithms. In Hobson, M. P. *et al.* (eds.) *Bayesian methods in cosmology*, Chapter 1. Cambridge, UK: Cambridge University Press.

Smeenk, C. 2014. Predictability crisis in early universe cosmology. *Studies in History and Philosophy of Modern Physics*, **46**, 122–33.

Steinhardt, P. J. 2011. The Inflation Debate: Is the theory at the heart of modern cosmology deeply flawed? *Scientific American*, **304**(4), 36–43.

Tegmark, M., Aguirre, A., Rees, M. J. and Wilczek, F. 2006. Dimensionless constants, cosmology, and other dark matters. *Physical Review D*, **73**(2), 023505.

Van Horn, K. S. 2003. Constructing a logic of plausible inference: a guide to Cox's theorem. *International Journal of Approximate Reasoning*, **34**(1), 3–24.

von Toussaint, U. 2011. Bayesian inference in physics. *Reviews of Modern Physics*, **83**(3), 943–99.

Whitrow, G. J. 1949. Cosmology and the a priori, Chapter IX. In *The structure of the universe*. London, UK: Hutchinson's University Library.

Wigner, E. 1932. On the Quantum Correction For Thermodynamic Equilibrium. *Physical Review*, **40**(June), 749–59.

Zinkernagel, H. 2011. Some Trends in the Philosophy of Physics. *Theoria-Revista De Teoria Historia Y Fundamentos De La Ciencia*, **26**(2), 215–41.

22

Testing the Multiverse: Bayes, Fine-Tuning and Typicality

LUKE A. BARNES

22.1 Introduction

Theory testing in the physical sciences has been revolutionized in recent decades by Bayesian approaches to probability theory. Here, I will consider Bayesian approaches to theory extensions, that is, theories like inflation which aim to provide a deeper explanation for some aspect of our models (in this case, the standard model of cosmology) that seem unnatural or fine tuned. In particular, I will consider how cosmologists can test the multiverse using observations of this universe.

Cosmologists will only ever get one horizon-full of data. Our telescopes will see so far, and no further. At any particular time, particle accelerators reach to a finite energy scale and no higher. And yet, it would be an unnatural constraint on our theories for them to fall silent beyond the edge of the observable universe and above a certain energy. Natural, simple theories need not confine themselves to the observable. How do we speculate beyond current data?

In particular, how do we evaluate (what I will call) theory extensions? That is, physical theories whose main attraction is that they provide a deeper, more natural understanding of some effective theory. For example, the appeal of cosmic inflation is its natural explanation of some of the "initial conditions" of the standard model of cosmology. The postulates of the standard model – a homogeneous and isotropic Robertson–Walker (RW) spacetime, a set of energy components and their densities (matter, radiation and a cosmological constant), and an initial set of adiabatic, Gaussian density and tensor perturbations – can explain all (or almost all) the cosmological data at our disposal: the expansion of the universe, big bang nucleosynthesis, the angular power spectrum of the cosmic microwave background (CMB), the galaxy and Lyman alpha forest power spectra, the baryon acoustic oscillation (BAO) scale, the luminosity distance-redshift relation of type Ia supernovae, and more.

So, why not simply declare cosmology to be finished? We have a model that explains all the data. Consider the following kind of reason for extending our cosmological theory. In the standard model of cosmology, photons in the CMB that are separated in the sky by more than ~ 1 degree were scattered by patches of gas that have never been in causal contact with each other. And yet the entire CMB is at the same temperature, to one part in 100,000.

If, alternatively, we propose that there was a period of accelerating expansion in the very early universe, then the regions we see in the CMB have been in causal contact, allowing them to come to thermal equilibrium. And thus, inflation solves the *horizon* problem, so the standard story goes.

Note well: the horizon problem does not involve a theory failing to predict an observation. Theories never predict their initial conditions. Rather, we argue that something about our model is open to a deeper explanation because it is unnatural, improbable, or an unexplained coincidence.

Examples could be multiplied. General relativity explains what to Newtonian gravity was a bare postulate: the equivalence of inertial and gravitational mass. Supersymmetry does not currently explain any data, but would explain why quantum corrections do not drive the Higgs Boson mass to the Planck scale, a fact which would otherwise be highly unnatural.

A calculation is required to make these arguments robust. Returning to inflation: how probable is an isotropic CMB given inflation, and not given inflation? And how simple is inflation as a hypothesis, given that we do not know what the inflaton is? How generic (probable?) are the initial conditions that lead to inflation? Observations can tell us something about the initial conditions of the observable universe; when should we accept a dynamical theory of those initial conditions, rather than simply postulating them?

Can we attack these questions with probability theory at all? Cosmology promises to stretch our interpretation of probabilities. It will be my contention here that objective Bayesian probabilities provide a consistent framework for extrapolating cosmological theories beyond our universe, and isolate the pertinent questions to ask such theories.

22.2 Objective Bayesian Probability

22.2.1 Probability from Uncertainty

We will start with an (oversimplified) overview of probability, and in particular my impressions of how it is used in the physical sciences. The interpretation of probability has a long and surprisingly turbulent history. In one corner stand the frequentists, for whom probabilities measure the relative frequencies of events in hypothetical infinitely repeated trials (or, for *finite* frequentists, in actual, known trials). When a scientist wants to test their ideas, they calculate the probability of the data given the theory. If this probability (known as a *likelihood*) passes certain tests, then we can announce that the theory is not disconfirmed.

The mathematical foundation of this approach was provided by Kolmogorov (1933), who builds probability theory from mathematical axioms, independent of any particular application to statistics. Probability, like tensor calculus or conic sections, is a tool that may or may not be useful to the scientist in the investigation of some physical systems.

If probabilities are frequencies of outcomes, it makes no sense to ask for the probability *of* a theory. We cannot compare the number of universes that obey Newtonian gravity with the number that obey Einstein's General Relativity. This is not a criticism of frequentism by its opponents. Ronald Fisher, the patron saint of frequentism, stated that "we can know nothing of the probability of hypotheses or hypothetical quantities" (Fisher, 1921).

In the other corner stand the Bayesians.[1] The basis of this approach is not abstract axioms but an attempt to start from the *desiderata* of rationality and develop probability theory as generalized logic. While classical logic is concerned with what follows deductively – if A then B – probability theory will include weaker degrees of certainty – if A then probably B. Probabilities such as $p(B|A)$ ("the probability of B given A") quantify the degree of certainty of the proposition B given the truth of the proposition A. Classical logic's implication $A \rightarrow B$ is the special case $p(B|A) = 1$; those two are the same statement. The goal is not merely to quantify subjective degrees of belief, that is, the psychological state of someone who believes A and is considering B. Just as classical logic's $A \rightarrow B$ says nothing about whether A is known by anyone, but instead denotes a connection between the truth values of the propositions A and B, so $p(B|A)$ quantifies a relationship between these propositions.[2]

How should degrees of certainty be assigned to certain propositions? Jaynes (2003) invites us to imagine a reasoning robot: insert a *given* proposition A in one slot, and the proposition of interest B in the other slot, and out comes a number indicating the degree of certainty. We program the robot according to the following desiderata:

D1. Probabilities are represented by real numbers. This ensures that degrees of plausibility can be compared on a single scale.
D2. Probabilities change in common sense ways. For example, if learning C makes B more likely, but does not change how likely A is, then learning C should make AB more likely.
D3. If a conclusion can be reasoned out in more than one way, then every possible way must lead to the same result.
D4. Information must not be arbitrarily ignored. All given evidence must be taken into account.
D5. Identical states of knowledge (except perhaps for the labeling of the propositions) should result in identical assigned probabilities.

Perhaps surprisingly, these desiderata are enough. Cox's theorem (Caticha, 2009; Jaynes, 2003) are required reading) shows that quantities assigned according to these desiderata obey the same rules as probabilities. In particular, we have a rule for each of the Boolean operations "and" (AB), "or" $(A + B)$ and "not" $(\bar{A})$,

$$p(AB|C) \equiv p(A|BC)\, p(B|C) \equiv p(B|AC)\, p(A|C) \qquad (22.1)$$

$$p(A + B|C) \equiv p(A|C) + p(B|C) - p(AB|C) \qquad (22.2)$$

$$p(\bar{A}|C) \equiv 1 - p(A|C) . \qquad (22.3)$$

---

[1] It is a simplification to speak of just two corners, but sufficient for our purposes.
[2] Neither are we considering degrees of truth; A and B are in fact either true or false.

These are *identities*, holding for any propositions A, B and C for which the relevant quantities are defined. In particular, from Eq. (22.1) we can derive Bayes' theorem,

$$p(A|BC) = \frac{p(B|AC)\, p(A|C)}{p(B|C)} \,.$$ (22.4)

Bayes's theorem often comes attached to a narrative about "prior" probabilities, which depend only on "known" "background" information (or worse, *temporally* prior information), that is updated with new "data" to produce revised "posterior" probabilities. None of this is essential to Bayesianism.

The goal of Bayesian probability theory is to calculate the probability of the proposition of interest A, given everything we know K. If you are handed $p(A|K)$ from the clouds, then your work is done. If, however, $p(A|K)$ is too much to handle then you will have to break it into smaller pieces. In particular, the sum total of everything you know K is likely to be expressible as a conjunction, $K = BC$, in which case Bayes's theorem is very useful. We use probability identities to write probabilities we *want* in terms of probabilities we *know*.

22.2.2 The Rise of Bayesianism

A revolution in the physical sciences over the last few decades has transformed what we do with data. New methods have been advanced because of a fundamental change in the way that scientists view probability. From these new foundations have come a new approach and a new set of tools, all marching under the banner of Bayes.

To underscore the dominance of Bayesian probability theory, a recent NASA Astrophysics Data System (ADS) search of the astronomy and physics literature for articles with the word "Bayesian" or "Bayes" in the title returned 7555 papers. A search for "frequentist" or "frequentism" in the title returned 71 papers, half of which also have "Bayes" in the title. Most of these are comparing methods. Frequentist methods are still used, and will not always be advertised as such. Nevertheless, this does show how few physicists and astronomers advertise their methods as frequentist. I have never seen frequentism defended in a scientific paper. On the rare occasions that the word appears, it is usually as a synonym for "oversimplified" or "archaic" or "wrong".

Why has Bayesianism risen so quickly in the physical sciences? I think that there are two main reasons.

Firstly, Bayesianism makes good sense of theory *testing*. Figure 22.1 shows the constraints from data from the Planck CMB satellite (Planck Collaboration *et al.*, 2015) on the average cosmic density of matter, relative to the critical density. The y-axis shows the probability (density) of a particular value of the parameter, normalized to the maximum value.

What *exactly* does the y-axis quantify? It is not a finite or hypothetical frequency – it's not saying that $\sim$95 per cent of universes we polled (or would hypothetically poll) have a mass density parameter between 0.27 and 0.36. The width of the peak is not an indication of the range of matter densities in different regions of the universe. It is not a chance, as

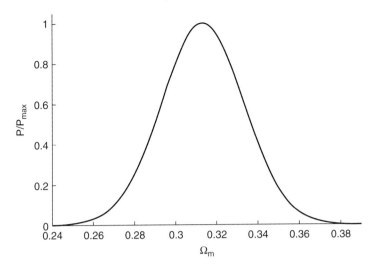

Figure 22.1 Constraints from the Planck CMB satellite on the average cosmic density of matter. The *y*-axis shows the probability (density) of a particular value of the parameter, normalized to the maximum value.

if the density of the universe is a stochastic property that every third Sunday of the month is less than 0.27. The universe only has one value of its average density, and so knows of only one point on the *x*-axis.

The *y*-axis of this plot most plausibly quantifies our degree of certainty. And yet, this is not a subjective credence. The Planck data analysis team is not reporting the effect that their satellite's instruments have had on their state of mind. What this plot reports is the implications of cosmological data for the knowledge of cosmological parameters.

More generally, science must be able to conclude, for example, that quantum mechanics is more likely to correctly describe atoms than classical electromagnetism. (Otherwise, what is the point? We would never learn anything.) This probability must be a statement about propositions, about states of knowledge. It cannot be a statement about frequencies or chances, because it is not a statement about the universe at all, or even a hypothetical ensemble of universes. Nature knows nothing of our incorrect theories.

This does not mean that frequencies and chances are useless. A frequency is a useful way to describe data. Chances are legitimate postulates of a physical theory, for example in describing the macroscopic state of a thermodynamic system or the indeterminacy of quantum systems. Bayesian probability theory does not imply that quantum probabilities are epistemic, or that statistic mechanics needs only human ignorance to link microphysics with thermodynamics. Rather, the claim is that frequencies and chances are insufficient for testing theories.

Secondly, the practice of Bayesian statistics exhibits a deep clarity and unity. The methods of orthodox statistics are a grab-bag of techniques, each intuitively reasonable but without any deeper insight into which is the best, or even of what "best" should mean. For example, Jaynes (2003) reports that, faced with linear regression (with both variables subject to an error of unknown variance), the orthodox textbook of Kempthorne & Folks (1971) formulates 16 different methods, and, being unable to choose between them, concludes with "It is all very difficult". A later survey of orthodox methods can "give only a long, somewhat dreary, list of one *adhockery* after another, with no firm final conclusions". In contrast, the Bayesian approach gives the scientist the impression of asking the right questions of the data, with no hidden assumptions and no black boxes.

22.2.3 Has Bayesianism Succeeded?

The claim of the Bayesian is that there are objective degrees of certainty or *credences* that can be modeled as probabilities. They are neither frequencies (actual or hypothetical), chances, nor merely subjective.

The reader might, and probably (!) should, be skeptical as to whether such an ambitious quantification of reasoning has indeed been achieved by the Bayesians. It might seem like alchemy, turning the base metal of ignorance into the gold of a precise probability distribution. Keep in mind, however, that Bayesian probabilities do not imply statistical frequencies: it does not follow from $p(B|A) = 0.5$ that there is a population of As that we could sample, half of whose members are Bs.

Further, Bayesian probabilities do not quantify everything that A says about B. Suppose that a mystery black box will flip a coin. What is the probability of heads H, given this information (A)? The Bayesian has no reason to prefer one side to the other; in particular, the coin and/or box might be biased towards one side, but we do not know which. To reflect this ignorance, we assign $p(H|A) = 0.5$. Now suppose that we examine the coin and box, and discover that the coin is (as best we can tell) perfectly symmetric and unbiased, and inside the box we find a mechanism that has shown no evidence of bias in the last billion flips. What is the probability of heads H, given this new, detailed information B? It has not changed: $p(H|B) = 0.5$. Should the Bayesian be worried that the probability does not reflect the vast difference in the information in A and B? Should we seek to expand probability to take into account this difference, using fuzzy probabilities or assigning distributions rather than numbers? Perhaps. But the unchanged probability is in some sense the right answer. Sure, we have learned a lot about the coin and the box, but this knowledge should not have changed our belief that heads will turn up.[3]

The assignment of probabilities is not derailed by ignorance. Ignorance is a state of knowledge, and probabilities describe states of knowledge. It may seem like assigning $p(H|A) = 0.5$ using the principle of indifference is misleadingly precise. We should reserve definite probability assignments for cases like B, and should instead say of A that "I do

---

[3] It will, appropriately, change the probability that the coin is biased, given a sequence of flips. Given A, a series of repeated heads will quickly convince us that the coin is biased. Given B, we will resist such a conclusion for longer, believing in the light of our examination of the coin and box that the repeated heads are mere chance.

not know". But this would sell ourselves short. "I do not know which one of these *two* statements is true" is a very different state of knowledge from "I don't know which one of these *trillion* statements is true". Our probabilities can and should reflect the size of the set of possibilities; the principle of indifference is invoked as a special case when this size is all we have. The assigned probabilities are only misleadingly precise if overinterpreted.

Nevertheless, Bayesian probability theory is not without worries. Some are pseudo-problems, such as the "problem of old evidence" (Glymour, 1980).[4] More troubling is the assignment of prior probabilities. Recall that prior probabilities are simply probabilities calculated using less than everything we know. So the problem is really: how do we assign probabilities when we do not know very much? The problem of the prior is particularly acute when faced with a continuum of possibilities, such as a probability distribution over a variable. We cannot say that each value is equally probable, or that each interval in an infinite range is equally probable, since these distributions do not sum (integrate) to one. The probabilities are worryingly shuffled by a change in variable. How do we model ignorance of an infinite number of possibilities?

Various methods have been advanced to solve this problem, including Jeffrey's prior and Jaynes *et al.*'s principle of maximum entropy. Whether these are successful is beyond the scope of this paper, but their failure would not sink Bayesianism. It would leave an open problem in the program. The most that the Bayesian might have to give up in light of these worries is that probabilities can be assigned to any proposition given any state of knowledge. For example, it seems absurd to suppose that there is such a thing as the probability that "the toilet paper is purple" given that "the plate is orange,"[5] that there is some *number* that uniquely captures the relationship between those propositions.

Faced with infinite possibilities, or vague statements about purple toilet paper, we might have to refrain from assigning a probability until more information is given. Jaynes (2003) argues that the problem of infinities is similar to the problem of vague statements – we have not really specified the problem until we know the limiting procedure that generates the infinity. Where one should draw the "too vague" line, however, is not clear.

22.3 Extending the Laws of Nature (As We Know Them)

22.3.1 Taking Stock

We want to apply Bayesian probability theory to the extension of the laws of nature, and then in particular to the multiverse. First, we must take stock of the laws of nature as we know them. We consider the somewhat idealized case in which we have identified the effective laws of nature that govern the physical regimes relevant to our observational evidence. Let,

4 Exercise for the reader. Hint: $p(E|B) = 1$ does not follow from "I know E".
5 Thanks to Eric Winsberg for this example.

- *U* = our observations of this universe;
- *B* = everything else we know;
- *L* = the laws of nature as we know them.

U represents the sum total of our observations of this universe, including every telescope observation and every experiment. *B* represents everything else we know, such that *UB* represents everything we know. *B* includes mathematical knowledge, and in particular all of theoretical physics. A statement such as "a bound test particle moving according to Newton's law of gravity would obey Kepler's laws of planetary motion" is true even if no particles actually obey Newton's law. It is not a statement about the actual world.

Regarding *L*, I am thinking here of the Lagrangian of the standard model of particle physics plus general relativity, but the details would not much matter. In a typically entertaining footnote, David Griffiths imagines "that God has a giant computer-controlled factory, which takes Lagrangians as input and delivers the universe they represent as output" (Griffiths, 2008, p. 373).

Actually, we need more than just the functional form of the Lagrangian. The equations of the laws of nature – as we know them – contain free parameters, numbers which are not predicted by the theory itself, but without which the laws are not fully specified. In addition, it is the *solutions* to the equations that describe a possible universe. We require further parameters to specify a particular universe from among the family of possible universes. These are usually specified as *initial conditions*, or more generally, boundary conditions.

We will represent the free parameters of the laws of nature, referred to as the *constants of nature*, as the set of numbers α_L. Similarly, we will represent the initial conditions required to specify a solution/universe by[6] β_L. The subscript *L* is a reminder that it is only in the context of a particular theory that a measurement of our universe becomes a *fundamental* constant.

We wish to evaluate the probability of our theory *L*, given the evidence we have $p(L|UB)$. We use Bayes' theorem:

$$p(L|UB) = \frac{p(L|B)}{p(U|B)} \, p(U|LB). \tag{22.5}$$

However, *L* is missing its parameters, and will not predict quantities until they are specified. We can introduce the free parameters α_L, β_L as *nuisance* parameters, to be integrated out:

$$p(L|UB) = \frac{p(L|B)}{p(U|B)} \int p(U|\alpha_L\beta_LLB)p(\alpha_L\beta_L|LB) \, d\alpha_L \, d\beta_L. \tag{22.6}$$

A few points to note. The first term on the top is the "prior" probability of the law *L*, $p(L|B)$. This is the probability that *L* describes this universe, given no information about this universe. Here is the place to formalize and implement Occam's razor – we expect simpler theories to be more probable (the interested reader is encouraged to consult MacKay, 2003, Chapter 28).

---

[6] The notation can be easily extended to functions or more advanced mathematical structures than lists of numbers.

The first term inside the integral (the likelihood) is where the theory, equipped with the appropriate constants and initial conditions, shows its predictive power by predicting observations. The second term inside the integral is the prior probability of the free parameters, that is, the probability of the parameters falling into a certain range, given no information about this universe. Note that this term takes L as given – the parameters have no law-independent meaning.

Our observations of the universe not only constrain L but its free parameters. We can, with a slight abuse of notation,[7] denote by $\boldsymbol{\alpha}_L^U$ and $\boldsymbol{\beta}_L^U$ the *set* of free parameters consistent with experiment, such that,

$$p(\boldsymbol{\alpha}_L^U \boldsymbol{\beta}_L^U | LUB) \gg p(\overline{\boldsymbol{\alpha}_L^U \boldsymbol{\beta}_L^U} | LUB). \tag{22.7}$$

Our goal as physicists is to identify the laws of nature that govern our observations of the universe. Ideally, L describes our observations better than any rival theory, $p(L|UB) \gg p(\bar{L}|UB)$, and while there exists a range of candidate deeper theories into which L could be embedded, none is significantly preferred by our data. We do not assume that L is *the* ultimate law of nature.

22.3.2 Why Extend the Laws?

One particular way in which we would like a deeper physical theory to differ from current theories is with regard to the constants of nature. In particular, we want them gone, and we can see why from Eq. (22.6). A sharply peaked $p(U|\boldsymbol{\alpha}_L\boldsymbol{\beta}_L LB)$ as a function of the free parameters $(\boldsymbol{\alpha}_L, \boldsymbol{\beta}_L)$ is precisely what physicists usually mean by "fine-tuned" – if the theory only adequately explains the data for a very narrow range of its free parameters, then we are suspicious. To illustrate in the one-dimensional case (Figure 22.2), suppose that the prior $p(\alpha|LB)$ is non-zero over a range $\sim R_\alpha$, and the likelihood $p(U|\alpha LB)$ is sharply peaked in a range of values $\Delta\alpha$, and negligible outside (i.e. remembering that the prior is normalized over α, but the likelihood is not). Then when we integrate over the nuisance parameter α, $p(U|LB) \sim \Delta\alpha/R_\alpha$. Unless the prior probability is fortuitously peaked in the same range, the likelihood $p(U|LB)$ will be very small.

The discovery that a theory is fine tuned opens the door for an alternative theory to replace it. This theory could have a broader likelihood, a narrower prior, or have no free parameters at all. Note that a preference for such a theory is not merely aesthetic, nor simply the desire to summarize the behaviour of nature as succinctly as possible.

---

[7] Specifically, there are two abuses of notation. We are using $\boldsymbol{\alpha}_L^U$ and $\boldsymbol{\beta}_L^U$ to refer to parameter regions, whereas before $\boldsymbol{\alpha}_L$ and $\boldsymbol{\beta}_L$ referred to particular values. Secondly, we should be placing propositions into our probability functions. We can think of $\boldsymbol{\alpha}_L^U$ as representing the proposition "the value of the fundamental constants of the theory L lie in the region $\boldsymbol{\alpha}_L^U$."

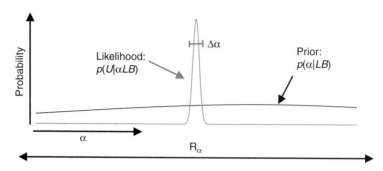

Figure 22.2 A Bayesian picture of a fine-tuned theory. $p(U|LB) = \int p(U|\alpha LB) \, p(\alpha|LB) \, d\alpha \sim \frac{\Delta\alpha}{R_\alpha} \ll 1.$

22.3.3 How to Evaluate a Theory Extension

So, we seek an extension to the laws of nature in which fewer arbitrary constants appear. How does the Bayesian evaluate theory extensions?

Consider, as an example, a detective entering a crime scene. She relies on background evidence B (what she knew before she entered the room), and inside the room she collects evidence E. The evidence clearly indicates that K, a local thug, is the killer: $p(K|EB) \gg p(\bar{K}|EB)$. Still, there may be puzzling or suspicious aspects of the hypothesis K; perhaps K did not know the victim. We thus are led to consider other propositions; not rival theories, but extensions to K. We might wonder whether K was (C) contracted to kill the victim by a local mob boss. We can evaluate this extended hypothesis in light of the data as follows:

$$p(CK|EB) = p(C|KEB)p(K|EB). \tag{22.8}$$

Now, we suppose that C does not explain the evidence of the crime scene beyond the hypothesis K, $p(E|CKB) = p(E|KB)$. That is, C seeks to explain K, and K explains E. For example, K's fingerprints at the scene are not rendered more or less probable by his status as a *contract* killer. We can then write,

$$p(CK|EB) = \frac{p(K|CB)}{p(K|B)} p(C|B)p(K|EB). \tag{22.9}$$

There are three factors of interest here. The first fraction denotes the probability of K being the killer given the contract hypothesis (and B), relative to the probability of K being the killer given background information alone. This is where the theory extension shows its worth, by leading us to expect that K would kill the victim. The second term is the prior probability of C, $p(C|B)$; the theory C is penalized if it is implausible given the background information. Thirdly, $p(K|EB)$ is the posterior probability of K, which by hypothesis is close to one.

22.3.4 Extending the Laws of Nature

Consider an extension to the laws of nature. We consider a deeper theory T, which aims to explain the laws and constants of nature as we observe them $L\boldsymbol{\alpha}_L^U\boldsymbol{\beta}_L^U$ (i.e. for convenience, we will write $L_{\alpha\beta}^U \equiv L\boldsymbol{\alpha}_L^U\boldsymbol{\beta}_L^U$ to denote the whole "laws + parameters" package). We assume that this deeper theory does not explain the data we observe beyond its ability to explain L, that is, $p(U|TL_{\alpha\beta}^U B) = p(U|L_{\alpha\beta}^U B)$. For example, let $L_{\alpha\beta}^U$ be the standard model of cosmology, beginning just prior to nucleosynthesis, and let T be inflation, which ends well before nucleosynthesis. Our prediction of the statistical properties of the CMB needs only $L_{\alpha\beta}^U$; inflation does not predict the properties of the CMB *beyond* predicting the "initial conditions" of the standard model of cosmology.

The formalism is then analogous to the crime scene case above:

$$p(TL_{\alpha\beta}^U|UB) = \frac{p(L_{\alpha\beta}^U|TB)}{p(L_{\alpha\beta}^U|B)}p(T|B)p(L_{\alpha\beta}^U|UB). \tag{22.10}$$

We can expand the fraction above,

$$p(TL_{\alpha\beta}^U|UB) = \frac{p(\boldsymbol{\alpha}_L^U\boldsymbol{\beta}_L^U|TLB)}{p(\boldsymbol{\alpha}_L^U\boldsymbol{\beta}_L^U|LB)}\frac{p(L|TB)}{p(L|B)}p(T|B)p(L_{\alpha\beta}^U|UB). \tag{22.11}$$

This is similar to the Bayesian formalism by which theories are tested with data, except that we are testing the theory extension T by using the effective theory and its measured constants $L_{\alpha\beta}^U$ as if they were *data*. Equation (22.11) highlights three questions to ask of *any* proposed extension to the laws of nature as we know them. Firstly, given the theory T, the effective laws of nature L and background information B, how probable are the constants and initial conditions of our universe? Secondly, given the theory T and background information B, how probable are the effective laws of nature L? Finally, given background information B, how probable is the theory T?

Let us look at some ways to do away with free parameters.

22.4 Extension 1: Replace Free Parameters with Mathematical Constants

To some, free parameters are a call to action, a hot poker in the Bayesian posterior. We are not satisfied, and we will not be satisfied until every physical measurement can be predicted from theory alone. Einstein (1949) dreamed of a set of equations such that "within these laws only rationally completely determined constants occur (not constants, therefore, whose numerical value could be changed without destroying the theory)".

In our formalism, this theory would set $p(L_{\alpha\beta}^U|TB) = 1$: given the deeper theory, there is only one low-energy effective theory with only one possible value of each "constant". Measuring the constants of nature would be akin to drawing a circle and determining its radius and circumference in order to "measure" π.

Unifying scientific theories can reduce the number of free parameters in physics. For example, Maxwell's unification of electricity and magnetism showed that $c = 1/\sqrt{\epsilon_0\mu_0}$ (c

speed of light, ϵ_0 vacuum permittivity, μ_0 vacuum permeability), thus reducing the number of free parameters of physics by one. This is a step in the right direction, but the progress of science can just as easily increase the number of constants by, for example, discovering a new fundamental particle.

Einstein's dream is not without its worries. A "perfect" unity likelihood is often a clue that the theory is *ad hoc* or jerry-rigged. For example, a theory with a large number of siblings – that is, mutually exclusive but similar theories that are equally probable given our background information – will only receive a small slice of the total prior probability of the family. This is, in essence, why theories with free parameters are suspicious in the first place. The theory can be thought of as a large family of theories, one for each value of the free parameter.

Thus, we need to worry about the prior probability of our deeper theory $p(T|B)$. It may have no free parameters, but if it is but one member of a large set of similar theories, the prior probability may still be small. In particular, while by hypothesis we cannot vary the parameters of the theory, this may merely indicate that we must look for fine-tuning at the next level deeper, as it were. Varying the effective parameters of our laws may require varying the deeper theory, leaving us no less at the mercy of a large set of possibilities.

This highlights one of Steven Weinberg's wishes in "Dreams of a Final Theory" (Weinberg, 1993), which he calls *logical isolation*. Weinberg argues that, while quantum mechanics is not logically inevitable, "any small change in quantum mechanics would lead to logical absurdities" (p. 70). In this sense, there is no obvious continuum of theories, of which quantum mechanics is just one. The Bayesian argument above fits nicely with Weinberg's intuition. Total logical isolation, however, seems too much to ask. Mathematical consistency is not trivial, but neither is it a rarity. There is no ultimate equation of our physical universe to which we can hope to say "mathematically, that's how things *must* be".

In addition, a theory that requires no initial conditions, or that somehow predicts its own initial conditions, would be rather strange. Rather than specifying the dynamical properties of physical objects in the form of counterfactuals, it would specify the state of the universe. For example, a Newtonian version of such a theory would not state that if two masses (m_1, m_2) are separated by distance r, then they would experience a force with magnitude Gm_1m_2/r^2. Rather, it would specify position as a function of time $r(t)$ for each particle in the universe. Rather than the complexity of the phenomena of the universe giving way to simple fundamental laws, a theory with no initial conditions would seem to require complexity all the way down.

22.5 Extension 2: Replace Free Parameters with Dynamical Entities

We have expounded the ingredients of physical theories as we know them: laws, constants and initial (or boundary) conditions. The laws describe dynamical entities – fields, particles, spacetime, etc. So, one way in which a constant could disappear in a deeper theory is by changing identity to become a dynamical quantity. The fine-structure constant, for example, could be the local value of a field. We can test this hypothesis by looking for changes in the value of the fine-structure constant over cosmic time and cosmic distances. To date, no convincing variation has been found (Cameron and Pettitt, 2012; King *et al.*, 2012; Webb *et al.*, 2011; Whitmore and Murphy, 2015).

Two problems immediately arise. Firstly, if the fine-structure constant is replaced by a quantum field, then it seems that we have merely replaced one constant with the parameters that describe the field (i.e. in fact, a field that varies so slowly over the observable universe requires a very low mass). Secondly, even if we could replace our constants with a totally constant-free field, this does not seem like progress. We have replaced a single number with a function: an infinite collection of numbers, one attached to each spacetime point. If we are in a typical place in the universe, then there is no further rationale for the value of the "constant" that we observe. There is some function that varies across spacetime, and we happen to be in the part that has $\alpha \approx 1/137$.

22.5.1 The Fine-Tuning of the Universe for Intelligent Life

However, there are good reasons to believe that we are not in a typical place in the universe. The universe is not an experiment. We are not Dr. Frankenstein, setting up our equipment, choosing the initial conditions, and observing the set-up at our leisure. We are the monster – we have awoken in a laboratory and are trying to figure out how it made us. Not all rooms can create a monster, so the fact that we are observing at all is a very stringent constraint on the contents of the room.

Similarly, not all laws of nature can be *scientific* laws, because not all laws of nature create scientists. There are certain equations that will not be written on a chalkboard in any universe that they describe. If the evolution of conscious observers shows a strong preference for certain laws or certain regions of parameter space, then an explanation for the values of the constants naturally arises. The reason why this set of constants exists at all is that there is a sufficiently large number of universe domains, with enough variation in their properties that at least one of them would hit on the right combination for life. The reason why we observe that we are in one of these rare regions is that we could not be anywhere else.

Beginning in the 1970s, a number of physicists have noticed the extreme sensitivity of the life-permitting qualities of our universe to the values of many of the physical constants and cosmological parameters of our universe. Seemingly small changes in the free parameters of the laws of nature as we know them have dramatic, uncompensated and detrimental effects on the ability of the universe to support the complexity needed by physical life forms. I have elsewhere reviewed the scientific literature on the fine-tuning of the universe for intelligent life (Barnes, 2012). Here are a few examples.

- The existence of structure in our universe *at all* places stringent bounds on the cosmological constant. Compared to the range of values for which our theories are well defined – roughly $\pm$ the Planck scale – the range of values that permit gravitationally bound structures is no more than one part in 10^{110}.

- A universe with structure also requires a fine-tuned value for the primordial density contrast Q. Too low, and no structure forms. Too high and galaxies are too dense to allow for long-lived planetary systems, as the time between disruption by a neighbouring star is too short. This places the constraint $10^{-6} \lesssim Q \lesssim 10^{-4}$ (Tegmark and Rees, 1998).
- The existence of long-lived stars, which produce and distribute chemical elements and are a stable source of energy that can power chemical reactions, requires an unnaturally small value for the "gravitational coupling constant" $\alpha_G = m_{\text{proton}}^2/m_{\text{Planck}}^2$; or, equivalently, that the proton mass be orders of magnitude smaller than the Planck mass. For stars to be stable at all, we require $\alpha_G \lesssim 10^{-33}$ (Adams, 2008).
- The existence of any atomic species and chemical processes whatsoever places tight constraints on the relative masses of the fundamental particles and the strengths of the fundamental forces. For example, Barr and Khan (2007) show the effect of varying the masses of the up and down quark, and find that star-and-chemistry permitting universes are huddled in a small shard of parameter space which has area $\Delta m_{\text{up}} \Delta m_{\text{down}}/m_{\text{Planck}}^2 \approx 10^{-42}$.

Note that these constraints are all multi-dimensional; I have quoted one-dimensional bounds for simplicity. See Barnes (2012) and references therein for plots demonstrating these and more constraints in multiple dimensions of parameter space. (It has never been the case that the fine-tuning literature has varied one variable at a time.)

These small numbers – 10^{-110}, 10^{-4}, 10^{-33}, 10^{-42} – are, in the Bayesian fashion, an attempt to quantify our ignorance. We are not assuming the existence of a random universe-generating machine, nor describing the properties of a real or imagined statistical sample. The laws of nature as we know them contain arbitrary constants, which are not constrained by anything in theoretical physics. As usual, we can react to small probabilities in a couple of ways. Perhaps, like the probability of a deck of cards falling on the floor in a particular order, something improbable has happened. Enough said. Alternatively, like the probability that the burglar correctly guessed the 12-digit code by chance on the first attempt, it may indicate that we have made an incorrect assumption. We should look for an alternative assumption (or theory), on which the fact in question is not so improbable.

22.5.2 *Making Predictions in a Multiverse*

Theories are tested by their predictions, and we saw above that theory extensions are tested by their ability to predict the effective laws and constants of nature. In practice, this means calculating likelihoods.

The multiverse is an example of a "population plus selection effect" explanation. There is some observed outcome X to be explained, and X is highly improbable on any single trial. We postulate a large, varied population to explain why any X exists at all, and a selection effect to explain why we observe X. For example, the front page of the newspaper reports correctly that Keith won the lottery. The probability of any particular person winning the lottery is very small. This occurrence is made more probable if we suppose that there is a large number of lottery players buying different tickets, and that only a lottery win would be considered newsworthy.

Where is the relevant selection effect when we are attempting to explain the statement that the effective laws of nature are L and the associated free parameters are $\boldsymbol{\alpha}_L^U$, $\boldsymbol{\beta}_L^U$? Recall that U represents everything that I know about *this* universe. Thus, to explain U, the proposition $L\boldsymbol{\alpha}_L^U\boldsymbol{\beta}_L^U$ must refer to this universe, the universe that I inhabit. $L\boldsymbol{\alpha}_L^U\boldsymbol{\beta}_L^U$ cannot simply state that "there is at least one universe in which the law L holds and in which the constants are $\boldsymbol{\alpha}_L^U$, $\boldsymbol{\beta}_L^U$", because this will not explain the fact that I observe U.

This highlights an important difference in probability between calculating the probability that "this X is Y" and "there is at least one X that is Y". Suppose I have just watched Alice deal herself five royal flushes in a row in a game of poker. The probability of *these* five hands being five royal flushes assuming a fair deal is 10^{-29}, making us wonder if Alice is cheating. The probability that *someone*, somewhere has fairly dealt five royal flushes depends on the number of poker deals there have ever been anywhere in the universe. If the universe is infinite, then this probability is one, making it useless for deciding whether Alice is cheating. As the Bayesian desiderata state, information must not be arbitrarily ignored. Reasoning as if we only knew that "there is at least one instance of five royal flushes" is to discard information.

Note that the correct distinction is *not* between first and third person probabilities, as is sometimes assumed in the multiverse literature. Third person probability can be as specific ("a particular X is Y") as first person probabilities. Also, there is nothing "mystical" about using indexical information in probabilities (Neal, 2006); "I" can successfully select a particular individual – in this case, the speaker of the sentence or calculator of the probability – without assuming that the individual is unique in reality on account of "some essence".

So, what is the likelihood that *this* universe has the observed constants, given a multiverse theory? We can calculate this in two pieces. We first calculate the probability that observers exist at all in the multiverse (O). So long as observer-permitting universes have non-zero chances and the universes in the multiverse are sufficiently varied, this probability will approach unity as the number of universes increases.

With an actual population of universes, the second probability piece is equal to a frequency: the fraction of observers (or observer moments) that observe our particular set of constants $\boldsymbol{\alpha}_L^U\boldsymbol{\beta}_L^U$. This will depend on two factors: the rate R_{obs} (per unit time and volume $\mathrm{d}x^\mu$) at which observers/observations are made at a particular point in spacetime, given the values of the "constants", and the probability of a particular set of constants at a particular spacetime point. Considering just the constants ($\boldsymbol{\alpha}_L$):

$$N_{\text{obs}} = \iint R_{\text{obs}}(x^\mu | \boldsymbol{\alpha}_L TLB)\, p(\boldsymbol{\alpha}_L | x^\mu TLB)\, \mathrm{d}x^\mu \mathrm{d}\boldsymbol{\alpha}_L \tag{22.12}$$

$$N_{\text{obs}}(\boldsymbol{\alpha}_L^U) = \iint_{\boldsymbol{\alpha}_L^U} R_{\text{obs}}(x^\mu | \boldsymbol{\alpha}_L TLB)\, p(\boldsymbol{\alpha}_L | x^\mu TLB)\, \mathrm{d}x^\mu \mathrm{d}\boldsymbol{\alpha}_L \tag{22.13}$$

$$\Rightarrow p(\boldsymbol{\alpha}_L^U | OTLB) = \frac{N_{\text{obs}}(\boldsymbol{\alpha}_L^U)}{N_{\text{obs}}} \tag{22.14}$$

The fine-tuning of the universe for intelligent life suggests that R_{obs} is strongly peaked in our neighborhood of parameter space, meaning that while regions of the universe with our constants are rare, they may be likely (or at least, not too unlikely) to be observed.

However, fine-tuning for life is not enough to ensure that a multiverse successfully predicts our constants of nature. The form of life with which we are familiar came about through biological evolution, via a gradual build up of complexity over timescales that are orders of magnitude longer than the lifetime of any particular individual. Such life forms require a stable planetary surface, a stable star producing usable photons, a ready supply of chemicals and so forth. However, observers could form without this history and environment as thermodynamic fluctuations. These *Boltzmann Brains* can cause problems for a multiverse theory because they mean that R_{obs} does not fall exactly to zero in seemingly hostile regions of parameter space.

In Eq. (22.12), we can write $R_{\text{obs}} = R_{\text{life}} + R_{\text{BB}}$ to represent the contribution of both biological life forms and Boltzmann Brains (BB) to the set of observers in a given multiverse. Thus, we can also write $N_{\text{obs}} = N_{\text{life}} + N_{\text{BB}}$. We have, then, a competition between whether most observers (or observations) are made by common observers in rare conditions (life) or rare observers in common conditions (BB).

In testing a multiverse, it matters what other hypothetical observers in the multiverse observe, since the likelihood is normalized over α_L. Theories must place their bets as to what data are to be expected; for the multiverse, this means predicting what an observer will observe. While our calculation of the posterior involves evaluating the likelihood at our particular value of the constants in our universe, the normalization of the likelihood means that the more observers there are that do not observe what we observe, the smaller the likelihood. Every observer counts, not just those who observe exactly what we observe.

Figure 22.3 presents a one-dimensional illustration of this *Boltzmann observer problem*. The problem is *not* that we might be Boltzmann Brains, or that most entities with my memories are fluctuation observers. We can call that the *Boltzmann Me* problem, and set it to one side. The Boltzmann observer problem is a straightforward case of a failed prediction. A multiverse, once the full range of observers is considered, can be strongly disconfirmed by the seemingly innocuous observation that I am not a brain floating in empty space. The problem is not that we might be Boltzmann Brains; the problem (for the theory) is that we are not.

Testing the multiverse thus requires an understanding of the conditions under which observers can fluctuate into existence. It is of particular interest whether quantum fluctuations in a vacuum can create observers; see Boddy *et al.* (2014) for the case against such observers. In broadly thermodynamic terms, the Boltzmann observer problem seems formidable. Biological life requires low entropy conditions in a large region; in fact, the entropy of this universe seems to be far lower than is required even by biological life forms (Eddington, 1931; Penrose, 2004). Boltzmann Brains, on the other hand, require only the smallest entropy fluctuation needed to create an observer. Given the usual connection between low entropy conditions and improbability, this would seem to make Boltzmann Brains far more numerous than biological life forms.

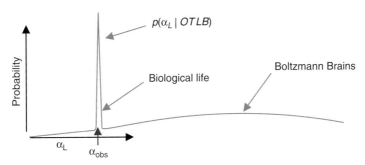

Figure 22.3 An illustration of the Boltzmann observer problem. The likelihood of the set of constants that we observe given a multiverse theory $p(\alpha_L|OTLB)$ is normalized over α_L. In evaluating the posterior probability of the multiverse, we evaluate the likelihood at the observed value of the constant α_{obs}. Boltzmann Brains can exist in universes which are hostile to biological life forms, and so can be found in a much larger region of parameter space. The larger the area under the broader Boltzmann Brain contribution, the smaller the (renormalized) likelihood of a biological life form observing α_{obs}.

We also face the *measure problem*, which in our formalism is the question of how to evaluate the likelihood of a multiverse theory when the number of observers is infinite. Jaynes (2003, p. 486–7) warns that "attempts to apply the rules of probability theory directly and indiscriminately on infinite sets" leads to paradoxes, and that the only cure for this disease is that "an infinite set should be thought of only as the limit of a specific (i.e. unambiguously specified) sequence of finite sets.... The mathematically generated paradoxes have been found only when we tried to depart from this policy by treating an infinite limit as something already accomplished, without regard to any limiting operation". The problem for an infinite multiverse is that there is no such limit – the infinity in question is "completed", an actually infinite set of universes and observers. In such circumstances, our probability assignments cannot be invariant under permutations of the labels on the observers (Olum, 2012). Infinite multiverse modellers could try to manufacture a limiting process – perhaps a sequence of spacetime volumes – or justify restricting attention to a finite subset.

22.5.3 Typicality and the Multiverse

Testing the multiverse has often focused on *typicality*: a theory is to be preferred if it predicts that human observers are typical in some class of objects in the universe (Hartle and Srednicki, 2007). For example, suppose we derive from a multiverse theory T the distribution of observed values of some constant α: $p(\alpha|TB)$. T predicts that, with 95 per cent certainty, our observed value of α falls inside the central 95 per cent of the distribution. If this prediction is correct, then the theory has passed this test.

This type of reasoning is transparently frequentist: the only probabilities that we can define are those of data with respect to theory, so we test theories by inventing a test for the likelihood. Should it pass, we try to think of another test, or else get more data. It ignores prior probabilities, and so cannot calculate the probability *of* a theory given the evidence.

As with other frequentists' methods, we can use Bayesian probability theory to expose the hidden assumptions. When is typicality – defined as closeness to the likelihood peak – a useful discriminant between models? Consider the simple case of two theories T_1 and T_2 competing to predict the value of some constant α. We calculate the likelihood distribution for α on each theory $p(\alpha|TB)$; suppose that it is roughly Gaussian. If (a) the prior probabilities of T_1 and T_2 are similar and, (b) if the widths of the likelihood distributions $p(\alpha|T_1B)$ and $p(\alpha|T_2B)$ are similar, then the theory for which the observed value of α lies closest to the peak of the distribution has the greater posterior probability.

Note that both conditions (a) and (b) are needed, and thus typicality is neither a necessary nor a sufficient condition for a multiverse theory to be a good theory. The problem with typicality is that it compares values of the likelihood at different values of α, when we should be comparing different theories by evaluating their likelihoods at the observed value of α.

Let us be clear of the status of typicality. It is not an assumption to be accepted or rejected at our leisure. It is not an assumption at all. Under certain conditions, it is a useful rule of thumb in evaluating competing multiverse theories. Bayesianism identifies these conditions.

22.6 Extension 3: Getting Metaphysical

In this book, George Ellis has invited us to think about not only cosmology, defined as the physics of the universe on large scales, but also cosmologia, which asks the great questions of existence, meaning and purpose that are raised by physical cosmology. Nothing in our formalism assumes that T is a *physical* theory. Indeed, if there is a final, ultimate physical theory of nature F, then whatever we think *about* that theory will have to be deeper than physics, so to speak. Even if all that remains is to state the definition of *naturalism*, that nothing other than the physical exists, we must acknowledge that this is a statement about physics, not of physics.

Further, we want to know whether or not naturalism is true. We can treat naturalism like any other theory, and consider its prior probability $p(N|B)$, and the probability of the final scientific laws on naturalism $p(F|NB)$. Even if we can not calculate these quantities, they point to the right questions to ask. Naturalism, as a hypothesis, is what statisticians call *non-informative* – it gives us no reason to prefer any particular F. In the case of naturalism, this is an *in principle* ignorance, since by hypothesis there are no true facts that explain why F rather than some other final law, why any law at all, why a mathematical law, what "breathes fire into the equations and makes a universe for them to describe?" (Hawking, 1988), what is existence, and so on?

Non-informative theories have likelihoods that are at the mercy of the size of their possibility space. For example, "the burglar guessed the 12-digit security code" gives us no reason to prefer any code over any other, and thus the likelihood of any particular code should reflect these trillion possibilities. The only thing in our background knowledge B that restricts the set of possible universes is internal (mathematical) consistency. Naturalism, then, is at the mercy of every possible way that concrete reality could consistently be. This places naturalism in an unenviable position.

Its competitors to explain F include axiarchism (Leslie, 1989) and theism (Swinburne, 2004), which argue that we should expect the existence of physical reality with significant *moral* value, including the moral good of embodied, free, conscious moral agents. Axiarchism and theism, then, bet heavily on the subset of possible laws that permit the existence of such life forms. Whether the fine-tuning of the laws *as we know them* ($L_{\alpha\beta}^{U}$) for life extends to final laws F, and their relative prior probabilities, will decide whether any of these theories is preferable to naturalism.

Acknowledgments

I would like to thank all the attendees of the Philosophy of Cosmology UK/US Conference, 2014, Tenerife for stimulating talks and discussions. Supported by a grant from the John Templeton Foundation. This publication was made possible through the support of a grant from the John Templeton Foundation. The opinions expressed in this publication are those of the author and do not necessarily reflect the views of the John Templeton Foundation.

References

Adams, F. C. (2008) Stars in other universes: stellar structure with different fundamental constants. *Journal of Cosmology and Astroparticle Physics*, **8**, 010.

Barnes, L. A. (2012) The Fine-Tuning of the Universe for Intelligent Life. *Publications of the Astronomical Society of Australia*, **29**, 529.

Barr, S. M. and Khan, A. (2007) Anthropic tuning of the weak scale and of mu/md in two-Higgs-doublet models. *Physical Review D*, **76**, 045002.

Boddy, K. K., Carroll, S. M. and Pollack, J. (2014) De Sitter Space Without Dynamical Quantum Fluctuations, arXiv:1405.0298.

Cameron, E., and Pettitt, T. (2012) On the Evidence for Cosmic Variation of the Fine Structure Constant (I): A Parametric Bayesian Model Selection Analysis of the Quasar Dataset. arXiv:1207.6223.

Caticha, A. (2009) Quantifying Rational Belief. *AIP Conf. Proc.* **1193**, 60.

Eddington, A. S. (1931) The End of the World: from the Standpoint of Mathematical Physics, *Nature* **127**, 3203.

Einstein, A. (1949) Autobiographical Notes. In Schilpp, P. A., ed. *Albert Einstein, Philosopher-Scientist*. Illinois: Open Court Publishing Company.

Fisher, R. A. (1921) On the 'Probable Error' of a Coefficient of Correlation Deduced from a Small Sample. *Metron*, **1**(332). 162, 164.

Glymour, C. (1980) *Theory and Evidence*. Princeton: Princeton University Press.

Griffiths, D. (2008) Introduction to Elementary Particles. New York: John Wiley & Sons.

Hartle, J. B. and Srednicki, M. (2007) Are we typical?, *Physical Review D*, **75**, 123523.

Hawking, S. W. (1988), *A brief history of time. From the Big Bang to Black Holes*. Toronto: Bantam Books.

Jaynes, E. T. (2003) *Probability Theory: The Logic of Science.* Cambridge, UK: Cambridge University Press.

King, J. A., Webb, J. K., Murphy, M. T., *et al.* (2012) Spatial variation in the fine-structure constant – new results from VLT/UVES. *MNRAS*, **422**, 3370.

Kempthorne, O. and Folks, L. (1971) *Probability, Statistics, and Data Analysis.* Ames, IA: The Iowa State University Press.

Kolmogorov, A. N. (1933) Translated as *Foundations of Probability* New York: Chelsea Publishing Company (1950).

Leslie, J. (1989) *Universes*, London, New York: Routledge.

MacKay, D. J. C. (2003) *Information Theory, Inference, and Learning Algorithms.* Cambridge: Cambridge University Press.

Neal, R. M. (2006) Puzzles of Anthropic Reasoning Resolved Using Full Non-indexical Conditioning. arXiv:math/0608592.

Olum, K. D. (2012) Is there any coherent measure for eternal inflation?, *Physical Review D*, **86**, 063509.

Penrose, R. (2004) *The Road to Reality: A Complete Guide to the Laws of the Universe.* London: Jonathan Cape.

Planck Collaboration, Ade, P. A. R., Aghanim, N., *et al.* (2015), arXiv:1502.01589.

Swinburne, R. (2004) *The Existence of God.* Oxford: Oxford University Press.

Tegmark, M. and Rees, M. J. (1998) Why Is the Cosmic Microwave Background Fluctuation Level 10^{-5}? *The Astrophysical Journal*, **499**, 526.

Webb, J. K., King, J. A., Murphy, M. T., *et al.* (2011) Indications of a Spatial Variation of the Fine Structure Constant, *Physical Review Letters*, **107**, 191101.

Weinberg, S. (1993) *Dreams of a Final Theory.* London: Vintage.

Whitmore, J. B. and Murphy, M. T. (2015) Impact of instrumental systematic errors on fine-structure constant measurements with quasar spectra. *MNRAS*, **447**, 446.

23

A New Perspective on Einstein's Philosophy of Cosmology

CORMAC O'RAIFEARTAIGH

23.1 Introduction

It has recently been discovered that Einstein once attempted – and subsequently abandoned – a 'steady-state' model of the expanding universe (Nussbaumer, 2014a; O'Raifeartaigh, 2014; O'Raifeartaigh *et al.*, 2014). An unpublished manuscript on the Albert Einstein Online Archive (Einstein, 1931a) demonstrates that Einstein explored the possibility of a universe that expands but remains essentially unchanged due to a continuous formation of matter from empty space (Figure 23.1). Several aspects of the manuscript indicate that it was written in the early months of 1931, during Einstein's first trip to California, and the work therefore probably represents Einstein's first attempt at a theoretical model of the cosmos in the wake of emerging evidence for an expanding universe (Nussbaumer, 2014a; O'Raifeartaigh *et al.*, 2014). It appears that Einstein abandoned the idea when he discovered that his steady-state model led to a null solution, as described below.

Many years later, steady-state models of the expanding cosmos were independently proposed by Fred Hoyle, Hermann Bondi and Thomas Gold (Bondi and Gold, 1948; Hoyle, 1948). The hypothesis formed a well-known alternative to 'big bang' cosmology for many years (Kragh, 1996, pp. 186–218; North, 1965, pp. 208–22; Nussbaumer and Bieri, 2009, pp. 161–3), although it was eventually ruled out by observations such as the distribution of the galaxies at different epochs and the cosmic microwave background (Kragh, 1996, pp. 318–80, 2007, pp. 201–6; Narlikar, 1988, p. 219). While it could be argued that steady-state cosmologies are of little practical interest now, we find it most interesting that Einstein conducted an internal debate between steady-state and evolving models of the cosmos decades before a similar debate engulfed the cosmological community. In particular, the episode offers several new insights into Einstein's cosmology, from his view of the role of the cosmological constant to his attitude to the question of cosmic origins. More generally, Einstein's exploration of steady-state cosmology casts new light on his philosophical journey from a static, bounded cosmology to the dynamic, evolving universe, and is indicative of a pragmatic, empiricist approach to cosmology.

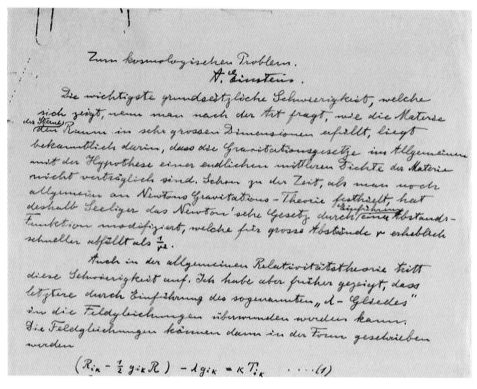

Figure 23.1 An excerpt from the first page of Einstein's steady-state manuscript (Einstein, 1931a). Reproduced by kind permission of the Hebrew University of Jerusalem.

23.2 Historical Context

Following the successful formulation of his general theory of relativity (Einstein, 1915, 1916), Einstein lost little time in applying his new theory of gravity, space and time to the universe as a whole. A major motivation was the clarification of the conceptual foundations of general relativity, i.e. to establish 'whether the relativity concept can be followed through to the finish, or whether it leads to contradictions' (Einstein, 1917a). Assuming a cosmos that was homogeneous, isotropic and static over time,[1] and that a consistent theory of gravitation should incorporate Mach's principle,[2] he found it necessary to add a new 'cosmological constant' term to the field equations of relativity in order to predict a universe with a non-zero density of matter (Einstein, 1917b). This approach led Einstein to a

---

[1] No empirical evidence for a non-static universe was known to Einstein at the time.

[2] Einstein's view of Mach's principle in these years was that space could not have an existence independent of matter and thus the spatial components of the metric tensor should vanish at infinity (Einstein, 1918a; Janssen, 2005).

finite, static cosmos of spherical spatial geometry whose radius was directly related to the density of matter.

That same year, the Dutch theorist Willem de Sitter noted that general relativity allowed another model of the cosmos, namely the case of a universe empty of matter (de Sitter, 1917). Einstein was greatly perturbed by de Sitter's solution as it suggested a spacetime metric that was independent of the matter it contained, in conflict with his understanding of Mach's principle (Einstein, 1918b). The de Sitter model became a source of some confusion among theorists for some years; it was later realised that the model was not static (Lemaître, 1925; Weyl, 1923). However, the solution attracted some attention in the 1920s because it predicted that the radiation emitted by objects inserted as test particles into the 'empty' universe would be red-shifted, a prediction that chimed with emerging astronomical observations of the spiral nebulae.[3]

In 1922, the young Russian physicist Alexander Friedmann suggested that non-stationary solutions to the Einstein field equations should be considered in relativistic models of the cosmos (Friedmann, 1922). With a second paper in 1924, Friedmann explored almost all the main theoretical possibilities for the evolution of the cosmos and its geometry (Friedmann, 1924). However, Einstein did not welcome Friedmann's time-varying models of the cosmos. His first reaction was that Friedmann had made a mathematical error (Einstein, 1922). When Friedmann showed that the error lay in Einstein's correction, Einstein duly retracted it (Einstein, 1923a); however, an unpublished draft of Einstein's retraction makes it clear that he considered time-varying models of the cosmos unrealistic (Einstein, 1923b; Stachel, 1977; Nussbaumer and Bieri, 2009, pp. 91–92).

Unaware of Friedmann's analysis, the Belgian physicist Georges Lemaître proposed an expanding model of the cosmos in 1927. A theoretician with significant training in astronomy, Lemaître was aware of V. M. Slipher's observations of the redshifts of the spiral nebulae (Slipher, 1915, 1917) and of Edwin Hubble's emerging measurements (Hubble, 1925) of the vast distances to the nebulae (Farrell, 2005, p. 90; Kragh, 1996, p. 29). Interpreting Slipher's redshifts as a relativistic expansion of space, Lemaître showed that a universe of expanding radius could be derived from Einstein's field equations, and estimated a rate of cosmic expansion from average values of the velocities and distances of the nebulae from Slipher and Hubble, respectively. This work received very little attention at first, probably because it was published in French in a little-known Belgian journal (Lemaître, 1927). However, Lemaître discussed the model directly with Einstein at the 1927 Solvay conference, only to have it dismissed with the forthright comment: 'Vos calculs sont corrects, mais votre physique est abominable' (Lemaître, 1958).

In 1929, Edwin Hubble published the first empirical evidence of a linear relation between the redshifts of the spiral nebulae (now known to be extra-galactic) and their radial distance (Hubble, 1929). By this stage, it had also been established that the static models of

---

[3] Observations of the redshifts of the spiral nebulae were published by V. M. Slipher in 1915 and 1917 (Slipher, 1915, 1917), and became widely known when they were included in a well-known book on relativity (Eddington, 1923).

Einstein and de Sitter presented problems of a theoretical nature: Einstein's universe was not stable against perturbation (Eddington, 1930; Lemaître, 1927) while de Sitter's universe was not static (Weyl, 1923; Lemaître, 1925). In consequence, theorists began to take Lemaître's model seriously, and a variety of time-varying relativistic models of the cosmos of the Friedmann–Lemaître type were advanced (de Sitter, 1930a, 1930b; Heckmann, 1931, 1932; Robertson, 1932, 1933; Eddington, 1930, 1931; Tolman, 1930a, 1930b, 1931, 1932).

By 1931, Einstein had accepted the dynamic universe. During a three-month sojourn at Caltech in Pasadena in early 1931, a trip that included discussions with the astronomers of Mount Wilson Observatory and with the Caltech theorist Richard Tolman,[4] Einstein made several public statements to the effect that he viewed Hubble's observations as likely evidence for a cosmic expansion. For example, the *New York Times* reported Einstein as commenting that 'New observations by Hubble and Humason concerning the redshift of light in distant nebulae makes the presumptions near that the general structure of the universe is not static' (Associated Press, 1931). Not long afterwards, Einstein published two distinct dynamic models of the cosmos, the Friedmann–Einstein model of 1931 and the Einstein–de Sitter model of 1932 (Einstein, 1931b; Einstein and de Sitter, 1932).

Written in April 1931, the Friedmann–Einstein model marked the first scientific publication in which Einstein formally abandoned the static universe. Citing Hubble's observations, he suggested that the assumption of a static universe was no longer justified (Einstein, 1931b). Adopting Friedmann's 1922 analysis of a universe of time-varying radius and positive spatial curvature, Einstein then removed the cosmological constant on the grounds that it was both unsatisfactory (it gave an unstable solution) and unnecessary. The resulting model predicted a cosmos that would undergo an expansion followed by a contraction, and Einstein made use of Hubble's observations to extract estimates for the current radius of the universe, the mean density of matter and the timespan of the expansion (Einstein, 1931b). Noting that the latter estimate was less than the ages of the stars estimated from astrophysics, Einstein attributed the problem to errors introduced by the simplifying assumptions of the model, notably the assumption of homogeneity.[5]

In early 1932, Einstein and Willem de Sitter proposed an alternative model of the expanding universe, based on Otto Heckmann's observation that a finite density of matter in a non-static universe does not necessarily demand a curvature of space (Heckmann, 1931). Mindful of a lack of empirical evidence for spatial curvature, Einstein and de Sitter set this parameter to zero (Einstein and de Sitter, 1932). With both the cosmological constant and spatial curvature removed, the resulting model described a cosmos of Euclidean geometry in which the rate of expansion h was related to the mean density of matter ρ by the simple relation $h^2 = \frac{1}{3}\kappa\rho$, where κ is the Einstein constant. Applying Hubble's value

---

[4] An account of Einstein's time in Pasadena can be found in Nussbaumer and Bieri (2009, pp. 144–6) and Bartusiak (2009, pp. 251–6).

[5] We have recently presented an analysis and first English translation of this work. We find that all of Einstein's estimates contain a systematic numerical error (O'Raifeartaigh and McCann, 2014).

of 500 km s^{-1} Mpc^{-1} for the recession rate of the galaxies, the authors calculated a value of 4×10^{-28} gcm^{-3} for the mean density of matter, a value that they found reasonably compatible with estimates from astronomy (Einstein and de Sitter, 1932).

The Einstein–de Sitter model became very well known and it played a significant role in the development of twentieth century cosmology (Kragh, 1996, p. 35; North, 1965, p. 134; Nussbaumer, 2014b; Nussbaumer and Bieri, 2009, p. 152). One reason was that it marked an important hypothetical case in which the expansion of the universe was precisely balanced by a critical density of matter; a cosmos of lower mass density would be of hyperbolic geometry and expand at an ever-increasing rate, while a cosmos of higher mass density would be of spherical geometry and eventually collapse. Another reason was the model's great simplicity; in the absence of any empirical evidence for spatial curvature or a cosmological constant, there was little reason to turn to more complicated models. However, the timespan of the expansion was not considered in the rather terse paper. We recently discovered a little-known paper by Einstein containing a review of the Einstein–de Sitter model (Einstein, 1933a; O'Raifeartaigh, *et al.*, 2015): as in the case of the Friedmann–Einstein model, it is noted that the time of expansion is less than the estimated ages of the stars and the problem is attributed to the simplifying assumptions of the model.

23.3 Einstein's Steady-State Manuscript

As pointed out in the introduction, it appears that Einstein's steady-state manuscript was written in early 1931, before the Friedmann–Einstein model of April 1931. The manuscript (Einstein, 1931a) opens with a brief discussion of what Einstein terms the 'cosmological problem', i.e. the problem of gravitational collapse in classical and relativistic models of the universe: 'It is well known that the most important fundamental difficulty that emerges when one asks how the stellar matter fills up space in very large dimensions is that the laws of gravity are not in general consistent with the hypothesis of a finite mean density of matter. Thus, at a time when Newton's theory of gravity was still generally accepted, Seeliger had already modified the Newtonian law by the introduction of a distance function that, for large distances r, diminishes considerably faster than 1/r^2'. Noting a similar problem in general relativity, Einstein recalls his introduction of the cosmological constant to the field equations in order to allow the prediction of a universe of constant radius and non-zero density of matter: 'This difficulty also arises in the general theory of relativity. However, I have shown in the past that this can be overcome by the introduction of the so-called "λ–term" to the field equations. The field equations can then be written in the form (see Figure 23.1)

$$\left(R_{ik} - \frac{1}{2}g_{ik}R\right) - \lambda g_{ik} = \kappa T_{ik}\dots \quad (23.1)$$

…At that time, I showed that these equations can be satisfied by a spherical space of constant radius over time, in which matter has a density ρ that is constant over space and time.'

Einstein then notes that his static model was invalidated on both theoretical and observational grounds. In the first instance, the static model was unstable, while dynamic solutions existed: 'On the one hand, it follows from investigations based on the same equations by [ ] and by Tolman [6] that there also exist spherical solutions with a world radius P that is variable over time, and that my solution is not stable with respect to variations of P over time.' Second, the astronomical observations of Edwin Hubble changed the playing field: 'On the other hand, Hubbel's [*sic*] exceedingly important investigations have shown that the extragalactic nebulae have the following two properties: 1) Within the bounds of observational accuracy they are uniformly distributed in space, 2) They possess a Doppler effect proportional to their distance.'

Einstein then points out that the time-varying solutions of the field equations proposed by de Sitter and Tolman are consistent with Hubble's observations, but predict a timespan for the expansion that is problematic: 'De Sitter and Tolman have already shown that there are solutions to equations (1) that can account for these observations. However the difficulty arose that the theory unvaryingly led to a beginning in time about 10^{10}–10^{11} years ago, which for various reasons seemed unacceptable.' The 'various reasons' in the quote is almost certainly a reference to the fact that the estimated timespan of dynamic models was not larger than the ages of stars as estimated from astrophysics. However, it is possible that Einstein's difficulty also concerns the very idea of a 'beginning in time' for the universe.

In the second part of the manuscript, Einstein suggests an alternative solution to the field equations that is also compatible with Hubble's observations – namely, an expanding universe in which the density of matter does not change over time: 'In what follows, I wish to draw attention to a solution to equation (1) that can account for Hubbel's [*sic*] facts, and in which the density is constant over time.'

Assuming a metric of flat space expanding exponentially,[7] Einstein derives two simultaneous equations from the field equations, eliminating the cosmological constant to solve for the matter density (see Figure 23.2):

'Equations (1) yield:

$$\frac{-3(9)}{4}\alpha^2 + \lambda c^2 = 0$$

$$\frac{3}{4}\alpha^2 - \lambda c^2 = \kappa \rho c^2$$

or

$$\alpha^2 = \frac{\kappa c^2}{3}\rho \ldots \tag{23.2}$$

From his equation (4), Einstein concludes that the density of matter ρ remains constant and is related to the expansion factor α: 'The density is therefore constant and determines

---

[6] The blank space representing theoreticians other than Tolman is puzzling as Einstein was unquestionably aware of the dynamic models of Friedmann and Lemaître.

[7] It is easily shown that assumptions of homogeneity and isotropy imply this metric for a steady-state model.

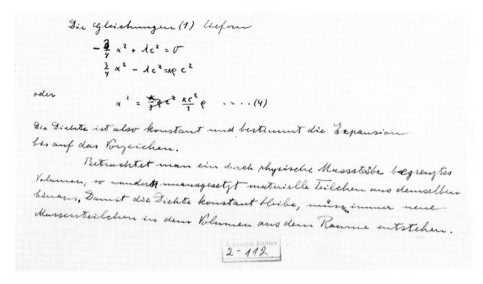

Figure 23.2 An excerpt from the last page of Einstein's steady-state manuscript (Einstein 1931a), reproduced by kind permission of the Hebrew University of Jerusalem. Equation (4) implies a direct relation between the expansion coefficient α and the mean density of matter ρ. However, the coefficient of α^2 in the first of the simultaneous equations was amended from 9/4 to -3/4 on revision, a correction that gives the null result $\rho = 0$ instead of equation (4).

the expansion apart from its sign.' This would be a stunning result, but it should be noted that equation (4) is incorrect, and arose from a numerical error in the derivation of the coefficient of α^2 in the first of the simultaneous equations. Careful study of the manuscript shows that Einstein later amended this coefficient from $+9/4$ to $-3/4$ (see Figure 23.2), an amendment that leads to the null solution $\rho = 0$ instead of equation (4).

In the final paragraph of the manuscript, Einstein proposes a physical mechanism to allow the density of matter to remain constant in an expanding universe, namely the continuous formation of matter from empty space: 'If one considers a physically bounded volume, particles of matter will be continually leaving it. For the density to remain constant, new particles of matter must be continually formed within that volume from space.' This proposal anticipates the later 'creation field' of Fred Hoyle in some ways (see Section 23.4). However, Einstein has not introduced a term representing the creation process into the field equations (unlike Hoyle). Instead, Einstein proposes that the cosmological constant assigns energy to empty space that can be associated with the creation of matter: 'The conservation law is preserved in that, by setting the λ-term, space itself is not empty of energy; its validity is well-known to be guaranteed by equations (1).' Thus, Einstein associates the continuous formation of matter from empty space with the cosmological constant. In reality, the lack of a specific term representing matter creation leads to a universe without matter in this model. It appears that Einstein recognised this problem on revision of the manuscript and set the model aside without pursuing the matter further.

23.4 On Steady-State Models of the Universe

The concept of a continuous creation of matter arose many times in twentieth century cosmology. In 1918, the American physicist William MacMillan proposed a continuous creation of matter from radiation in order to avoid a gradual 'running down' of the universe due to the conversion of matter into energy in stellar processes (MacMillan, 1918, 1925). The proposal was welcomed by Robert Millikan, who suggested that the process might be the origin of cosmic rays (Millikan, 1928). The idea of a continuous creation of matter from radiation was also briefly considered by Richard Tolman as a means of introducing matter into the empty de Sitter universe, although he found the idea improbable (Tolman, 1929).

Other physicists considered the possibility of a continuous creation of matter from empty space. In 1928, James Jeans speculated that matter was continuously created in the centre of the spiral nebulae (Jeans, 1928) and similar ideas of continuous creation were explored by Svante Arrhenius and Walther Nernst (Arrhenius, 1908; Nernst, 1928).[8]

Following the discovery of the systematic recession of the spiral nebulae, Richard Tolman suggested that a continuous annihilation of matter into radiation might be responsible for an expansion of space (Tolman, 1930a). While Eddington took the view that this process would retard expansion (Eddington, 1930), it is possible that Tolman's paper provided the inspiration for Einstein's steady-state model. As pointed out by Harry Nussbaumer, Einstein had many conversations with Tolman at the relevant time and Einstein's steady-state manuscript bears some mathematical similarities to Tolman's model – if not matter annihilation, why not matter creation? (Nussbaumer, 2014a).

The concept of an expanding universe that remains in a steady state due to a continuous creation of matter from empty space is most strongly associated with the Cambridge physicists Fred Hoyle, Hermann Bondi and Thomas Gold (Bondi and Gold, 1948; Hoyle, 1948). In the late 1940s, these physicists became concerned with well-known problems associated with evolving models of the cosmos. In particular, they noted that the evolving models predicted a timespan for expansion that was problematic and disliked Lemaître's hypothesis of a universe with a fireworks beginning (Lemaître's, 1931a, 1931b, 1931c). Another concern was philosophical in nature; if the universe was truly different in the past, was it not inconsistent to assume that the present laws of physics applied? In order to circumvent these problems, the trio explored the idea of an expanding universe that does not evolve over time, i.e., an expanding cosmos in which the mean density of matter is maintained constant by a continuous creation of matter from the vacuum (Bondi and Gold, 1948; Hoyle, 1948).

In the case of Bondi and Gold, the proposal of a steady-state model took as a starting point the 'perfect cosmological principle', a philosophical principle that stated that the universe should appear essentially the same to all observers in all locations at all times. This principle demanded a continuous creation of matter in order to maintain a constant

---

[8] A review of steady-state cosmologies in the early twentieth century can be found in Kragh (1996, pp. 143–62).

density of matter in the expanding universe. The resulting model bore some similarities to Einstein's steady-state model, but it is difficult to compare the theories directly as the Bondi–Gold theory was not formulated in the framework of general relativity. On the other hand, Fred Hoyle constructed a steady-state model of the cosmos by means of a daring modification of the Einstein field equations (Hoyle, 1948; Mitton, 2005, pp. 118–19). Replacing Einstein's cosmological constant with a new 'creation-field' term C_{ik} to represent a continuous formation of matter from the vacuum, Hoyle obtained the equation

$$\left(R_{ik} - \frac{1}{2} g_{ik} R \right) - C_{ik} = \kappa T_{ik}$$

Hoyle's creation-field term allowed for an unchanging universe but was of importance only on the largest scales, in the same manner as the cosmological constant. In this model, the expansion of space was driven by the creation of matter, and the perfect cosmological principle emerged as a consequence rather than a starting assumption. A more sophisticated formulation of the model, based on the principle of least action, was proposed in later years (Hoyle and Narlikar, 1962).

As is well known, a significant debate was waged between steady-state and evolving models of the cosmos during the 1950s and 1960s (Kragh, 1996, pp. 252–68, 2007, pp. 187–90; Mitton, 2005, pp. 167–96). Eventually, the steady-state universe was effectively ruled out by observation, not least by the study of the distribution of the galaxies at different epochs and by the discovery of the cosmic microwave background (Kragh, 1996, pp. 318–80, 2007, pp. 201–6; Narlikar, 1988, pp. 218–19). There is no evidence that any of the steady-state theorists were aware of Einstein's attempt; indeed, it is likely that they would have been greatly intrigued to learn that Einstein had once considered a steady-state model.

23.5 On Einstein's Philosophy of Cosmology

It should come as no great surprise that, when confronted with empirical evidence for an expanding universe, Einstein considered a steady-state or 'stationary' model of the expanding cosmos. Such a model fits well with his lack of interest in non-static solutions to the field equations in 1917, and his hostility to the dynamic models of Friedmann and Lemaître when they were first proposed (see Section 23.2). Indeed, a model of the expanding cosmos in which the mean density of matter remains unchanged over time seems a natural successor to Einstein's static model of 1917 from a philosophical point of view.

However, Einstein's attempt at a steady-state model led to a null solution, and it appears that he abandoned the idea rather than pursue it further. (One possibility would have been to introduce a matter-creation term to the field equations in the manner of Hoyle; another to consider a fluid of negative pressure (McCrea, 1951).) Instead, Einstein turned to expanding models of varying matter density that could be described 'naturally' by the field equations, i.e. without the use of the cosmological constant term (Einstein, 1931b;

Einstein and de Sitter, 1932). It therefore seems very likely that Einstein abandoned steady-state cosmology on the grounds that it was more contrived than evolutionary models of the cosmos.

Taken together, Einstein's abandonment of steady-state cosmology, his removal of the cosmological constant term in the Friedmann–Einstein model (Einstein, 1931b), and the removal of spatial curvature in the Einstein–de Sitter model (Einstein and de Sitter, 1932), suggest a simple, pragmatic approach to cosmology. Where theorists such as Friedmann, Heckmann and Robertson considered all possible universes (see Section 23.2), Einstein sought the simplest model of the universe that could account for observation. It is worth asking whether this practical 'Occam's razor' approach was in fact characteristic of Einstein's cosmology all along, as considered below.

23.5.1 Einstein's Journey From the Static to the Evolving Universe

Einstein's journey from a static, bounded cosmology to the evolving universe is traditionally characterised as that of a reluctant convert; a conservative Einstein, hidebound by philosophical prejudice until overwhelmed by irrefutable evidence (Giulini and Straumann, 2006; Kragh, 1996, p. 26; Nussbaumer, 2014b; Nussbaumer and Bieri, 2009, pp. 92; Smeenk, 2014). We suggest that Einstein's steady-state manuscript provides a useful clue that this narrative may be somewhat inaccurate.

Considering first Einstein's cosmic model of 1917, it is often asserted that the cosmological constant was introduced to the field equations in order to predict a static rather than a contracting universe. In fact, it is more accurate to say that the purpose of the cosmological constant was to allow the prediction of a finite density of matter in a universe that was assumed *a priori* to be static. No evidence for a dynamic universe was known at the time, and the notion of an expanding or contracting universe would have seemed very far-fetched. (Indeed, Einstein refers to the model as 'making possible a quasi-static distribution of matter, as required by the fact of the small velocities of the stars' (Einstein, 1917b).) When Friedmann explored time-varying solutions of the field equations as a hypothetical possibility in 1922, Einstein was one of the few who paid attention; however, he found non-static solutions 'suspicious' due to a lack of supporting evidence (see Section 23.2). In 1927, Lemaître's expanding model of the universe was inspired by observations at the cutting edge of astronomical research; Einstein's rejection of this model can probably be attributed to a lack of familiarity with advances in astronomy. Lemaître certainly thought so, commenting later that Einstein did not seem to be aware of recent astronomical measurements (Lemaître, 1958).

With the publication of astronomical observations suggestive of an expanding cosmos in 1929, Einstein lost little time in abandoning the static universe. It seems that he had no difficulty changing his viewpoint once such a change was warranted by the evidence. One is reminded of a famous comment attributed to John Maynard Keynes: 'When the facts change, I change my mind – what do you do, Sir?' It now seems that at this point, Einstein's first guess was an expanding universe that remains essentially unchanged over time – the obvious next step after his static model. However, when this attempt led to an empty universe, Einstein turned to evolving models instead. Noting that expanding models did not necessarily require a cosmological constant, he removed this term (Einstein, 1931b). When he realised that spatial curvature was also no longer a given in dynamic cosmologies, this parameter was removed in turn (Einstein and de Sitter, 1932). This sequence of ever simpler models suggests an approach to cosmology that was not conservative but pragmatic – a minimalist, empirical approach to the study of the universe. Tellingly, Einstein did not propose any major cosmic models beyond this point; as he explained later, he saw little point in speculating further in the absence of empirical data on cosmological parameters such as spatial curvature and the density of matter (Einstein, 1945, pp. 133–4).

We note that this approach to cosmology is very typical of Einstein's general approach to physics, at least in his younger years. Sometimes described as positivist, Einstein's approach is more accurately described as a philosophy of *logical empiricism* – he embraced the central importance of observations in the testing of a theoretical hypothesis, at least in a holistic sense, but also assigned great importance to the construction of consistent theories from analytic principles of logic (Einstein, 1949, pp. 680–681; Frank, 1948, pp. 259–63, 1949, pp. 271–86; Reichenbach, 1949, pp. 309–11). This is a very different approach to that of Compte or Mach, who suggested that the fundamental laws of physics should only contain concepts that could be defined by direct observations, or at least be connected to observation by a short chain of thought. It is also different to that of empiricists such as Moritz Schlick or Rudolf Carnap because it contained both positivist and metaphysical elements.[9] An insight into Einstein's philosophy of science in these years can be found in his 1933 Herbert Spencer Lecture at Oxford: 'Experience remains, of course, the sole criterion of the physical utility of a mathematical construction. But the creative principle resides in mathematics' (Einstein, 1933b, 1934, p. 36).

23.5.2 *On the Cosmological Constant and Dark Energy*

Until recently, it was universally assumed that, with the emergence of the first empirical evidence for an expanding universe, Einstein immediately abandoned the cosmological constant along with the static universe (Kragh, 1996, p. 34; Nussbaumer, 2014b; Nussbaumer and Bieri, 2009, p. 147; North, 1965, p. 132; Straumann, 2002). Certainly, Einstein made it clear on several occasions that he disliked the term, at least from the perspective of the general theory of relativity. (For example, in 1919 he described the term as 'gravely detrimental to the formal beauty of the theory' (Einstein, 1919).) However, Einstein's steady-state manuscript demonstrates that he retained the cosmological constant in at least one cosmic model he attempted *after* Hubble's observations, albeit for a new purpose. It appears that when presented with evidence for a cosmic expansion, Einstein's

[9] See Howard (2014) for an overview of this point.

attraction to an unchanging universe at first outweighed his dislike of the cosmological constant, just as it did in 1917.

It will not have escaped the reader's attention that Einstein's association of the cosmological constant with an energy of space in his steady-state model is not unlike the current hypothesis of dark energy, at least from a philosophical standpoint. Where Einstein attempted to associate a continuous creation of matter with the cosmological constant, we now currently assume an energy for an accelerated expansion.[10] More generally, it has often been noted that the cosmological constant term of 1917 anticipates the notion of dark energy in some ways. It is less well known that Einstein also considered – and dismissed – the possibility of a time-varying energy of space, a concept not unlike the modern hypothesis of quintessence. Within a few months of the publication of Einstein's static model of 1917, Erwin Schrödinger suggested that the cosmological term could be placed on the right hand side of the field equations (a negative energy density term in the matter-energy tensor) and that the term could be time-varying (Schrödinger, 1918). Einstein's response was that, if constant, placing the term in the matter-energy tensor was equivalent to his original formulation. If not constant, the term would necessitate undesirable speculation on the nature of its variation over time: 'The course taken by Herr Schrödinger does not appear passable to me because it leads too deeply into the thicket of hypotheses' (Einstein, 1918c). Once again, this attitude indicates a strong dislike of complicated solutions unless necessitated by observation.[11]

We note that a great deal has been written over the years about Einstein's evolving view of the cosmological constant. For example, the well-known Russian physicist George Gamow stated that Einstein once declared the term 'my greatest blunder' (Gamow, 1956, 1970, p. 44), while others have cast doubt on this statement (Livio, 2013, pp. 233–41; Straumann, 2002). We will not enter this debate here, but simply note that Einstein soon dispensed with the term in his non-static cosmology. His considered view is probably best summed up in a footnote to his 1945 review of cosmology: 'If Hubble's expansion had been discovered at the time of the creation of the general theory of relativity, the cosmologic member would never have been introduced. It seems now so much less justified to introduce such a member into the field equations, since its introduction loses its sole original justification – that of leading to a natural solution of the cosmologic problem' (Einstein, 1945, p. 130). This stance should be contrasted with Einstein's attitude to spatial curvature. While the Einstein–de Sitter model was based on the fact that the presence of matter in a dynamic universe does not automatically imply spatial curvature, the authors were careful not to rule it out: 'It is possible to represent the facts without assuming a curvature of three-dimensional space. The curvature is, however, essentially determinable, and an increase in the precision of the data derived from observations will enable us in the future to fix its sign and to determine its value' (Einstein and de Sitter, 1932).

[10] See Peebles and Ratra (2003) for a review of dark energy.
[11] See Harvey (2012) for a fuller discussion of this episode.

23.5.3 On the Question of Cosmic Origins

To modern eyes, a striking aspect of Einstein's steady-state manuscript is the lack of reference to the problem of the singularity for the case of evolving models, or to the related question of an origin for the universe. Indeed, the manuscript is the only steady-state model of the expanding universe known to us that is not motivated (at least in part) by a desire to circumvent such problems. While Einstein is clearly conscious of the puzzle of the short timespan of evolving models, there is no reference to the problem of origins (see Section 23.3).

One explanation might be that Einstein's steady-state manuscript almost certainly pre-dated Lemaître's proposal of a 'fireworks beginning' for the universe (Lemaître, 1931b, 1931c). However, the issue of cosmic origins for evolving models was recognised before these papers were published (de Sitter, 1932; Eddington, 1930, 1931). We note instead that Einstein's silence on the question is very typical of his cosmology – there is no reference to the problem in either of his evolving models (Einstein, 1931b; Einstein and de Sitter, 1932) or in a contemporaneous review of relativistic cosmology (Einstein, 1933a). In later years, Einstein made it clear that this silence did not stem from a philosophical difficulty with the notion of a physical origin for the cosmos, but from doubts concerning the validity of relativistic models at early epochs: 'For large densities of field and of matter, the field equations and even the field variables which enter into them will have no real significance. One may not therefore assume the validity of the equations for very high density of field and of matter' (Einstein, 1945, pp. 132–3).

23.5.4 On Einstein's Philosophy of Relativity

We note in passing that Einstein's steady-state manuscript does not contain any considerations of philosophical issues associated with the theory of relativity, as opposed to cosmology. Reading the opening section of the work, the professional philosopher may be somewhat disappointed by the lack of reference to problems such as the use of idealised clocks and rulers in relativity,[12] or the question of the geometrisation of gravity.[13] This silence is once again very typical of Einstein's cosmology; such issues are not discussed in any of Einstein's static or dynamic models of the cosmos, although he did consider them elsewhere (Einstein, 1948). This suggests once more that Einstein's approach to cosmology was essentially pragmatic; general relativity was a useful tool to describe the universe, but by no means the ultimate answer. As we have argued elsewhere (O'Raifeartaigh and McCann, 2014), it is likely that Einstein's search for a unified field theory in these years made him very conscious of the limitations of relativistic models of the cosmos.

---

[12] See Brown (2014) for a review.
[13] A longstanding question was whether the spacetime metric of relativity was a mathematical tool to describe gravity, or whether gravity 'was' geometry (Lehmkuhl, 2014).

23.5.5 On Paradigm Shifts in Cosmology

We note finally that Einstein's steady-state manuscript does not support a view that his acceptance of the evolving universe occurred as an abrupt change to a new worldview. As described above, the model appears as an intermediate step in a long journey from the static universe to an expanding, evolving cosmology. Indeed, the manuscript provides a new piece of evidence that today's 'big bang' cosmology did not emerge as an abrupt 'paradigm shift' in the manner envisioned by Thomas Kuhn (Kuhn, 1962), but rather as a slow dawning in both theory and observation within a single paradigm, the relativistic universe.

It is unfortunate that Einstein's cosmology papers of the 1930s are not better known, as the pragmatic, empirical approach we have discussed above is very different to Einstein's work on unified field theory in these years (Einstein and Mayer, 1930, 1931, 1932). Indeed, we find the cosmology papers quite reminiscent of the young Einstein's approach to emerging phenomena (Einstein, 1905a, 1905b, 1905c). One wonders whether the familiar narrative that Einstein became more and more attached to a formal mathematical approach to physics in his later years is entirely accurate. Could it be that Einstein's philosophical approach to science did not truly change but that the intense level of mathematical abstraction one associates with Einstein's later work was simply a facet of the great technical challenge posed by unified field theory?

23.6 Conclusions

Einstein's attempt at a steady-state model was abandoned before publication but it offers many insights into his philosophy of cosmology. His hypothesis of a universe of expanding radius and constant matter density is very different to his static model of 1917 or his evolving models of 1931 and 1932, and anticipates in some ways the well-known steady-state cosmology of Hoyle, Bondi and Gold. The model was almost certainly written in early 1931, when Einstein first learnt of observational evidence for a cosmic expansion, but was quickly abandoned when it led to a null solution. The steady-state manuscript is nevertheless of interest because it offers new evidence that Einstein's philosophical journey from a static, bounded cosmology to the dynamic, evolving universe was that of a pragmatic empiricist, rather than a reluctant conservative.

We note finally that Einstein's steady-state model finds an echo in current theories of cosmic inflation. In particular, the de Sitter metric of flat, exponentially expanding space used in inflationary models[14] recalls the steady-state models of Einstein and Hoyle. Indeed, many scholars have noted that inflationary models are effectively steady-state cosmologies over an extremely limited timespan (Barrow, 2005; Hoyle, 1994, p. 271; Narlikar, 1988, pp. 223–5, 2005). Furthermore it has been suggested (Linde 1986a, 1986b; Vilenkin, 1983)

---

[14] See Liddle (1999) for a review of inflationary cosmology.

that the inflationary process inevitably creates the conditions for further inflation in a never-ending cycle. This concept of 'eternal inflation' raises the possibility that the observed, evolving universe is a local anomaly in a global ensemble that is in a steady state (Barrow, 2005), a scenario that is not dissimilar to Hoyle's later proposal of a steady-state universe permeated with local 'little bangs' (Hoyle and Narlikar, 1966; Hoyle *et al.*, 1993; Narlikar, 2005). Thus it can be concluded that, like the cosmological constant, the concept of the steady-state universe is proving hard to banish from modern cosmology.

23.7 Acknowledgements

The author would like to thank the Hebrew University of Jerusalem for permission to display the excerpts shown in Figures 23.1 and 23.2. He also thanks the Dublin Institute of Advanced Studies for access to the Collected Papers of Albert Einstein (Princeton University Press).

References

Arrhenius, S., 1908. *World in the Making; the Evolution of the Universe.* (transl. H. Borns). Michigan: Harper.

Associated Press. 1931. Prof. Einstein begins his work at Mt. Wilson. *New York Times*, Jan 3, p. 1.

Barrow, J. 2005. Worlds without end or beginning. In Gough, J, ed. *The Scientific Legacy of Fred Hoyle.* Cambridge: Cambridge University Press.

Bartusiak, M. 2009. *The Day We Found the Universe.* New York: Vintage Books.

Bondi, H. and Gold, T. 1948. The steady-state theory of the expanding universe *Mon. Not. Roy. Ast. Soc.*, **108**, 252–70.

Brown, H. 2014. The behaviour of rods and clocks in general relativity, and the meaning of the metric field. To be published in the Einstein Studies Series. ArXiv preprint 0911.4440. http://arxiv.org/abs/0911.4440.

de Sitter, W. 1917. On Einstein's theory of gravitation and its astronomical consequences. *Month. Not. Roy. Astron. Soc.*, **78**, 3–28.

de Sitter, W. 1930a. On the distances and radial velocities of the extragalactic nebulae , and the explanation of the latter by the relativity theory of inertia. *Proc. Nat. Acad. Sci.*, **16**, 474–88.

de Sitter, W. 1930b. The expanding universe. Discussion of Lemaître's solution of the equations of the inertial field. *Bull. Astron. Inst. Neth.*, **5**(193), 211–18.

de Sitter, W, 1932. *Kosmos: A Course of Six Lectures on the Development of our Insight into the Structure of the Universe, Delivered for the Lowell Institute in Boston in November 1931.* Cambridge, MA: Harvard University Press.

Eddington, A. S. 1923. *The Mathematical Theory of Relativity.* Cambridge: Cambridge University Press.

Eddington, A. S. 1930. On the instability of Einstein's spherical world. *Month. Not. Roy. Astron. Soc.*, **90**, 668–78.

Eddington A. S. 1931. The recession of the extra-galactic nebulae. *Mon. Not. Roy. Astr. Soc.*, **92**, 3–6.

Einstein, A. 1905a. Über einen die Erzeugung und Verwandlung des Lichtes betreffenden heuristischen Gesichtspunkt. *Annal. Physik*, **17**, 132–48. Available in English translation in CPAE **2** (Doc. 14).

Einstein, A. 1905b. Über die von der molekularkinetischen Theorie der Wärme geforderte Bewegung von in ruhenden Flüssigkeiten suspendierten Teilchen. *Annal. Physik*, **17**, 549–60. Available in English translation in CPAE **2** (Doc. 16).

Einstein, A. 1905c. Zur Elektrodynamik bewegter Körper. *Annal. Physik*, **17**, 891–921. Available in English translation in CPAE **2** (Doc. 23).

Einstein, A. 1915. Die Feldgleichungen der Gravitation. *Sitzungsb. König. Preuss. Akad.*, 844–7. Available in English translation in CPAE **6** (Doc. 25).

Einstein, A. 1916. Die Grundlage der allgemeinen Relativitätstheorie. *Ann. Physik.*, **49**, 769–822. Available in English translation in CPAE **6** (Doc. 30).

Einstein, A. 1917a. Letter to de Sitter, March 12th. Available in English translation in CPAE **8** (Doc. 311).

Einstein, A. 1917b. Kosmologische Betrachtungen zur allgemeinen Relativitätstheorie. *Sitzungsb. König. Preuss. Akad.*, 142–52. Available in English translation in CPAE **6** (Doc. 43).

Einstein, A. 1918a. Prinzipielles zur allgemeinen Relativitätstheorie. *Ann. Physik*, **55**, 241–44. Available in English translation in CPAE **7** (Doc. 4).

Einstein, A. 1918b. Kritisches zur einer von Hrn. De Sitter gegebenen Lösung der Gravitationsgleichungen. *Sitzungsb. König. Preuss. Akad.*, 270–2. Available in English translation in CPAE **7** (Doc. 5).

Einstein, A. 1918c. Bemerkung zu Herrn Schrödingers Notiz "Uber ein Lösungssystem der allgemein kovarianten Gravitationsgleichungen", *Physikalische Zeitschrift*, **19**, 165–6. Available in English translation in CPAE **7** (Doc. 3).

Einstein, A. 1919. Spielen Gravitationsfelder im Aufbau der materiellen Elementarteilchen eine wesentliche Rolle? *Sitzungsb. König. Preuss. Akad.*, 349–56. Available in English transl. in CPAE **7** (Doc. 17).

Einstein, A. 1922. Bemerkung zu der Arbeit von A. Friedmannn "Über die Krümmung des Raumes". *Zeit. Phys.*, **11**, 326.

Einstein, A.1923a. Notiz zu der Arbeit von A. Friedmannn "Über die Krümmung des Raumes". *Zeit. Phys.*, **16**, 228.

Einstein, A. 1923b. Notiz zu der Arbeit von A. Friedmannn "Über die Krümmung des Raumes". The Albert Einstein Archives, Document **1**–26. http://alberteinstein.info/vufind1/Record/EAR000034026.

Einstein, A. 1931a. Zum kosmologischen Problem. Albert Einstein Archive Online, Document **2**–112, http://alberteinstein.info/vufind1/Record/EAR000034354. Available in English translation in (O'Raifeartaigh *et al.*, 2014).

Einstein, A. 1931b. Zum kosmologischen Problem der allgemeinen Relativitätstheorie. *Sitzungsb. König. Preuss. Akad.* 235–7. Available in English translation in (O'Raifeartaigh and McCann, 2014).

Einstein, A. 1933a. Sur la Structure Cosmologique de l'Espace (transl. M. Solovine). In *'Les Fondaments de la Théorie de la Relativité Générale'*, Hermann et Cie, Paris. Available in English translation in (O'Raifeartaigh *et al.*, 2015).

Einstein, A. 1933b. *On the Method of Theoretical Physics*. The Herbert Spencer Lecture, Oxford, 10 June, 1933. Oxford: Clarendon Press. Reprinted in (Einstein, 1934).

Einstein, A. 1934. *The World As I See It*. New York: Cocivi-Friede.

Einstein, A. 1945. On the cosmological problem. Appendix to *The Meaning of Relativity*, 2nd edn. Princeton: Princeton University Press. Also available in later editions.

Einstein, A. 1948. Letter to Lincoln Barnett. June 19th. Document no. **1**–154, Einstein Online Archive.

Einstein, A. 1949. Autobiographical Notes. In P. A. Schilpp, ed. *Albert Einstein: Philosopher-Scientist*. The Library of Living Philosophers **VII**, Ed. P. A. Schilpp. Wisconsin: George Banta Publishing, pp. 3–105.

Einstein, A. and de Sitter, W. 1932. On the relation between the expansion and the mean density of the universe. *Proc. Nat. Acad. Sci.*, **18**, 213–14.

Einstein, A. and Mayer, W. 1930. Zwei strenge statische Lösungen der Feldgleichungen der einheitlichen Feldtheorie. *Sitzungsb. König. Preuss. Akad.*, 110–20.

Einstein, A. and Mayer, W. 1931. Einheitliche Feldtheorie von Gravitation und Elektrizität. *Sitzungsb. König. Preuss. Akad.*, 541–57.

Einstein, A. and Mayer, W. 1932. Einheitliche Feldtheorie von Gravitation und Elektrizität, 2. Abhandlung. *Sitzungsb. König. Preuss. Akad.*, 130–7.

Farrell, J. 2005. *The Day Without Yesterday: Lemaître, Einstein and the Birth of Modern Cosmology*. New York: Thunder's Mouth Press.

Frank, P. 1948. *Einstein: His Life and Times*. London: Jonathan Cape.

Frank, P. 1949. Einstein, Mach and Logical Positivism. In *Albert Einstein: Philosopher Scientist*. The Library of Living Philosophers **VII**, Ed. P. A. Schilpp. Wisconsin: George Banta Publishing, pp. 269–287.

Friedmann, A. 1922. Über die Krümmung des Raumes. *Zeit. Physik.*, **10**, 377–86.

Friedmann, A. 1924. Über die Möglichkeit einer Welt mit konstanter negativer Krümmung des Raumes. *Zeit. Physik.*, **21**, 326–32.

Gamow, G. 1956. The evolutionary universe. *Scientific American*, **192**, 136–154.

Gamow, G. 1970. *My Worldline*. New York: Viking Press.

Giulini, D. and Straumann, N. 2006. Einstein's impact on the physics of the twentieth century. *Studies in History and Philosophy of Modern Physics*, **37**, 115–73.

Harvey, A. 2012. How Einstein discovered dark energy. http://arxiv.org/abs/1211.6338.

Heckmann, O. 1931. Über die Metrik des sich ausdehnenden Universums. *Nach. Gesell. Wiss. Göttingen, Math.-Phys., Klasse*: 1–5.

Heckmann, O. 1932. Die Ausdehnung der Welt in ihrer Abhängigkeit von der Zeit. *Nach. Gesell. Wiss. Göttingen, Math.-Phys., Klasse*: 97–106.

Hoyle, F. 1948. A new model for the expanding universe. *Mon. Not. Roy. Ast. Soc.*, **108**, 372–82.

Hoyle, F. 1994. *Home Is Where The Wind Blows*. Mill Valley, CA: University Science Books.

Hoyle, F. and Narlikar, J. 1962. Mach's principle and the creation of matter. *Proc. Roy. Soc.*, **270**, 334–41.

Hoyle, F. and Narlikar, J. 1966. A Radical Departure from the 'Steady-State' Concept in Cosmology. *Proc. Roy. Soc.*, **290**, 162–76.

Hoyle, F., Burbidge, G. and Narlikar, J. 1993. A quasi-steady state cosmological model with creation of matter. *Ast. J.*, **410**, 437–57.

Howard, D. 2014. Einstein and the development of twentieth century philosophy of science. In Janssen, M. and Lehner, C., eds. *The Cambridge Companion to Einstein*. Cambridge: Cambridge University Press, pp. 354–76.

Hubble, E. 1925. Cepheids in spiral nebulae. *Observatory*, **48**, 139–42.

Hubble, E. 1929. A relation between distance and radial velocity among extra-galactic nebulae. *Proc. Nat. Acad. Sci.*, **15**, 168–73.

Janssen, M. 2005. Of pots and holes: Einstein's bumpy road to general relativity. *Ann. Physik*, **14** (Suppl), 58–85.

Jeans, J. 1928. *Astronomy and Cosmogony.* Cambridge: Cambridge University Press.

Kragh, H. 1996. *Cosmology and Controversy.* Princeton, NJ: Princeton University Press.

Kragh, H. 2007. *Conceptions of Cosmos.* Oxford: Oxford University Press.

Kuhn, T. 1962. *The Structure of Scientific Revolutions.* Cambridge: Cambridge University Press.

Lehmkuhl, D. 2014. Why Einstein did not believe that general relativity geometrizes gravity. *Studies in History and Philosophy of Modern Physics*, **46**, 316–26.

Lemaître, G. 1925. Note on de Sitter's universe. *J. Math. Phys.*, **4**, 188–92.

Lemaître, G. 1927. Un univers homogène de masse constant et de rayon croissant, rendant compte de la vitesse radiale des nébeleuses extra-galactiques. *Annal. Soc. Sci. Brux.* Série A. **47**, 49–59.

Lemaître, G. 1931a. A homogeneous universe of constant mass and increasing radius, accounting for the radial velocity of the extra-galactic nebulae. *Mon. Not. Roy. Astr. Soc.*, **91**, 483–90.

Lemaître, G. 1931b. The beginning of the world from the point of view of quantum theory. *Nature,* **127**, 706.

Lemaître, G. 1931c. L'expansion de l'espace. *Rev. Quest. Sci.,* **17**, 391–440.

Lemaître, G. 1958. Recontres avec Einstein. *Rev. Quest. Sci.,* **129**, 129-132.

Liddle. A. R. 1999. An Introduction to Cosmological Inflation. In Masiero, A., Senjanovic, G. and Smirnov, A., eds. *High Energy Physics and Cosmology* . Singapore: World Scientific Publishers.

Linde, A. 1986a. Eternal chaotic inflation. *Mod. Phys. Lett. A*, **1**(02), 81–85.

Linde, A. 1986b. Eternally existing self-reproducing chaotic inflationary universe. *Phys. Lett. B.* **175**, 395–400.

Livio, M. 2013. *Brilliant Blunders: from Darwin to Einstein.* Simon & Schuster.

MacCrea, W. H. 1951. Relativity theory and the creation of matter. *Proc. Roy. Soc.* **206**, 562–75.

MacMillan, W. D. 1918. On stellar evolution. *Astrophys. J.*, **48**, 35–49.

MacMillan, W. D. 1925. Some mathematical aspects of cosmology. *Science* **62**: 63–72, 96–9, 121–7.

Millikan R. 1928. Available energy. *Science*, **68**, 279–84.

Mitton, S. 2005. *Fred Hoyle: A Life in Science.* London: Aurum Press.

Narlikar, J. 1988. *The Primeval Universe.* Oxford: Oxford University Press.

Narlikar, J. 2005. Alternative ideas in cosmology. In Gough, J, ed. *The Scientific Legacy of Fred Hoyle.* Cambridge: Cambridge University Press.

Nernst, W. 1928. Physico-chemical considerations in astrophysics. *J. Franklin Inst.* **206**, 135–42.

North, J. D. 1965. *The Measure of the Universe: A History of Modern Cosmology.* Oxford: Oxford University Press.

Nussbaumer, H. 2014a. Einstein's aborted model of a steady-state universe. To appear in the volume "In memoriam Hilmar W. Duerbeck", *Acta Historica Astronomiae.* Ed. W. Dick, R. Schielicke and C. Sterken. http://arxiv.org/abs/1402.4099.

Nussbaumer, H. 2014b. Einstein's conversion from his static to an expanding universe. *Eur. Phys. J (H)*, **39**(1), 37–62.

Nussbaumer, H. and Bieri, L. 2009. *Discovering the Expanding Universe*, Cambridge: Cambridge University Press.

O'Raifeartaigh, C. 2014. Einstein's steady-state cosmology. *Physics World*, **27**(9), 30–33.

O'Raifeartaigh, C. and McCann, B. 2014. Einstein's cosmic model of 1931 revisited; an analysis and translation of a forgotten model of the universe. *Eur. Phys. J (H)*, **39**(1), 63–85.

O'Raifeartaigh, C., McCann, B., Nahm, W. and Mitton, S. 2014. Einstein's steady-state theory: an abandoned model of the cosmos. *Eur. Phys. J (H)*, **39**(3), 353–67.

O'Raifeartaigh, C., O'Keeffe, M., Nahm, W. and Mitton, S. 2015. Einstein's cosmology review of 1933: a new perspective on the Einstein–de Sitter model of the cosmos. *Eur. Phys. J. (H)*, **40**(3), 301–35.

Peebles and Ratra. 2003. *Rev. Mod. Phys.* **75**(2), 559–606.

Reichenbach, H. 1949. The philosophical significance of the theory of relativity. In Schilpp, P. A., ed. *Albert Einstein: Philosopher Scientist*. The Library of Living Philosophers **VII**. Wisconsin: George Banta Publishing, pp. 287–313.

Robertson, H. P. 1932. The expanding universe. *Science*, **76**, 221–6.

Robertson, H. P. 1933. Relativistic cosmology. *Rev. Mod. Phys.*, **5**, 62–90.

Schrödinger E. 1918. Uber ein Lösungssystem der allgemein kovarianten Gravitations-gleichungen. *Physikalische Zeitschrift*, **19**, 20–22.

Slipher, V. M. 1915. Spectrographic observations of nebulae. *Pop. Ast.*, **23**, 21–4.

Slipher, V. M. 1917. Nebulae, *Proc. Am. Phil. Soc.*, **56**, 403–9.

Smeenk, C. 2014. Einstein's role in the creation of relativistic cosmology. In Janssen, M. and Lehner, C., eds. *The Cambridge Companion to Einstein*. Cambridge: Cambridge University Press, pp. 228–70.

Stachel, J. 1977. Notes on the Andover Conference. In Earman, J. S., Glymour, C. N. and Stachel, J., eds. *Foundations of Space-Time Theories*, Minnesota Studies in the Philosophy of Science **VIII**, pp. viii–xii.

Straumann, N. 2002. The history of the cosmological constant problem. In Brax, P., Martin, J. and Uzan, J. P., eds. *On the Nature of Dark Energy: Proceedings of the 18th IAP Astrophysics Colloquium*. Frontier Group. ArXiv preprint no. 0208027.

Tolman, R. 1929. On the astronomical implications of the de Sitter line element for the universe. *Ast. J.*, **69**, 245–74.

Tolman, R. 1930a. The effect of the annihilation of matter on the wave-length of light from the nebulae. *Proc. Nat. Acad. Sci.*, **16**, 320–37.

Tolman, R.C. 1930b. More complete discussion of the time-dependence of the non-static line element for the universe. *Proc. Nat. Acad. Sci.*, **16**, 409–20.

Tolman, R.C. 1931. On the theoretical requirements for a periodic behaviour of the universe. *Phys. Rev.*, **38**, 1758–71.

Tolman, R.C. 1932. On the behaviour of non-static models of the universe when the cosmological term is omitted. *Phys. Rev.*, **39**, 835–43.

Vilenkin, A. 1983. Birth of inflationary universes. *Phys. Rev. D* **27**, 2848–55.

Weyl, H. 1923. Zur allgemeinen Relativitätstheorie. *Physik. Zeitschrift.* **24**, 230–32.

24

The Nature of the Past Hypothesis

DAVID WALLACE

There is a narrative about the nature of asymmetry in time which can be caricatured like this. Firstly, there is our fundamental physics – if we are being really careful about it, this is the standard model or our preferred post-standard-model physics; in reality, in the practical cases we think about, it is more likely to be ordinary quantum mechanics or maybe even classical Hamiltonian dynamics. In any case, it is supposed to be the physics of the micro-constituents of the world. And it is time-reversal invariant and shows no particular direction of time.

And secondly there is the observed world, which is full of various kinds of observed asymmetries: dynamical asymmetries, entropic asymmetries, causal, psychological asymmetries, and so on. And the general way we set the problem up is as a contradiction between what our physics says, which is that the world is time-reversal invariant, and what we see around us, which is not time-reversal invariant.

I want to suggest the advantages of a slightly more nuanced way of thinking about the problem. It is really not the case that all of physics, or even most of physics, or even very much of physics, frankly, is "fundamental" physics. In the middle – between the fundamental physics at the bottom, and the directly observed macro-world at the top – we have a huge range of what we might call higher-level (or "emergent") dynamical systems, governed by higher-level dynamical equations. I am thinking of the equations of fluid dynamics; I am thinking of the Boltzmann equation that governs dilute gases and many similar systems; I am thinking of the Langevin equation and the Fokker–Planck equation that govern Brownian motion; I am thinking of the equations of radioactive decay. (In principle I am also thinking about all the various equations and rules of the higher sciences, but we can confine our attention here just to the panoply of different systems and different equations that we study in physics.)

Actual physics is a plethora of different dynamical systems governed by different sorts of laws. And if you look at those higher-level laws you find – not universally, but very frequently – that *they* have a whole range of properties which are not shared by the fundamental physics. I want to focus on two particular properties like this.

This chapter is a transcript of a talk and discussion at the Tenerife conference on the Philosophy of Cosmology.

Firstly, they tend to have a lack of time symmetry. It is worth pausing on what that means for a second, because in one sense the standard model (and its plausible successors) do not have time symmetry either: they are symmetric under CPT but not under T alone. But higher-level theories are time-reversal-non-invariant in a more important way that the standard model does not share: they are generally *irreversible*, which is to say that their dynamical maps are many-to-one, either in the literal sense that they take different initial states to the *same* later state or in the sense that they take different initial states to final states that get closer and closer together over time, so that to any given grain of resolution they might as well be taken as the same final state. And frequently these higher-level theories are not just irreversible in general, but have attractors: particular points in their state space such that all the states which share the same conserved quantities as that attractor end up at that attractor, according to the equations of those theories.

And secondly – somewhat less importantly for my purposes, but not irrelevantly – the dynamical equations of these higher-level theories tend to be probabilistic. Which is to say: sometimes they are stochastic equations; sometimes they are equations for the evolution of classical probability distributions; quantum-mechanically they tend to be equations for the evolution of mixed states. In general we tend to recover determinism for these kinds of theories only in a law-of-large-numbers sense, and only when we are talking about systems with a lot of degrees of freedom. Something like the Boltzmann equation will do as an example; characteristically [1] that is set out probabilistically as an evolution equation for a one-particle marginal (whether that is a density operator, as in quantum theory, or a classical probability distribution), but the multi-body correlations are weak enough and the particle numbers are large enough that at the end of the process we can treat the predictions as deterministic. That is not always how it goes – Boltzmann famously derived the classical Boltzmann equation directly without going through a probability route [2] – but it tends to be the general pattern.

Pause for a second – and set aside for a second how any of this higher-level physics is linked to the underlying fundamental physics – and ask just how we would go about making claims about the past and the future of a system governed by these kinds of laws. In a theory governed by *reversible* dynamical laws, there is going to be a fairly obvious symmetry in how we do it. How do you use the theory to learn about the future? You look at what the present state is, turn the handle of the dynamics, and out comes a prediction about the future state that can be checked against experiment. How do you use the theory to learn about the past? Pretty much the same. You plug in the present state, you run the dynamical equations backwards, that tells you what the past state is supposed to be, and then you compare it with what it actually was. And there is a reason we call this by the neologism "*retrodiction*": it is to suggest that we are doing the same kind of thing as a *pre*diction about the future.

In a theory governed by *irreversible* dynamical laws, prediction is going to be a similar kind of game: you plug in the present state of the world, evolve it forward in time, and out comes a statement – possibly probabilistic – about the future state of the world. We are not in a position to retrodict in the same way, because in an irreversible theory, the present state is typically *not* going to determine a past state. That could be because the dynamics is deterministic but irreversible, so that many past states are compatible with the present state; it could also be because the dynamics is probabilistic and just is not really in the business of telling us *anything* about what the past looked like.

What we do in practice is a kind of "guess and check". At the crudest level, we take a guess as to what looks like a reasonable past state, we evolve it forward and see what the state *would* be now, we compare with what it actually *is*, and we iterate that process. If we want to be slightly more careful, more systematic, more formal about that, we could put something like a Bayesian prior distribution over our collective initial states, use that Bayesian prior to work out a probability distribution over possible present states, conditionalize on the actual present state and see where that leads us. However this process is made precise, let us call it *historical inference*, and distinguish it sharply from retrodiction. It is our normal means by which we learn about the past in these kinds of theories.

It is worth reminding ourselves of the cases where we actually use retrodiction, just to see the contrast with historical inference. If we want to work out the dates of some eclipse that is mentioned in some fifth century BC history we really do carry out pretty much the time-reverse of the calculations that we use to work out the dates of an eclipse in the next century – that is, we really do just run the equations of the solar system backwards. But this is very much the exception which proves the rule, and even then it only applies approximately.

In a world which was, hypothetically, really governed by these kind of irreversible (and often probabilistic) laws, it would not be particularly mysterious that we had a whole bunch of psychological asymmetries, causal asymmetries, entropic asymmetries, and so forth. If our underlying physics contained a whole bunch of irreversibility and violation of time-reversal invariance, we would expect that the observed world would also have those features.

All this suggests a different way of setting up the problem of asymmetry in time from the way it is normally set up, the way we started with. In that approach, typically we say: how do we reconcile the fundamental-physics level directly with the observed world? My suggestion, as a friendly amendment to this way of thinking, is that there are advantages in keeping the discussion internal to the equations of physics, and asking: how can we reconcile the bottom level, the fundamental physics level, with the emergent higher levels, the levels of irreversible dynamical equations? If we can sort this out, the remaining step – reconciling that higher level with the observed-macro-world asymmetries – looks rather tractable.

Why do I make this suggestion? There are three reasons. The first is just division of labour. It is quite a job to say: ok, how am I to reconcile the microscopic physics of the world with the fact that I remember the past but not the future? There are so many different layers, and so many different bits of science, filtering into that kind of story that trying to do the whole thing all in one go inevitably involves radically simplifying models, and dynamical assumptions that we are not remotely in a position to check. On the other hand, if we can outsource the problem of understanding why we remember the past but not the future to those with relevant expertise in memory and cognition and evolution and so forth while handing them on a plate a whole bunch of underlying physics that has time irreversibility in it already, then the task looks a bit more tractable than trying to do the whole thing ourselves.

The second reason is that we need to understand the fundamental/emergent relation anyway. An account of why we remember the past and not the future, or an account of why ice melts in qualitative terms, that does not also tell us quantitatively how it is that we can have the Langevin equation or something similar holding compatibly with our micro physics has not finished the job, because that equation is demonstrably correct for some physical systems and we need to understand why. Conversely, if we can account for the latter, if we can get a grip on how to reconcile higher-level irreversible dynamical systems with bottom-level reversible dynamical systems, we shall get a lot of the rest more or less for free.

The third reason, and the one I want to dwell on, is: there is something funny about saying it is a deep mystery how we can reconcile time reversible microphysics with time irreversible higher-level physics. In one sense it *is* a mystery: we can prove a whole bunch of formal incompatibilities and show that no time-asymmetric physics can be derived from a time-symmetric starting point. But on the other hand, mostly we do not get our higher-level dynamical physics purely from phenomenology, purely from experiment. To a very large extent we actually do derive it from the microphysics – or perhaps, to avoid begging the question, we at least *construct* it from the microphysics. We have a collection of thoroughly used and highly successful techniques for starting with micro-level physics and getting out equations governing the higher-level physics. In fact the great bulk of what we call the evidence for our low-level physics is actually mediated through this kind of process. And we do not just get the qualitative form of the higher-level equations out, we actually get the coefficients.

For instance, think about the decay rates of particles. Those are governed by a time-irreversible decay equation, but we get that equation out from quantum field theory. And we get out the decay coefficient while doing it. So at some level we clearly know how to do this, or at least we have a trick that very reliably works, and which works, as philosophers would say, *projectably*, in two senses. It allows us to work out dynamical equations which seem to be laws for the systems in question, in that they apply to those systems wherever they are in space and time. And the fact that we can do this also turns out to be projectable: if we take a novel physical system where we have not yet tried to work out what the macro-dynamics are, and if we use these kinds of techniques, we tend to get out empirically correct laws.

So that suggests that if we want to understand where the time asymmetry comes into physics, we ought to be looking at what ingredient we are actually putting in when we construct the Langevin equation, or the radioactive decay equation, or any of these higher-level equations. And indeed, this is a sanity check on extant claims of where the asymmetry is. If, for instance, it is claimed that the origin of temporal asymmetry is a specific low entropy boundary condition for the Universe, we ought to be able to see how it is that this low entropy boundary condition, perhaps indirectly, underpins whatever is actually being put in to our derivational process to get out the Langevin equation and the like from our microphysics.

And that moves me on to the second half of what I want to talk about: let us actually have a look in a little bit more detail at these derivations, and see what is going on. For technical details here, see Wallace [3] chapter 9 and references therein.

I am going to be very qualitative, and as general as I can manage, but of course the activity of deriving higher-level dynamical equations from lower-level dynamical equations is enormously wide and varied – indeed, in a sense it is the great bulk of physics – and I only know small corners of it. So let us stipulate: I am talking about a certain subclass of such processes. That subclass is not empty; conjecturally I would say that that subclass contains an awful lot of what we do, but it is not particularly my brief to say that it covers every low-level/high-level derivation known.

Here is a generic model of how I am going to think about things. To a large extent what we are doing when we construct higher-level physics is some kind of coarse-graining. At the kinematic level, there is a state space S_L of the low-level theory, and there is a state-space S_H of the high-level theory, and there is what we might call a *reduction map* that takes us from points of S_L to points of S_H. That map is typically going to be many-to-one: it is going to take a whole group of low-level states and associate them with a single high-level state.

The paradigmatic example of this is something like the way we do fluid or gas dynamics. Here the low-level theory, at least classically, is going to be the Hamiltonian dynamics of 10^{23} point particles interacting under some force laws, and the states in the high-level theory are going to be specified by giving the pressure, density and velocity of the gas averaged over, say, one-cubic-micron cells. This means that S_H is still a large dimensional state space, but it is a lot smaller than the low-level space S_L: A whole bunch of different particle arrangements are going to be associated to single fluid states. And it is going to follow, of course, just from that reduction map, that if I take a trajectory in S_L defined by the low-level dynamics, that trajectory is going to be mapped to some path in S_H.

If we move outside the particular case I have just discussed and ask generally how we might try to set up high-level to low-level correspondences in physics, there is a slightly unfortunate tendency at this point to go into pragmatics and epistemology. Sometimes you will hear people say that what we are doing when we go from the low-level to the high-level theory is: we are keeping only those features of the system that we are interested in and discarding those that do not interest us. For instance, maybe we are not interested in the precise positions of the gas: we just want to know its bulk state.

That is not going to work. I am not actually very interested in the cubic micron of gas over there in that corner; I am not really very interested in the gas in this room at all. I could perfectly happily go through the rest of my life knowing nothing about it, and I imagine so could the rest of you. But it is still true that its dynamics is governed by the equations of fluid dynamics; that generality is no less true because I do not care about it.

Conversely, I am actually quite interested in what the stock market looks like tomorrow. I would love to be able to do a dynamical reduction where I course-grained over everything except the leading numbers in the price index tomorrow and kept those. But, however fascinated I might be about that, I cannot do it, not in a way that is predictively useful. What is going on here is that we are not really asking: when is the high-level theory defined by the degrees of freedom we are interested in? We are actually asking: when is the high-level theory defined by degrees of freedom for which we can write down autonomous dynamical rules?

What do I mean by that? The low-level theory has dynamics, which determine low-level trajectories, and each low-level trajectory determines a high-level trajectory, but there is no *a priori* guarantee that the low-level *dynamics* determines a high-level *dynamics*. There is no guarantee, for instance, that there cannot be two trajectories in S_L whose images in S_H are identical up till some moment of time and then diverse; which is to say that there is no guarantee that in moving from the low level to the high level I have not discarded some information which is actually necessary to predict the future evolution of the high-level states.

So what we actually want are reduction maps that do not have this feature: that actually do generate a high-level dynamics from a low-level dynamics. In this situation we have something we can call a *dynamical reduction* process. Another way to put it is that we can find a dynamics on S_H such that evolving a state in S_L forward in time under the lower-level dynamics and then mapping it to S_H, or mapping it to S_H immediately and evolving it under the dynamics on S_H, gives the same results. In mathematical terminology, dynamical evolution and the coarse-graining reduction map commute.

There does not *need* to be any such high-level dynamics, but often there is. And in particular, often that high-level dynamics is pretty robust against the fine-grained details of how we define the reduction map. (I talked about my gas on cubic micron cells, but clearly I could have chosen ten cubic microns or half a cubic micron.)

The existence of these high-level dynamical laws is, or ought to be, kind of surprising. After all, step back from the fact that we know it works empirically, and ask how is it that we can find equations for just a small number of degrees of freedom in a huge system that are autonomous, that can be studied dynamically while discarding the information at the other degrees of freedom? As with the stock market case, it is not generally true that that can be done: generally we cannot just pick a subset of degrees of freedom and work out what they do, given that they are dynamically coupled to all the other degrees of freedom.

So why do high-level laws ever exist? Very broadly, we can see two reasons. Firstly, sometimes it happens because there really is a dynamical decoupling of some degrees of freedom from others, and generally that happens when we have got a symmetry of some kind in play. So it really is the case that, under appropriate approximations, the centre of mass degree of freedom of the Earth decouples from the zillions of other degrees of freedom of the Earth. And so I really can write down an evolution equation for the centre of mass of the Earth treated as a point with just three degrees of freedom. Even that is not perfect: if the gravitational field varies quickly on scales comparable to the length of the Earth, I am going to have trouble, but to a first approximation I can do it. And I can do it basically because the symmetry structure of the dynamics lets me decouple the centre-of-mass degrees of freedom from the rest.

Much more commonly, as for instance in the gas, what is going on is not that there is complete decoupling of this kind: it is that the residual degrees of freedom, those discarded when we apply a coarse-graining reduction map, are very large in number and very random in the fine details of their dynamics, and each one of them is contributing only to a very small extent to the overall dynamics. So we can do a statistical trick: rather than keep track of them in bulk we can just keep track of their statistical averages; we can sum all their contributions up and treat the whole thing as a sort of generalized noise term. (And you see this in some of the ways one actually derives these kinds of equations in detail.) Using that method is implicitly taking a bet that actually those fine-grain details do not matter and that we really can treat them as a sort of averaged-over noise.

That sounds great, but there is a sense in which we know that it cannot really be true. And the sense is the following: if it really were the case that some reduction map from low-level to high-level state space was compatible with a completely accurate, robust reduced dynamics for the high-level theory, then if the dynamics of the low-level theory is time-reversal invariant the dynamics of the high-level theory had better be time-reversal invariant as well. And it is not. So something went wrong.

And if we look at what went wrong – if we look at the mathematics of what we are doing here – what is going wrong is that it is not true that *any* distribution of the microscopic degrees of freedom with such and such average will behave in such and such a way. There will be ways of tuning and setting up the microscopic degrees of freedom, so that they are aligned in just the right way to break our assumptions about how the dynamics is going to work.

Let me give an example here – partly to help see how general this discussion is; it is not a statistical mechanical example in the usual sense. Think about radioactive decay. We have simple higher-level dynamics for decay: the probability of an unstable particle not decaying is exponential in time, and there are also terms for particles being kicked into undecayed states by absorbing the sort of particles that comprise the decay products.

Now ask how all this works quantum-mechanically. If I take a particle on my desk which has in fact not decayed and evolve it forward in time, it will evolve into a superposition of the undecayed particle and a whole bunch of decay products at different times. If I try evolving it backwards under Schrödinger's equation, I will erroneously predict the time reverse of decay: after all, the equation has a time-reversal symmetry. How do I reconcile that with the fact that the particle has just been sitting on my desk undecayed?

The answer is that the full quantum state at the present time contains not just a term describing the undecayed particle but all the terms corresponding to the particle having decayed at various times in the *past*. And properly running the dynamics backwards means allowing for all of these other terms – which all have just the right mod-squared-amplitudes, and just the right phases, that they continually cancel out the decay terms that are produced from the backward evolution of the undecayed term. It is because of that set-up being aligned just right – in Everettian terms, [3, 4] it is because of all the branches going backwards in time interfering in just the right way – that I get the wrong answer if I try to retrodict using the normal radioactive decay equations.

And for the same reason, if I take the time reverse of the current state (including all the decay terms as well as the undecayed term corresponding to our current experience), that state is *forward* time evolution will not match the normal prediction of radioactive decay. More or less any old generic state of the way the fields could be would not do this but this very carefully prepared one, prepared by time evolving the system forward and then time-reversing, is set up in just the right way to do it.

This state is a *counter-example* state, a state for which our methods of statistical averaging fail to generate the right higher-level evolution. And in general, there will exist counter-example states whenever we try to derive irreversible higher-level dynamics from reversible lower-level dynamics.

What can we say in general about the counter-example states? Just that they are going to be very delicately and carefully structured. It is tempting to say that they are very low probability states, in some sense, but I want to warn against that, because we are mostly here talking about theories that are already formulated as probability theories or theories of mixed states, so our space of states here is a space of probability distributions or mixed states anyway. And talk of "low-probability probability distributions" is at least *prima facie* not well defined. So it is not that the counter-example states are low probability, exactly, but that heuristically, and in some cases demonstrably, they are going to be very delicate, complicated, carefully specified states. (Indeed, the only way we really know to write down any such states is the one we used in the case of radiation: take a simple state, evolve it forwards through time, and then time-reverse it.)

So then the question of why high-level dynamics of this kind in general works (and in particular, the question of why non-equilibrium statistical mechanics works) is going to be the question of why we are allowed to assume that initial state is not like that, is not one of the counter-example states. And of course if we take any given physical system – some small, mundane system, like the glass of water on my desk - it is not surprising that its initial state is not like that: its initial state has been prepared by a whole bunch of other dynamical processes, so unless those dynamical processes are really, really carefully set up in a certain way, or unless the initial state that we feed into *those* processes was very carefully set up in a certain way, we would be very surprised to find that the initial state of our mundane system was one of the very delicately assembled states which count as the exceptions to the usual reduction rule.

What we have done there is push back the requirement that the state is not careful and special in this sense back to an earlier system: the system which led dynamically to our mundane system having the initial state it had. And of course - and this is my justification in talking about this material in a conference on cosmology - if you keep pushing these things back and back, following from our initial system, to the system that generated it, to the system that generated *that* system, … where you are going to get to is a requirement that the initial state of the Universe as a whole is not one of these very special counter-example states.

To sum up: what we seem to get out if we look at the actual structure of reduction and emergence in physics is a claim about the initial state of the universe – something that is hardly novel in discussions of time asymmetry - but a claim that is slightly different from the usual claims. Here I have in mind the neo-Boltzmannian approach espoused by, for example, Albert [5], Penrose [6], Carroll [7] or Goldstein [8]. We are not requiring anything of the *macro*-state of the early universe: in particular, we are not requiring it to have low entropy. Instead, we are requiring something about its *micro* structure: we are requiring it *not* to be set up in the kind of delicately correlated ways that invalidate our general averaging moves.

From this perspective, far from introducing a condition which makes the initial universe very special (the usual ways "past-hypothesis" claims about the early Universe are phrased), we are asking that it should *not* be very special. We need to be a bit careful, though, because of course if we suppose that the universe is closed and has a final state as well as an initial state, and if we take this "not very special" initial state and run it forward, then - precisely because this licenses us to apply higher-level, emergent dynamics – the *final* state is going to be really, really special. It is going to have built into it just the right correlations such that if time-reversed and run backwards it *will not* be governed by the macro-level dynamics, but rather by their time reverse. So we have not somehow dissolved away the need for an asymmetric assumption. But that assumption looks a bit different, conceptually and mathematically, from what we are used to in these discussions: we are not supposing that the state at one end of time is particularly low-entropy, but that it is relatively simple, relatively free of complicated and delicate correlative structure.

What *should* we say about the fact that the early universe has a very low entropy? Well, it certainly seems to need explanation, but it does not seem to be very different from other facts about the initial condition of the universe. How do we learn about the fact that the early universe has low entropy? By historical inference: we take our possible guesses as to what the initial state is, we evolve them forward and we compare with what we have. And since our macro-level dynamical theories are entropy-increasing, given that they are correct, the fact that the early universe has a low entropy compared to the present day universe is a claim that is not very difficult to get out historically.

So: my suggested way of proceeding gives us, in David Albert's terms [5], a different sort of transcendental assumption that we need to make in order that our physics works. But it is not a transcendental assumption of the kind we can actually empirically check in cosmology; it is a transcendental assumption about the fine-grained micro-level delicate structure of the early universe. And it is basically the assumption that it has not got much of it.

24.1 Discussion

David Albert: Thank you first as usual for a beautiful talk. I guess I have two small comments: one about the issue of division of labour that you were talking about. Here is the thing: presumably what we are interested in showing, among the asymmetries between past and future. For example, we are interested in showing not merely that epistemic access to the past is different from epistemic access to the future for human beings or for mammals, or for terrestrial organisms, or biological organisms in general. It is something much more general and much more fundamental than that. So that in that sense there is a sort of natural expectation that it should have some fairly direct link to the fundamental physics. If we were to encounter Martians that had time-reversed epistemic access relative to us we would be flabbergasted. So actually there is a sense in which it does not feel like a job for neurologists or psychologists or experts in human cognition. There is a sense in which it feels like a job for the fundamental physics. That is one thing.

A related point: The people who talk about features of the initial macro-state of the universe are trying to do a job slightly different from the one you describe here. That is, they are not just trying to justify the macro-dynamics or to explain the macro-dynamics. What such people are usually trying to do is something in a way more ambitious. I mean, maybe that is a foolhardy task, but they are trying to *both* justify the macro-dynamics *and* sort of systematize the whole process of inference towards past and future. So, for example, on the way of reasoning that you are describing, there is a bunch of earlier states that historical inference could lead to. One is the state we think pertained five minutes ago, the macro-state of the world. Another is the macro-state of the world we think pertained ten minutes ago. And so on and so forth, all the way back to the Big Bang. So I think I completely agree with you that if the task at hand is just to explain the macro-dynamics you need much less than that. If the task you are looking to do is more ambitious than that, you may need more.

David Wallace: It is absolutely right of course that we want a very general explanation of the epistemic asymmetries, and I agree with you it is a job for physics, but it is not necessarily a job for *fundamental* physics. We live in a world where the degrees of freedom and the macroscopic functionals are governed by time-asymmetric dynamical laws. In a situation like that it is unsurprising to find these kinds of general epistemic asymmetries; there is further work to try and understand them and get them out but it is not a mystery in that sense.

Having said that, in some sense this is spoils to the victor: if we can give that explanation directly in terms of fundamental physics, why not?

As to the other point: I do not disagree there at all. But it is sometimes said that the past hypothesis is specifically something we need to explain the fact that statistical mechanics techniques work. My claim is: that is not really what is doing the work. Put in your framework, what is actually doing all the work is the probability distribution over the initial macro-condition, not the choice of macro-condition.

Carlo Rovelli: If I had to cover both talks, David [Wallace] and David [Albert], my comment would be that according to my own understanding everything that has been said is exactly right, as far as I understand. (Which, of course, might be due to the fact that I am a theoretical physicist, I do not get saddled with philosophy.) But it seems to me: did not we know that; is that not the way we understand things; have you not just put clearly what was understood? And in fact was not it all understood by Boltzmann, especially in what he wrote *after* he got all the criticisms to his H theorem?

I want to make a point related to that. First, let me say that I talked about the problem of the special initial condition in my first talk this morning; perhaps my talk should have been *after* yours because the point of departure of my talk was the conclusion that you arrived at. But what I want to say is that after getting the criticism for his H theorem Boltzmann got to this point, which I think is not often appreciated: he was thinking in probabilistic terms, he was thinking in terms of equilibrium fluctuations, and what he proved – or rather, what he stated was the case, and in fact what has been proved recently by people in statistical theory, is the following. That if you take a statistical system and you look at the fluctuations of its entropy (which of course is not going to be maximal, it is going to fluctuate) you can ask why, if I am at a certain point away from the maximum, I find that it goes up in one direction in time and down in one direction in time, while my theory tells me that it goes down in both directions. And here is an answer to that: it is a beautiful answer. Given a value the most probable situation is that it is a peak of a fluctuation. So this makes it clear that his own result is valid not because it breaks time-reversal invariance; it confirms time-reversal invariance and it shows that given a situation in which you are out of equilibrium the most likely solution is going to be in *both* directions. And of course (in the context of our observations) the only way to make sense of that is to go back to the origin of the universe, so making it a cosmological posit. The theorem has been proved quite recently in statistical mechanics rigorously; Boltzmann just guessed it.

David Wallace: I want to pick up on the first point. The job of the philosopher of physics is to tell people the physics they know and then say that it is profound. I am actually almost serious about that, not just being self-deprecating: a lot of the point of this kind of work is to take things that are tacitly grasped by people, and understand clearly and consistently and explicitly how those things should be understood. It almost shades in to pedagogy at some level.

But I do think there is a degree of mismatch between what is said in general and verbal terms about why systems equilibrate, and what you actually find if you go down into the weeds and look at the way we construct and derive our detailed quantitative understanding of equilibration. I think that is of some interest, and in particular I think it is somewhat striking that the role of low entropy assumptions is more indirect and much less clear than it can seem in the general discussions.

I have been interested in this stuff partly because I thought: okay, I buy the general claim that the way to understand why statistical mechanics works is because the initial universe had a low entropy, so let us actually have a look at how those equations work and see where the low entropy assumption in particular is coming in. And then it turned out to be more complicated than that.

Simon Saunders: I think the value of your talk, David, for me anyway, is that it seems to undercut a sense of mystery or some sense of strangeness or weirdness about how special the initial condition has to be. It seems – and Penrose has made a lot of this [6] – that by virtue of the fact that the final state of the universe, given black holes and eventually black hole evaporation, has a fabulously higher entropy than it has now, one gets to the picture that the early state of the universe had to be amazingly, precisely fine-tuned so as to have low entropy. And your punchline, in a way is: no, that it is not that special.

Don Page: That is very controversial, I did not think you [Wallace] said that. Did you really say that?

Simon Saunders: Let me carry on, because I am presenting a gloss on what David is saying and maybe David will tell me I have got the gloss wrong. My gloss is that the initial state of the universe, far from being special, precisely *is not* at the microscopic level carefully calibrated so as to have the sorts of coincidental relations that could bring together what would look to us like time-reversed macroscopic phenomenology.

And I just want to push that, in that if you imagine Penrose, or some other physicist or mathematician, doing some other calculation which shows how yet more extraordinarily special the initial state seems to have to be, because of anthropic considerations, perhaps the right way to think of that is that what they have really shown is how extraordinarily high the final entropy of the universe can be, how extraordinarily large entropy can grow through physical processes. And perhaps black hole thermodynamics is an example of that, that prior to the understanding of black holes we did not realize how large the entropy could get to be.

So I wonder if you would embrace that way of thinking about it. I think the comeback on that would be to say, well, okay, the initial state of the universe has to be amazingly non-special at the microscopic level, it must not have all of these careful, finely tweaked relations among the, whatever it is, particles and so forth, and you might say that what we are really learning is: you think you have got a non-special state because it does not have those finely-tuned correlations, well, guess what, it has got to be even more non-special than we thought it had to be.

Just to summarize then, suppose we learn of some new mechanism as we did in black hole thermodynamics, such that the final state of the universe could be even more extraordinarily high entropy, is that the right way to be surprised? Or is it Penrose's way? No, the surprise is to find how extraordinarily fine tuned the state of the universe has to be so as to be low entropy. Or, the third way that I am offering as well: is it that we learn how extraordinarily non-special the initial state has to be?

David Wallace: I think this partly says what we mean by 'special'. So (and I take this to be what is worrying Don), I am obviously *not* claiming that the initial macro-state does not occupy an extremely small region of phase space. Clearly it does. What I am claiming is that this fact about the macro-state is not the thing which is doing the explanatory work of saying why the law-like regularities of high-level statistical mechanics hold.

Now of course if we imagine a fictional world that (unlike in real cosmology) actually had an equilibrium state – a box-like world of some kind – then if it is initial state was unspecial in the macro sense it would be at equilibrium. The claim that the macro-dynamical laws held would then be boring because they just say: you are at the attractor, stay there. But nonetheless the thing that is doing the work in explaining *why* those (boring) law-like regularities hold is the non-specialness of the micro-structure within that macro-state. And you could perfectly well imagine a universe that was actually at equilibrium but whose micro-state *was* extremely special in just the right way that it went away from equilibrium. (Take a system that has just equilibrated, and time-reverse its micro-state.)

On the broader question of the specialness of the initial *macro* state: I have to confess I am not entirely clear, and I am less clear the more I think about it, quite what it is that concerns us about it. Granted, the theory that God created the universe by picking a point uniformly at random with respect to the Liouville measure on phase space is falsified by the data. But that was not a very plausible creation myth in the first place.

I think what is going on is something like this. In general, a really good way of studying a large system in a given macro-state is to assume its micro-state is chosen at random with respect to the Liouville measure over that macro-state. And we have reasonable dynamical grounds to explain why that is a good thing to do. But we are in danger of extrapolating this back to a more transcendental principle that it is *a priori* the case that a system is equally likely to be anywhere in phase space. And that is what gets you into skeptical catastrophes, where we say that somehow our memory of the past is completely unreliable because it is much, much more likely we have fluctuated in from a higher entropy state. But I do not think that there are any good reasons on the table to think that the right *a priori* probability distribution is uniform across phase space. There plausibly *are* good general *a priori* assumptions, coming from general epistemology or general philosophy of science, for thinking that the initial macro-state should be a relatively simple, or relatively easily describable state, but in gravitational systems that simplistic criterion tends to pick out low entropy states rather than high entropy states. For discussion of the conceptual features entropy in self-gravitating systems see Wallace [9] and references therein.

So in conclusion, I am not entirely sure what the fuss is about, so I am not quite sure therefore what I should be saying in response to your trilemma at the end.

Cormac O'Raifeartaigh: Many thanks to both Davids for talks which were extremely clear - that is not easy when you are going from discipline to discipline. If I could ask in that spirit a very simple question which for the philosophers is probably kindergarten. What do you make of Lee Smolin's view that there is not *necessarily* a tension between the fact that some of our equations in particle physics do not include time, and the way that we see time in the observable universe? His simple answer to that [tension] is that since the Dirac equation physicists have been haunted by the notion that every solution *has* to represent the real world, whereas in fact mathematics is simply a representation of nature. That we fall into the old problem of confusing reality itself with our representation of reality? What do you make of people who duck the whole question by saying "there is not necessarily a conflict there, this is simply a facet of the way we represent nature"?

David Wallace: The move of saying "our physics is not fully representing the world" is always available, but I think the only really good way to tell if our physics *cannot* represent the world is to try really, really hard to represent the world and see if we fail. I am a bit nervous about *assuming* that it is the case. That is a general philosophical nervousness about Smolin's move.

Let me use that as a sort of advert for some of the virtues of the approach in my talk. The question of how we reconcile time-symmetric micro-dynamics with the panoply of phenomena in the observed universe is so general, has so many different facets, that there is all sorts of space for more philosophical attempts to make the problem go away. The question of how it is that we can derive the Fokker–Planck equation from Newtonian dynamics, given that Newtonian dynamics is time-reversal symmetric and the Fokker–Planck equation is not, and that the Fokker–Planck equation is probabilistic and Newtonian mechanics is not; that question is much tighter and sharper. The problem is not some kind of philosophical paradox; it is a straight mathematical contradiction, so some of the assumptions in it must be wrong, and you can push at where those are.

References

[1] Balescu, R. (1977) *Statistical Dynamics: Matter Out of Equilibrium*. London: Imperial College Press.

[2] Brown, H. R., Myrvold, W. and Uffink, J. (2009) Boltzmann's H-theorem, its discontents, and the birth of statistical mechanics. *Stud. Hist. Phil. Mod. Phys*, **40**, 174–91.

[3] Wallace, D. (2012) *The Emergent Multiverse: Quantum Theory According to the Everett Interpretation*. Oxford: Oxford University Press.

[4] Everett, H. (1957) Relative State Formulation of Quantum Mechanics. *Rev. Mod. Phys.*, **29**, 454–62.

[5] Albert, D. (1999) *Time and Chance*, chapter 6. Cambridge, MA: Harvard University Press.

[6] Penrose, R. (1989) *The Emperor's New Mind: Concerning Computers, Minds, and the Laws of Physics*, chapter 7. Oxford: Oxford University Press.

[7] Carroll, A. (2010) *From Eternity to Here: The Quest for the Ultimate Theory of Time*. New York: One World.

[8] Goldstein, S. (2001) Boltzmann's approach to statistical mechanics. In Bricmont, J., Durr, D., Galavotti, M., Petruccione, F. and Zanghi, N., eds. *Chance in Physics: Foundations and Perspectives*. Berlin: Springer. Available online at http://arxiv.org/abs/cond-mat/0105242) or Price, H. (1996) *Time's Arrow and Archimedes' Point*. Oxford: Oxford University Press.

[9] Wallace, D. (2010) Gravity, Entropy and Cosmology: In Search of Clarity. *Br. J. Phil. Sci.*, **61**, 513–40.

25

Big and Small

DAVID Z. ALBERT

Our everyday macroscopic experience of being in the world is saturated with asymmetries – thermodynamic asymmetries, and radiative asymmetries, and epistemic asymmetries, and phenomenological asymmetries, and asymmetries of over-determination, and asymmetries of influence, and what have you – between the past and the future.

And there is a long-cherished hope – something that has its origins in the work of Boltzmann, and which has been pursued, by any number of other investigators, through any number of fits and starts and revelations and wrong turns, ever since – that all of those asymmetries can ultimately be traced back to some relatively simple characteristic of the initial macrocondition of the universe. The thought (as people put it now) is that all we need to do, in order to account for these asymmetries, is to add to the fundamental time-reversal-symmetric dynamical *laws*, and to the standard statistical-mechanical probability-measure over the space of possible fundamental physical *states*, a simple postulate – a so-called *past-hypothesis* – to the effect that the world first came into being in whatever particular low-entropy macrocondition it is that the normal inferential procedures of cosmology are eventually going to present to us.

The business of working this thought out in detail is a large undertaking, which is still very much in its infancy, and which is still very much under debate – and I do not want to attempt anything along the lines of an overview of that project here. All I want to talk about in this chapter is a widespread and fundamental and perennial sort of puzzlement about how such a project could even seriously be *entertained* – a puzzlement (that is) about how it is that the macrocondition of the universe 14 billion years ago – all by itself – could even *imaginably* be up to the job of explaining so much about the feel, now and on earth, of the passing of time.

This puzzlement takes a number of different forms, and arises in a number of different contexts.

On the most trivial level, there is a question of how the lowness of the entropy of the world 14 billion years ago can impose any genuinely profound and vivid constraints *whatever* on what the world is doing *now*. And all that needs to be said, in order to make *that* sort of puzzlement go away, is that although 14 billion years is a long time, the entropy of the universe at that time was very, very low – and that (in particular) 14 billion years is a great deal *shorter* than the expected *relaxation* time of the state in which our universe seems to have started out.

There is a somewhat more *interesting* question about how the lowness of the entropy of the world 14 billion years ago can have any genuinely profound and vivid effects, or impose any genuinely profound and vivid constraints, on what particular, localized, human-scale, quasi-isolated *sub-systems* of the world are doing now. There is a worry (in particular) that runs like this: Suppose we grant that the standard Boltzmannian arguments do indeed establish that the overall entropy of a universe which starts out in a low-entropy past-hypothesis sort of macrostate is overwhelmingly likely to rise towards the future. That does nothing at all (so the worry goes) to show that the entropies of quasi-isolated *sub*-systems of the world are likely to rise in the same direction – and it is the behaviors of those *latter* systems (after all), and not of the universe as a whole, that is the central and paradigmatic topic of thermodynamics!

Jennan Ismael (for example) has written, in precisely this connection, that "It should be clear that you can't in general move from a property of the whole to properties of its parts. The fact that dogs can run doesn't mean that dog heads can run. Forests expand and babies grow, but not because the trees or cells that make them up do. The global story isn't just the local story, writ large. There are all kinds of ways in which a system could have global dynamical properties that are not properties of any of its parts." (Ismael, 2016). And there is a paper of Eric Winsberg's (2016) in which he worries that the conditions that are required in order to run the standard Boltzmannian argument to the effect that the entropy of an isolated system will increase towards the future may somehow be *altogether different* from the conditions that are required in order to run the *analogous* argument on whatever quasi-isolated *sub*-systems of that larger system happen to emerge as the larger system evolves in time.

It is not clear exactly what this worry is about, or where it comes from (although I may be missing something). The Boltzmannian tradition has given us an argument that the overwhelming majority of microconditions compatible with any non-equilibrium macro-condition of any isolated thermodynamic system are sitting on trajectories on which *all of its thermodynamic properties* – and not merely its overall entropy – are going to evolve towards the future, in the general direction of equilibrium, *in accord with our usual thermodynamic expectations*. If (for example) the isolated system in question consists of two gases, which are initially at different temperatures, and which are in thermal contact with one another, what the Boltzmannian arguments make plausible is not merely that the overwhelming majority of microconditions compatible with this initial macrocondition are sitting on trajectories on which the overall entropy of this system is going to increase towards the future, but (in addition) that the overwhelming majority of those microconditions are sitting on trajectories on which the entropy of the *hotter* gas is going to *decrease* towards the future. And if the isolated system in question consists of (say) 12 isolated gases, in 12 separated containers, each of which is initially far from its own individual equilibrium state, *then* what the Boltzmannian arguments are going to make plausible is not merely that the overwhelming majority of microconditions compatible with this initial macrocondition are sitting on trajectories on which the overall entropy of this system will rise towards the future, but (in addition) that the overwhelming majority of those micro-conditions are sitting on trajectories on which the separate entropies of each of those 12 gases are *all*, *individually*, going to rise towards the future. Of course, the number of micro-conditions on which the entropy of some particular one of those gases goes down towards the future will be much larger than the number on which the overall entropy of the 12 of them together goes down towards the future – but both of those numbers are going to be fantastically small, and in both cases the microconditions in question are going to be scattered, more or less at random, in unimaginably tiny clumps, all over the phase space, and the likelihood of the microstate of the system ever wandering into *either one* of these two different sorts of clumps seems, on the face of it, extremely remote.

But let me go on to the question I really want to address – which is the third and the deepest and the most interesting and the most amorphous and the most phenomenological form of the general puzzlement about the Boltzmannian project that I mentioned at the outset of the chapter.

Let me put it in four increasingly concrete and increasingly simple and increasingly tractable ways:

1. How can it seriously be imagined that my own sense of the passage of time, how can it seriously be imagined (for example) that my own sense – right here and right now – of whether some particular baseball happens to be flying towards me or away from me, is somehow anchored in the lowness of the entropy of the universe 14 billion years ago?

2. How can it *be*, how can it *work*, that the increase of the entropy of the world, or of myself, somehow constitutes the standard or the yardstick against which I judge the direction in which events are unfolding? How is it (that is) that the entropy gradient of anything ever comes into the picture? I am certainly not aware of *checking* on the entropy gradient of anything in the course of deciding whether the baseball is flying towards me or away from me. No comparison with anything else, so far as I am aware, is involved. I simply, directly, unmediatedly *see* that the baseball is flying either towards me or away from me.

3. Consider the sense of the direction of time that is implicit in the operations of (for example) a simple mechanical realization of a Turing machine. Can anyone seriously believe that *thermodynamical* characteristics of the world somehow play a role in the way a machine like that distinguishes between what it has just done and what it is to do *next*? How so? How can that be? How would that work? Machines like that can apparently function perfectly well, machines like that apparently have no trouble at all distinguishing between what they have just done and what they are to do next, without the aid of special devices for measuring the entropy-gradient of the world, or themselves, or anything else!

4. Consider (finally) a simple mechanical device which has *no other business* than distinguishing between what it has just done and what it is to do next – the *paradigmatic* distinguisher, the distinguisher *par excellence*, between what it has just done and what

it is to do next. Think (that is) of a clock. And think (for the sake of concreteness, for the sake of simplicity) of an old-fashioned, fully mechanical, *pendulum*-clock.

Good. Now we have our hands on something that we are in a position to analyze in detail.

Note that in the course of the normal and intended operations of a clock like that, there are going to be moments – the moments (in particular) when the pendulum is precisely at the apogee of its swing – when every last one of its macroscopic moving parts is fully *at rest*. Note (to put it slightly differently) that in the course of the normal and intended operations of a clock like that, there are going to be moments – the moments (again) when the pendulum is precisely at the apogee of its swing -when the macrocondition of the clock, in its entirety, is *invariant under time-reversal*. And consider how it is, at such moments, that the clock manages to distinguish between what it has just done and what it is to do next.

The macrocondition of the clock, together with the microscopic dynamical equations of motion, together with the statistical postulate, is manifestly not going to do the trick. For if the present macrocondition of the clock together with the microscopic dynamical equations of motion and the statistical postulate makes it likely that the clock is going to read (say) 3:05 five minutes from now, and if the present macrocondition of the clock is invariant under time-reversal, then the present macrocondition of the clock, together with the microscopic dynamical equations of motion and the statistical postulate – both of which are invariant under time-reversal as well – is necessarily *also* going to make it likely, and to exactly the same degree, that the clock read 3:05 five minutes *ago*.

And all there is to break the symmetry, all there is that *stands in the way* of the clock's having read 3:05 five minutes ago, is the past-hypothesis. The clock's ability to distinguish between what it did last and what it does next, and your ability to distinguish between a baseball's flying towards you and a baseball's flying away from you, are anchored in the entropy-gradient of the universe. If we were to hold the present macrocondition of the world fixed, and move the past-hypothesis from the beginning of time to its end, *the clock would run backwards*.

Period.

People sometimes find it hard to take this in. Consider (for example) the following reaction, from a well-known theoretical physicist, which is worth quoting exactly, and in its entirety:

It's uncharacteristic of Albert to pass over details. He could've described how a pendulum clock works (e.g. falling weight version) in a couple of minutes, but he didn't. The mechanism CANNOT run backwards. What drives it is the falling of the weight, pulling on the cord. If there's no pull on the cord the clock stops. If there's PUSH on the cord the clock stops. It'll ONLY work if there's tension on the cord and that will make the hands move clockwise because of the way the cord is wound around the drive axle. And that would be true even if the weight rose into the air and started pulling upwards. Putting the low entropy in the future of the universe can't make the clock run backwards, as he claims. The clock "knows" which way to go, if it's going to go at all, because the information is built into it in the way that the cord is wound around the axle.

It is hard to know exactly what to say about a worry like that – except to repeat, maybe a little louder, that it is simply and radically mistaken. Absolutely no such additional details about how the clock works can make the slightest bit of difference. Consider (again) a moment when the macroscopic state of the clock is stationary. Consider (that is) a moment when the macroscopic state of the clock is invariant under a reversal of all of the velocities of its microscopic particulate constituents. In the absence of a past-hypothesis, the predictions of statistical mechanics about the evolution of the macroscopic condition of this clock away from that moment towards the future are going to be *identical* to its predictions about the evolution of its macroscopic condition away from that moment towards the past, for the simple reason that *there is nothing whatever in the situation to distinguish between them.* In the absence of a past-hypothesis, the predictions of statistical mechanics are that *as we proceed away from the present*, in *either* temporal direction, *the cord is always going to be unwinding, and the weight is always going to be going down, and the hands of the clock are always going to be turning in the clockwise direction*. And it is only because of the truth of the past-hypothesis that (as a matter of actual fact) those hands turn *counter*-clockwise, as we proceed away from the present, in the direction of the past. And if we were to switch the low-entropy condition from the beginning of the universe to its end, then everything would be the other way around.

It turns out (then) to be *essential* to the intended functioning of a pendulum clock, or of a Turing machine, or of a human brain – it turns out (that is) to be precisely the *opposite* of an irrelevancy or an inconvenience or a potential source of error – that it be in thermal *disequilibrium*. A pendulum clock – no less than a puff of smoke or a block of ice – is (among other things) an instrument for measuring *the entropy-gradient of the world*. A pendulum clock is (more particularly) an instrument whose hands cannot help but to turn clockwise – for exactly the sorts of reasons that are spelled out in the quotation from the physicist above – *in the temporal direction that points away from the past-hypothesis.* That is how they differ from (say) projectiles, or gyroscopes: a projectile can just as easily move this way as that, and a gyroscope can just as easily turn clockwise as counter-clockwise, as the entropy of the world increases – and so they are no good at all (unlike a pendulum clock, or a block of ice, or me, or you) at telling past from future.

References

Ismael, J. (forthcoming) Do Statistical-Mechanical Probabilities Need Shuffling: Albert's Boltzmann Story and GRW Collapses. In *Essays on Time and Chance*. Cambridge, MA: Harvard University Press.

Winsberg, E. (forthcoming) The Metaphysical Foundations of Statistical Mechanics: The Status of PROB and PH. In *Essays on Time and Chance*. Cambridge, MA: Harvard University Press.

Index